Microcontroller Theory

and Applications with the PIC18F

Microcontroller Theory
and Applications with the PIC18F

First Edition

M. RAFIQUZZAMAN, Ph.D.
Professor
California State Polytechnic University, Pomona
and
President
Rafi Systems, Inc.
Diamond Bar, California

Vice President and Executive Publisher	Don Fowley
Associate Publisher	Daniel Sayre
Marketing Manager	Chris Ruel
Production Manager	Micheline Frederick
Cover Designer	Wendy Lai
Editorial Assistant	Katie Singleton

Published by John Wiley & Sons, Inc., Hoboken, New Jersey.
Published simultaneously in Canada.

This title was set by the author and printed and bound by Hamilton Printing.

Library of Congress Cataloging-in-Publication Data:

ISBN 13 978-0470-94769-2

Printed in the United States of America.

10 9 8 7 6 5 4 3 2 1

To my wife, Kusum; my son, Tito; and my brother, Elan

Contents

PREFACE

Microcontrollers play an important role in the design of digital systems. They are found in a wide range of applications including office automation systems (copiers and fax machines), consumer electronics (microwave ovens), digital instruments, and robotics.

This book is written in a very simplified manner to present the fundamental concepts of assembly and C language programming and interfacing techniques associated with typical microcontrollers. Microchip Technology's PIC18F4321 is used for this purpose. The PIC18F family continues to be popular. The PIC18F family is an excellent educational tool for acquiring an understanding of both hardware and software aspects of typical microcontrollers.

The PIC18F uses Harvard architecture with a RISC-based CPU. Conventional CPUs complete fetch, decode, and execute cycles of an instruction in sequence. However, the PIC18F uses pipelining, in which instruction fetch and execute cycles are overlapped. This speeds up the instruction execution time of the PIC18F. A brief coverage of CPU architectures, RISC vs. CISC, pipelining, assembly/C language programming, and I/O techniques associated with typical microcontrollers is provided in the first part of the book. These topics are then related to a popular member of the PIC18F family such as the PIC18F4321 in the second part of the book.

As far as the programming is concerned, assembly language programming is mostly covered in this book using the PIC18F. An adequate coverage of C is also provided. Although writing programs using C is easier than using assembly language, assembly language programming will provide an exposure to the internal architecture of microcontrollers. Furthermore, programming in assembly language may sometimes be useful for real-time systems.

Several assembly and some C language programs along with I/O examples are developed using Microchip's MPLAB and PICkit™3. The MPLAB software package includes a text editor, PIC18F assembler, C compiler, and a simulator. The PICkit™3 is a programmer provided by Microchip. One can build an inexpensive PIC18F-based system on a breadboard using one of the PIC18F devices such as the PIC18F4321. The programmer can download the compiled or assembled programs using the PICkit™3 from the personal computer or laptop, and then perform meaningful experiments. This is the most inexpensive way of implementing laboratory experiments using a typical microcontroller such as the PIC18F4321. Note that Appendix F provides a tutorial showing step-by-step procedure for assembling and debugging a PIC18F assembly language program using Microchip MPLAB PIC18F assembler/debugger. Appendix G, on the other hand, includes a tutorial showing step-by-step procedure for compiling and debbuging a C program using the MPLAB C18 compiler/debugger.

The book is self-contained and includes a number of basic topics. A background in basic digital logic and C language programming is assumed. Characteristics and principles common to typical microcontrollers are emphasized and basic microcontroller interfacing techniques are demonstrated via examples using the simplest possible devices, such as switches, LEDs, A/D and D/A converters, the hexadecimal keyboard, and seven-segment and LCD displays. Most of the examples are implemented successfully in the laboratory.

The text is divided into 10 chapters. In Chapter 1, we provide a review of terminology, number systems, and evolution of microcontrollers. A comparison of the basic features of some members of the PIC18F family and typical microcontroller applications are also included.

Chapters 2 through 5 provide basic concepts needed to understand the material presented in Chapters 6 though 10. Chapter 2 covers typical microcontroller architectures. The concepts of CPU architecture, program and data memory units, pipelining, and RISC vs. CISC are included.

Chapter 3 contains programming concepts associated with typical microcontrollers. Topics include machine, assembly, and C language programming, typical addressing modes, and instruction sets.

Chapter 4 is focused on the memory organization and I/O (Input / Output) techniques associated with typical microcontrollers. The basic concepts associated with main memory array design, including memory maps, are also covered. Typical microcontroller input/output techniques including programmed I/O and interrupt I/O are included.

Chapter 5 includes PIC18F architecture and addressing modes. The PIC18F pipelining, register architecture, memory maps, and addressing modes are provided.

Chapters 6 through 9 form the nucleus of the book. The concepts of assembly language programming covered in Chapter 3 are demonstrated in Chapters 6 and 7 by means of a typical 8-bit microcontroller. A specific device from the PIC18F family such as the PIC18F4321 is used to illustrate the concepts. Several PIC18F assembly language programming examples are included.

The I/O techniques covered in Chapter 4 are demonstrated in Chapters 8 and 9 using the PIC18F4321. Several I/O examples using PIC18F assembly language are also included. These chapters also demonstrate how the software and hardware work together by interfacing simple I/O devices such as switches, LEDs, and seven-segment displays to more advanced devices such as LCDs (Liquid Crystal Displays), hexadecimal keyboard, and A/D and D/A converters. The PIC18F timers and CCP (Compare/Capture/PWM) module along with Serial I/O are also covered. Typical examples include designing a PIC18F4321-based voltmeter using both programmed and interrupt I/O.

The concepts of C language programming covered in Chapter 3 are demonstrated in Chapter 10 using the PIC18F4321 microcontroller from an introductory point-of-view. Chapter 10 starts with a brief coverage of basics of C language, and then implements most of the assembly language programming examples in Chapters 8 and 9 using C. Typical C programs include I/O examples with LEDs and switches, PIC18F-based voltmeter, A/D and D/A converters, LCD displays, timers, and motor control using PWM (Pulse Width Modulation).

The book can easily be adopted as a text for a one- semester or one-quarter course in microcontrollers taught at the undergraduate level in electrical/computer engineering and computer science departments. The students are expected to have a background in C language and digital logic (both combinational and sequential) design. The book will also be useful for practicing microcontroller system designers. Practitioners of microcontroller-based applications will find more simplified explanations, together with examples and comparison considerations, than are found in manufacturers' manuals.

As mentioned before, emphasis is given in this book on assembly language programming using a typical microcontroller such as the PIC18F4321. Adequate coverage of I/O and interfacing using C is included.

Since C language programming is prerequisite for this course on microcontrollers, coverage of I/O and interfacing using C would suffice. A basic coverage of assembly language programming using a typical microcontroller such as the PIC18F is provided.

The author is especially indebted to his colleague, Dr. R. Chandra, of California State Poly University, Pomona; to his student, Luke Stankiewicz; and to others for their valuable comments and for making constructive suggestions. The author also wishes to express his sincere appreciation to his student, Michael Nguyen for drawing several figures in the book, and to CJ Media of California for preparing the final version of the manuscript. The author is also grateful to his student, Sevada Isayan, and to Marc McComb and Rob Stransky of Microchip Technology, Inc. for their inspiration and support throughout the writing effort. Finally, the author is indebted especially to his deceased parents, who were primarily responsible for his accomplishments.

Pomona, California M. RAFIQUZZAMAN

CREDITS

The material cited here is used by permission of the sources listed below.

Copyright of Microchip Technology, Inc. 2009, Used by Permission: PIC18F4321 Family Data Sheet, DS39689F. All mnemonics of Microchip PIC18F Microcontroller Family are courtesy of Microchip Technology, Inc.

1 INTRODUCTION TO MICROCONTROLLERS

Digital systems are designed to store, process, and communicate information in digital form. They are found in a wide range of applications, including process control, communication systems, digital instruments, and consumer products. A digital computer, more commonly called simply a *computer*, is an example of a typical digital system.

A computer manipulates information in digital or, more precisely, binary form. A *binary number* has only two discrete values: zero or one. Each discrete value is represented by the OFF and ON status of an electronic switch called a *transistor*. All computers understand only binary numbers. Any decimal number (base 10, with ten digits from 0 to 9) can be represented by a binary number (base 2, with digits 0 and 1).

The basic blocks of a computer are the central processing unit (CPU), the memory, and the input/output (I/O). The CPU of a computer is basically the same as the brain of a human being; so computer memory is conceptually similar to human memory. A question asked of a human being is analogous to entering a program into a computer using an input device such as a keyboard, and a person answering a question is similar in concept to outputting the program result to a computer output device such as a printer. The main difference is that human beings can think independently, whereas computers can answer only questions for which they are programmed. Computer *hardware* includes such components as memory, CPU, transistors, nuts, bolts, and so on. Programs can perform a specific task, such as addition, if the computer has an electronic circuit capable of adding two numbers. Programmers cannot change these electronic circuits but can perform tasks on them using instructions.

Computer *software* consists of a collection of programs that contain instructions and data for performing a specific task. All programs, written using any programming language (e.g., C), must be translated into binary prior to execution by a computer because the computer understands only binary numbers. Therefore, a translator is necessary to convert such a program into binary, and this is achieved using a translator program called a *compiler*. Programs in the binary form of 1's and 0's are then stored in the computer memory for execution. Also, as computers can only add and compare, all operations, including subtraction, multiplication, and division, are performed by addition.

Due to advances in semiconductor technology, it is possible to fabricate a CPU on a single chip. The result is a *microprocessor*. Both metal-oxide semiconductor (MOS) and bipolar technologies are used in the fabrication process. The CPU can be placed on a single chip when MOS technology is used. However, several chips are required with bipolar technology. At present, HCMOS (high-speed complementary MOS) or BICMOS (combination of bipolar and HCMOS) technology is normally used to fabricate a microprocessor on a single chip. Along with the microprocessor chip, appropriate memory and I/O chips can be used to design a *microcomputer*. The pins on each one of these chips can be connected to the proper lines on a system bus, which consists of address, data, and control lines. In the past, some manufacturers designed a complete microcomputer (CPU, memory, and I/O) on a single chip with limited capabilities. Single-chip microcomputers such as the Intel 8048 were used in a wide range of industrial and home applications.

Microcontrollers evolved from single-chip microcomputers. Microcontrollers are normally used for dedicated applications such as automotive systems, home appliances, and home entertainment systems. Typical microcontrollers include a CPU, memory, I/O, along with peripheral functions such as timers, A/D (analog-to-digital), and serial I/O all on a single chip. Microchip Technology's PIC (peripheral interface controller) is an example of a typical microcontroller.

In this chapter we first define some basic terms associated with microcontrollers. We then describe briefly the evolution of microcontrollers. Finally, typical microcontroller-based applications are included.

1.1 Explanation of Terms

Before we go on, it is necessary to understand some basic terms.

- *Address* is a pattern of 0's and 1's that represents a specific location in memory or a particular I/O device. An 8-bit microcontroller with 16 address bits can produce 2^{16} unique 16-bit patterns from 0000000000000000 to 1111111111111111, representing 65,536 different address combinations (addresses 0 to 65,535).

- *Addressing mode* is the manner in which the microcontroller determines the operand (data) and destination addresses during execution of an instruction.

- *Arithmetic-logic unit* (ALU) is a digital circuit that performs arithmetic and logic operations on two n-bit digital words. The value of n for microcontrollers can be 8-bit or 16-bit. Typical operations performed by an ALU are addition, subtraction, ANDing, ORing, and comparison of two n-bit digital words. The size of the ALU defines the size of the microcontroller. For example, an 8-bit microcontroller contains an 8-bit ALU.

- *Big endian* convention is used to store a 16-bit number such as 16-bit data in two bytes of memory locations as follows: the low memory address stores the high byte while the high memory address stores the low byte. The Motorola/Freescale HC11 8-bit microcontroller follows the big endian format.

- *Bit* is an abbreviation for the term *binary digit.* A binary digit can have only two values, which are represented by the symbols 0 and 1, whereas a decimal digit can have 10 values, represented by the symbols 0 through 9. The bit values are easily implemented in electronic and magnetic media by two-state devices whose states portray either of the binary digits 0 and 1. Examples of such two-state devices are a transistor that is conducting or not conducting, a capacitor that is charged or discharged, and a magnetic material that is magnetized north to south or south to north.

- *Bit size* refers to the number of bits that can be processed simultaneously by the basic arithmetic circuits of a microcontroller. A number of bits taken as a group in this manner is called a *word.* For example, an 8-bit microcontroller can process an 8-bit word. An 8-bit word is referred to as a *byte* , and a 4-bit word is known as a *nibble.*

- *Bus* consists of a number of conductors (wires) organized to provide a means of communication among different elements in a microcontroller system. The conductors in a bus can be grouped in terms of their functions. A microcontroller normally has an address bus, a data bus, and a control bus. Address bits are sent to memory or to an external device on the *address bus*. Instructions from memory, and data to/from memory or external devices, normally travel on the *data bus*. Control signals for the other buses and among system elements are transmitted on the *control bus*. Buses are sometimes *bidirectional*; that is, information can be transmitted in either direction on the bus, but normally in only one direction at a time.

- *Clock* is analogous to human heart beats. The microcontroller requires synchronization among its components, and this is provided by a *clock* or timing circuits.

- *CPU* (Central Processing Unit) contains several registers (memory elements), an ALU, and a control unit. Note that the control unit translates instructions and performs the desired task. The number of peripheral devices depends on the particular application involved and may even vary within an application.

- *EEPROM* or *E^2PROM* (Electrically Erasable Programmable ROM) is nonvolatile. EEPROMs can be programmed without removing the chip from the socket. EEPROMs are called Read Most Memories (RMMs), because they have much slower write times than read times. Therefore, these memories are usually suited for applications when mostly reading rather than writing is performed. An example of EEPROM is the 2864 (8K x 8).

- *EPROM* (Erasable Programmable ROM) is nonvolatile. EPROMs can be programmed and erased. The EPROM chip must be removed from the socket for programming. This memory is erased by exposing the chip to ultraviolet light via a lid or window on the chip. Typical erase times vary between 10 and 30 minutes. The EPROM is programmed by inserting the chip into a socket of the EPROM programmer, and providing proper addresses and voltage pulses at the appropriate pins of the chip. An example of EPROM is the 2764 (8K x 8).

- *Flash memory* is designed using a combination of EPROM and EEPROM technologies. Flash memory was invented by Toshiba in the mid 1980s and is nonvolatile. Flash memory can be programmed electrically while embedded on the board. One can change multiple bytes at a time. An example of flash memory is the Intel 28F020 (256K x 8). Flash memory is typically used in cell phones and digital cameras.

- *Harvard architecture* is a type of CPU architecture that uses separate instruction and data memory units along with separate buses for instructions and data. This means that these processors can execute instructions and access data simultaneously. Processors designed with this architecture require four buses for program memory and data memory. These are one data bus for instructions, one address bus for addresses of instructions, one data bus for data, and one address bus for addresses of data. The sizes of the address and data buses for instructions may be different from the address and data buses for data. Several microcontrollers including the PIC18F are designed using the Harvard architecture. This is because it is inexpensive to implement these buses inside the chip since both program and data memories are internal to the chip.

- *Instruction set* of a microcontroller is a list of commands that the microcontroller is designed to execute. Typical instructions are ADD, SUBTRACT, and STORE. Individual instructions are coded as unique bit patterns that are recognized and executed by the microcontroller. If a microcontroller has three bits allocated to the representation of instructions, the microcontroller will recognize a maximum of 2^3, or eight, different instructions. The microcontroller will then have a maximum of eight instructions in its instruction set. It is obvious that some instructions will be more suitable than others to a particular application. For example, in a control application, instructions inputting digitized signals to the processor and outputting digital control variables to external circuits are essential. The number of instructions necessary in an application will directly influence the amount of hardware in the chip set and the number and organization of the interconnecting bus lines.

- *Little endian* convention is used to store a 16-bit number such as 16-bit data in two bytes of memory locations as follows: the low memory address stores the low byte while the high memory address stores the high byte. The PIC18F microcontroller follows the little endian format.

- *Microcomputer* typically consists of a microprocessor (CPU) chip, input and output chips, and memory chips in which programs (instructions and data) are stored.

- *Microcontroller* is implemented on a single chip containing a CPU, memory, and IOP (I/O and peripherals). Note that a typical IOP contains the I/O unit of a microcomputer, timers, an A/D (analog-to-digital) converter, analog comparators, serial I/O, and other peripheral functions (to be discussed later).

- *Microprocessor* is the CPU of a microcomputer contained on a single chip, and must be interfaced with peripheral support chips in order to function.

- *Pipelining* is a technique that overlaps instruction fetch (instruction read) with execution. This allows a microcontroller's processing operation to be broken down into several steps (dictated by the number of pipeline levels or stages) so that the individual step outputs can be handled by the microcontroller in parallel. Pipelining is often used to fetch the microcontroller's next instruction while executing the current instruction, which speeds up the overall operation of the micro controller considerably. Microchip technology's PIC18F (8-bit microcontroller) uses a two-stage instruction pipeline in order to speed up instruction execution.

- *Program* contains instructions and data. Two conventions are used to store a 16-bit number such as 16-bit data in two bytes of memory locations. These are called little endian and big endian byte ordering. In little endian convention, the low memory address stores the low byte while the high memory address stores the high byte. For example, the 16-bit hexadecimal number 2050 will be stored as two bytes in two 16-bit locations (Hex 5000 and Hex 5001) as follows: address 5000 will contain 50 and address 5001 will store 20. In big endian convention, on the other hand, the low memory address stores the high byte while the high memory address stores the low byte. For example, the same 16-bit hexadecimal number 2050 will be stored as two bytes in two 16-bit locations (Hex 5000 and Hex 5001) as follows: address 5000 will contain 20 and address 5001 will store 50. Motorola / Freescale HC11 (8-bit microcontroller) follows big endian convention. Microchip PIC18F (8-bit microcontroller), on the other hand, follows the little endian format.

- *Random-access memory* (RAM) is a storage medium for groups of bits or words whose contents cannot only be read but can also be altered at specific addresses. A RAM normally provides *volatile storage*, which means that its contents are lost in case power is turned off. There are two types of RAM: static RAM (SRAM), and dynamic RAM (DRAM). *Static RAM* stores data in flip-flops. Therefore, this memory does not need to be refreshed. An example of SRAM is 6116 (2K x 8). *Dynamic RAM*, on the other hand, stores data in capacitors. That is, it can hold data for a few milliseconds. Hence, dynamic RAMs are refreshed typically by using external refresh circuitry. Dynamic RAMs (DRAMs) are used in applications requiring large memory. DRAMs have higher densities than SRAMs. Typical examples of DRAMs are the 4464 (64K x 4), 44256 (256K x 4), and 41000 (1M x 1). DRAMs are inexpensive, occupy less space, and dissipate less power than SRAMs.

- *Read-only memory* (ROM) is a storage medium for the groups of bits called *words*, and its contents cannot normally be altered once programmed. A typical ROM is fabricated on a chip and can store, for example, 2048 eight-bit words, which can be accessed individually by presenting to it one of 2048 addresses. This ROM is referred to as a 2K by 8-bit ROM. 10110111 is an example of an 8-bit word that might be stored in one location in this memory. A ROM is a *nonvolatile storage* device, which means that its contents are retained in case power is turned off. Because of this characteristic, ROMs are used to store permanent programs (instructions and data).

- *Reduced Instruction Set Computer* (RISC) contains a simple instruction set. In contrast, a *Complex Instruction Set Computer* (CISC) contains a large instruction set. The PIC18F is a RISC-based microcontroller whereas Motorola/Freescale HC11 is a CISC-based microcontroller.

- *Register* can be considered as volatile storage for a number of bits. These bits may be entered into the register simultaneously (in parallel) or sequentially (serially) from right to left or from left to right, 1 bit at a time. An 8-bit register storing the bits 11110000 is represented as follows:

1	1	1	1	0	0	0	0

- *von Neumann (Princeton) architecture* uses a single memory unit and the same bus for accessing both instructions and data. Although CPUs designed using this architecture are slower compared to Harvard architecture, since instructions and data cannot be accessed simultaneously because of the single bus, typical microprocessors such as the Pentium use this architecture. This is because memory units such as ROMs, EPROMs, and RAMs are external to the microprocessor. This will require almost half the number of wires on the mother board because address and data pins for only two buses rather than four buses (Harvard architecture) are required. This is the reason Harvard architecture would be very expensive if utilized in designing microprocessors. Note that microcontrollers using Harvard architecture internally will have to use von Neumann architecture externally. Texas Instrument's MSP 430 uses the von Neumann architecture.

1.2 Microcontroller Data Types

In this section we discuss data types used by typical microcontrollers: unsigned and signed binary numbers, ASCII (American Standard Code for Information Interchange), EBCDIC (extended binary coded decimal interchange code) and binary-coded decimal (BCD).

1.2.1 Unsigned and Signed Binary Numbers

An *unsigned binary number* has no arithmetic sign and therefore is always positive. Typical examples are your age or a memory address, which are always positive numbers. An 8-bit unsigned binary integer represents all numbers from 00_{16} through FF_{16}(0_{10} through 255_{10}).

A *signed binary number*, on the other hand, includes both positive and negative numbers. It is represented in the microcontroller in two's complement form. For example, the decimal number +15 is represented in 8-bit two's complement form as 00001111 (binary) or 0F (hexadecimal). The decimal number -15 can be represented in 8-bit two's complement form as 11110001 (binary) or F1 (hexadecimal). Also, the most significant bit (MSB) of a signed number represents the sign of the number. For example, bit 7 of an 8-bit number, bit 15 of a 16-bit number, and bit 31 of a 32-bit number represent the signs of the respective numbers. A "0" at the MSB represents a positive number; a "1" at the MSB represents a negative number. Note that the 8-bit binary number 11111111 is 255_{10} when represented as an unsigned number. On the other hand, 11111111_2 is -1_{10} when represented as a signed number.

One can convert an unsigned binary number from lower to higher length using zero extension. For example, an 8-bit unsigned number FF (hex) can be converted to a 16-bit unsigned number 00FF (hex) by extending 0's to the upper byte of 00FF (hex). Both FF (hex) and 00FF (hex) have the same decimal value of 255. This is called zero extension. Zero extension is useful for performing arithmetic operations between two unsigned binary numbers of different lengths.

A signed binary number, on the other hand, can be converted from lower to higher length using sign extension. For example, an 8-bit signed number FF (hex) can be converted to a 16-bit signed number FFFF (hex) by extending the sign bit ('1' in this case) to the upper byte of FFFF (hex). Both FF (hex) and FFFF (hex) have the same decimal value of -1. Sign extension is useful for performing arithmetic operations between two signed binary numbers of different lengths.

Sign extension is useful when one wants to perform an arithmetic operation on two signed numbers of different lengths. For example, the 16-bit signed number 0020 (hex) can be added with the 8-bit signed number E1 (hex) by sign-extending E1 as follows:

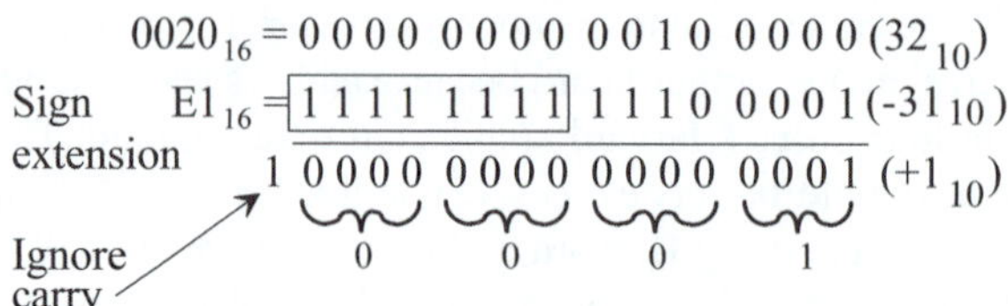

An error (indicated by overflow in a microcontroller) may occur while performing two's complement arithmetic. The microcontroller automatically sets an overflow bit to 1 if the result of an arithmetic operation is too big for the microcontroller's maximum word size; otherwise it is reset to 0. For signed arithmetic operations such as addition, the overflow $V = C_f \oplus C_p$, where C_f is the final carry and C_p is the previous carry. This can be illustrated by the following examples.

Consider the following examples for 8-bit numbers. Let C_f be the final carry (carry out of the most significant bit or sign bit) and C_p be the previous carry (carry out of bit 6 or seventh bit). We will show by means of numerical examples that as long as C_f and C_p are the same, the result is always correct. If, however, C_f and C_p are different, the result is incorrect and sets the overflow bit to 1. Now, consider the following cases.

Case 1: C_f and C_p are the same.

```
       0 0 0 0 0 1 1 0      06₁₆
       0 0 0 1 0 1 0 0     +14₁₆
     0 0 0 0 1 1 0 1 0      1A₁₆
Cf = 0    Cp = 0

       0 1 1 0 1 0 0 0      68₁₆
       1 1 1 1 1 0 1 0     -06₁₆
     1 0 1 1 0 0 0 1 0      62₁₆
Cf = 1    Cp = 1
```

Therefore, when C_f and C_p are either both 0 or both 1, a correct answer is obtained.

Case 2: C_f and C_p are different.

```
       0 1 0 1 1 0 0 1      59₁₆
       0 1 0 0 0 1 0 1     +45₁₆
     0 1 0 0 1 1 1 1 0     -62₁₆ ?
Cf = 0    Cp = 1
```

$C_f = 0$ and $C_p = 1$ give an incorrect answer because the result shows that the addition of two positive numbers is negative.

```
       1 0 1 1 0 1 1 0     -4A₁₆
       1 0 0 0 0 0 0 1     -7F₁₆
     1 0 0 1 1 0 1 1 1     +37₁₆ ?
Cf = 1    Cp = 0
```

$C_f = 1$ and $C_p = 0$ provide an incorrect answer because the result indicates that the addition of two negative numbers is positive. Hence, the overflow bit (V) will be set to zero if the carries C_f and C_p are the same, that is, if both C_f and C_p are either 0 or 1. On the other hand, the overflow bit (V) will be set to 1 if carries C_f and C_p are different. The relationship among C_f, C_p, and V can be summarized in a truth table as follows:

Inputs		Output
C_f	C_p	V
0	0	0
0	1	1
1	0	1
1	1	0

From the truth table, overflow $V = \overline{C_f} C_p + C_f \overline{C_p} = C_f \oplus C_p$

Note that the symbol ⊕ represents exclusive-OR logic operation. Exclusive-OR means that when two inputs are the same (both one or both zero), the output is zero. On the other hand, if two inputs are different, the output is one. The overflow can be considered as an output while C_f and C_p are the two inputs. The answer is incorrect when the overflow bit is set to 1; the answer is correct if the overflow bit is 0.

Typical 16- and 32-bit microprocessors such as Motorola/Freescale 68000/68020 have separate unsigned and signed multiplication and division instructions as follows: MULU (multiply two unsigned numbers), MULS (multiply two signed numbers), DIVU (divide two unsigned numbers), and DIVS (divide two signed numbers). It is important for the programmer to understand clearly how to use these instructions.

For example, suppose that it is desired to compute $X^2/255$. If X is a signed 8-bit number, the programmer should use the MULS instruction to compute $X * X$ which is always unsigned (the square of a number is always positive), and then use DIVU to compute $X^2/255$ (16-bit by 8-bit unsigned divide) since 255_{10} is positive. But if the programmer uses DIVS, both $X *X$ and $255_{10}(FF_{16})$ will be interpreted as signed numbers. FF_{16} will be interpreted as -1_{10}, and the result will be wrong. On the other hand, if X is an unsigned number, the programmer needs to use MULU and DIVU to compute $X^2/255$.

The PIC18F microcontroller includes unsigned multiplication instruction. However, the PIC18F does not provide any signed multiplication and signed or unsigned division instructions. However, as shown in chapter 7, these instructions can be achieved by writing subroutines using PIC18F instructions.

1.2.2 ASCII and EBCDIC Codes

If it is to be very useful, a microcontroller must be capable of handling nonnumeric information. In other words, a microcontroller must be able to recognize codes that represent numbers, letters, and special characters. These codes are classified as alphanumeric or character codes. A complete and adequate set of necessary characters includes the following:

- 26 lowercase letters
- 26 uppercase letters
- 10 numerical digits (0-9)
- Approximately 25 special characters, which include +, /, #, %, and others.

This totals 87 characters. To represent 87 characters with some type of binary code would require at least 7 bits. With 7 bits, there are $2^7 = 128$ possible binary numbers; 87 of these combinations of 0 and 1 bits serve as the code groups representing the 87 different characters.

Two alphanumerical codes are the American Standard Code for Information Interchange (ASCII) and the extended binary-coded-decimal interchange code (EBCDIC). ASCII is typically used with microcontrollers; IBM uses EBCDIC code. Eight bits are used to represent characters, although 7 bits suffice, because the eighth bit is frequently used to test for errors and is referred to as a *parity bit*. It can be set to 1 or 0 so that the number of 1 bits in the byte is always odd or even.

Note that decimal digits 0 through 9 are represented by 30_{16} through 39_{16} in ASCII. On the other hand, these decimal digits are represented by $F0_{16}$ though $F9_{16}$ in EBCDIC. Note that ASCII and unicode are widely used these days. EBCDIC is outdated. However, ASCII and EBCDIC are used in the following example merely for illustrative purposes.

A microcontroller program is usually written for code conversion when input/output devices of different codes are connected to the microcontroller. For example, suppose that it is desired to enter the number 5 into a computer via an ASCII keyboard and to print this number on an EBCDIC printer. The ASCII keyboard will generate 35_{16} when the number 5 is pushed. The ASCII code 35_{16} for the decimal digit 5 enters the microcontroller and resides in the memory. To print the digit 5 on the EBCDIC printer, a program must be written that will convert the ASCII code 35_{16} for 5 to its EBCDIC code, $F5_{16}$. The output of this program is $F5_{16}$. This will be input to the EBCDIC printer. Because the printer understands only EBCDIC codes, it inputs the EBCDIC code $F5_{16}$ and prints the digit 5. Typical microprocessors such as the Intel Pentium include instructions to provide correct unpacked BCD after performing arithmetic operations in ASCII. The Pentium instruction AAA (ASCII adjust for addition) is such an instruction. The PIC18F does not provide such an instruction.

1.2.3 Unpacked and Packed Binary-Coded-Decimal Numbers

The 10 decimal digits 0 through 9 can be represented by their corresponding 4-bit binary numbers. The digits coded in this fashion are called *binary-coded-decimal digits* in 8421 code, or BCD digits. Two unpacked BCD bytes are usually packed into a byte to form *packed BCD*. For example, two unpacked BCD bytes 02_{16} and 05_{16} can be combined as a packed BCD byte 25_{16}.

Let us consider entering data decimal 24 via an ASCII keyboard into a microcontroller. Two keys (2 and 4) will be pushed on the ASCII keyboard. This will generate 32 and 34 (32 and 34 are ASCII codes in hexadecimal for 2 and 4, respectively) inside the microcontroller. A program can be written to convert these ASCII codes into unpacked BCD 02_{16} and 04_{16}, and then to convert to packed BCD 24 or to binary inside the microcontroller to perform the desired operation. Unpacked BCD 02_{16} and 04_{16} can be converted into packed BCD 24 (00100100_2) by logically shifting 02_{16} four times to the left to obtain 20_{16}, then logically ORing with 04_{16}. On the other hand, to convert unpacked BCD 02_{16} and 04_{16} into binary, one needs to multiply 02_{16} by 10 and then add 04_{16} to obtain 00011000_2 (the binary equivalent of 24).

Note that BCD correction (adding 6) is necessary for the following:

i) if the binary sum is greater than or equal to decimal 16 (this will generate a carry of 1)

ii) if the binary sum is 1010 through 1111

For example, consider adding packed BCD numbers 97 and 39:

	111←Intermediate Carries		
97	1001	0111	BCD for 97
+39	0011	1001	BCD for 39
136	1101	0000	invalid sum
	+0110	+0110	add 6 for correction
0001	0011	0110	← correct answer 136
1	3	6	

Typical 32-bit microprocessors such as the Motorola 68020 include PACK and UNPK instructions for converting an unpacked BCD number to its packed equivalent, and vice versa. The PIC18F microcontroller contains an instruction called DAW, which provides the correct BCD result after binary addition of two packed BCD numbers.

1.3 Evolution of the Microcontroller

The Intel Corporation is generally acknowledged as the company that introduced the first microprocessor successfully into the marketplace. Its first microprocessor, the 4004, was introduced in 1971 and evolved from a development effort while a calculator chip set was being made. The 4004 microprocessor was the central component in the chip set, which was called the MCS-4. The other components in the set were a 4001 ROM, a 4002 RAM, and a 4003 shift register.

Shortly after the 4004 appeared in the commercial marketplace, three other general-purpose microprocessors were introduced: the Rockwell International 4-bit PPS-4, the Intel 8-bit 8008, and the National Semiconductor 16-bit IMP-16. Other companies, such as General Electric, RCA, and Viatron, also made contributions to the development of the microprocessor prior to 1971.

The microprocessors introduced between 1971 and 1972 were first-generation systems designed using PMOS technology. In 1973, second-generation microprocessors such as the Motorola 6800 and the Intel 8080 (8-bit microprocessors) were introduced. The second-generation microprocessors were designed using NMOS technology. This technology resulted in a significant increase in instruction execution speed over PMOS and higher chip densities. Since then, microprocessors have been fabricated using a variety of technologies and designs. NMOS microprocessors such as the Intel 8085, the Zilog Z80, and the Motorola 6800/6809 were introduced based on second-generation microprocessors. A third-generation HMOS microprocessor, introduced in 1978, is typically represented by the Intel 8086 and the Motorola 68000, which are 16-bit microprocessors.

During the 1980s, fourth-generation HCMOS and BICMOS (a combination of bipolar and HCMOS) 32-bit microprocessors evolved. Intel introduced the first commercial 32-bit microprocessor, the problematic Intel 432, which was eventually discontinued. Since 1985, more 32-bit microprocessors have been introduced. These include Motorola's 68020, 68030, 68040, 68060, and PowerPC; Intel's 80386, 80486, and Pentium family, Core Duo, and Core2 Duo microprocessors.

The performance offered by the 32-bit microprocessor is more comparable to that of superminicomputers such as Digital Equipment Corporation's VAX11/750 and VAX11/780. Intel and Motorola also introduced RISC microprocessors: the Intel 80960 and Motorola 88100/PowerPC, which had simplified instruction sets. Note that the purpose of RISC microprocessors is to maximize speed by reducing clock cycles per instruction. Almost all computations can be obtained from a simple instruction set. Note that, in order to enhance performance significantly, Intel Pentium Pro and other succeeding members of the Pentium family and Motorola 68060 are designed using a combination of RISC and CISC.

Single-chip microcomputers such as the Intel 8048 evolved during the '80s. Soon afterward, based on the concept of single-chip microcomputers, Intel introduced the first 8-bit microcontroller—the Intel 8051, which uses Harvard architecture. The 8051 is designed using CISC. The 8051 contains a CPU, memory, I/O, A/D and D/A converters, a timer, and a serial communication interface—all in a single chip. The microcontrollers became popular during the '80s.

The 8-bit microcontrollers are small enough for many embedded applications, but also powerful enough to allow a lot of complexity and flexibility in the design process of an embedded system. Several billion 8-bit microcontrollers were sold during the last decade. Several contemporary microcontroller manufacturers use RISC architecture, and thus provide a cost effective approach. In addition, typical 8-bit microcontrollers such as the PIC18F implemented several on-chip enhanced peripheral functions including PWM (pulse-width modulation) and flash memories. Note that the Motorola/Freescale popular 8-bit microcontroller HC11 does not have on-chip flash memory and PWM functions. PWM is a very desirable feature for applications such as automotive and motor control. These applications may include driving servo motors. In HC11, a timer section is used to generate PWM signals. However, Motorola/Freescale implemented these features in the HC12, which is a 16-bit microcontroller. Note that the HC11 has been popular because of its rich instruction set.

Like EEPROM, flash memory can be programmed and erased electrically. Flash memory is very popular these days compared to EEPROM. Note that EEPROM can be erased one byte at a time while flash memory can be erased only in blocks.

Table 1.1 provides a comparison of some of the basic features of some of the typical microcontrollers.

Microchip has introduced several different versions of the PIC18F microcontroller over the years. All members of the PIC18F family basically contain the same instruction set. However, certain features such as memory sizes, number of I/O ports, A/D channels, and PWM modules may vary from one version to another. In this book, a specific PIC18F chip such as the PIC18F4321 will be considered in detail.

1.4 Typical Microcontroller Applications

Some of the typical microcontroller applications include the following:

- automotive systems
- operation of devices such as a microwave oven, a radiator fan in a car, or servo motors used to move the handles on a foosball table
- barcode readers
- hotel card key writers
- robotics

In the following discussion, a microcontroller-based temperature control system is first described. Since microcontrollers are widely used as "embedded controllers" in embedded applications, the basic concepts associated with embedded controllers are then considered.

TABLE 1.1 Comparison of basic features of typical microcontrollers

	PIC18F	MSP 430	HC11	AVR
Manufacturer	Microchip Technology	Texas Instruments	Motorola / Freescale	Atmel
Introduced	2000; the first PIC in 1989	Late 1990s	1985	1996
Size	8-bit	16-bit	8-bit	8-bit
Architecture	Harvard	von Neumann	von Neumann	Harvard
Design approach	RISC	RISC	CISC	RISC
On-chip flash memory	Yes	Yes	No	Yes. First to offer on-chip flash
On-chip PWM	Yes.	Yes	No	Yes
CPU Clock	40-MHz (maximum)	1-MHz (maximum)	4-MHz (maximum)	20-MHz (maximum)
Total Instructions	75	27	144	123
Total Addressing modes	6	7	6	5

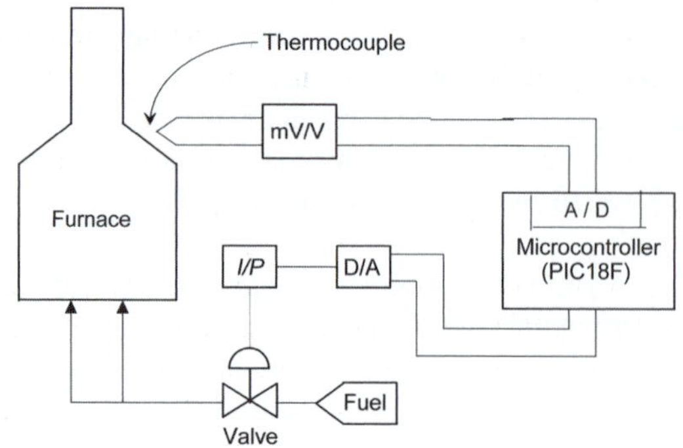

FIGURE 1.1 Furnace temperature control

1.4.1 A Simple Microcontroller Application

To put microcontrollers into perspective, it is important to explore a simple application. For example, consider the microcontroller-based dedicated controller shown in Figure 1.1. Suppose that it is necessary to maintain the temperature of a furnace to a desired level to maintain the quality of a product. Assume that the designer has decided to control this temperature by adjusting the fuel. This can be accomplished using a typical microcontroller such as the PIC18F along with the interfacing components as follows. Temperature is an analog (continuous) signal. It can be measured by a temperature-sensing (measuring) device such as a thermocouple. The thermocouple provides the measurement in millivolts (mV) equivalent to the temperature.

Since microcontrollers only understand binary numbers (0's and 1's), each analog mV signal must be converted to a binary number using the microcontroller's on-chip analog-to-digital (A/D) converter. Note that the PIC18F contains an on-chip A/D converter. The PIC18F does not include an on-chip digital-to-analog (D/A) converter. However, the D/A converter chip can be interfaced to the PIC18F externally.

First, the millivolt signal is amplified by a mV/V amplifier to make the signal compatible for A/D conversion. A microcontroller such as the PIC18F can be programmed to solve an equation with the furnace temperature as an input. This equation compares the temperature measured with the temperature desired, which can be entered into the microcontroller using the keyboard. The output of this equation will provide the appropriate opening and closing of the fuel valve to maintain the appropriate temperature. Since this output is computed by the microcontroller, it is a binary number. This binary output must be converted into an analog current or voltage signal.

The D/A (digital-to-analog) converter chip inputs this binary number and converts it into an analog current (*I*). This signal is then input into the current/pneumatic (*I/P*) transducer for opening or closing the fuel input valve by air pressure to adjust the fuel to the furnace. The furnace temperature desired can thus be achieved. Note that a transducer converts one form of energy (electrical current in this case) to another form (air pressure in this example).

1.4.2 Embedded Controllers

Embedded microcontroller systems, also called embedded controllers, are designed to manage specific tasks. Once programmed, the embedded controllers can manage the functions of a wide variety of electronic products. In embedded applications, the microcontrollers are embedded in the host system; their presence and operation are basically hidden from the host system.

Typical embedded control applications include office automation products such as copiers, laser products, fax machines, and consumer electronics such as VCRs and microwave ovens. Applications such as printers typically utilize a microcontroller. The RISC microcontrollers are ideal for these types of applications. Note that the personal computer interfaced to the printer is the host.

RISC microcontrollers such as the PIC18F are well suited for applications including robotics, controls, instrumentation, and consumer electronics. The key features of the RISC microcontrollers that make them ideal for these applications are their relatively low level of integration in the chip, and instruction pipeline architecture. These characteristics result in low power consumption, fast instruction execution, and fast recognition of interrupts.

Although microcontrollers including PIC18F are considered ideal for many embedded applications, sometimes they might not be able to perform certain tasks. For example, applications such as laser printers require a high performance microprocessor with on-chip floating-point hardware. The PowerPC RISC microprocessor with on-chip floating-point hardware is ideal for these types of applications. Note that the personal computer interfaced to the laser printer is the host. The PIC18F will not be suitable for such an application since it does not provide floating-point instructions.

2
MICROCONTROLLER BASICS

In this chapter we describe the fundamental material needed to understand the basic characteristics of microcontrollers. It includes topics such as typical microcontroller architectures, timing signals, CPU organization, and status flags. An overview of pipelining and RISC vs. CISC is included. Finally, an introduction to the functional characteristics of the PIC18F is included.

2.1 Basic Blocks of a Microcomputer

In order to understand the functions performed by typical modules contained in a microcontroller, it is necessary to cover the basic blocks of a microcomputer.

A microcomputer has three basic blocks: a microprocessor (CPU on a chip), a memory unit, and an input/output (I/O) unit. Figure 2.1 shows the basic blocks of a microcomputer. A system bus (comprised of several wires) connects these blocks. The CPU executes all the instructions and performs arithmetic and logic operations on data. The CPU of the microcomputer contains all the registers and the control unit, as well as arithmetic-logic circuits of the microcomputer.

A *memory unit* stores both data and instructions. The memory section typically contains ROM and RAM chips. The ROM can only be read and is nonvolatile; that is, it retains its contents when the power is turned off. A ROM is typically used to store instructions and data that do not change. For example, it might store a table of seven-segment codes for outputting data to a display external to the microcomputer for turning on a digit from 0 through 9.

One can read from and write into a RAM. The RAM is volatile; that is, it does not retain its contents when the power is turned off. A RAM is used to store programs and data that are temporary and might change during the course of executing a program. An *I/O unit* transfers data between the microcomputer and the external devices via I/O ports (registers). The transfer involves data, status, and control signals.

In a single-chip microcomputer, these three elements are on one chip, whereas in a single-chip microprocessor, separate chips are required for memory and I/O. Microcontrollers, which evolved from single-chip microcomputers, are typically used for dedicated applications such as automotive systems, home appliances, and home entertainment systems. Microcontrollers include a CPU, memory, and IOP (I/O and Peripherals) on a single chip. Note that a typical IOP contains I/O unit of a microcomputer, timers, A/D (analog-to-digital) converter, analog comparators, serial I/O, and other peripheral functions (to be discussed later). Two popular microcontrollers are Microchip Technology's 8-bit PIC (peripheral interface controller) microcontroller and Motorola's HC11 (8-bit).

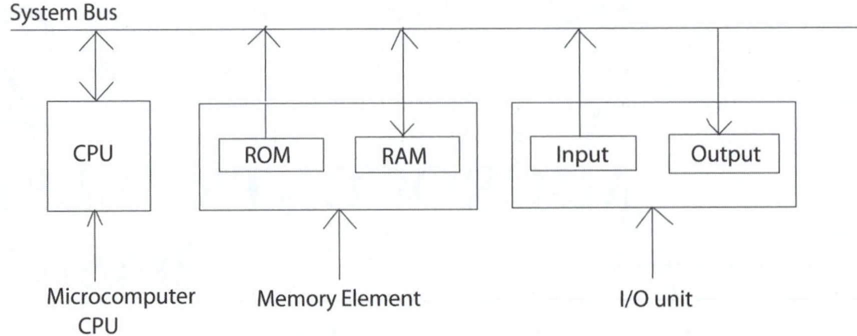

FIGURE 2.1 Basic blocks of a microcomputer

Since the microcomputer is an integral part of a microcontroller, it is necessary to investigate a typical microcomputer in detail. Once such a clear understanding is obtained, it will be easier to work with any specific microcontroller. Figure 2.2 illustrates a very simplified version of a typical microcomputer and shows the basic blocks of a microcomputer system. The various buses that connect these blocks are also shown. Although this figure looks very simple, it includes all of the main elements of a typical microcomputer system.

2.1.1 System Bus

The microcomputer's system bus (internal to the microcontroller) contains three buses, which carry all of the address, data, and control information involved in program execution. These buses connect the CPU to each of the ROM, RAM, and I/O chips so that information transfer between the CPU and any of the other elements can take place. In a microcomputer, typical information transfers are carried out with respect to the memory or I/O. When a memory or an I/O chip receives data from the microprocessor, it is called a *WRITE operation*, and data are written into a selected memory location or an I/O port (register). When a memory or an I/O chip sends data to the microprocessor, it is called a *READ operation*, and data are read from a selected memory location or an I/O port.

In the *address bus*, information transfer takes place in only one direction, from

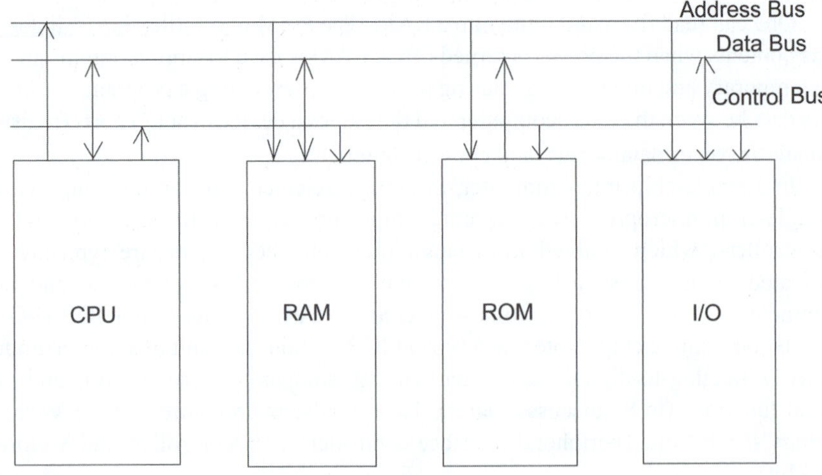

FIGURE 2.2 Simplified version of a typical microcomputer

the microprocessor to the memory or I/O elements. This is therefore called a *unidirectional bus*. The size of the address bus determines the total number of memory addresses available in which programs can be executed by the microprocessor. The address bus is specified by the total number of address bits required by the CPU. This also determines the direct addressing capability or the size of the main memory of the microcontroller. The microcontroller's CPU can execute programs located only in the main memory. For example, a CPU with 16 address bits can generate 2^{16} = 64,536 bytes [64 kilobytes (kB)] of different possible addresses (combinations of 1's and 0's) on the address bus. The CPU includes addresses from 0 to $65{,}535_{10}$ (0000_{16} through $FFFF_{16}$). A memory location can be represented by each of these addresses. For example, an 8-bit data item $2B_{16}$ can be stored at 16-bit address 0200_{16}.

When a CPU with a 16-bit address bus wants to transfer information between itself and a certain memory location, it generates the 16-bit address from an internal register on its 16 address pins, A_0–A_{15}, which then appears on the address bus. These 16 address bits are decoded to determine the desired memory location. The decoding process normally requires hardware (decoders) not shown in Figure 2.2.

In the *data bus*, data can flow in both directions, that is, to or from the CPU. This is therefore a bidirectional bus.

The *control bus* consists of a number of signals that are used to synchronize operation of the individual microcomputer elements. The CPU sends some of these control signals to the other elements to indicate the type of operation being performed. Each microcontroller has a unique set of control signals. However, some control signals are common to most microcontrollers. We describe some of these control signals later in this section.

2.1.2 Clock Signals

The system clock signals are contained in the control bus. These signals generate the appropriate clock periods during which instruction executions are carried out by the CPU. Typical microcontrollers have an internal clock generator circuit to generate a clock signal. Figure 2.3 shows a typical clock signal.

The number of cycles per second (Hertz, abbreviated as Hz) is referred to as the *clock frequency*. The CPU clock frequencies of typical microcontrollers vary from 1MHz (1×10^6Hz) to 40MHz (40×10^6Hz) . The clock defines the speed of the microcontroller. Note that one clock cycle = $1/f$, where f is the clock frequency. The execution times of microcontroller instructions are provided in terms of the number of clock cycles.

For example, suppose that execution time for the addition instruction by a microcontroller is one cycle. This means that a microcontroller with a 40MHz clock will execute the ADD instruction in 25 nanoseconds [clock cycle = $1/(40 \times 10^6)$ = 25 nanoseconds]. On the other hand, for a 4MHz microcontroller, the addition instruction will be executed in 250 nanoseconds [clock cycle = $1/(4 \times 10^6)$ = 250 nanoseconds]. This implies that the higher the clock frequency, the faster the microcontroller can execute the instructions.

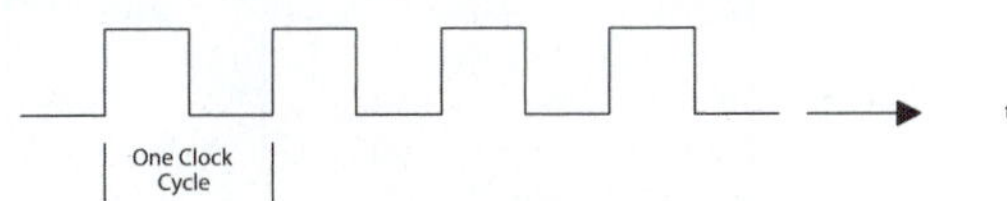

FIGURE 2.3 Typical clock signal

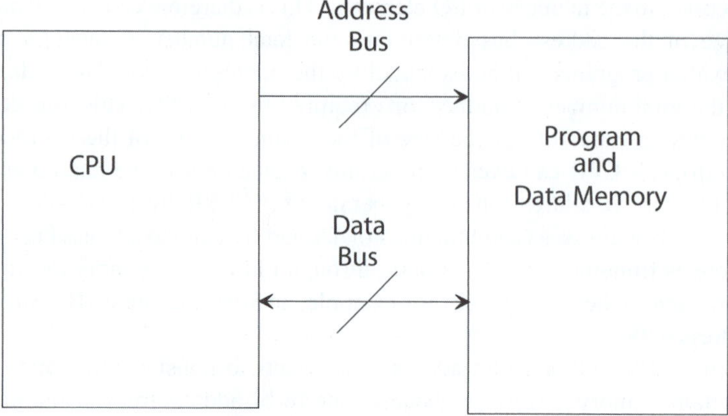

FIGURE 2.4 von Neumann architecture

2.2 Microcontroller Architectures

The microcontroller requires memory to store programs and data. The various microcontrollers available today are basically the same in principle. The main variations are in the number of memory units, and the address and data buses they use. As mentioned in Chapter 1, two types of CPU architectures are used for designing microcontrollers. They are von Neumann (Princeton) and Harvard.

In von Neumann architecture, a single memory system with the same address and data buses is used for accessing both programs and data. This means that programs and data cannot be accessed simultaneously. This may slow down the overall speed. Texas Instrument's MSP 430 uses von Neumann architecture. Figure 2.4 shows a block diagram of the von Neumann architecture.

Harvard architecture is a type of computer architecture that uses separate program and data memory units along with separate buses for instructions and data. This means that these processors can execute instructions and access data simultaneously. Processors designed with this architecture require four buses for program memory and data memory. These are: one data bus for instructions, one address bus for addresses of instructions, one data bus for data, and one address bus for addresses of data. The sizes of the address and data buses for instructions may be different from those of the address and data buses

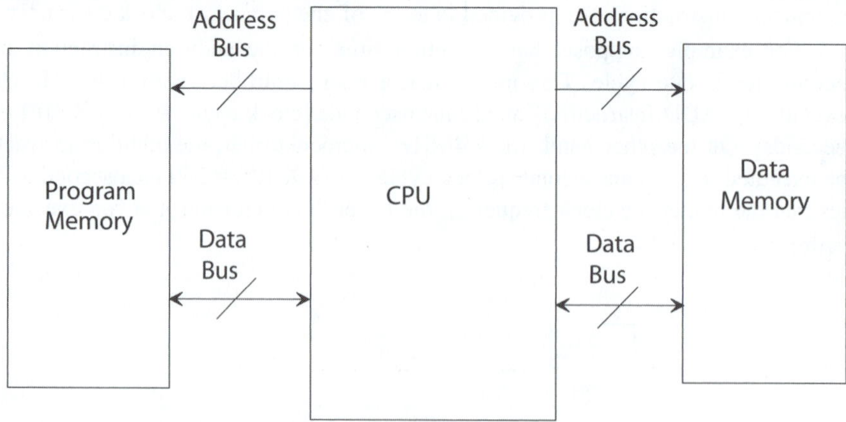

FIGURE 2. 5 Harvard architecture

for data. Several microcontrollers including the PIC18F are designed using the Harvard architecture. Figure 2.5 shows a block diagram of the Harvard architecture.

Most microcontrollers use the Harvard architecture. This is because it is inexpensive to implement these buses inside the chip since both program and data memories are internal to the chip.

Although processors designed using the von Neumann architecture are slower compared to the Harvard architecture since instructions and data cannot be accessed simultaneously because of the single bus, typical microprocessors such as the Pentium use this architecture. This is because memory units such as ROMs and RAMs are external to the microprocessor. This will require almost half the number of wires on the mother board since address and data pins for only two buses rather than four buses (Harvard architecture) are required. This is the reason Harvard architecture would be very expensive if utilized in designing microprocessors. Note that microcontrollers using Harvard architecture internally will have to use von Neumann architecture externally.

2.3 Central Processing Unit (CPU)

As mentioned earlier, the CPU is the brain of the microcontroller. Therefore, the power of the microcontroller is determined by the capabilities of the CPU. Its clock frequency determines the speed of the microcontroller. The number of data and address bits on the CPU make up the microcontroller's word size and maximum memory size. The microcontroller's I/O and interfacing capabilities are determined by the control bus of the CPU.

The logic inside the CPU can be divided into three main areas: the register section, the control unit, and the arithmetic-logic unit (ALU). A CPU chip with these three sections is shown in Figure 2.6.

2.3.1 Register Section

The number, size, and types of registers vary from one CPU to another. However, the various registers in all CPUs carry out similar operations. The register structures of CPUs play a major role in the design of a microcontroller. Also, the register structures for a specific CPU determine how convenient and easy it is to program the microcontroller. We first describe the most basic types of CPU registers, their functions, and how they are used. We then consider other common types of registers.

Basic CPU Registers There are four basic CPU registers: instruction register, program counter, memory address register, and accumulator.

- *Instruction register* (IR). The instruction register stores instructions. The contents of an instruction register are always decoded by the CPU as an instruction. After fetching an instruction code from memory, the CPU stores it in the instruction register. The instruction is decoded (translated) internally by the CPU, which then performs the operation required. The word size of the CPU normally determines the size of the instruction register.

Registers
ALU
Control Unit

FIGURE 2.6 CPU with the main functional elements

- *Program Counter* (PC). The program counter contains the address of the instruction or operation code (op-code), normally the address of the next instruction to be executed. Note the following features of the program counter:
  1. Upon activating the CPU's RESET input, the address of the first instruction to be executed is normally loaded into the program counter.
  2. To execute an instruction, the CPU typically places the contents of the program counter on the address bus and reads ("fetches") the contents of this address (i.e., instruction) from memory. The program counter contents are incremented automatically by the CPU's internal logic. The CPU thus executes a program sequentially, unless the program contains an instruction such as a JUMP instruction, which changes the sequence.
  3. The size of the program counter is determined by the size of the address bus.
  4. Many instructions, such as JUMP and conditional JUMP, change the contents of the program counter from its normal sequential address value. The program counter is loaded with the address specified in these instructions.
- *Memory Address Register* (MAR). The memory address register contains the address of data. The CPU uses the address as a direct pointer to memory. The contents of the address are the actual data that are being transferred.
- *Accumulator* (A). The accumulator is typically an 8-bit register. It stores the results after most ALU operations. These 8-bit CPUs have instructions to shift or rotate the accumulator one bit to the right or left through the carry flag. The accumulator is typically used for inputting a byte into the accumulator from an external device or for outputting a byte to an external device from the accumulator. The accumulator in the PIC18F is called the working register (WREG).

Depending on the register section, the CPU can be classified either as an accumulator- or general-purpose register-based machine. In an accumulator-based microcontroller (PIC18F), the data are assumed to be held in a register called the accumulator. All arithmetic and logic operations are performed using this register as one of the data sources. The result of the operation is stored in the accumulator. Microchip Technology's PIC18F (8-bit microcontroller) is accumulator-based.

Texas Instrument's MSP430, on the other hand, is a general-purpose register-based 16-bit microcontroller. The term *general-purpose* comes from the fact that these registers can hold data, memory addresses, or the results of arithmetic or logic operations. The number, size, and types of registers vary from one microcontroller to another. Most registers are general-purpose, but some, such as the program counter (PC), are provided for dedicated functions. The PC normally contains the address of the next instruction to be executed.

As mentioned before, upon activating the CPU's RESET input pin, the PC is normally initialized with the address of the first instruction. For example, the PIC18F, upon hardware reset, reads the first instruction from address 0. Note that the PC in PIC18F is 21-bit wide. Hence, upon hardware reset, the PC in PIC18F will contain 21 zeros. To execute the instruction, the PIC18F places the PC contents (0 in this case) on the address bus and reads (fetches) the first instruction from internal memory. The program counter contents are then incremented automatically by the ALU. The PC normally points to the next instruction.

Use of Basic CPU Registers To provide a clear understanding of how the basic registers in an accumulator-based CPU are used, a binary addition program will be considered. The program logic will be explained by showing how each instruction changes the contents of the four basic CPU registers (PC, IR, MAR, A). In PIC18F, the MAR is called "File Register"or "Register" and is located in data memory which is external to the CPU. Note that there are several File registers in the data memory in the PIC18F. Assume that the address of program memory is 16-bit wide with 16-bit contents while the address of data memory is 8-bit wide with 8-bit contents. Suppose that the contents of the MAR with data memory address 0x20 are to be added to the contents of the accumulator. Assume that [NNNN] represents the contents of the memory location NNNN. Now, assume that [0x20] = 0x05.

The steps involved in adding [0x20] with the contents of the accumulator can be summarized as follows:

1. Load 'A' with the first data (0x02) to be added.
2. Add the contents of the accumulator 'A' to [0x20], and store the result in address 0x20.

   The following instructions for the PIC18F will be used to achieve the above addition:

0x0E02	Load 0x02 into 'A'
0x2620	Add [A] with [0x20] and store result in MAR with address 0x20.

The complete program in hexadecimal, starting at location 0x200 (arbitrarily chosen), is given in Figure 2.7. Note that program memory address stores 16 bits. Hence, memory addresses are shown in increments of 2. Data memory, on the other hand, stores 8-bit. Hence, data addresses are shown in increments of 1. Assume that the CPU can be instructed that the starting address of the program is 0x0200. This means that the program counter can be initialized to contain 0x0200, the address of the first instruction to be executed. Note that the contents of the other three registers (IR, MAR. A) are not known at

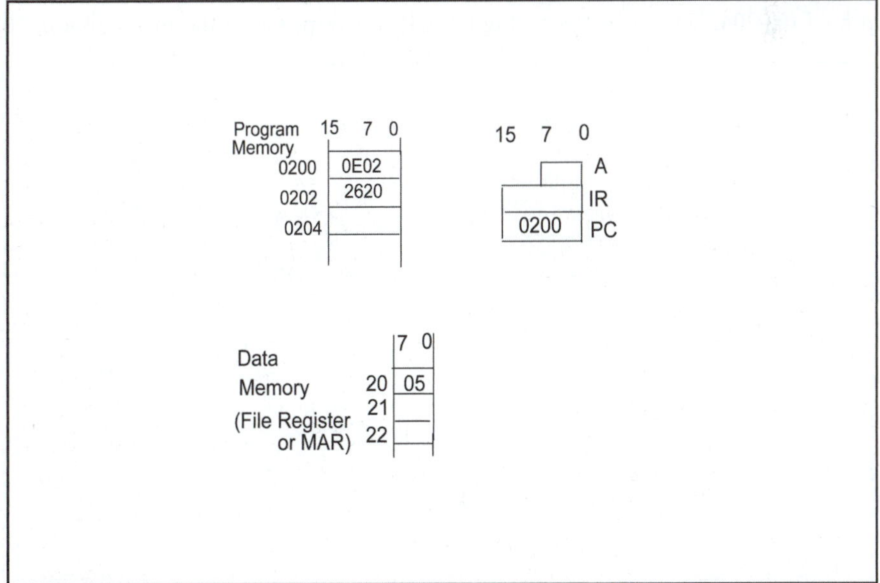

FIGURE 2.7 Addition program with initial register and memory contents

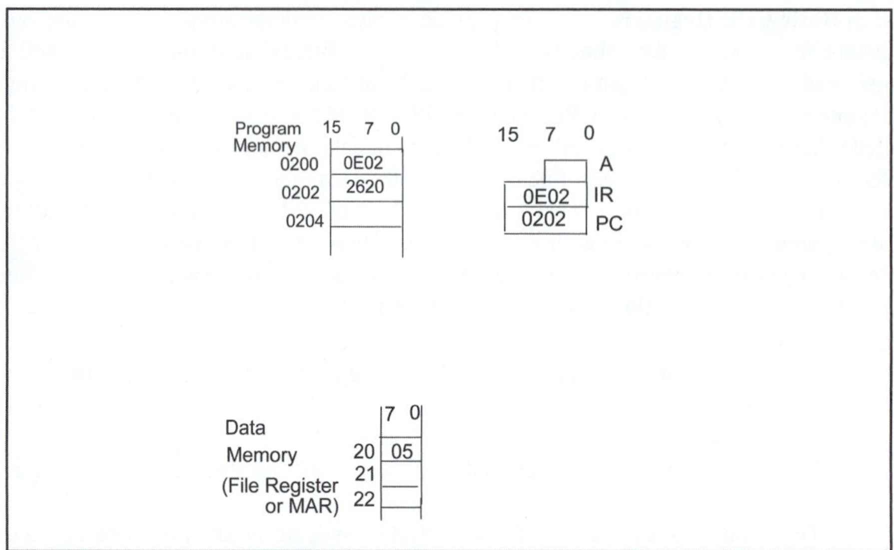

FIGURE 2.8 Addition program (modified during execution)

this point. The CPU loads the contents of the memory location addressed by the program counter into IR. Thus, the first instruction, $0E02_{16}$, stored in address 0x200, is transferred into IR.

The program counter contents are then incremented by 2 by the ALU to hold 0x0202. The register contents along with the program are shown in Figure 2.8.

The binary code 0x0E02 in the IR is executed by the CPU. The CPU then takes appropriate actions. Note that the instruction 0x0E02 loads 0x02 into 'A' register. This is shown in Figure 2.9.

Next, the CPU loads the contents of the memory location addressed by the PC into the IR; thus, 0x2620 is loaded into the IR. The PC contents are then incremented by 2 to hold 0x0204. This is shown in Figure 2.10. In response to the instruction 0x2620,

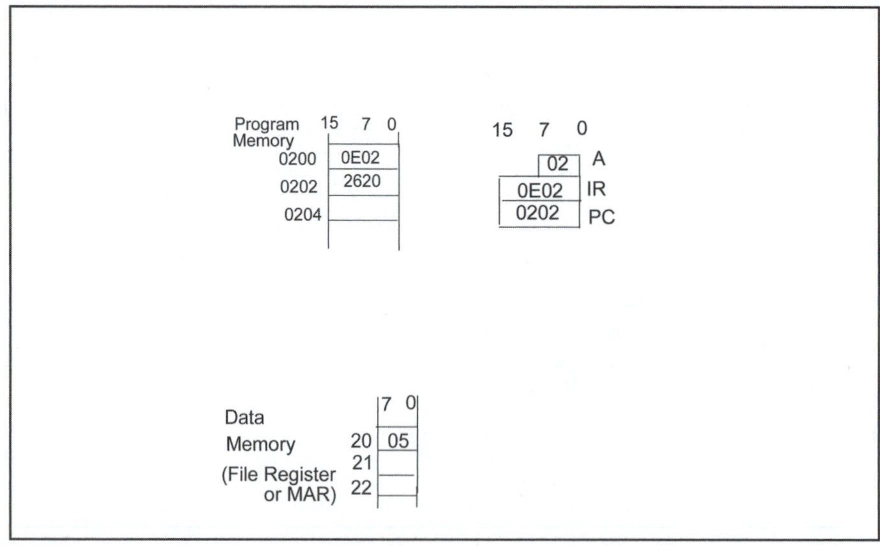

FIGURE 2.9 Addition program (modified during execution)

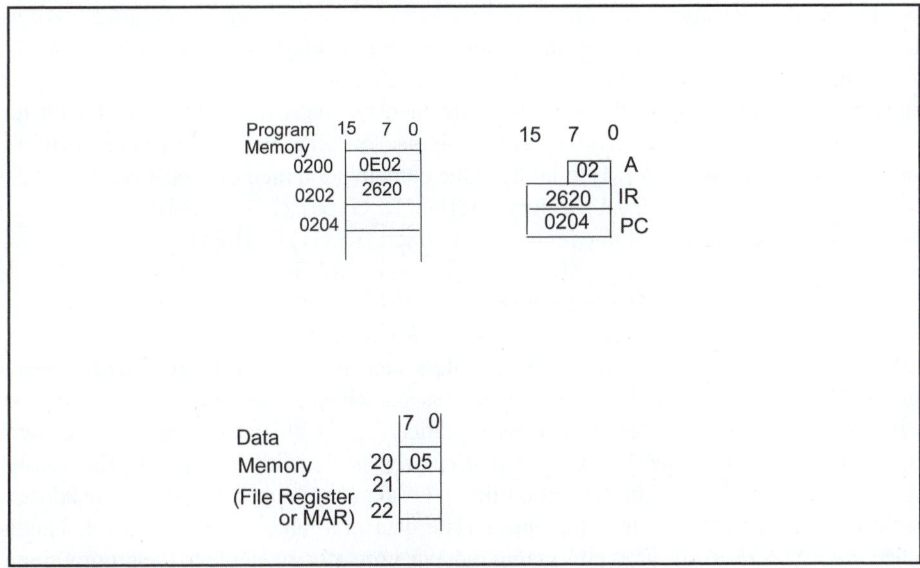

FIGURE 2.10 Addition program (modified during execution)

the contents of the data memory location addressed by the MAR (0x20) are added to the contents of the accumulator A; thus, 0x05 is added to 0x02. The result 0x07 is stored in data memory address 0x20. Note that the previous contents (0x05) of data memory address 0x20 are lost. The contents of the PC are not incremented this time. This is because 0x05 is obtained from data memory. Figure 2.11 shows the details.

Other CPU Registers In the following, we describe other CPU registers such as general-purpose registers, index register, status register, and stack pointer register.

General-Purpose Registers Some microcontrollers such as the Texas Instrument's MSP430 have a number of general-purpose registers for storing temporary data or for

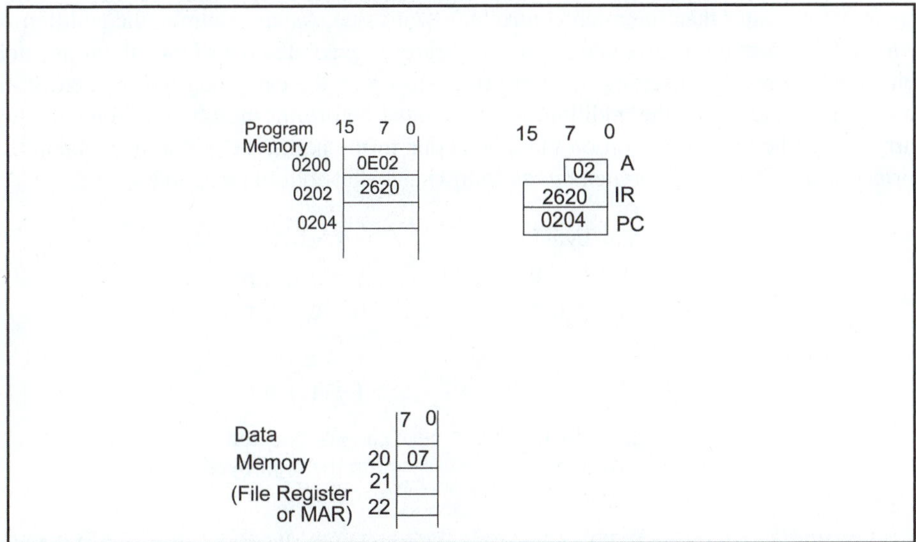

FIGURE 2.11 Addition program (modified during execution)

carrying out data transfers between various registers. The use of general-purpose registers speeds up the execution of a program because the microcontroller does not have to read data from external memory via the data bus if data are stored in one of its general-purpose registers. Some of the typical functions performed by instructions associated with the general-purpose registers are given here. We will use [REG] to indicate the contents of the general-purpose register and [M] to indicate the contents of a memory location.

1. Move [REG] to or from memory: [M] ← [REG] or [REG] ← [M].
2. Move the contents of one register to another: [REG1] ← [REG2].
3. Increment or decrement [REG] by 1: [REG] ← [REG] + 1 or [REG] ← [REG] - 1.
4. Load 16-bit data into a register [REG] : [REG] ← 16-bit data.

Index Register Some microcontrollers such as the PIC18F provide an indexed addressing mode using an index register to access an element in an array. An *index register* is typically used as a counter in address modification for an instruction or for general storage functions. The index register is particularly useful with instructions that access tables or arrays of data. In this operation the index register is used to modify the address portion of the instruction. Thus, the appropriate data in a table can be accessed. This is called *indexed addressing*. This addressing mode is normally available to the programmers of PIC18F. The effective address for an instruction using the indexed addressing mode is determined by adding the address portion of the instruction to the contents of the index register. Note that the accumulator is used as the index register in the PIC18F.

Status Register A *status register,* also known as a *processor status word register* or *condition code register*, contains individual bits, with each bit having special significance. The bits in the status register are called *flags*. The status of a specific microcontroller operation is indicated by a flag, which is set or reset by the microcontroller's internal logic to indicate the status of certain operations such as arithmetic and logic operations. The status flags are also used in conditional JUMP instructions. We describe some of the common flags in the following.

A *carry flag* is used to reflect whether or not the result generated by an arithmetic operation is greater than the microcontroller's word size. As an example, the addition of two 8-bit numbers might produce a carry. The carry is generated out of the 8th bit position (bit 7), which results in setting the carry flag. However, the carry flag will be zero if no carry is generated from the addition. As mentioned before, in multibyte arithmetic, any carry out of the low-byte addition must be added to the high-byte addition to obtain the correct result. This can illustrated by the following 16-bit addition example:

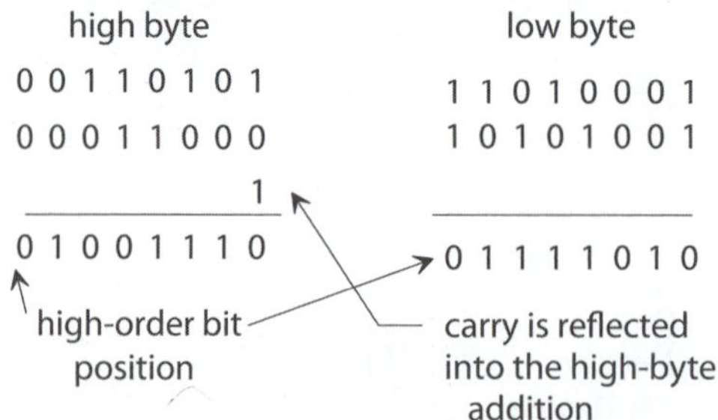

While performing BCD arithmetic with microcontrollers, the carry out of the low nibble (4 bits) has a special significance. Because a BCD digit is represented by 4 bits, any carry out of the low 4 bits must be propagated into the high 4 bits for BCD arithmetic. This carry flag is known as *"Digit Carry (DC)"* flag in the PIC18F and is set to 1 if the carry out of the low 4 bits is 1; otherwise, it is 0. A *zero flag* is used to show whether the result of an operation is zero. It is set to 1 if the result is zero, and it is reset to 0 if the result is nonzero. A *sign flag* (sometimes called a negative flag) is used to indicate whether the result of the last operation is positive or negative. If the most significant bit of the last operation is 1, this flag is set to 1 to indicate that the result is negative. This flag is reset to 0 if the most significant bit of the result is zero: that is, if the result is positive.

As mentioned earlier, an *overflow flag* arises from representation of the sign flag by the most significant bit of a word in signed binary operation. The overflow flag is set to 1 if the result of an arithmetic operation is too big for the microcontroller's maximum word size; otherwise it is reset to 0. Let C_f be the final carry out of the most significant bit (sign bit) and C_p be the previous carry. It was shown in Chapter 1 that the overflow flag is the exclusive- OR of the carries C_p and C_f:

$$\text{overflow} = C_p \oplus C_f$$

For 8-bit signed arithmetic operations, the overflow flag will be set to 1 if the result is greater than $+127_{10}$ or less than or equal to - 128_{10}.

Stack Pointer Register A *stack* consists of a number of RAM locations set aside for reading data from or writing data into these locations and is typically used by subroutines (a *subroutine* is a program that performs operations frequently needed by the main or calling program). The address of the stack is contained in a register called a *stack pointer*. The size of the stack memory is normally the same as that of the program counter. For example, since the program counter in PIC18F is 21-bit wide, the stack memory is also 21-bit. However, the size of the stack pointer in PIC18F is 5-bit which provides 32 (2^5) locations for the stack.

Two instructions, PUSH and POP, are usually available with a stack. The *PUSH operation* is defined as writing to the top or bottom of the stack, whereas the *POP operation* means reading from the top or bottom of the stack. Some microcontrollers access the stack from the top; others access via the bottom. When the stack is accessed from the bottom, the stack pointer is incremented after a PUSH and decremented after a POP operation. On the other hand, when the stack is accessed from the top, the stack pointer is decremented after a PUSH and incremented after a POP. Microcontrollers typically use internal registers for performing PUSH or POP operations. The incrementing or decrementing of a stack pointer depends on whether the operation is PUSH or POP and on whether the stack is accessed from the top or the bottom.

We now illustrate stack operations in more detail. We use 16-bit registers and 16-bit addresses in Figures 2.12 through 2.15. All data (hex) are chosen arbitrarily. In Figure 2.12, the stack pointer is incremented by 2 (16-bit register) after the PUSH to contain the value 20CA. Now, consider the POP operation of Figure 2.13. The stack pointer is decremented by 2 after the POP. The contents of address 20CA are assumed to be empty conceptually after the POP operation. Next, consider the PUSH operation of Figure 2.14. The stack is accessed from the top. The stack pointer is decremented by 2 after a PUSH. Finally, consider the POP operation of Figure 2.15. The stack pointer is incremented by 2 after the POP. The contents of address 20C6 are assumed to be empty conceptually after a POP operation.

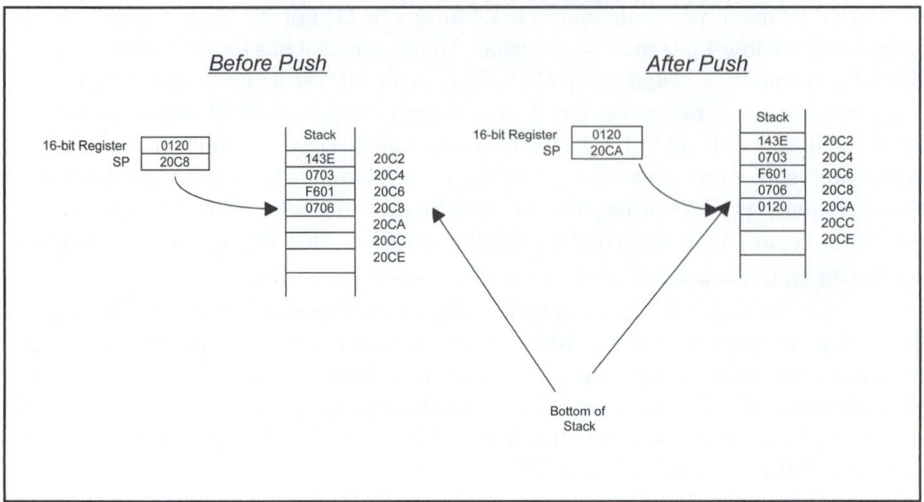

FIGURE 2.12 PUSH operation when accessing a stack from the bottom

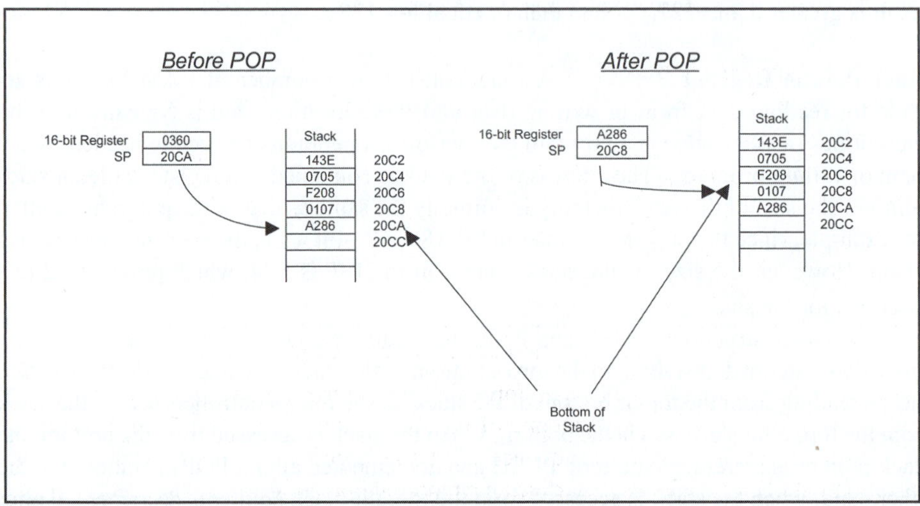

FIGURE 2.13 POP operation when accessing a stack from the bottom

Note that the stack is a LIFO (last in, first out) memory. As mentioned earlier, a stack is typically used during subroutine CALLs. The CPU automatically PUSHes the return address onto a stack after executing a subroutine CALL instruction in the main program. After executing a RETURN from a subroutine instruction (placed by the programmer as the last instruction of the subroutine), the CPU automatically POPs the return address from the stack (previously PUSHed) and then returns control to the main program. Note that the PIC18F accesses the stack from the top. This means that the stack pointer in the PIC18F holds the address of the top of the stack. Hence, in the PIC18F, the stack pointer is incremented after a PUSH, and decremented after a POP.

2.3.2 Control Unit

The main purpose of the control unit is to read and decode instructions from the program memory. To execute an instruction, the control unit steps through the appropriate blocks of the ALU based on the op-codes contained in the instruction register. The

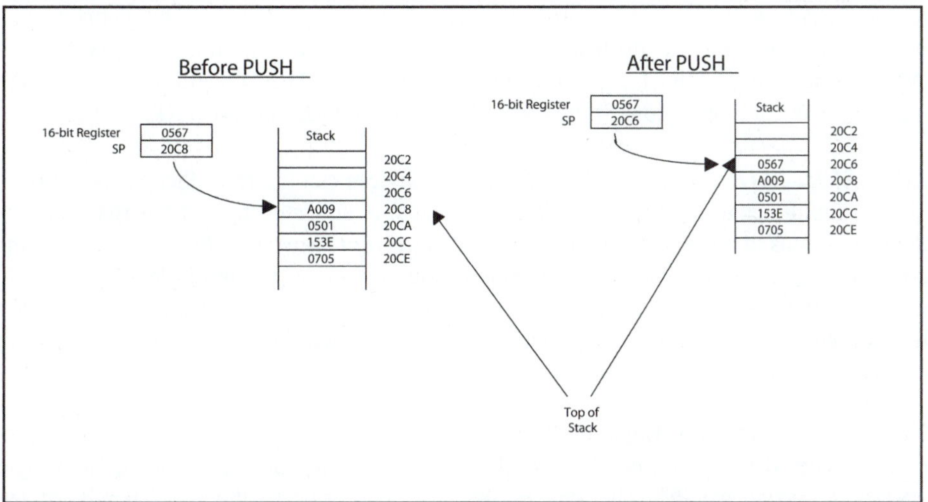

FIGURE 2.14 PUSH operation when accessing a stack from the top

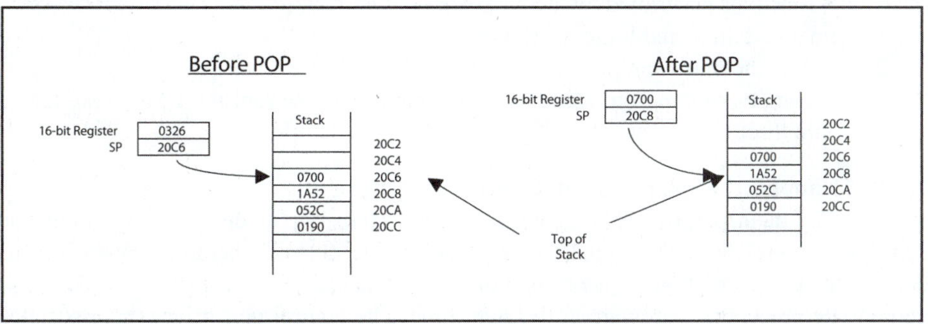

FIGURE 2.15 POP operation when accessing a stack from the top

op-codes define the operations to be performed by the control unit to execute an instruction. The control unit interprets the contents of the instruction register and then responds to the instruction by generating a sequence of enable signals. These signals activate the appropriate ALU logic blocks to perform the required operation.

Control Bus Signals The control unit generates the *control signals*, which are output to the other microcontroller elements via the control bus. The control unit also takes appropriate actions in response to the control signals on the control bus provided by the other microcontroller elements. The control signals vary from one CPU to another. For each specific CPU, these signals are described in detail in the manufacturer's manual. It is impossible to describe all of the control signals for various manufacturers. However, we cover some of the common ones in the following discussion.

$\overline{\text{RESET}}$. This input is common to all CPUs. When this input pin is driven HIGH or LOW (depending on the CPU), the program counter is loaded with a predefined address specified by the manufacturer. As mentioned before, the PIC18F, upon hardware reset, loads the 21-bit program counter with 0's. This means that the instruction stored at memory location 0 is executed first.

READ/$\overline{WRITE}$ (R/$\overline{W}$). This output line is common to all CPUs. The status of this line tells the other microcontroller elements whether the CPU is performing a READ or a WRITE operation. A HIGH signal on this line indicates a READ operation, and a LOW indicates a WRITE operation. Some CPUs have separate READ and WRITE inputs.

INTERRUPT REQUEST. The external I/O devices can interrupt the microcontroller via this input signal on the CPU. When this signal is activated by the external devices, the CPU jumps to a special program called the *interrupt service routine*. This program is normally written by the user for performing tasks that the interrupting device wants the CPU to carry out. After completing this program, the CPU returns to the main program it was executing when the interrupt occurred. This topic will be covered in more detail in Chapters 3, 8, 9, and 10.

2.3.3 Arithmetic and Logic Unit (ALU)

The ALU performs all of the data manipulations, such as arithmetic and logic operations, inside a CPU. The size of the ALU conforms to the word length of the microcontroller. This means that an 8-bit microcontroller will have an 8-bit ALU. Some of the typical functions performed by the ALU are

1. binary addition and logic operations
2. finding the one's complement of data
3. shifting or rotating the contents of a general-purpose register 1 bit to the left or right through a carry

2.3.4 Simplified Explanation of Control Unit Design

The main purpose of the control unit is to translate or decode instructions and generate appropriate enable signals to accomplish the desired operation. Based on the contents of the instruction register, the control unit sends the data items selected to the appropriate processing hardware at the right time. The control unit drives the associated processing hardware by generating a set of signals that are synchronized with a master clock.

The control unit performs two basic operations: instruction interpretation and instruction sequencing. In the interpretation phase, the control unit reads (fetches) an instruction from the memory addressed by the contents of the program counter into the instruction register. The control unit inputs the contents of the instruction register. It recognizes the instruction type, obtains the necessary operands, and routes them to the appropriate functional units of the execution unit (registers and ALU). The control unit then issues the necessary signals to the execution unit to perform the desired operation and routes the results to the destination specified. In the sequencing phase, the control unit generates the address of the next instruction to be executed and loads it into the program counter.

There are two methods for designing a control unit: hardwired control and microprogrammed control. In the hardwired approach, synchronous sequential circuit design procedures are used in designing the control unit. Note that a control unit is a clocked sequential circuit. The name *hardwired control* evolved from the fact that the final circuit is built by physically connecting components such as gates and flip-flops. In the microprogrammed approach, on the other hand, all control functions are stored in a ROM inside the control unit. This memory is called the *control memory*. The words in this memory, called *control words*, specify the control functions to be performed by the control unit. The control words are fetched from the control memory and the bits are routed to appropriate functional units to enable various gates. An instruction is thus executed. The PIC18F uses the hardwired approach for designing its control unit for the RISC-based CPU.

Design of control units using microprogramming (sometimes called *firmware* to distinguish it from hardwired control) is more expensive than using hardwired controls. To execute an instruction, the contents of the control memory in microprogrammed control must be read, which reduces the overall speed of the control unit.The most important advantage of microprogramming is its flexibility; alterations can be made simply by changing the microprogram in the control memory. A small change in the hardwired approach may lead to redesigning the entire system.

Microprogramming is typically used by a CPU designer to program the logic performed by the control unit. On the other hand, assembly language programming is a popular programming language used for programming a microcontroller to perform a desired function. A microprogram is stored in the control unit. An assembly language program is stored in the program memory. The assembly language program is called a *macroprogram*. A macroinstruction (or simply, an instruction) initiates execution of a complete microprogram.

2.4 Basic Concept of Pipelining

To execute a program, a conventional CPU repeats the following three steps for completing each instruction:

1. *Fetch.* The CPU fetches (instruction read) the instruction from the program memory (external to the CPU) into the instruction register.

2. *Decode.* The CPU decodes or translates the instruction using the control unit. The control unit inputs the contents of the instruction register, and then decodes (translates) the instruction to determine the instruction type.

3. *Execute.* The CPU executes the instruction using the control unit. To accomplish the task, the control unit generates a number of enable signals required by the instruction.

For example, suppose that it is desired to add the contents of two registers, X and Y, and store the result in register Z. To accomplish this, the conventional CPU performs the following steps:

1. The CPU fetches the instruction into the instruction register.
2. The control unit (CU) decodes the contents of the instruction register.
3. The CU executes the instruction by generating enable signals for the register and ALU sections to perform the following:
    a. The CU transfers the contents of registers X and Y from the Register section into the ALU.
    b. The CU commands the ALU to ADD.
    c. The CU transfers the result from the ALU into register Z of the register section.

Hence, the conventional CPU executes a program by completing one instruction at a time and then proceeds to the next. This means that the control unit would have to

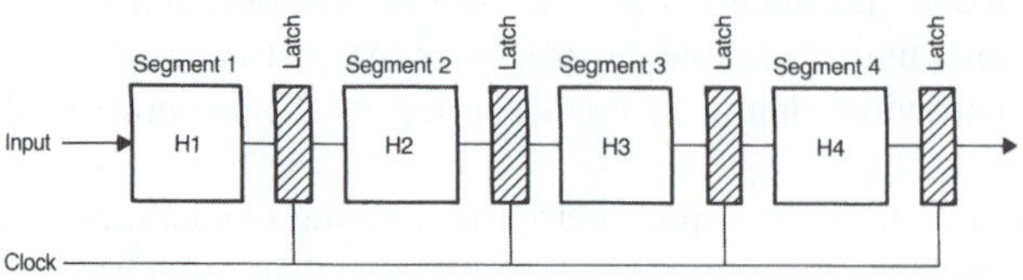

FIGURE 2.16 Four-segment pipeline

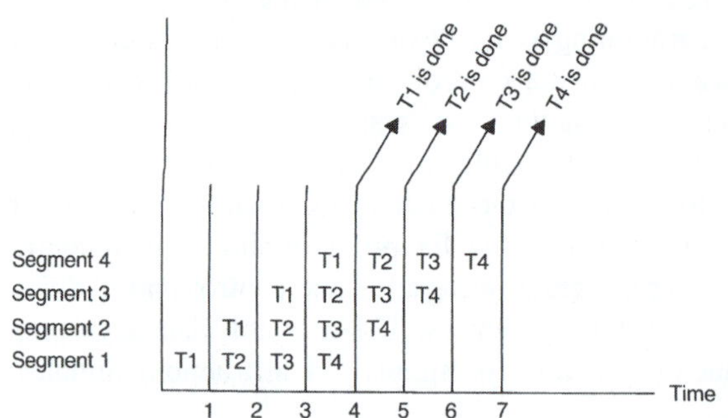

FIGURE 2.17 Overlapped execution of four tasks using a pipeline

wait until the instruction is fetched from memory. Also, the ALU would have to wait until the required data are obtained. Since the speeds of microcontrollers are increasing at a more rapid rate than memory speeds, the control unit and ALU will be idle while the conventional CPU fetches each instruction and obtains the required data. Typical microcontrollers such as the PIC18F utilize the control unit and ALU efficiently by prefetching the next instruction(s) and the required data before the control unit and ALU require them. As mentioned earlier, conventional CPUs execute programs in sequence; typical microcontrollers such as the PIC18F, on the other hand, implement the feature called "pipelining" to prefetch the next instruction while the control unit is busy executing the current instruction. Hence, PIC18F implements pipelining to increase system throughput.

The basic concepts associated with pipelining will be considered next. Assume that a task T is carried out by performing four activities: Al, A2, A3, and A4, in that order. Hardware Hi is designed to perform activity Ai. Hi is referred to as a segment, and it essentially contains combinational circuit elements. Consider the arrangement shown in Figure 2.16. In this configuration, a latch is placed between two segments so the result computed by one segment can serve as input to the following segment during the next clock period.

The execution of four tasks Tl, T2, T3, and T4 using the hardware of Figure 2.16 is described using the space-time chart shown in Figure 2.17.

Initially, task Tl is handled by segment 1. After the first clock, segment 2 is busy with Tl while segment 1 is busy with T2. Continuing in this manner, task Tl is completed at the end of the fourth clock. However, following this point, one task is shipped out per clock. This is the essence of the pipelining concept. A pipeline gains efficiency for the same reason as an assembly line does. Several activities are performed but not on the same material.

The PIC18F implements a two-stage pipeline. As mentioned earlier, the execution of an instruction by a typical CPU is completed in two stages. During the first stage, the instruction is fetched from program memory. During the second stage, the task specified in the instruction is accomplished. Note that the PIC18F CPU, fetches the instruction during the first stage like a typical CPU. However, during the second stage, the PIC18F CPU while executing the instruction, fetches the next instruction. This is called "two-stage instruction pipelining," and is used by the PIC18F to increase the speed of instruction execution. It should be mentioned that when the PIC18F fetches a branch instruction, it clears or flushes the pipeline and executes a new sequence of instructions starting at the new branch address.

2.5 RISC vs. CISC

There are two types of CPU architectures: RISC (reduced instruction set computer) and CISC (complex instruction set computer). A RISC microcontroller such as the PIC18F emphasizes simplicity and efficiency. RISC designs start with a necessary and sufficient instruction set. The purpose of using RISC architecture is to maximize speed by reducing clock cycles per instruction. Almost all computations can be obtained from a few simple operations. The goal of RISC architecture is to maximize the effective speed of a design by performing infrequent operations in software and frequent functions in hardware, thus obtaining a net performance gain. The following list summarizes the typical features of a RISC CPU:

1. The RISC CPU is designed using hardwired control with little or no microcode. Note that variable-length instruction formats generally require microcode design. All RISC instructions have fixed formats, so microcode design is not necessary.
2. A RISC CPU executes most instructions in a single cycle.
3. The instruction set of a RISC CPU typically includes only register, load, and store instructions. All instructions involving arithmetic operations use registers, and load and store operations are utilized to access memory.
4. The instructions have a simple fixed format with few addressing modes.
5. A RISC CPU processes several instructions simultaneously and thus includes pipelining.
6. Software can take advantage of more concurrency. For example, jumps occur after execution of the instruction that follows. This allows fetching of the next instruction during execution of the current instruction.

RISC CPUs are suitable for embedded applications. *Embedded controllers* are embedded in the host system. This means that the presence and operation of these controllers are basically hidden from the host system. Typical embedded control applications include office automation systems such as printers.

RISC CPUs are well suited for applications such as image processing, robotics, and instrumentation. The key features of the RISC CPUs that make them ideal for these applications are their relatively low level of integration in the chip and instruction pipeline architecture. These characteristics result in low power consumption, fast instruction execution, and fast recognition of interrupts.

CISC CPUs such as the Motorola /Freescale HC11 CPU contain a large number of instructions and many addressing modes. In contrast, RISC CPUs such as the PIC18F include a simple instruction set with a few addressing modes. Almost all computations can be obtained from a few simple operations. RISC basically supports a small set of commonly used instructions that are executed at a fast clock rate compared to CISC, which contains a large instruction set (some of which are rarely used) executed at a slower clock rate. To implement the fetch/execute cycle for supporting a large instruction set for CISC, the clock is typically slower.

In CISC, most instructions can access memory whreas RISC contains mostly load/ store instructions. The complex instruction set of CISC requires a complex control unit, thus requiring microprogrammed implementation. RISC utilizes hardwired control which is faster. CISC is more difficult to pipeline; RISC provides more efficient pipelining. An advantage of CISC over RISC is that complex programs require fewer instructions in CISC with fewer fetch cycles, while RISC requires a large number of instructions to accomplish the same task with several fetch cycles. However, RISC can significantly improve its performance with a faster clock, more efficient pipelining, and compiler optimization.

2.6 Functional Representation of a Typical Microcontroller—The PIC18F4321

Figure 2.18 depicts the functional block diagram of the PIC18F4321 microcontroller. The block diagram can be divided into three sections, namely, CPU, memory, and I/O (input/output). A brief description of these blocks will be provided in the following.

The PIC18F4321 CPU contains registers, ALU, an instruction decode and control unit, along with the oscillator blocks. Typical CPU registers include IR (instruction register), W (accumulator), program counter (PC), three memory address registers (FSR0 through FSR2), and stack pointer (STKPTR). An on-chip hardware multiplier is also included for performing unsigned multiplication. These registers are described in more detail in Chapter 5.

The on-chip memory contains program memory and data memory. As mentioned before, the PIC18F4321 is designed using the Harvard architecture; a separate 21-bit address bus for program memory and a separate 12-bit address bus for data memory are shown in the figure.

The on-chip I/O block includes five I/O ports (Port A through Port E), four hardware timers, a 10-bit ADC (analog-to-digital Converter), and CCP (capture, compare, PWM) and associated modules. As mentioned before, the PIC18F4321 can perform functions such as capture, compare, and pulse width modulation (PWM) using the timers and CCP modules. The PIC18F4321 can compute the period of an incoming signal using the capture module. The PIC18F4321 can produce a periodic waveform or time delays using the compare module. The PIC18F4321's on-chip PWM can be used to obtain pulse waveforms with a particular period and duty cycle, ideal for applications such as motor control.

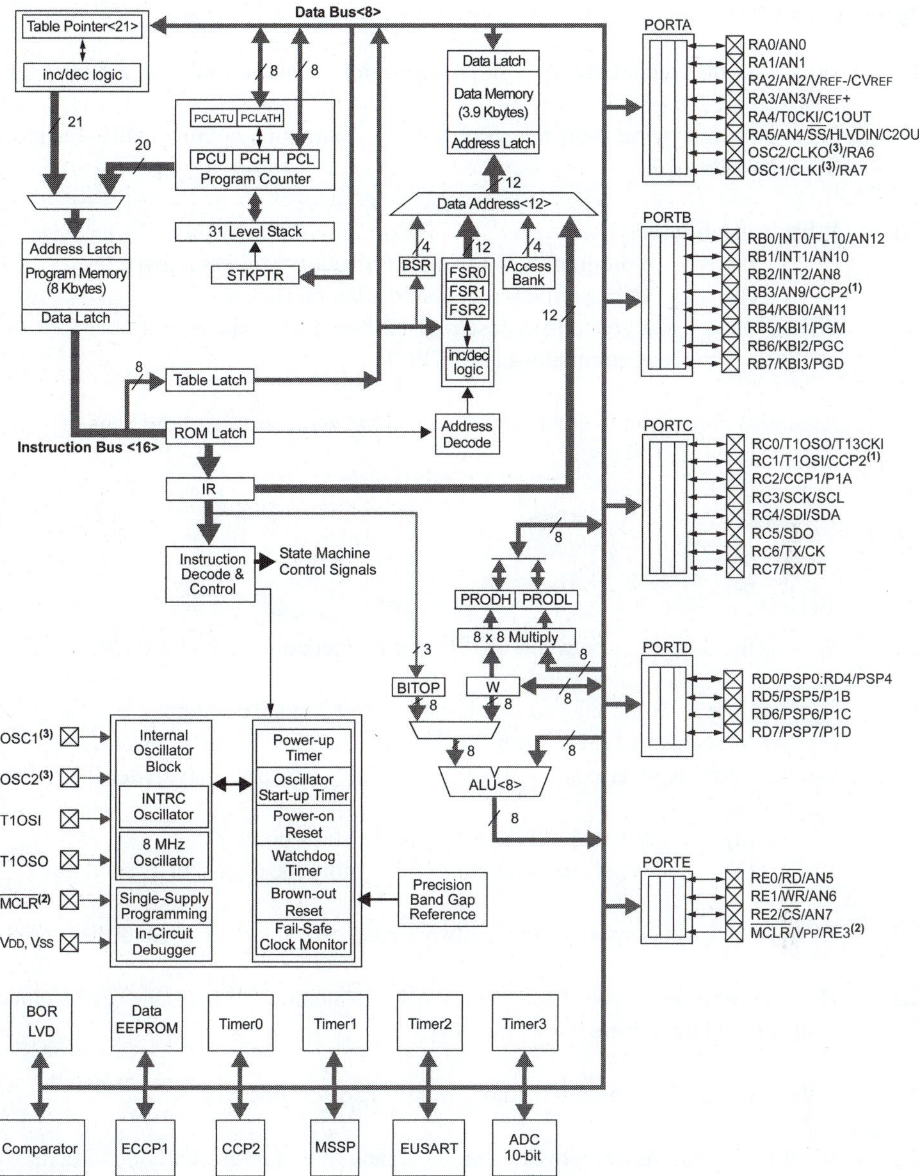

Note 1: CCP2 is multiplexed with RC1 when configuration bit, CCP2MX, is set, or RB3 when CCP2MX is not set.

2: RE3 is available only when MCLR functionality is disabled.

3: OSC1/CLKI and OSC2/CLKO are available only in select oscillator modes and when these pins are not being used as digital I/O. Refer to **Section 2.0 "Oscillator Configurations"** for additional information.

Figure 2.18 PIC18F4321 block diagram

Questions and Problems

2.1 What is the difference between a single-chip microcomputer and a microcontroller?

2.2 What is meant by an 8-bit microcontroller? Name one commercially available 8-bit microcontroller.

2.3 What is the difference between
- (a) a program counter and a memory address register?
- (b) an accumulator and an instruction register?
- (c) a general-purpose register-based CPU and an accumulator-based CPU?

2.4 Assuming signed numbers, find the sign, carry, zero, and overflow flags of
- (a) $09_{16} + 17_{16}$
- (b) $A5_{16} - A5_{16}$
- (c) $71_{16} - A9_{16}$
- (d) $6E_{16} + 3A_{16}$
- (e) $7E_{16} + 7E_{16}$

2.5 What is the difference between PUSH and POP operations in the stack?

2.6 Suppose that an 8-bit microcontroller has a 16-bit stack pointer and uses a 16-bit register to access the stack from the top. Assume that initially the stack pointer and the 16-bit register contain $20C0_{16}$ and 0205_{16}, respectively. After the PUSH operation
- (a) What are the contents of the stack pointer?
- (b) What are the contents of memory locations $20BE_{16}$ and $20BF_{16}$?

2.7 What is the main purpose of the hardware reset pin on the microcontroller chip?

2.8 How many bits are needed to access a 4 MB data memory? What is the hexadecimal value of the last address in this memory?

2.9 If the address of an on-chip memory is 0x7FF, determine its size.

2.10 What is the difference between von Neumann and Harvard CPU architectures? Provide an example of a commercially available microcontroller using each type of CPU.

2.11 What is the basic difference between program execution by a conventional CPU and the PIC18F CPU?

2.12 Discuss the basic features of RISC and CISC.

2.13 Discuss briefly the purpose of the functional units (CCP, A/D, serial communication) implemented in the PIC18F.

2.14 What is meant by pipelining?

2.15 Summarize the basic features of PIC18F pipelining.

3

INTRODUCTION TO PROGRAMMING LANGUAGES

In this chapter we provide the fundamental concepts of programming languages. Typical programming characteristics such as programming languages, basics of assembly language programming, instruction formats, microcontroller instruction sets, and addressing modes are discussed.

3.1 Basics of Programming Languages

Microcontrollers are typically programmed using semi-English-language statements (assembly language). In addition to assembly languages, micrococontrollers use a more understandable human-oriented language called *high-level language*. No matter what type of language is used to write programs, microcontrollers understand only binary numbers. Therefore, all programs must eventually be translated into their appropriate binary forms. The principal ways to accomplish this are discussed later.

Programming languages can typically be divided into three main types: machine language, assembly language, and high-level language. A *machine language program* consists of either binary or hexadecimal op-codes. Programming a microcontroller with either one is relatively difficult, because one must deal only with numbers. The CPU architecture of the microcontroller determines all of its instructions. These instructions are called the microcontroller's *instruction set*. Programs in *assembly* and *high-level languages* are represented by instructions that use English-language-type statements. The programmer finds it relatively more convenient to write programs in assembly or high-level language than in machine language. However, a translator must be used to convert such programs into binary machine language so that the microcontroller can execute the programs. This is shown in Figure 3.1.

An *assembler* translates a program written in assembly language into a machine language program. A *compiler* or *interpreter*, on the other hand, converts a high-level language program such as C into a machine language program. Assembly or high-level language programs are called *source codes*. Machine language programs are known as *object codes*. A *translator* converts source codes to object codes. Next, we discuss the three main types of programming language in more detail.

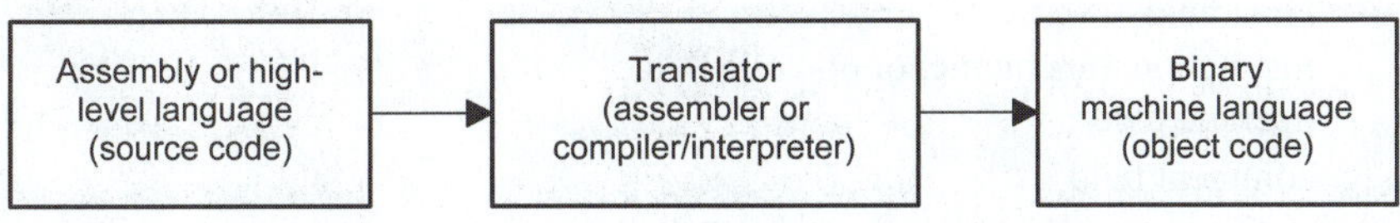

FIGURE 3.1 Translating assembly or high-level language into binary machine language

3.2 Machine Language

A microcontroller has a unique set of machine language instructions defined by its manufacturer. No two microcontrollers by different manufacturers have the same machine language instruction set. For example, Microchip Technology's PIC18F microcontroller uses the code $D7FF_{16}$ for the assembly language statement "HERE BRA HERE" (branch always to HERE), whereas the Motorola/Freescale HC11 microcontroller uses the code $20FE_{16}$ for the same statement with its BRA instruction. Therefore, a machine language program for one microcontroller will not run on the microcontroller of a different manufacturer.

At the most elementary level, a microcontroller program can be written using its instruction set in binary machine language. As an example, the following program adds two numbers using the PIC18F machine language:

```
0000111000000010
0000111100000011
0110111001000000
1110111100000011
```

Obviously, the program is very difficult to understand unless the programmer remembers all of the PIC18F codes, which is impractical. Because one finds it very inconvenient to work with 1's and 0's, it is almost impossible to write an error-free program on the first try. Also, it is very tiring for a programmer to enter a machine language program written in binary into the microcontroller's RAM. For example, the programmer needs a number of binary switches to enter the binary program. This is definitely subject to error.

To increase the programmer's efficiency in writing a machine language program, hexadecimal numbers rather than binary numbers are used. The following is the same addition program in hexadecimal using the PIC18F instruction set:

```
0E02
0F03
6E40
EF03
```

It is easier to detect an error in a hexadecimal program, because each byte contains only two hexadecimal digits. One would enter a hexadecimal program using a hexadecimal keyboard. A keyboard monitor program in ROM provides interfacing of the hexadecimal keyboard with the microcontroller. This program must convert each key actuation into binary machine language in order for the microcontroller to understand the program. However, programming in hexadecimal is not normally used.

3.3 Assembly Language

The next programming level uses assembly language. Each line in an assembly language program includes four fields:

- label field
- instruction, mnemonic, or op-code field
- operand field
- comment field

As an example, a typical program for adding two 8-bit numbers written in PIC18F assembly language is as follows:

Label	Mnemonic	Operand	Comment
	MOVLW	1	;Move 1 into accumulator
	ADDLW	2	;Add 2 with 3 , store result in accumulator
	SLEEP		;Halt

Obviously, programming in assembly language is more convenient than programming in machine language, because each mnemonic gives an idea of the type of operation it is supposed to perform. Therefore, with assembly language, the programmer does not have to find the numerical op-codes from a table of the instruction set, and programming efficiency is improved significantly.

An assembly language program is translated into binary via a program called an *assembler*. The assembler program reads each assembly instruction of a program as ASCII characters and translates them into the respective binary op-codes. For example, the PIC18F assembler translates the SLEEP (places the PIC18F in sleep mode; same as HALT instruction in other processors) instruction into its 16-bit binary op-code as 0000 0000 0000 0011 (0003 in hex), as depicted in Table 3.1.

An advantage of the assembler is address computation. Most programs use addresses within the program as data storage or as targets for jumps or calls. In machine language programming, these addresses must be calculated by hand. The assembler solves this problem by allowing the programmer to assign a symbol to an address. The programmer may then reference that address elsewhere by using the symbol. The assembler computes the actual address for the programmer and fills it in automatically. One can obtain hands-on experience with a typical assembler for a microcontroller by downloading it from the Internet.

3.3.1 Types of Assemblers

Most assemblers use two passes to assemble a program. This means that they read the input program text twice. The first pass is used to compute the addresses of all labels in the program. To find the address of a label, it is necessary to know the total length of all of the binary code preceding that label. Unfortunately, however, that address may be needed in that preceding code. Therefore, the first pass computes the addresses of all labels and stores them for the next pass, which generates the actual binary code. Various types of assemblers are available today:

- *One-pass assembler.* This assembler goes through an assembly language program once and translates it into a machine language program. This assembler has the problem of defining forward references. This means that a JUMP instruction using an address that appears later in the program must be defined by the programmer after the program is assembled.

TABLE 3.1 Conversion of PIC18F SLEEP instruction into its binary op-code

Assembly code	Binary form of ASCII codes as seen by the assembler	Binary op-code created by the MPLAB PIC18F assembler
S	0101 0011	
L	0100 1000	0000 0000 0000 0011
E	0100 0101	
E	0100 0101	
P	0101 0000	

- *Two-pass assembler.* This assembler scans an assembly language program twice. In the first pass, this assembler creates a symbol table. A symbol table consists of labels with addresses assigned to them. This way, labels can be used for JUMP statements and no address calculation has to be done by the user. On the second pass, the assembler translates the assembly language program into machine code. The two-pass assembler is more desirable and much easier to use. Note that the MPLAB PIC18F assembler is a two-pass assembler.

- *Macroassembler.* This type of assembler translates a program written in macro language into machine language. This assembler lets the programmer define all instruction sequences using macros. Note that by using macros, the programmer can assign a name to an instruction sequence that appears repeatedly in a program. The programmer can thus avoid writing an instruction sequence that is required many times in a program by using macros. The macroassembler replaces a macroname with the appropriate instruction sequence each time it encounters a macroname.

  It is interesting to see the difference between a subroutine and a macroprogram. A specific subroutine occurs once in a program. A subroutine is executed by CALLing it from a main program. The program execution jumps out of the main program and executes the subroutine. At the end of the subroutine, a RET instruction is used to resume program execution following the CALL SUBROUTINE instruction in the main program. A macro, on the other hand, does not cause the program execution to branch out of the main program. Each time a macro occurs, it is replaced by the appropriate instruction sequence in the main program. Typical advantages of using macros are shorter source programs and better program documentation. A typical disadvantage is that effects on registers and flags may not be obvious.

  Conditional macroassembly is very useful in determining whether or not an instruction sequence is to be included in the assembly, depending on a condition that is true or false. If two different programs are to be executed repeatedly based on a condition that can be either true or false, it is convenient to use conditional macros. Based on each condition, a particular program is assembled. Each condition and the appropriate program are typically included within IF and ENDIF pseudoinstructions.

- *Cross assembler.* This type of assembler is typically resident in a processor and assembles programs for another processor for which it is written. The cross assembler program is written in a high-level language so that it can run on different types of processors that understand the same high-level language.

- *Resident assembler.* This type of assembler assembles programs for a processor in which it is resident. The resident assembler may slow down operation of the processor on which it runs.

- *Meta-assembler.* This type of assembler can assemble programs for many different types of processors. The programmer usually defines the particular processor being used.

3.3.2 Assembler Delimiters

As mentioned before, each line of an assembly language program consists of four fields: label, mnemonic or op-code, operand, and comment. The assembler ignores the comment field but translates the other fields. The label field must start with an uppercase alphabetic character. The assembler must know where one field starts and another ends. Most assemblers allow the programmer to use a special symbol or delimiter to indicate the beginning or end of each field. Typical delimiters used are spaces, commas, semicolons, and colons:

- Spaces are used between fields.
- Commas (,) are used between addresses in an operand field.
- A semicolon (;) is used before a comment.
- A colon (:) or no delimiter is used after a label.

3.3.3 Specifying Numbers by Typical Assemblers

To handle numbers, some assemblers consider all numbers as decimal numbers unless specified otherwise. All assemblers will also specify other number systems, including hexadecimal numbers. The user must define the type of number system used in some way. This is generally done by using a letter or a symbol before or after the number. For example, Intel uses the letter H after a number to represent it as a hex number, whereas Microchip's MPLAB assembler uses several ways to represent a hex number. Three of the most common ways to represent a number as a hex number are 0x before the number, H after the number, or default (without using a letter or a symbol before or after the number). As an example, 60 in hexadecimal is represented by the MPLAB assembler as either 0x60 or 60H, or simply 60.

Some assemblers such as the MASM 32 assembler for the Pentium microprocessor require hexadecimal numbers to start with a digit (0 through 9). A 0 is typically used if the first digit of the hexadecimal number is a letter. This is done to distinguish between numbers and labels. For example, typical assemblers such as MASM32 will normally require the number F3H to be represented as 0F3H; otherwise, the assembler will generate an error. However, the MPLAB assembler used in this book for assembling PIC18F assembly language programs does not require, '0' to be used if the first digit of a hexadecimal number is a letter.

The MPLAB uses D before a 'number' to specify a decimal number. For example, decimal number 60 can be represented as D'60'. A binary number is specified by the MPLAB by using B before the 'number'. For example, the 8-bit binary number 01011100 can be represented by the MPLAB as B'01011100'.

3.3.4 Assembler Directives or Pseudoinstructions

Assemblers use pseudoinstructions or directives to make the formatting of the edited text easier. Pseudoinstructions are not translated directly into machine language instructions. They equate labels to addresses, assign the program to certain areas of memory, or insert titles, page numbers, and so on. To use the assembler directives or pseudoinstructions, the programmer puts them in the op-code field, and if the pseudoinstructions require an address or data, the programmer places them in the label or data field. Typical pseudoinstructions are Origin (ORG), Equate (EQU), Define Byte (DB), and Define Word (DW).

Origin (ORG) The directive ORG lets a programmer place programs anywhere in memory. Internally, the assembler maintains a program counter type of register called an *address counter*. This counter maintains the address of the next instruction or data to be processed.

An ORG directive is similar in concept to a JUMP instruction. Note that the JUMP instruction causes a processor to place a new address in the program counter. Similarly, the ORG pseudoinstruction causes the assembler to place a new value in the address counter.

Typical ORG statements are

```
ORG 0x100
SLEEP
```

The MPLAB assembler will generate the following code for these statements:

```
100 0003
```

Most assemblers assign a value of zero to the starting address of a program if the programmer does not define this by means of an ORG.

Equate (EQU). The directive EQU assigns a value in its operand field to an address in its label field. This allows the user to assign a numerical value to a symbolic name. The user can then use the symbolic name in the program instead of its numeric value. This reduces errors.

A typical example of EQU is START EQU 0x0200, which assigns the value 0200 in hexadecimal to the label START.

Note that if a label in the operand field is equated to another label in the label field, the label in the operand field must have been defined previously. For example, the EQU statement

```
BEGIN EQU    START
```

will generate an error unless START is defined previously with a numeric value.

Define Byte (DB) The directive DB is generally used to set a memory location for a certain byte value. For example,

```
START DB      0x45
```

will store the data value 45 hex to the address START. With some assemblers, the DB pseudoinstruction can be used to generate a table of data as follows:

```
      ORG     0x200
TABLE DB      0x20, 0x30, 0x40, 0x50
```

In this case, 20 hex is the first bit of data in the memory location 0x200; 30 hex, 40 hex, and 50 hex occupy the next three memory locations. Therefore, the data in memory will look like this:

```
200     20
300     30
400     40
500     50
```

Define Word (DW) The directive DW is typically used to assign a 16-bit value to two memory locations. For example,

```
      ORG     0x100
START DW      0x4AC2
```

will assign C2 to location 100 and 4A to location 101. It is assumed that the assembler will follow little endian; that is, it will assign the low byte first (C2) and then the high byte (4A). With some assemblers, the DW directive can be used to generate a table of 16-bit data as follows:

```
                ORG 0x80
POINTER         DW  0x5000, 0x6000, 0x7000
```

In this case, the three 16-bit values 0x5000, 0x6000, and 0x7000 are assigned to memory locations starting at the address 0x80. That is, the array would look like this:

```
80              00
```

81	50
82	00
83	60
84	00
85	70

END This directive indicates the end of the assembly language source program.

3.3.5 Assembly Language Instruction Formats

In this section, assembly language instruction formats available with typical microcontrollers are discussed. Depending on the number of addresses specified, the following instruction formats can be used: three-address, two-address, one-address, or zero-address. Because all instructions are stored in the main memory, instruction formats are designed in such a way that instructions take less space and have more processing capabilities. It should be emphasized that the microcontroller architecture has considerable influence on a specific instruction format. The following are some important technical points that have to be considered while designing an instruction format:

- The size of an instruction word is chosen such that it facilitates the specification of more operations by a designer. For example, with 4- and 8-bit op-code fields, we can specify 16 and 256 distinct operations, respectively.

- Instructions are used to manipulate various data elements, such as integers, floating-point numbers, and character strings. In particular, all programs written in a symbolic language such as C are stored internally as characters. Therefore, memory space will not be wasted if the word length of the machine is some integral multiple of the number of bits needed to represent a character. Because all characters are represented using typical 8-bit character codes such as ASCII, it is desirable to have 8 or 16 bit as the word length for typical microcontrollers.

- The size of the address field is chosen such that high resolution is guaranteed. Note that in any microcontroller, the ultimate resolution is a bit. Memory resolution is a function of the instruction length, and, in particular, short instructions provide less resolution. For example, in a microcontroller with 32K 16-bit memory words, at least 19 bits are required to access each bit of the word. (This is because 2^{15} = 32K and 2^4 = 16.)

The general form of a *three-address instruction* is

<op-code> Addr1,Addr2,Addr3

Some typical *three-address instructions* are

```
MUL   A,B,C        ;     C <- A * B
ADD   A,B,C        ;     C <- A + B
SUB   R1,R2,R3     ;     R3 <- R1 - R2
```

In this specification, all alphabetic characters are assumed to represent memory addresses, and the string that begins with the letter R indicates a register. The third address of this type of instruction is usually referred to as the *destination address*. The result of an operation is always assumed to be saved in the destination address.

Typical programs can be written using three-address instructions. For example, consider the following sequence of three-address instructions:

```
MUL   A, B, R1     ;     R1 <- A * B
```

```
MUL    C, D, R2      ;    R2 <- C * D
MUL    E, F, R3      ;    R3 <- E * F
ADD    R1,R2,R1      ;    R1 <- R1 + R2
SUB    R1,R3,Z       ;    Z <- R1 - R3
```

This sequence implements the statement Z = A * B + C * D - E * F. The three-address format, in addition to the other formats, is normally used by typical microcontrollers such as the PIC18F.

If we drop the third address from the three-address format, we obtain the two-address format, whose general form is

<op-code> Addr1,Addr2

Some typical *two-address instructions* are

```
MOV    A,R1     ;    R1 <- A
ADD    C,R2     ;    R2 <- R2 + C
SUB    R1,R2    ;    R2 <- R2 - R1
```

In this format, the addresses Addr1 and Addr2 represent source and destination addresses, respectively.

The following sequence of two-address instructions is equivalent to the program using three-address format presented earlier:

```
MOV    A,R1     ;    R1 <- A
MUL    B,R1     ;    R1 <- R1 * B
MOV    C,R2     ;    R2 <- C
MUL    D,R2     ;    R2 <- R2 * D
MOV    E,R3     ;    R3 <- E
MUL    F,R3     ;    R3 <- R3 * F
ADD    R2,R1    ;    R1 <- R1 + R2
SUB    R3,R1    ;    R1 <- R1 - R3
MOV    R1,Z     ;    Z <- R1
```

Some typical *one-address instructions* are

```
LDA    B    ;    Acc <- B
ADD    C    ;    Acc <- Acc + C
MUL    D    ;    Acc <- Acc * D
STA    E    ;    E <- Acc
```

The following program illustrates how we can translate the C language statement z = (a * b) + (c * d) - (e * f) into a sequence of one-address instructions:

```
lda    e     ;    Acc <- e
mul    f     ;    Acc <- e * f
sta    t1    ;    t1 <- Acc
lda    c     ;    Acc <- c
mul    d     ;    Acc <- c * d
sta    t2    ;    t2 <- Acc
lda    a     ;    Acc <- a
mul    b     ;    Acc <- a * b
add    t2    ;    Acc <- Acc + t2
sub    t1    ;    Acc <- Acc - t1
sta    z     ;    Z <- Acc
```

In this program, t1 and t2 represent the addresses of memory locations used to store temporary results. Instructions that do not require any addresses are called *zero-address instructions*. All microcontrollers include some zero-address instructions in the instruction set. Typical examples of zero-address instructions are CLC (clear carry) and NOP (no operation).

3.3.6 Typical Instruction Set

An instruction set of a specific microcontroller consists of all the instructions that it can execute. The capabilities of a microcontroller are determined to some extent by the types of instructions it is able to perform. Each microcontroller has a unique instruction set designed by its manufacturer to do a specific task. We discuss some of the instructions that are common to all microcontrollers. We group together chunks of these instructions that have similar functions. These instructions typically include:

- ***Arithmetic and Logic Instructions.*** These operations perform actual data manipulations. The instructions typically include arithmetic/logic, increment/decrement, and rotate/shift operations. Typical arithmetic instructions include ADD, SUBTRACT, COMPARE, MULTIPLY, and DIVIDE. Note that the SUBTRACT instruction provides the result and also affects the status flags, whereas the COMPARE instruction performs subtraction without any result and affects the flags based on the result.

  Typical microcontrollers utilize common hardware to perform addition and subtraction operations for both unsigned and signed numbers. The instruction set for a microcontroller typically includes the same ADD and SUBTRACT instructions for both unsigned and signed numbers. The interpretations of unsigned and signed ADD and SUBTRACT operations are performed by the programmer. For example, consider adding two 8-bit numbers, A and B ($A = FF_{16}$ and $B = FF_{16}$), using the ADD instruction by a microcontroller as follows:

$$
\begin{array}{rl}
 & 1111111 \leftarrow \text{Intermediate carries} \\
 & FF_{16} = 11111111 \\
+ & FF_{16} = 11111111 \\
 & \text{---------------------------} \\
\text{final carry} \rightarrow & 111111110 = FE_{16}
\end{array}
$$

  When the addition above is interpreted by the programmer as an unsigned operation, the result will be $A + B = FF_{16} + FF_{16} = 255_{10} + 255_{10} = 510_{10}$, which is FE_{16} with a carry, as shown above. However, if the addition is interpreted as a signed operation, then $A + B = FF_{16} + FF_{16} = (-1_{10}) + (-1_{10}) = -2_{10}$, which is FE_{16}, as shown above, and the final carry must be discarded by the programmer. Similarly, the unsigned and signed subtraction can be interpreted by the programmer.

  The unsigned and signed multiplication and division algorithms will be discussed in the following.

  Unsigned Multiplication Several unsigned multiplication algorithms are available. Multiplication of two unsigned numbers can be accomplished via repeated addition. For example, to multiply 4_{10} by 3_{10}, the number 4_{10} can be added twice to itself to obtain the result, 12_{10}.

  Signed Multiplication Signed multiplication can be performed using various algorithms. A simple algorithm follows. Assume that M (multiplicand) and Q (multiplier) are in two's complement form. Also assume that M_n and Q_n are the most significant bits (sign bits) of the multiplicand (*M*) and the multiplier (*Q*), respectively. To perform signed multiplication, proceed as follows:

1. If $M_n = 1$, compute the two's complement of *M*.
2. If $Q_n = 1$, compute the two's complement of *Q*.
3. Multiply the $n - 1$ bits of the multiplier and the multiplicand using unsigned multiplication.
4. The sign of the result $S_n = M_n \oplus Q_n$.
5. If $S_n = 1$, compute the two's complement of the result obtained in step 3.

Next, consider a numerical example. Assume that *M* and *Q* are two's complement numbers. Suppose that $M = 1100_2$ and $Q = 0111_2$. Because $M_n = 1$, take the two's complement of $M = 0100_2$; because $Q_n = 0$, do not change *Q*. Multiply 0111_2 and 0100_2 using the unsigned multiplication method discussed before. The product is 00011100_2. The sign of the product $S_n = M_n \oplus Q_n = 1 \oplus 0 = 1$. Hence, take the two's complement of the product 00011100_2 to obtain 11100100_2, which is the final answer: -28_{10}.

Unsigned Division Unsigned division can be accomplished via repeated subtraction. For example, consider dividing 7_{10} by 3_{10} as follows:

Dividend	Divisor	Subtraction Result	Counter
7_{10}	3_{10}	$7 - 3 = 4$	1
		$4 - 3 = 1$	$1 + 1 = 2$

Quotient = counter value = 2
Remainder = subtraction result = 1

Here, '1' is added to a counter whenever the subtraction result is greater than the divisor. The result is obtained as soon as the subtraction result is smaller than the divisor

Signed Division Signed division can be performed using various algorithms. A simple algorithm follows. Assume that DV (dividend) and DR (divisor) are in two's complement form. For the first case, perform unsigned division using repeated subtraction of the magnitudes without the sign bits. The sign bit of the quotient is determined as $DV_n \oplus DR_n$, where DV_n and DR_n are the most significant bits (sign bits) of the dividend (DV) and the divisor (DR), respectively. To perform signed division, proceed as follows:

1. If $DV_n = 1$, compute the two's complement of DV, else keep DV unchanged.
2. If $DR_n = 1$, compute the two's complement of DR, else keep DR unchanged.
3. Divide the $n - 1$ bits of the dividend by the divisor using unsigned division algorithm (repeated subtraction).
4. The sign of the quotient $Q_n = DV_n \oplus DR_n$. The sign of the remainder is the same as the sign of the dividend unless the remainder is zero. The following numerical examples illustrate this:

   The general equation for division can be used for signed division. Note that the general equation for division is *dividend = quotient * divisor + remainder*. For example, consider dividend = – 9, divisor = 2. Three possible solutions are shown below:

   (a) $-9 = -4 * 2 - 1$, quotient = – 4, remainder = – 1.
   (b) $-9 = -5 * 2 + 1$, quotient = – 5, remainder = +1.

(c) $-9 = -6 * 2 + 3$, quotient = – 6, remainder = +3.

However, the correct answer is shown in (a), in which, the quotient = – 4 and the remainder = – 1. Hence, for signed division, the sign of the remainder is the same as the sign of the dividend, unless the remainder is zero.

5. If $Q_n = 1$, compute the two's complement of the quotient obtained in step 3, else keep the quotient unchanged.

The above algorithm will be verified using numerical examples provided in the following:

i) <u>Signed division with zero remainder</u>

Assume 4-bit numbers.

Dividend = +6 = 0110_2 divisor = -2 = two's complement of 2 = 1110_2.

Since the sign bit of dividend is 0, do not change dividend. Because the sign bit of divisor is 1, take two's complement of 1110, which is 0010. Now, divide 0110 by 0010 using repeated subtraction as follows:

DIVIDEND	DIVISOR	SUBTRACTION RESULT USING TWO'S COMPLEMENT	COUNTER
0110	0010	0110-0010=0100	0001
		0100-0010=0010	0010
		0010-0010=0000	0011

Result of unsigned division: Quotient = counter value = 0011_2

Remainder = subtraction result = 0000_2

Result of signed division 6 (0110) divided by -2 (1110):

Sign of the quotient = (sign of dividend) $\oplus$ (sign of divisor) = $0 \oplus 1 = 1$.

Hence, quotient = two's complement of $0011_2 = 1101_2 = -3_{10}$, remainder = 0000_2

ii) <u>Signed division with nonzero remainder</u>

Assume 4-bit numbers.

Dividend = -5 = two's complement of $0101_2 = 1011_2$ divisor = -2 = two's complement of 2 = 1110_2.

Since the sign bit of dividend is 1, take two's complement of 1011, which is 0101. Because the sign bit of divisor is 1, take two's complement of 1110 which is 0010. Now, divide 0101 by 0010 using repeated subtraction as follows:

DIVIDEND	DIVISOR	SUBTRACTION RESULT USING TWO'S COMPLEMENT	COUNTER
0101	0010	0101-0010=0011	0001
		0011-0010=0001	0010

Result of unsigned division: Quotient = Counter value = 0010_2

Remainder = Subtraction result = 0001_2

Result of signed division -5 (1011) divided by -2 (1110)

Sign of the quotient = (sign of dividend) $\oplus$ (sign of divisor) = $1 \oplus 1 = 0$. Hence, do not take two's complement of quotient.

Quotient = 0010_2 = $+2_{10}$; Remainder has the same sign as the dividend, which is negative (bit 3 = 1). Hence, remainder = two's complement of $0001_2 = 1111_2 = -1_{10}$.

Note that the sign of the quotient = (sign of dividend) $\oplus$ (sign of divisor). However, the sign of the remainder is the same as the sign of the dividend unless the remainder is 0. This can be verified by the following numerical examples using decimal numbers:

Case 1: when the remainder is 0

i) Assume both dividend and divisor are positive

Dividend = +6, divisor = +2
Result: quotient = +3, remainder = 0

ii) Assume dividend is negative and divisor is positive.
Dividend = -6, divisor = +2
Result: quotient = -3, remainder = 0

iii) Assume dividend is positive and divisor is negative.
Dividend = +6, divisor = -2
Result: quotient = -3, remainder = 0

iv) Assume both dividend and divisor are negative.
Dividend = -6, divisor = -2
Result: quotient = +3, remainder = 0

Case 2: when the remainder is nonzero
Since dividend = quotient x divisor + remainder
Hence, remainder = dividend - quotient x divisor.

i) Assume both dividend and divisor are positive.
Dividend = +5, divisor = +2
Result: quotient = +2, Remainder can be obtained from the equation, remainder = dividend - quotient x divisor. Hence, remainder = +5 - (+2 x +2) = +1.

ii) Assume dividend is negative and divisor is positive.
Dividend = -5, divisor = +2
Result: quotient = -2, Remainder can be obtained from the equation, remainder = dividend - quotient x divisor. Hence, remainder = -5 - (-2 x +2) = - 1.

iii) Assume dividend is positive and divisor is negative.
Dividend = +5, divisor = -2
Result: quotient = -2, Remainder can be obtained from the equation, remainder = dividend - quotient x divisor. Hence, remainder = +5 - (-2 x --2) = + 1.

iv) Assume both dividend and divisor are negative.
Dividend = -5, divisor = -2
Result: quotient = +2, Remainder can be obtained from the equation, remainder = dividend - quotient x divisor. Hence, remainder = -5 - (+2 x -2) = - 1.

From above, the sign of the remainder is the same as the sign of the dividend unless the remainder is zero.

RISC microcontrollers such as the PIC18F include the unsigned multiplication instruction. The PIC18F instruction set does not contain instructions for signed multiplication, unsigned and signed division. However, subroutines using PIC18F assembly language can be written to obtain them using the above algorithms.

- ***Logic Instructions.*** Typical logic instructions perform traditional Boolean operations such as AND, OR, and Exclusive-OR. The AND instruction can be used to perform a masking operation. If the bit value in a particular bit position is desired in a word, the word can be logically ANDed with appropriate data to accomplish this. For example, the bit value at bit 2 of an 8-bit number 0100 1Y10 (where an unknown bit value of Y is to be determined) can be obtained as follows:

```
        0 1 0 0 1 Y 1 0 -- 8-bit number
AND     0 0 0 0 0 1 0 0 -- masking data
        ---------------------
        0 0 0 0 0 Y 0 0 -- result
```

If the bit value Y at bit 2 is 1, the result is nonzero (flag Z = 0); otherwise, the result is zero (Flag Z = 1). The Z flag can be tested using typical conditional JUMP instructions such as JZ (Jump if Z=1) or JNZ (Jump if Z = 0) to determine whether Y is 0 or 1. This is called a masking operation. The AND instruction can also be used to determine whether a binary number is ODD or EVEN by checking the least significant bit (LSB) of the number (LSB = 0 for even and LSB = 1 for odd). The OR instruction can typically be used to insert a 1 in a particular bit position of a binary number without changing the values of the other bits. For example, a 1 can be inserted using the OR instruction at bit 3 of the 8-bit binary number 0 1 1 1 0 0 1 1 without changing the values of the other bits:

```
        0 1 1 1 0 0 1 1 -- 8-bit number
OR      0 0 0 0 1 0 0 0 -- data for inserting a 1 at bit 3
        -------------------
        0 1 1 1 1 0 1 1 -- result
```

The Exclusive-OR instruction can be used to find the one's complement of a binary number by XORing the number with all 1's as follows:

```
        0 1 0 1 1 1 0 0 - - 8-bit number
XOR     1 1 1 1 1 1 1 1 - - data
        -------------------
        1 0 1 0 0 0 1 1 -- result (one's complement of the 8-bit
                                   number 0 1 0 1 1 1 0 0)
```

- ***Shift and Rotate Instructions.*** Next, the concept of logic and arithmetic shift and rotate operations is reviewed. In a logical shift operation, a bit that is shifted out will be lost, and the vacant position will be filled with a 0. For example, if we have the number

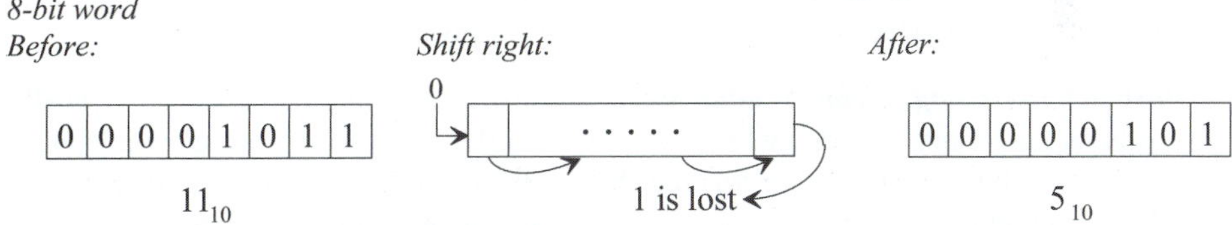

FIGURE 3.2 Logical right shift operation

TABLE 3.2 Typical logic/arithmetic and shift/rotate operations

Shift type	Logic	Arithmetic	Rotate
Right	0 Lost	MSB Lost	
Left	0 Lost	0 Lost	

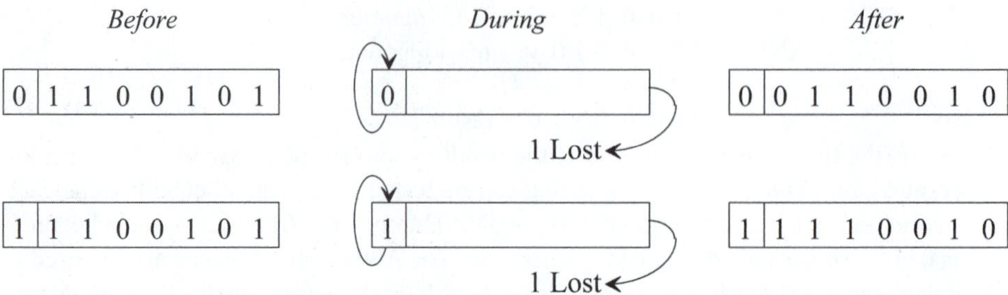

FIGURE 3.3 True arithmetic right shift operations

$(11)_{10}$ after a logical right shift operation, the register contents shown in Figure 3.2 will occur. Typical examples of logic/arithmetic and shift/rotate operations are given in Table 3.2.

It must be emphasized that a logical left or right shift of an unsigned number by n positions implies multiplication or division of the number by 2^n, respectively, provided that a 1 is not shifted out during the operation.

In the case of true arithmetic left or right shift operations, the sign bit of the number to be shifted must be retained. However, in computers, this is true for right shift and not for left shift operation. For example, if a register is shifted right arithmetically, the most significant bit (MSB) of the register is preserved, thus ensuring that the sign of the number will remain unchanged. This is illustrated in Figure 3.3.

There is no difference between arithmetic and logical left shift operations. If the most significant bit changes from 0 to 1, or vice versa, in an arithmetic left shift, the result is incorrect and the CPU sets the overflow flag to 1.

- ***Instructions for controlling microcontroller operations.*** These instructions typically include those that set the reset specific flags and halt or stop the CPU.
- ***Data movement instructions.*** These instructions move data from a register to memory, and vice versa, between registers, and between a register and an I/O device.
- ***Instructions using memory addresses.*** An instruction in this category typically contains a memory address, which is used to read a data word from memory into a microcontroller register or for writing data from a register into a memory location. Many instructions under data processing and movement fall in this category.
- ***Conditional and unconditional JUMP.*** These instructions typically include one of the following:

  1. An unconditional JUMP, which always transfers the memory address specified in the instruction into the program counter
  2. A conditional JUMP, which transfers the address portion of the instruction into the program counter based on the conditions set by one of the status flags in the flag register

3.3.7 Typical Addressing Modes

One of the tasks performed by a microcontroller during execution of an instruction is the determination of the operand and destination addresses. The manner in which a microcontroller accomplishes this task is called the "addressing mode." Now, let us present the typical microcontroller addressing modes, relating them to the instruction set of the PIC18F.

An instruction is said to have "implied or inherent addressing mode" if it does not have any operand. For example, consider the following instruction: SLEEP, which is equivalent to the HALT instruction in other microcontrollers. The SLEEP instruction is a no-operand instruction.

Whenever an instruction/operand contains data, it is called an "immediate mode" instruction. For example, consider the following PIC18F instruction:

```
ADDLW      3      ;      [WREG]←[WREG] + 3
```

Note that the accumulator in the PIC18F is called the WREG register. This instruction adds 3 to the contents of the WREG and then stores the result in WREG.

An instruction is said to have an *absolute* or *direct addressing mode* if it contains a memory address in the operand field. For example, consider the PIC18F instruction:

```
MOVWF      0x20   ;      [0x20]←[WREG]
```

The MOVWF 0x20 instruction moves the contents of the WREG register into a memory location whose address is 0x20. The contents of WREG are unchanged. MOVWF 0x20 uses direct address mode since address 0x20 is directly specified in the MOVWF instruction.

When an instruction specifies a microcontroller register to hold the address, the resulting addressing mode is known as the *register indirect mode*. For example, consider the PIC18F instruction:

```
MOVWF      INDF0  ; Move contents of WREG into a data RAM
                  ; address pointed to by FSR0 since INDF0
                  ; is associated with FSR0
```

The above instruction moves the 8-bit contents of WREG to a data memory location whose address is in PIC18F's FSR0 register. This instruction uses the contents of the FSR0 register as a pointer to data memory. Also, INDF0 in the instruction MOVWF IND0 means that the FSR0 register will hold the address of data memory.

Conditional branch instructions are used to change the order of execution of a program based on the conditions set by the status flags. Some microcontrollers use conditional branching with the absolute mode. The op-code verifies a condition set by a particular status flag. If the condition is satisfied, the program counter is changed to the value of the operand address (defined in the instruction). If the condition is not satisfied, the program counter is incremented, and the program is executed in its normal order.

Typical microcontrollers such as the PIC18F use conditional branch instructions. Some conditional branch instructions are 16 bits wide. The first byte is the op-code for checking a particular flag. The second byte is an 8-bit offset, which is added to the contents of the program counter if the condition is satisfied to determine the effective address. This offset is considered as a signed binary number with the most significant bit as the sign bit. It means that the offset can vary from -128_{10} to $+127_{10}$ (0 being positive). This is called the *relative mode*.

This means that for forward branching, the range of the offset value is from 0x00 to 0x7F. For backward branching, this range varies from 0x80 to 0xFF. Since conditional branch instructions are 16 bits wide in the PIC18F, the PC (program counter) is incremented by 2 to point to the next instruction while executing the conditional branch instruction. The offset is multiplied by 2 and then added to PC+2 to find the branch address if the condition is true. Note that the offset is multiplied by 2 since the contents of the PC must always be an even number for 16- and 32-bit instruction lengths.

As an example, consider BNZ 0x03. Note that BNZ stands for "Branch if not zero." If the Z (zero flag) in the Status register is 0, then the PC is loaded with the (PC + 2 + 03H x 2). When the PIC18F executes the BNZ instruction, the PC points to the next instruction. This means that if BNC is located at address 0050H in program memory, the PC will contain 0052H (PC + 2) when the PIC18F executes BNZ. Hence, if Z = 0, then after execution of the BNZ 0x03 instruction, the PC will be loaded with address 0058H (0052H + 03H x 2). Hence, the program will branch to address 0058H which is six steps forward relative to the current contents of PC. This is called the "relative addressing mode." Note that the relative mode is useful for developing position independent code.

3.3.8 Subroutine Calls in Assembly Language

It is sometimes desirable to execute a common task many times in a program. Consider the case when the sum of squares of numbers is required several times in a program. One could write a sequence of instructions in the main program for carrying out the sum of squares every time it is required. This is all right for short programs. For long programs, however, it is convenient for the programmer to write a small program known as a *subroutine* for performing the sum of squares, and call this program each time it is needed in the main program. Therefore, a subroutine can be defined as a program carrying out a particular function that can be called by another program, known as the *main program*. The subroutine needs to be placed only once in memory starting at a particular memory location. Each time the main program requires this subroutine, it can branch to it, typically by using PIC18F's CALL to subroutine (CALL) instruction along with its starting address. The subroutine is then executed. At the end of the subroutine, PIC18F's RETURN from subroutine instruction takes control back to the main program.

3.4 High-Level Language

As mentioned earlier, a programmer's efficiency increases significantly with assembly language compared to machine language. However, the programmer needs to be well acquainted with the CPU architecture and its instruction set. Further, the programmer has to provide an op-code for each operation that the CPU has to carry out in order to execute a program. As an example, for adding two numbers, the programmer would instruct the CPU to load the first number into a register, add the second number to the register, and then store the result in memory. However, the programmer might find it tedious to write all the steps required for a large program. Also, to become a reasonably good assembly language programmer, one needs to have a lot of experience.

High-level language programs composed of English-language-type statements rectify all these deficiencies of machine and assembly language programming. The programmer does not need to be familiar with the internal microcontroller structure or its instruction set. Also, each statement in a high-level language corresponds to a number of assembly or machine language instructions. For example, consider the statement "f = a + b;" written in a high-level language called C. This single statement adds the contents of 'a' with 'b' and stores the result in f. This is equivalent to a number of steps in machine or assembly language, as mentioned before. It should be pointed out that the letters a, b, and f do not refer to particular registers within the CPU. Rather, they are memory locations.

C is widely used a very popular language used with microcontrollers. A high-level language is a problem-oriented language. The programmer does not have to know the details of the architecture of the microcontroller and its instruction set. Basically, the programmer follows the rules of the particular language being used to solve the problem at hand. A second advantage is that a program written in a particular high-level language can be executed by two different microcontrollers, provided that they both understand that language. For example, a program written in C for a PIC18F microcontroller will run on a Texas Instrument's MSP 430 microcontroller because both microcontrollers have a compiler to translate the C language into their particular machine language; minor modifications are required for I/O programs.

Typical microcontrollers are also provided with a program called an "interpreter." This is provided as part of the software development package. The interpreter reads each high-level statement such as F = A + B and directs the microcontroller to perform the operations required to execute the statement. The interpreter converts each statement into machine language codes but does not convert the entire program into machine language codes prior to execution. Hence, it does not generate an object program. Therefore, an interpreter is a program that executes a set of machine language instructions in response to each high-level statement in order to carry out the function. A compiler, however, converts each statement into a set of machine language instructions and also produces an object program that is stored in memory. This object program must then be executed by the CPU to perform the required task in the high-level program.

In summary, an interpreter executes each statement as it proceeds, without generating an object code, whereas a compiler converts a high-level program into an object program that is stored in memory. This program is then executed. Compilers normally provide inefficient machine codes because of the general guidelines that must be followed for designing them. Note that C is a high-level language that includes input/output instructions. However, the compiled codes generate many more lines of machine code than an equivalent assembly language program. Therefore, the assembled program will take up less memory space and will execute much faster compared to the compiled C code.

3.5 Choosing a Programming Language

Compilers used to provide inefficient machine codes because of the general guidelines that must be followed for designing them. However, modern C compilers generate very tight and efficient codes. Hence, C is widely used these days. Assembly language programming, on the other hand, is important in the understanding of the internal architecture of a microcontroller, and may sometimes be useful for writing programs for real-time applications.

3.6 Flowcharts

Before an assembly language program is written for a specific operation, it is convenient to represent the program in a schematic form called a *flowchart.* A brief listing of the basic shapes used in a flowchart and their functions is given in Table 3.3.

Note that the flowchart symbols of Table 3.3 are used for writing some of the PIC18F assembly language programming examples in Chapters 6 and 7.

TABLE 3.3 Flowchart symbols

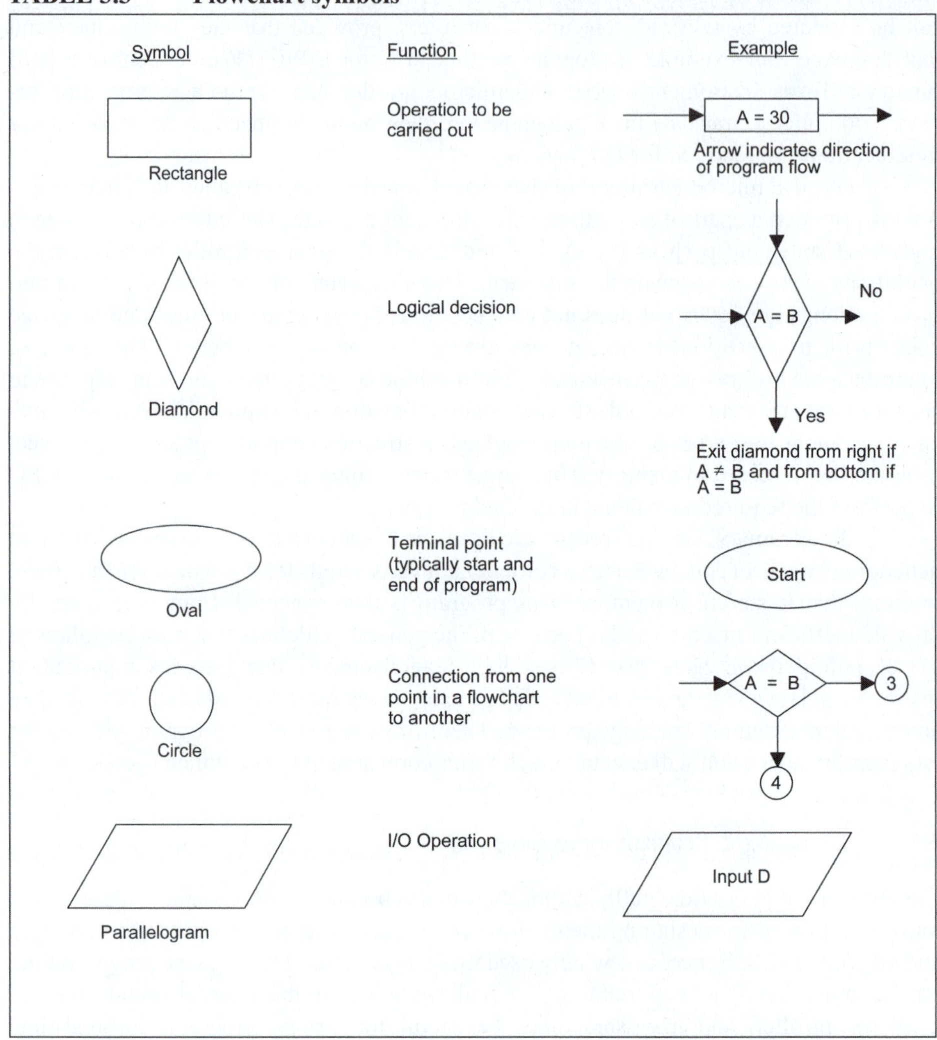

Questions and Problems

3.1 What is the basic difference between assembly and high-level languages? Why would you choose one over the other?

3.2 Assume that two microcontrollers, the PIC18F and the HC12, have C compilers. Will a program written in C language run on both microcontrollers?

3.3 Will a program written in Microchip's PIC18F assembly language run on microcontrollers from other manufacturers?

3.4 Determine the contents of address 0x23 after assembling the following:

(a)
```
ORG   0x20
DB    00H, 05H, 07H, 00H, 03H
```
(b)
```
ORG   0x20
DW    0702H, 123FH, 7020H, 0000H
```

3.5 What is the difference between

(a) a cross assembler and a resident assembler?

(b) a two-pass assembler and a meta-assembler?

3.6 Write a program equivalent to the C language assignment statement

z = a + (b * c) + (d * e) – (f / g) - (h * i);

Use only

(a) three-address instructions

(b) two-address instructions

3.7 Assume that a microcontroller has only two registers, R1 and R2, and that only the following instruction is available:

```
XOR   Ri,Rj   ; Rj <- Ri ⊕ Rj
              ; i,j = 1,2
```

Using this XOR instruction, find an instruction sequence to exchange the contents of registers R1 and R2.

3.8 Assume 2 two's complement signed numbers, $M = 11111111_2$ and $Q = 11111100_2$. Perform signed multiplication using the algorithm described in Section 3.3.6.

3.9 Using the signed division algorithm described in Section 3.3.6, find the quotient and remainder of (-25)/3.

3.10 Find the logic operation and 8-bit data for clearing bits 2 and 4 of an 8-bit number, $7E_{16}$, to 0's without changing the other bits.

3.11 Find the logic operation and 8-bit data for setting bits 0 and 7 of an 8-bit number, $3A_{16}$, to 1's without changing the other bits.

3.12 Find the overflow bit after performing an arithmetic shift on $B6_{16}$ three times to the left.

3.13 Describe the meaning of each of the following addressing modes.

(a) Immediate (b) Absolute
(c) Register indirect (d) Relative
(e) Implied

3.14 What are the advantages of subroutines?

3.15 Explain the use of a stack in implementing subroutine calls.

4 MICROCONTROLLER MEMORY AND INPUT/OUTPUT (I/O)

In this chapter we describe basic concepts of memory organization and Input/Output techniques associated with typical microcontrollers. We will also discuss signals common to CPU, memory, and I/O circuits inside typical microcontroller chips. Topics include main memory array design, and programmed and interrupt I/O.

4.1 Introduction to Microcontroller Memory

A memory unit is an integral part of any microcontroller, and its primary purpose is to hold instructions and data. The major design goal of on-chip memory inside the microcontroller is to allow it to operate at a speed close to that of the CPU. In a broad sense, a microcontroller memory system can be divided into two groups:

1. CPU registers
2. Primary or main memory

Microcontroller CPU registers are used to hold temporary results when a computation is in progress. Also, there is no speed disparity between these registers and the CPU because they are fabricated using the same technology. However, the cost involved in this approach limits a microcontroller architect to include only a few registers in the CPU.

In an accumulator-based CPU such as the PIC18F microcontroller, typical registers inside the CPU include the accumulator, program counter, stack pointer, and status register. In a general purpose register-based microcontroller such as Texas Instrument's MSP430 contain both dedicated registers and general purpose registers. Typical dedicated registers include program counter, stack pointer, and status register. In addition, several general-purpose registers are also provided and any of these registers can be used as an accumulator.

Primary or main memory is the storage area in which all programs are executed. The microcontroller can directly access only those items that are stored in main memory. Therefore, all programs must be in the main memory prior to execution. CMOS technology is normally used in main memory design. The size of the main memory is usually much larger than the number of registers, and its operating speed is slower than that of processor registers. Typically, microcontrollers such as the PIC18F contain main memory consisting of Flash memory (program memory) and SRAM (data memory).

4.1.1 Main Memory

As mentioned before, the main memory (or simply, the memory) stores both instructions and data. For 8-bit microcontrollers, the memory is divided into a number

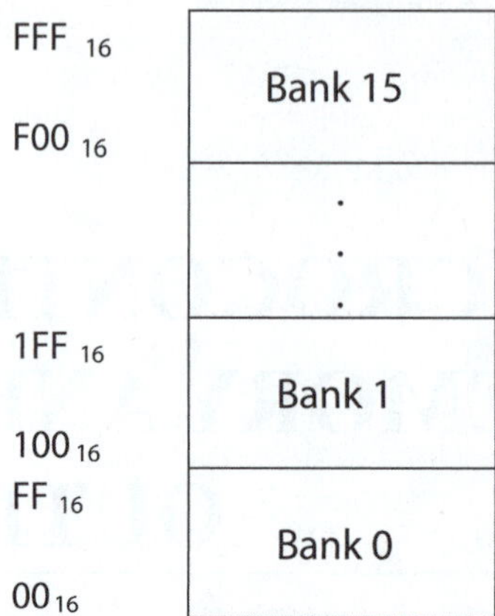

FIGURE 4.1 PIC18F data memory

of 8-bit units called *memory words*. An 8-bit unit of data is termed a *byte*. Therefore, for an 8-bit microcontroller, *memory word* and *memory byte* mean the same thing. For 16-bit microcontrollers, a word contains two bytes (16 bits). A memory word is identified in the memory by an address. For example, the PIC18F 4321 is an 8-bit microcontroller, and can directly address a maximum of two megabytes (2^{21}) of program memory space. The data memory address, on the other hand, is 12 bits wide. Hence, the PIC18F family members can directly address data memory of up to 4 Kbytes (2^{12}). This provides a maximum of 2^{12} = 4096 bytes of data memory addresses, ranging from 000 to FFF in hexadecimal.

An important characteristic of a memory is whether it is volatile or nonvolatile. The contents of a volatile memory are lost if the power is turned off. On the other hand, a nonvolatile memory retains its contents after power is switched off. ROM is a typical example of nonvolatile memory. RAM is a volatile memory unless backed up by batteries.

Large areas of data memory require an efficient addressing scheme to make rapid access to any address possible. Ideally, this means that an entire address does not need to be provided for each read or write operation. For PIC18F, this is accomplished with a RAM banking scheme. This divides the memory space into 16 contiguous banks (bank 0 through 15) of 256 bytes. Depending on the instruction, each location can be addressed directly by its full 12-bit address, or an 8-bit low-order address and a 4-bit bank pointer.

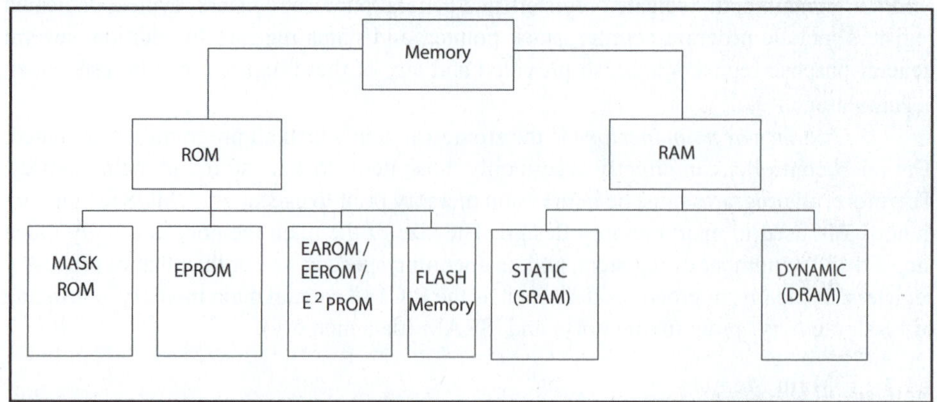

FIGURE 4.2 Summary of available semiconductor memories for microcontroller systems

Figure 4.1 shows a simplified data memory layout of the PIC18F. In the figure, the high 4 bits of an address specify the bank number. As an example, consider address 0x105 of segment 1. The high 4 bits, 0001, of this address define the location as in bank 1, and the low 8 bits, 0x05, specify the particular address in bank 1.

Memories can be categorized into two main types: read-only memory (ROM) and random-access memory (RAM). As shown in Figure 4.2, ROMs and RAMs are then divided into a number of subcategories, which is discussed next.

Read-Only Memory ROMs can only be read, so these are nonvolatile memory. CMOS technology is used to fabricate ROMs. ROMs are divided into two common types: mask ROM and erasable programmable ROM (EPROM) such as the 2732, and EAROM (electrically alterable ROM) [also called EEPROM or E^2PROM (electrically erasable PROM)] such as the 2864.

Mask ROMs are programmed by a masking operation performed on a chip during the manufacturing process. The contents of mask ROMs are permanent and cannot be changed by the user. EPROMs can be programmed, and their contents can also be altered by using special equipment, called an *EPROM programmer*. When designing a microcontroller for a particular application, permanent programs are stored in ROMs. Control memories used to microprogram the control unit are ROMs.

EPROMs can be reprogrammed and erased. The EPROM chip must be removed from the system before programming. This memory is erased by exposing the chip to ultraviolet light via a lid or window on the chip. Typical erase times vary between 10 and 20 minutes. The EPROM can be programmed by inserting the EPROM chip into a socket of the EPROM programmer and providing proper addresses and voltage pulses at the appropriate pins of the chip.

EEPROMs can be programmed without removing the memory from the ROM's sockets. These memories are also called *read-mostly memories* (RMMs), because they have much slower write times than read times. Therefore, these memories are usually suited for operations when mostly reading rather that writing will be performed.

Flash Memory Another type of memory, called *Flash memory* (nonvolatile), invented in the mid-1980s by Toshiba, is designed using a combination of EPROM and E^2PROM technologies. Flash memory can be reprogrammed electrically while embedded on the board. One can change multiple bytes at a time. An example of flash memory is the Intel 28F020 (256K x 8-bit). Flash memory is typically used in cellular phones and digital cameras. Note that the PIC18F uses flash memory as its program memory.

Random-Access Memory There are two types of RAM: static RAM (SRAM) and dynamic RAM (DRAM). *Static RAM* stores data in flip-flops. Therefore, this memory does not need to be refreshed. RAMs are volatile unless backed up by battery. The PIC18F uses SRAM for its data memory.

Dynamic RAM stores data in capacitors. That is, it can hold data for a few milliseconds. Hence, dynamic RAMs are refreshed typically by using external refresh circuitry. DRAMs are used in applications requiring large memory. DRAMs have higher densities than SRAMs. Typical examples of DRAMs are the 4464 (64K x 4-bit), 44256 (256K x 4-bit), and 41000 (1M x 1-bit). DRAMs are inexpensive, occupy less space, and dissipate less power than SRAMs. Two enhanced versions of DRAM are EDO DRAM (extended data output DRAM) and SDRAM (synchronous DRAM).

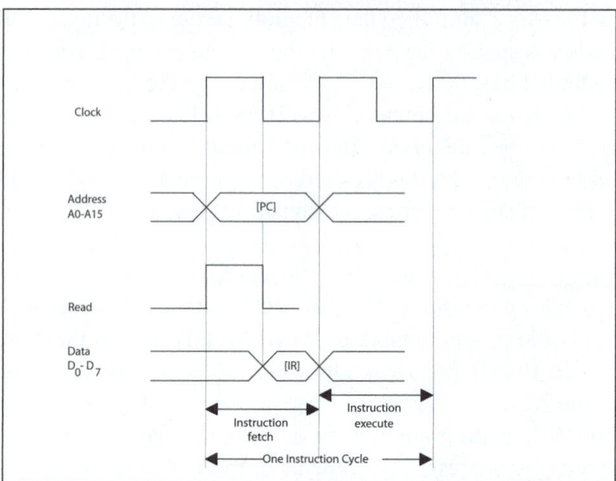

FIGURE 4.3 Typical instruction fetch timing diagram for an 8-bit microprocessor

The EDO DRAM provides fast access by allowing the DRAM controller to output the next address at the same time the current data are being read. An SDRAM contains multiple DRAMs (typically, four) internally. SDRAMs utilize the multiplexed addressing of conventional DRAMs. That is, like DRAMs, SDRAMs provide row and column addresses in two steps. However, the control signals and address inputs are sampled by the SDRAM at the leading edge of a common clock signal (133 MHz maximum). SDRAMs provide higher densities than conventional DRAMs by further reducing the need for support circuitry and faster speeds. The SDRAM has been used in PCs (personal computers).

4.1.2 READ and WRITE Timing Diagrams

To execute an instruction, the CPU of the microcontroller reads or fetches the op-code via the data bus from a memory location in the ROM/RAM external to the CPU. It then places the opcode (instruction) in the instruction register. Finally, the CPU executes the instruction. Therefore, the execution of an instruction consists of two portions, instruction fetch and instruction execution. We consider the instruction fetch, memory READ, and memory WRITE timing diagrams in the following paragraphs using a single clock signal. Figure 4.3 shows a typical instruction fetch timing diagram.

In Figure 4.3, to fetch an instruction, when the clock signal goes to HIGH, the CPU places the contents of the program counter on the address bus via address pins A_0–A_{15} on the chip. Note that since each of lines A_0–A_{15} can be either HIGH or LOW, both transitions are shown for the address in Figure 4.3. The instruction fetch is basically a memory READ operation. Therefore, the CPU raises the signal on the READ pin to HIGH. As soon as the clock goes to LOW, the logic external to the CPU gets the contents of the memory location addressed by A_0–A_{15} and places them on the data bus D_0–D_7. The CPU then takes and stores the data in the instruction register so that the data get interpreted as an instruction. This is called *instruction fetch*. The CPU performs this sequence of operations for every instruction.

We now describe the READ and WRITE timing diagrams. A typical READ timing diagram is shown in Figure 4.4. Memory READ is basically loading the contents of a memory location of the main ROM/RAM into an internal register of the CPU. The address of the location is provided by the contents of the memory address register (MAR).

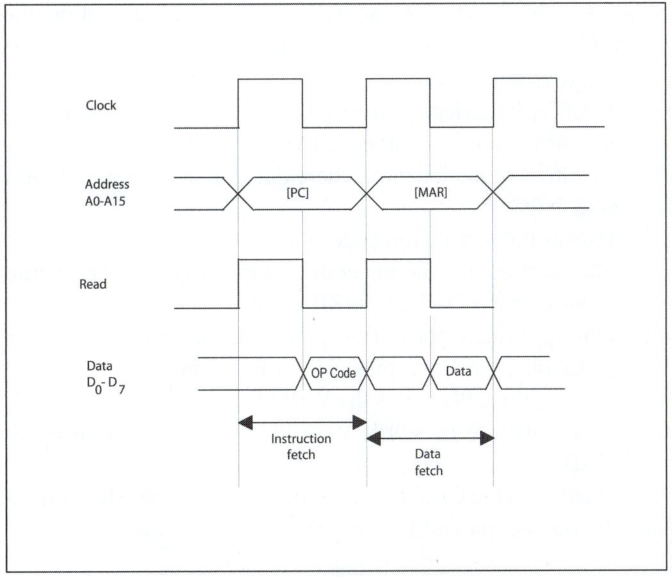

FIGURE 4.4 Typical memory READ timing diagram

Let us now explain the READ timing diagram of Figure 4.4.

1. The CPU performs the instruction fetch cycle as before to READ the opcode.
2. The CPU interprets the opcode as a memory READ operation.
3. When the clock pin signal goes HIGH, the CPU places the contents of the memory address register on the address pins A_0–A_{15} of the memory module.
4. At the same time, the CPU raises the READ pin signal to HIGH.
5. The logic external to the CPU gets the contents of the location in the main ROM/

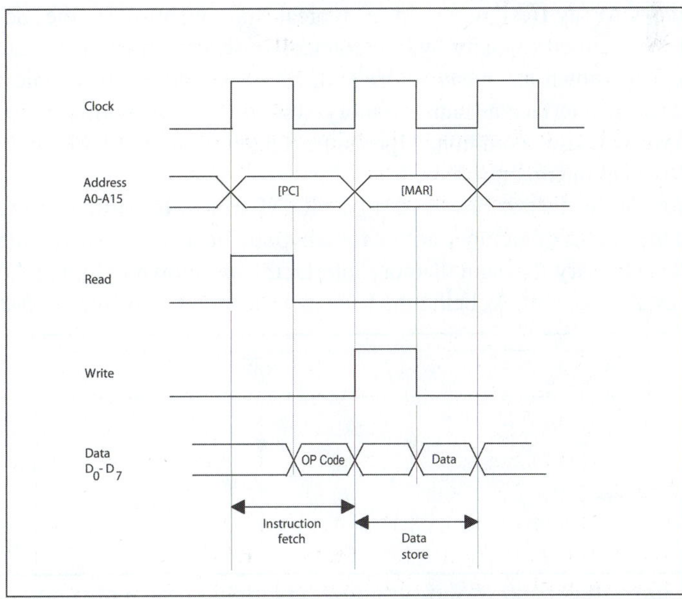

FIGURE 4.5 Typical memory WRITE timing diagram

RAM addressed by the memory address register and places it on the data bus.

6. Finally, the CPU gets these data from the data bus via pins $D_0 - D_7$ and stores them in an internal register.

Memory WRITE is basically storing the contents of an internal register of the CPU into a memory location of the main RAM. The contents of the memory address register provide the address of the location where data are to be stored. Figure 4.5 shows a typical WRITE timing diagram.

1. The CPU fetches the instruction code as before.
2. The CPU interprets the instruction code as a memory WRITE instruction and then proceeds to perform the DATA STORE cycle.
3. When the clock pin signal goes HIGH, the CPU places the contents of the memory address register on the address pins A_0–A_{15} of the memory module.
4. At the same time, the CPU raises the WRITE pin signal to HIGH.
5. The CPU places data to be stored from the contents of an internal register onto data pins D_0–D_7.
6. The logic external to the CPU stores the data from the register into a RAM location addressed by the memory address register.

4.1.3 Main Memory Organization

Typical microcontroller on-chip main memory, also called “memory module,” may include ROM/EPROM/E^{2}PROM, and SRAM. As mentioned earlier, the PIC18F main memory (program memory and data memory) consists of Flash memory and SRAMs. A microcontroller system designer is normally interested in how the microcontroller memory is organized or, in other words, how to connect the memory units to the CPU, and then determine the memory map of the microcontroller. That is, the PIC18F designer would be interested in finding out what memory locations are assigned to the Flash memory and SRAMs.

Main Memory Array Design In a typical microcontroller, the designer has to implement the required capacity by interconnecting several memory circuits to the CPU. This concept is known as *memory array design.* We address this topic in this section and show how to interface a data memory system (SRAM) with a typical CPU. In the following, we will use common signals associated with the CPU and memory units internal to typical microcontrollers.

Now let us discuss how to design SRAM arrays. In particular, our discussion is focused on the design of memory arrays for a hypothetical CPU. The pertinent signals of a typical CPU necessary for main memory interfacing are shown in Figure 4.6. There are 16 address lines, A_{15}-A_0, with A_0 being the least significant bit. This means that this CPU can

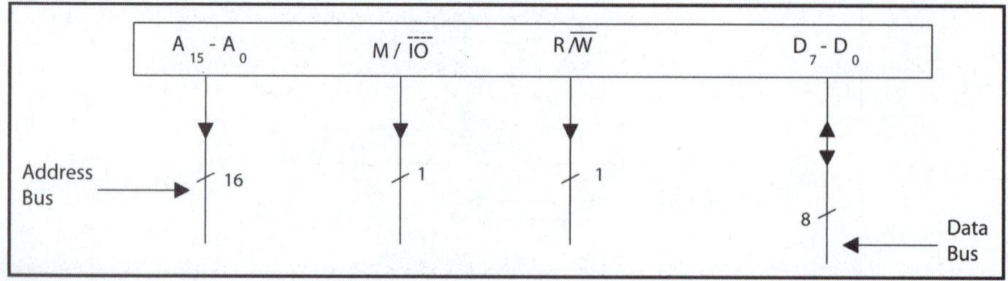

FIGURE 4.6 Pertinent signals of a typical CPU required for main memory interfacing

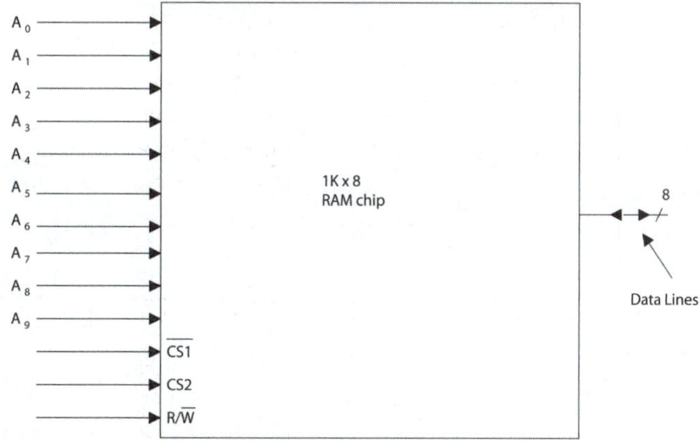

FIGURE 4.7 Typical 1K × 8 SRAM unit

address directly a maximum of 2^{16} = 65,536 or 64K bytes of memory locations.

The control line $M/\overline{IO}$ goes LOW if the CPU executes an I/O instruction; it is held HIGH if the CPU executes a memory instruction. Similarly, the CPU drives control line $R/\overline{W}$ HIGH for READ operation; it is held LOW for WRITE operation. Note that all 16 address lines and the two control lines ($M/\overline{IO}$, $R/\overline{W}$) described so far are unidirectional in nature; that is, information can always travel on these lines from the processor to external units. Eight bidirectional data lines, D_7-D_0 (with D_0 being the least significant bit), are also shown in Figure 4.6. These lines are used to allow data transfer from the CPU to memory module, and vice versa.

The block diagram of a typical 1K × 8 RAM SRAM unit is shown in Figure 4.7. In this circuit, there are 10 address lines, A_9-A_0, so one can read or write 1024 (2^{10} = 1024) different memory words. Also, in this chip there are eight bidirectional data lines, D_7-D_0, so that information can travel back and forth between the CPU and the memory module. The three control lines $\overline{CS1}$, CS2, and $R/\overline{W}$ are used to control the SRAM unit according to the truth table shown in Table 4.1, from which it can be concluded that the RAM chip is enabled only when $\overline{CS1}$ = 0 and CS2 = 1. Under this condition, $R/\overline{W}$ = 0 and $R/\overline{W}$ = 1 imply write and read operations, respectively.

To connect a CPU to the memory module, two address decoding techniques are commonly used for each memory type: linear decoding and full decoding. Let us discuss first how to interconnect a CPU with a 4K SRAM array comprised of the four 1K SRAM units of Figure 4.7 using the linear decoding technique. Figure 4.8 uses linear decoding to accomplish this. In this approach, address lines A_9-A_0 of the CPU are connected to all SRAM units. Similarly, the control lines $M/\overline{IO}$ and $R/\overline{W}$ of the CPU are connected to control lines CS2 and $R/\overline{W}$, respectively, to each of the SRAM units. The high-order

TABLE 4.1 Truth table for controlling SRAM unit

$\overline{CS1}$	CS2	$R/\overline{W}$	Function
0	1	0	Write operation
0	1	1	Read operation
1	X	X	The chip is not selected.
X	0	X	The chip is not selected.

X means "don't care."

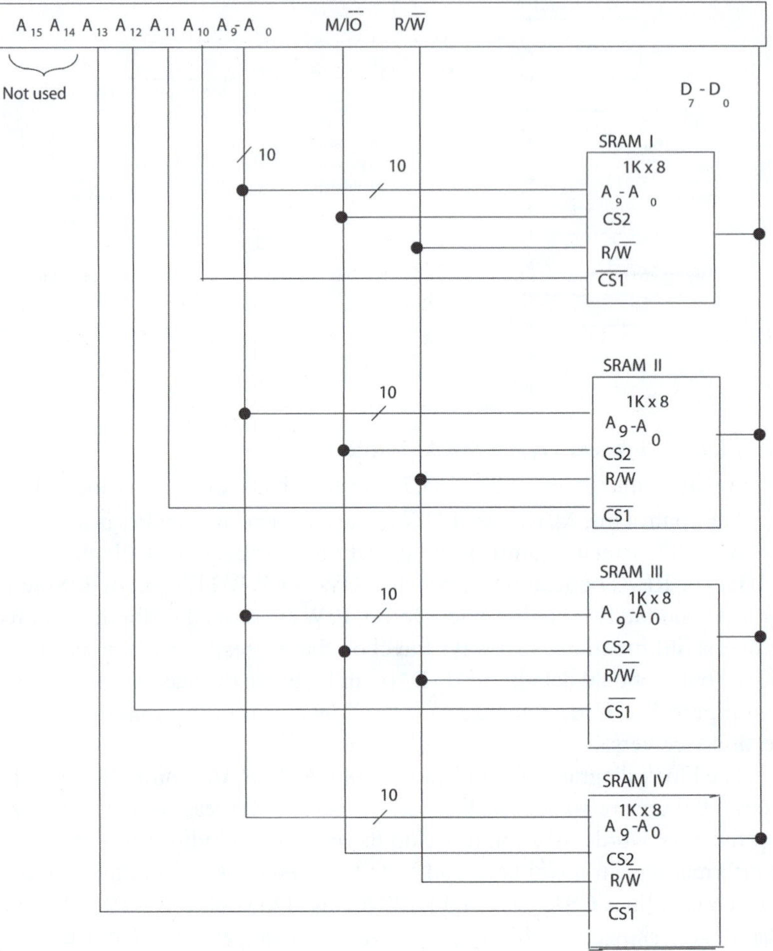

FIGURE 4.8 CPU connected to 4K SRAM using the linear select decoding technique

address lines A_{10}-A_{13} are used for selecting memory units. In particular, address lines A_{10} and A_{11} select SRAM units I and II, respectively. Similarly, the address lines A_{12} and A_{13} select the SRAM units III and IV, respectively. A_{15} and A_{14} are don't cares and are assumed to be zero.

Table 4.2 describes how the addresses are distributed among the four 1K SRAM units. The primary advantage this method, known as linear select decoding, is that it does not require decoding hardware. However, if two or more of lines A_{10}-A_{13} are low at the

TABLE 4.2 Address map of the memory organization of Figure 4.8

Address range (hex)	SRAM number
3800-3BFF	I
3400-37FF	II
2C00-2FFF	III
1C00-1FFF	IV

TABLE 4.3 Decoding guide

A_{12}	A_{11}	A_{10}	SRAM number
0	0	0	I
0	0	1	II
0	1	0	III
0	1	1	IV

same time, more than one SRAM unit is selected, and this causes a bus conflict. Because of this potential problem, the software must be written such that it never reads into or writes from any address in which more than one of bits A_{13}-A_{10} are low. Another disadvantage of this method is that it wastes a large amount of address space. For example, whenever the address value is B800 or 3800, SRAM chip I is selected. In other words, address 3800 is the mirror reflection of address B800 (this situation is also called memory foldback). This technique is therefore limited to a small system. The system of Figure 4.8 can be expanded up to a total capacity of 6K using A_{14} and A_{15} to select two more 1K SRAM units.

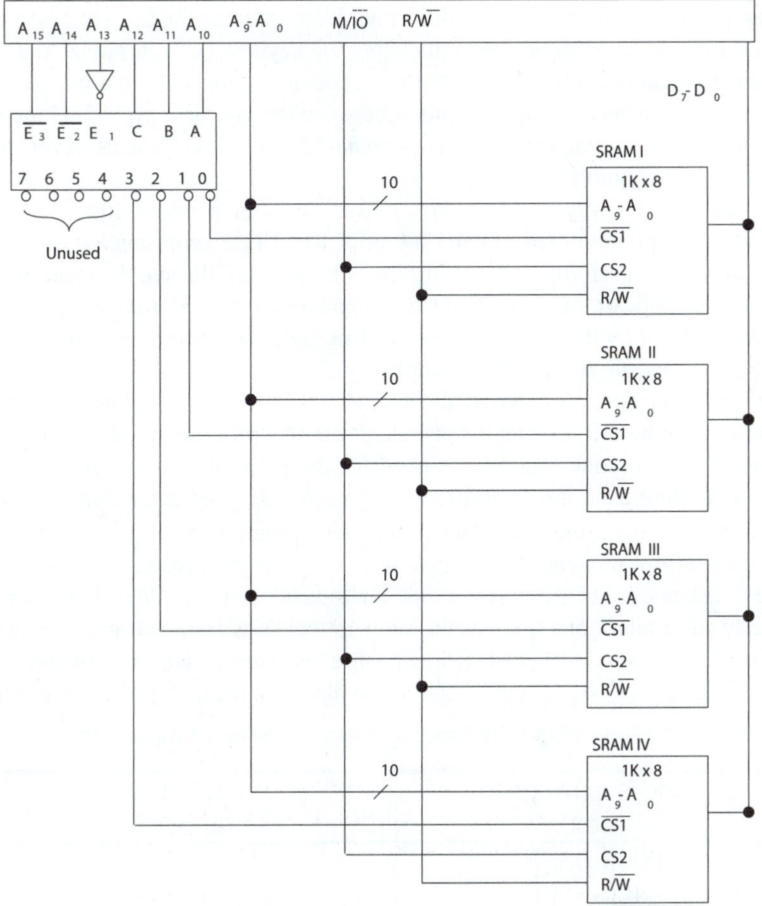

FIGURE 4.9 **Interconnecting a CPU with a 4K RAM using full decoded memory addressing**

To resolve problems with linear decoding, we use full decoded memory addressing. In this technique we use a decoder. The 4K memory system designed using this technique is shown in Figure 4.9. In Figure 4.9 the decoder output selects one of the four 1K SRAMs, depending on the values of A_{12}, A_{11}, and A_{10} (Table 4.3).

Note that the decoder output will be enabled only when $\overline{E3} = \overline{E2} = 0$ and E1 = 1. Therefore, in the organization of Figure 4.9, when any one of the high-order bits A_{15}, A_{14}, or A_{13} is 1, the decoder will be disabled, and thus none of the SRAMs will be selected. In this arrangement, the memory addresses are assigned as shown in Table 4.4.

This approach does not waste any address space since the unused decoder outputs (don't cares) can be used for memory expansion. For example, the 3-to-8 decoder of Figure 4.9 can select eight 1K SRAMs. Also, this method does not generate any bus conflict. This is because the decoder output selected ensures enabling of one memory unit at a time.

4.2 Microcontroller Input/Output (I/O)

The technique of data transfer between a microcontroller and an external device is called *input/output* (I/O). One communicates with a microcontroller via the I/O devices interfaced to it. The user can enter programs and data using the keyboard on a terminal and execute the programs to obtain results. Therefore, the I/O devices connected to a microcontroller provide an efficient means of communication between the microcontroller and the outside world. These I/O devices, commonly called *peripherals,* include keyboards, seven-segment displays, and LCDs (liquid crystal displays).

There are two ways of transferring data between a microcontroller and I/O devices. These are programmed I/O and interrupt I/O. Using *programmed I/O*, the CPU executes a program to perform all data transfers between the CPU and the external device. The main characteristic of this type of I/O technique is that the external device carries out the functions dictated by the program contained in the microcontroller memory. In other words, the CPU controls all transfers completely.

In *interrupt I/O*, an external device can force the CPU to stop executing the current program temporarily so that it can execute another program known as an *interrupt service routine*. This routine satisfies the needs of the external device. After completing this program, a return from interrupt instruction can be executed at the end of the service routine to return control at the right place in the main program.

The interrupt procedure is similar in concept to the procedure associated with subroutine CALL and RETURN instructions. The subroutine CALL /RETURN includes a main program and a subroutine whereas the interrupt contains a main program and a service routine. The subroutine CALL instruction pushes the current contents of the program counter onto the stack. The RETURN instruction placed at the end of the subroutine pops

TABLE 4.4 Address map of the memory organization of Figure 4.9

Address range (hex)	SRAM number
0000-03FF	I
0400-07FF	II
0800-0BFF	III
0C00-0FFF	IV

the previously pushed program counter, and returns control to the main program.

The interrupt, on the other hand, is initiated externally via hardware or internally via occurrence of events such as completion of ADC (analog-to-digital converter). Once the interrupt is recognized, the microcontroller normally pushes the program counter (PC) and the status register (SR) onto the stack, and automatically branches to an address predefined by the manufacturer. The user writes a program called "interrupt service routine" at this address. This program is similar to the subroutine. A "Return from Interrupt" instruction placed by the user at the end of the interrupt service routine will pop the previously pushed PC and SR , and will return control to the main program at the proper location.

4.2.1 Overview of Digital Output Circuits

For simplicity, a basic background in TTL outputs will be provided next. Since the CMOS technology is used in designing typical microcontrollers, these concepts will then be related to CMOS outputs.

TTL Outputs There are three types of output configurations for TTL. These are open-collector output, totem-pole output, and tristate (three-state) output. The open-collector output means that the TTL output is a transistor with nothing connected to the collector. The collector voltage provides the output of the gate. For the open-collector output to work properly, a resistor (called the pullup resistor), with a value of typically 1 Kohm, should be connected between the open collector output and a +5 V power supply.

If the outputs of several open-collector gates are tied together with an external resistor (typically 1 Kohm) to a +5 V source, a logical AND function is performed at the connecting point. This is called wired-AND logic.

Figure 4.10 shows two open-collector outputs (*A* and *B*) connected together to a common output point *C* via a 1 KΩ resistor and a +5 V source.

The common-output point *C* is HIGH only when both transistors are in cutoff (OFF) mode, providing *A* = HIGH and *B* = HIGH. If one or both of the two transistors is turned ON, making one (or both open-collector outputs) LOW, this will drive the common output *C* to LOW. Note that a LOW (ground, for example) signal when connected to a HIGH (+5 V, for example) signal generates a LOW. Thus, *C* is obtained by performing a logical AND operation of the open collector outputs *A* and *B*.

Let us briefly review the totem-pole output circuit shown in Figure 4.11. The circuit operates as follows:

When transistor Q_1 is ON, transistor Q_2 is OFF. When Q_1 is OFF, Q_2 is ON. This

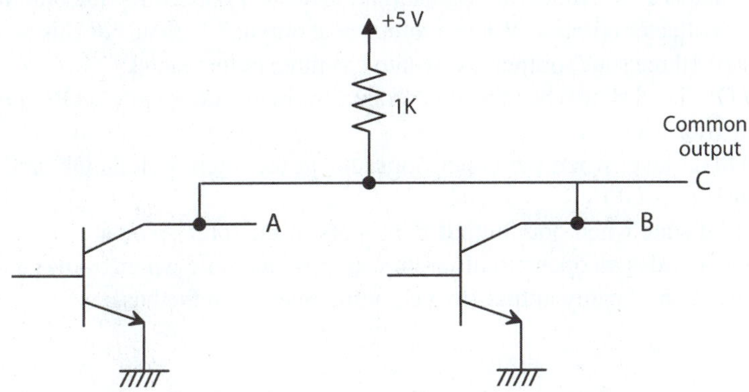

FIGURE 4.10 Two open-collector outputs A and B tied together

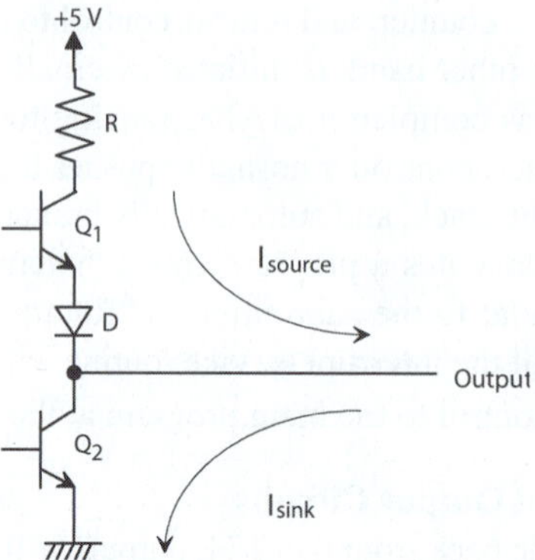

FIGURE 4.11 TTL totem-pole output

is how the totem-pole output is designed. The complete TTL gate connected to the bases of transistors Q_1 and Q_2 is not shown; only the output circuit is shown.

In the figure, Q_1 is turned ON when the logic gate circuit connected to its base sends a HIGH output. The switches in transistor Q_1 and diode D close while the switch in Q_2 is open. A current flows from the +5 V source through R, Q_1, and D to the output. This current is called I_{source} or output high current, I_{OH}. This is typically represented by a negative sign in front of the current value in the TTL data book, a notation indicating that the chip is losing current. For a low output value of the logic gate, the switches in Q_1 and D are open and the switch in Q_2 closes. A current flows from the output through Q_2 to ground. This current is called I_{sink} or output low current, I_{OL}. This is represented by a positive sign in front of the current value in the TTL data book, indicating that current is being added to the chip. Either I_{source} or I_{sink} can be used to drive a typical output device such as an LED. I_{source} (I_{OH}) is normally much smaller than I_{sink} (I_{OL}). I_{source} (I_{OH}) is typically −0.4 mA (or −400 μA) at a minimum voltage of 2.7 V at the output. I_{source} is normally used to drive devices that require high currents. A current amplifier (buffer) such as a transistor or an inverting buffer chip such as 74LS368 needs to be connected at the output if I_{source} is used to drive a device such as an LED requiring high current (10 to 20 mA). I_{sink} is normally 8 mA.

The totem-pole outputs must not be tied together. When two totem-pole outputs are connected together with the output of one gate HIGH and the output of the second gate LOW, the excessive amount of current drawn can produce enough heat to damage the transistors in the circuit.

Tristate is a special totem-pole output that allows connecting the outputs together like the open-collector outputs. When a totem-pole output TTL gate has this property, it is called a tristate (three state) output. A tristate has three output states:

1. A LOW level state when the lower transistor in the totem-pole is ON and the upper transistor is OFF
2. A HIGH level when the upper transistor in the totem-pole is ON and the lower transistor is OFF
3. A third state when both output transistors in the totem-pole are OFF. This third state provides an open circuit or high-impedance state which allows a direct wire connection of many outputs to a common line called the bus.

CMOS Outputs Like TTL, the CMOS logic offers three types of outputs. These are

FIGURE 4.12 Typical switch for a microcontroller's input

push-pull (totem-pole in TTL), open drain (open collector in TTL), and tristate outputs. For example, the 74HC00 contains four independent 2-input NAND gates and includes push-pull output. The 74HC03 also contains four independent 2-input NAND gates, but has open drain outputs. The 74HC03 requires a pull-up resistor for each gate. The 74HC125 contains four independent tristate buffers in a single chip. Note that CMOS technology is normally used in designing microcontrollers.

4.2.2 Simple I/O Devices

A simple input device such as a DIP switch can be connected to a microcontroller's I/O port, as shown in Figure 4.12. The figure shows a switch circuit that can be used as a single bit input into an I/O port. When the DIP switch is open, V_{IN} is HIGH. When the switch is closed, V_{IN} is LOW. V_{IN} can be used as an input bit for performing laboratory experiments. Note that unlike TTL, a 1 Kohm resistor is connected between the switch and the input of the MOS gate. This provides protection against static discharge.

For performing simple I/O experiments using programmed I/O, light-emitting diodes (LEDs) and seven-segment displays can be used as output devices. An LED is typically driven by low voltage and low current, which makes it a very attractive device for use with microcontrollers.

Table 4.5 provides the current and voltage requirements for red, yellow, and green LEDs. Basically, an LED will be ON, generating light, when its cathode is sufficiently negative with respect to its anode. A microcontroller can therefore light an LED either by grounding the cathode (if the anode is tied to +5 V) or by applying +5 V to the anode (if the cathode is grounded) through an appropriate resistor value. A typical hardware interface

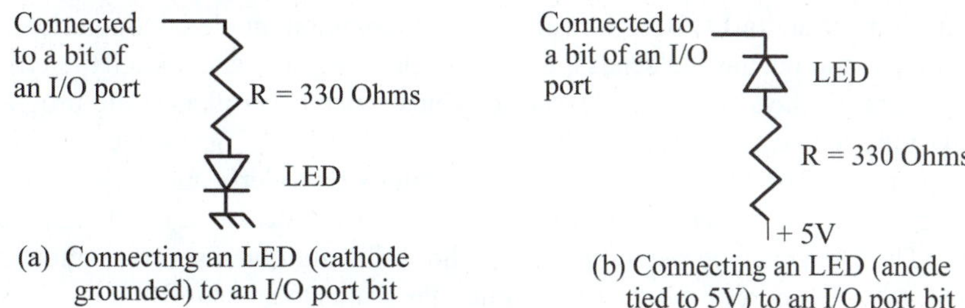

(a) Connecting an LED (cathode grounded) to an I/O port bit

(b) Connecting an LED (anode tied to 5V) to an I/O port bit

FIGURE 4.13 Interfacing LED to PIC18F

TABLE 4. 5 Current and voltage requirements of LEDs

LEDs	Red	Yellow	Green
Current	10 mA	10 mA	20 mA
Voltage	1.7 V	2.2 V	2.4 V

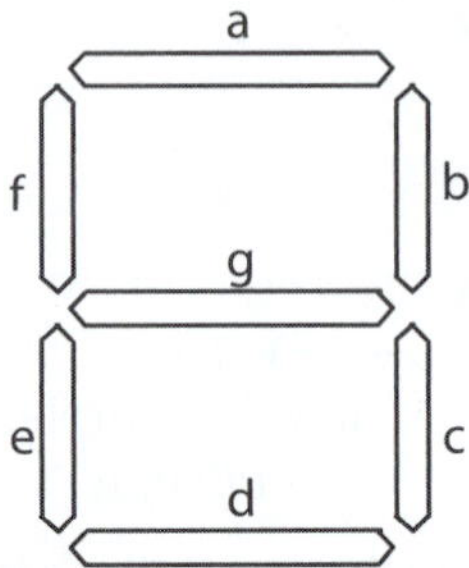

FIGURE 4.14 A seven-segment display

between a microcontroller and an LED is depicted in Figure 4.13.

A typical microcontroller such as the PIC18F outputs adequate current to turn an LED ON or OFF. In Figure 4.13 (a), a '1' from the microcontroller will turn the LED ON while a '0' will turn it OFF. In Figure 4.13 (b), on the other hand, a '0' from the microcontroller will turn the LED ON while a '1' will turn it OFF.

From Table 4.5, a red LED requires 10 mA current at 1.7 V. In Figure 4.13 (a), a HIGH at the microcontroller output will turn the LED ON. This will allow a path of current to flow from the +5 V source through R and the LED to the ground. In Figure 4.13 (b), a LOW at the microcontroller output will turn the LED ON. This will allow a path of current to flow from the +5 V source through R and the LED to the ground (microcontroller I/O port). The appropriate value of R needs to be calculated to satisfy the voltage and current requirements of the LED. The value of R can be calculated as follows:

$$R= \frac{5 - 1.7}{10\text{ mA}} = \frac{5 - 1.7}{10\text{ mA}} = 330\ \Omega$$

Therefore, the interface design is complete, and a value of $R = 330\ \Omega$ is required. A seven-segment display can be used with programmed I/O to display, for example, decimal numbers from 0 to 9. The name *seven segment* is based on the fact that there are seven LEDs, one in each segment of the display. Figure 4.14 shows a typical seven-segment display. In the figure, each segment contains an LED. All decimal numbers from 0 through 9 can be displayed by turning the appropriate segment ON or OFF. For example, a '0' can be displayed by turning the LED in segment *g* OFF and turning the other six LEDs in segments *a* through *f* ON. There are two types of seven-segment displays: common-cathode and common-anode. In common-cathode arangement, the microcontroller sends a HIGH to light a segment and a LOW to turn it off. In a common-anode configuration, on the other hand, the microcontroller sends a LOW to light a segment and a HIGH to turn it off. Seven-segment displays can be interfaced to typical microcontrollers using programmed I/O. BCD to seven-segment code converter chips such as 7447 or 7448 can be replaced by a lookup table. This table can be stored in a microcontroller's memory. An assembly language program can be written to read the appropriate code for a BCD digit stored in this table. These data can be output to display the BCD digit on a seven-segment display connected to an I/O port of the microcontroller. Programs to accomplish this are written in PIC18F assembly language (Chapter 8).

4.2.3 Programmed I/O

A microcontroller communicates with an external device via one or more registers called *I/O ports* using programmed I/O. Each bit in the port can be configured individually as either input or output. Each port can be configured as an input or output port by another register usually called the *Data Direction Register (DDR)*. The port contains the actual

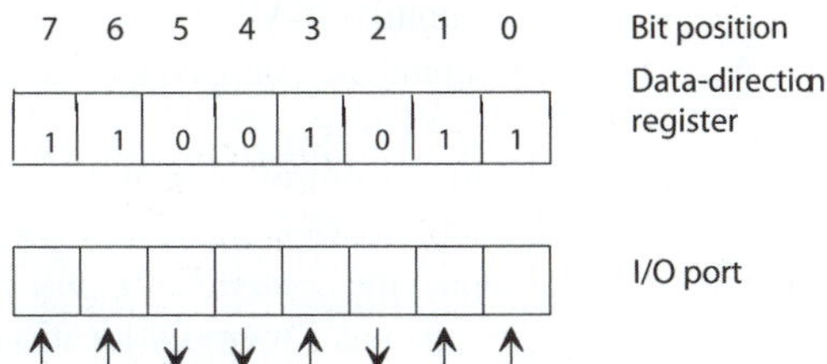

FIGURE 4.15 I/O port with the corresponding data direction register

input or output data. The data direction register is an output register and can be used to configure the bits in the port as inputs or outputs.

Each bit in the port can be set up as an input or output, normally by writing a '0' or a '1' in the corresponding bit of the DDR. The PIC18F microcontroller makes an I/O port bit an input by writing a '1' in the corresponding bit in DDR. On the other hand, writing a '0' in a particular bit in DDR will configure the corresponding bit in the port as an output.

For example, if an 8-bit DDR in the PIC18F contains 0xCB (CB Hex), the corresponding port is defined as shown in Figure 4.15. In this example, because 0xCB (1100 1011) is stored in the data direction register, bits 0, 1, 3, 6, and 7 of the port are set up as inputs, and bits 2, 4, and 5 of the port are defined as outputs. The microcntroller can then send output to external devices, such as LEDs, connected at bits 2, 4, and 5 through a proper interface.

Similarly, the microcontroller can input the status of external devices, such as switches, through bits 0, 1, 3, 6, and 7. To input data from the input switches, the microcontroller inputs the complete byte, including the bits to which LEDs are connected. While receiving input data from an I/O port, however, the microcontroller places a value, probably 0, at the bits configured as outputs and the program must interpret them as "don't cares." At the same time, the microcontroller's outputs to bits configured as inputs are disabled.

I/O ports are addressed using either standard I/O or memory-mapped I/O techniques. Using *Standard I/O* {sometimes called *port I/O* (also called *isolated I/O* by Intel)}, the CPU outputs an internal signal such as the M/$\overline{\text{IO}}$ for memory and I/O units on the microcontroller chip. The CPU outputs a HIGH on M/$\overline{\text{IO}}$ to indicate to memory and the I/O that a memory operation is taking place. A LOW output from the CPU to M/$\overline{\text{IO}}$ indicates an I/O operation. Execution of an IN or OUT instruction makes the M/$\overline{\text{IO}}$ LOW, whereas memory-oriented instructions, such as MOVE, drive the M/$\overline{\text{IO}}$ to HIGH.

In standard I/O, the CPU uses the M/$\overline{\text{IO}}$ output signal to distinguish between I/O and memory. Intel microcontrollers such as the 8051 use standard I/O.

In *memory-mapped I/O*, the CPU does not use the M/$\overline{\text{IO}}$ control signal. Instead, the CPU uses an unused address pin to distinguish between memory and I/O. The CPU uses a portion of the memory addresses to represent I/O ports. The I/O ports are mapped as part of the microcontroller's main memory addresses which may not exist physically, but are used by the microcontroller's memory-oriented instructions, such as MOVE, to generate the necessary control signals to perform I/O. The PIC18F uses memory-mapped I/O.

When standard I/O is used, microcontrollers normally use an IN or OUT instruction with 8-bit ports as follows:

```
IN      A, PORTA        ; Inputs 8-bit data from PORTA into the 8-bit
```

```
                        ; accumulator A
OUT   PORTA,A           ; Outputs the contents of the 8-bit accumulator A
                        ; into PORTA
```

With memory-mapped I/O, the microcontroller normally uses an instruction such as MOV as follows:

```
MOV  mem, reg           ; Inputs the contents of a port called "mem"
                        ; mapped as a memory location into a register
MOV reg,mem             ; outputs the contents of a register to a port called
                        ; "mem"  mapped as a memory location
```

4.2.4 Unconditional and Conditional Programmed I/O

There are typically two ways in which programmed I/O can be utilized: unconditional I/O and conditional I/O. The microcontroller can send data to an external device at any time using *unconditional programmed I/O*. The external device must always be ready for data transfer. A typical example is that of a microcontroller outputting a 7-bit code through an I/O port to drive a seven-segment display connected to this port.

In *conditional programmed I/O*, the microcontroller waits for a particular condition to occur, and then outputs data to an external device based on the condition. Conditional programmed I/O is sometimes called *polled I/O*.

As an example of conditional programmed I/O, consider Figure 4.16. Suppose that a comparator is connected to bit 0 of Port C , and an LED is connected to bit 1 of Port D of the PIC18F4321 microcontroller. It is desired to turn the LED ON when the comparator output becomes HIGH (Vx > Vy). In a situation such as this, the microcontroller needs to wait in a loop until the condition "Vx > Vy" occurs. The microcontroller will send a HIGH to bit 1 of Port D as soon as the condition occurs.

Note that TRISC is the DDR for Port C and TRISD is the DDR for Port D. A '1' in a particular position will make the corresponding bit in each of these ports as an input while a '0' will make it an output. Also, note that the PIC18F4321 uses memory-mapped I/O. Hence, the following assembly language program starting at address 0x200 for the PIC18F4321 microcontroller will accomplish this:

```
                ORG     0x200
                SETF    TRISC       ; Make Port C as input by setting all bits of
                                    ; TRISC to 1's
                CLRF    TRISD       ; Make Port C as output by clearing all bits
                                    ; of TRISD to 0's
WAIT            MOVF    PORTC, W    ; Input commparator output into WREG
                                    ; (ACCUMULATOR) via bit 0 of PORTC
                ANDLW   0x01        ; AND to check bit 0 (comparator output)
                                    ; of WREG is 1
                BZ      WAIT        ; Wait in loop if comparator output is 0 or
                                    ; Z flag is 1
                MOVLW   0x02        ; Move 1 to bit 1 of WREG (Accumulator)
                                    ; register
                MOVWF   PORTD       ; Turn the LED ON
                SLEEP               ; HALT
```

The PIC18F instructions used in the above program will now be explained. The "SETF TRISC" in the above program sets all bits of the TRISC (DDR for Port C) to 1's

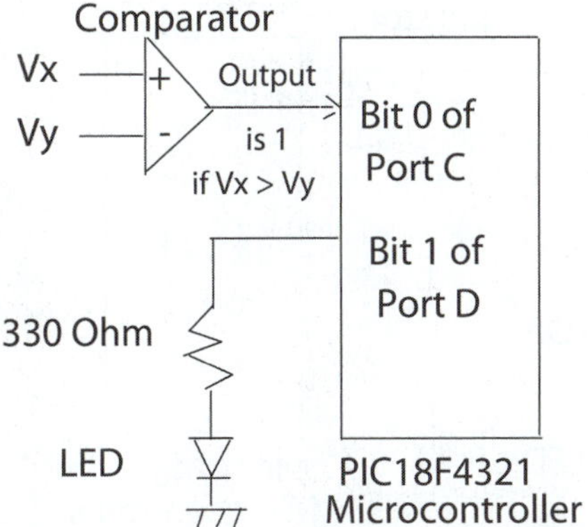

FIGURE 4.16 Example illustrating conditional or polled I/O

and thus configures bit 0 of Port C as an input bit. The "CLRF TRISD," on the other hand, clears all bits in TRISD (DDR for Port D) to 0's and configures bit 1 of Port D as an output bit. The "MOVF PORTC, W" instruction moves (inputs) the contents of PORTC into the WREG register (accumulator of the PIC18F). Thus, the comparator output connected to bit 0 of Port C is input into bit 0 of the WREG register.

The "ANDLW 0x01" logically ANDs the contents of WREG with 0x01, and stores the result in WREG. The contents of WREG will be zero (Z = 1) if the comparator output at bit 0 of Port C is 0; the contents of WREG will be one (Z = 0) if the comparator output is 1. The "BZ WAIT" instruction checks the Z flag. If the Z flag is 1 (comparator output is 0), the program branches back to WAIT, and stays in the loop until the comparator output is 1. As soon as the comparator output is 1 (Z = 0), the "MOVLW 0x02" moves 0x02 into WREG, and thus, the bit 1 of WREG is a '1'. The MOVWF PORTD" instruction moves the contents of WREG to Port D. Thus, a '1' is output to bit 1 of Port D, and the LED is turned ON. The "SLEEP" instruction then halts the microcontroller.

Note that in the program, the PIC18F4321 has to wait in a loop indefinitely for the comparator output to become one (Vx > Vy). This is called "conditional" or "polled I/O," and is obviously inefficient because of the wait loop.

4.2.5 Interrupt I/O

As mentioned before, a disadvantage of conditional programmed I/O is that the CPU needs to check the status bit (output of the comparator) by waiting in a loop. This type of I/O transfer is dependent on the occurrence of the external condition. This waiting may slow down the CPU's ability to process other data. The interrupt I/O technique is efficient in this type of situation.

Interrupt I/O is a device-initiated I/O transfer. The external device is connected to a pin called the *interrupt* (INT) pin on the microcontroller chip. When the device needs an I/O transfer with the microcontroller, it activates its interrupt pin. The microcontroller usually completes execution of the current instruction and saves the contents of the current program counter and the status register onto the stack.

The microcontroller then loads an address into the program counter automatically to branch to a subroutine-like program called the *interrupt service routine*. This program is written by the user. The external device wants the microcontroller to execute this program to transfer data. The last instruction of the service routine is a RETURN, which is typically similar in concept to the RETURN instruction used at the end of a subroutine. The RETURN from interrupt instruction typically restores the program counter and the status

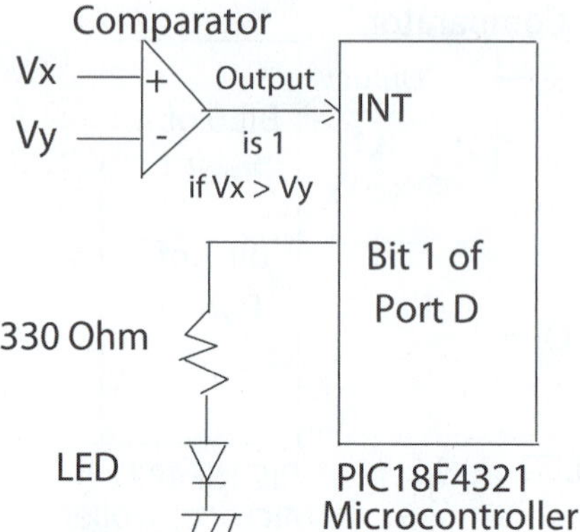

FIGURE 4.17 Example illustrating interrupt I/O

register with the information saved in the stack before going to the service routine.

Figure 4.17 provides a simple example for illustrating the concept interrupt I/O. This is the same example used to illustrate polled I/O of Figure 4.16 except that the comparator output is connected to the microcontroller's interrupt (INT) pin instead of bit 0 of Port C.

Assume that the PIC18F4321 microcontroller is executing the following main program:

```
        ORG     0x200
        SETF    TRISC    ; Make Port C as input by setting all
                         ; bits of TRISC to 1's
        CLRF    TRISD    ; Make Port C as output by clearing all
                         ; bits of TRISD to 0's
        MOVLW   0x15
        MOVWF   STKPTR   ; Initialize STKPTR to 0x15
        MOVLW   3        ; Move 3 into WREG register

BEGIN   MOVWF   0x30     ; Move WREG into 0x30
        -
        -
        -
```

Note that the last two instructions, MOVLW and MOVWF, are chosen arbitrarily. In the above program, the SETF and CLRF instructions configure Port C and Port D of Figure 4.17. The "MOVLW 0x15" and "MOVWF STKPTR" initializes the PIC18F stack pointer (STKPTR) to 0x15. The value of the STKPTR is chosen arbitrarily.

Since interrupt I/O uses stack to save the return address, the stack pointer should be initialized in the main program. The PIC18F4321 then continues with execution of the "MOVLW 3" instruction. Suppose that during execution of the "MOVLW 3" instruction, the output of the comparator becomes HIGH, indicating that Vx is greater than Vy. This drives the INT signal to HIGH, interrupting the microcontroller. The microcontroller completes execution of the current instruction, "MOVLW 3." It then saves the current contents of the program counter (address BEGIN) and the status register automatically onto the stack and executes a subroutine-like program called the *service routine*. This program is usually written by the user. The microcontroller manufacturer normally specifies the starting address of the service routine. This address is 0x000008 in the PIC18F. The user writes a service routine at this address to turn the LED ON, and then returns to the main program as follows:

```
ORG       0x000008  ; Starting address of the service routine
MOVLW     0x02      ; Move 1 to bit 1 of WREG (accumulator) register
MOVWF     PORTD     ; Turn the LED ON
RETFIE              ; Restore PC and SR, and return from interrupt
```

In this service routine, using the MOVLW and MOVWF instructions, the microcontroller turns the LED ON. The return instruction RETFIE, at the end of the service routine loads the address BEGIN and the previous status register contents from the stack, and loads the program counter and status register with them. The microcontroller executes the "MOVWF 0x30" instruction at the address BEGIN and continues with the main program. The basic characteristics of interrupt I/O have been discussed so far. The main features of interrupt I/O provided with a typical microcontroller are discussed next.

Interrupt Types There are typically two types of interrupts: external interrupts and internal interrupts. *External interrupts* are initiated through a microcontroller's interrupt pins by external devices such as the comparator in the previous example. External interrupts can be divided further into two types: maskable and nonmaskable. The nonmaskable interrupt cannot be enabled or disabled by instructions, whereas a microcontroller's instruction set typically contains instructions to enable or disable maskable interrupt. A nonmaskable interrupt has a higher priority than a maskable interrupt. If maskable and nonmaskable interrupts are activated at the same time, the processor will service the nonmaskable interrupt first.

A nonmaskable interrupt is typically used as a power failure interrupt. Microcontrollers normally use +5 V dc, which is transformed from 110 V ac. If the power falls below 90 V ac, the DC voltage of +5 V cannot be maintained. However, it will take a few milliseconds before the ac power drops below 90 V ac. In these few milliseconds, the power-failure-sensing circuitry can interrupt the processor. The interrupt service routine can be written to store critical data in nonvolatile memory such as battery-backed CMOS RAM, and the interrupted program can continue without any loss of data when the power returns.

Internal interrupts are usually nonmaskable, and cannot be disabled by instructions. They are activated internally by conditions such as completion of analog-to-digital conversion, timer interrupt, or interrupt due to serial I/O. Internal interrupts are handled in the same way as external interrupts. The user writes a service routine to take appropriate action to handle the interrupt. Some microcontrollers include software interrupt instructions. When one of these instructions is executed, the microcontroller is interrupted and serviced similarly to external or internal interrupts.

Some microcontrollers such as the Motorola/Freescale HC11/HC12 provide both external (maskable and nonmaskable) and internal (exceptional conditions and software instructions). The PIC18F provides external maskable interrupts only. The PIC18F does not have any external nonmaskable interrupts. However, the PIC18F provides internal interrupts. The internal interrupts are activated internally by conditions such as timer interrupts, completion of analog-to-digital conversion, and serial I/O.

Interrupt Address Vector The technique used to find the starting address of the service routine (commonly known as the *interrupt address vector*) varies from one processor to another. The microcontroller manufacturers typically define the fixed starting address for each interrupt.

Saving the Microcontroller Registers When a microcontroller is interrupted, it normally saves the program counter (PC) and the status register (SR) onto the stack so that the microcontroller can return to the main program with the original values of PC and SR after executing the service routine. The user should know the specific registers the microcontroller saves prior to executing the service routine. This will allow the user to use the appropriate return instruction at the end of the service routine to restore the original conditions upon return to the main program.

Questions and Problems

4.1 What is the basic difference between main memory and secondary memory?

4.2 A microcontroller has 24 address pins. What is the maximum size of the main memory?

4.3 What is the basic difference between (a) EPROM and EEPROM? (b) SRAM and DRAM?

4.4 What is flash memory?

4.5 Given a memory with a 14-bit address and an 8-bit word size:
  (a) How many bytes can be stored in this memory?
  (b) If this memory were constructed from 1K × 1 RAMs, how many memory chips would be required?
  (c) How many bits would be used for chip select?

4.6 Draw a block diagram showing the address and data lines for the 2732 and 2764 EPROM chips.

4.7 (a) How many address and data lines are required for a 1M × 16 memory chip?
  (b) What is the size of a decoder with one chip enable ($\overline{CE}$) to obtain a 64K × 32 memory from 4K × 8 chips? Where are the inputs and outputs of the decoder connected?

4.8 A microcontroller with 24 address pins and eight data pins is connected to a 1K × 8 memory with one enable. How many unused address bits of the microcontroller are available for interfacing other 1K × 8 memory units? What is the maximum directly addressable memory available with this microcontroller?

4.9 Name the methods used in main memory array design. What are the advantages and disadvantages of each?

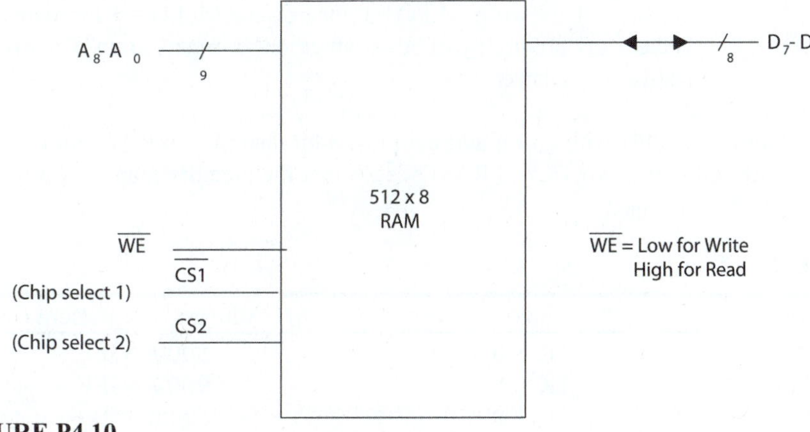

FIGURE P4.10

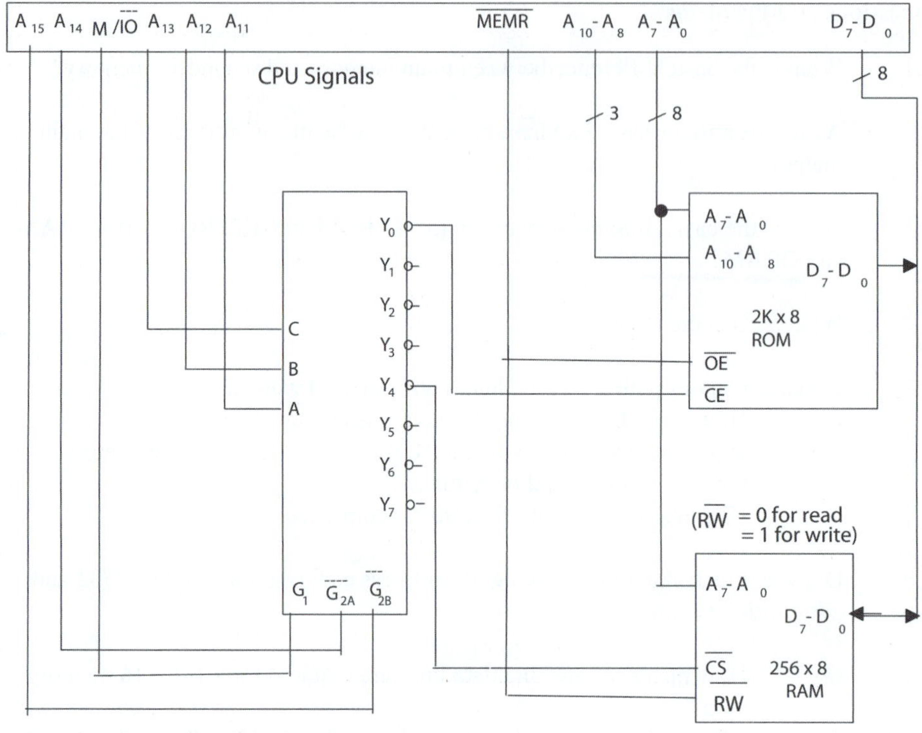

FIGURE P4.11

4.10 The block diagram of a 512 × 8 RAM is shown in Figure P4.10. In this arrangement the memory unit is enabled only when $\overline{CS1}$ = L and CS2 = H. Design a 1K × 8 RAM system using the 512 × 8 RAM as the building block. Draw a neat logic diagram of your implementation. Assume that the CPU can directly address 64K with an R/$\overline{W}$ and eight data pins. Using linear decoding and don't-care conditions as 1's, determine the memory map in hexadecimal.

4.11 Consider the hardware schematic shown in Figure P4.11.

(a) Determine the address map of this system. *Note:* $\overline{MEMR}$ = 0 for read, $\overline{MEMR}$ = 1 for write, M/$\overline{IO}$ = 0 for I/O, and M/$\overline{IO}$ = 1 for memory.

(b) Is there any possibility of bus conflict in this organization? Clearly justify your answer.

4.12 Interface a CPU with 16-bit address pins, 8-bit data pins, a R/$\overline{W}$ pin to a 1K × 8 EPROM, and two 1K × 8 RAM's such that the memory map shown in Table P4.12 is obtained:

TABLE P4.12

Device	Size	Address assignment (hex)
EPROM	1K × 8	8000–83FF
RAM chip 0	1K × 8	9000–93FF
RAM chip 1	1K × 8	C000–C3FF

TABLE P4.13

Device	*Size*	*Address assignment in hex*
EPROM	1K × 8	7000–73FF
RAM 0	1K × 8	D000–D3FF
RAM 1	1K × 8	F000–F3FF

Assume that both EPROM and RAM contain two enable pins: $\overline{CE}$ and $\overline{OE}$ for the EPROM, and $\overline{CE}$ and $\overline{WE}$ for each RAM. Note that $\overline{WE} = 1$ and $\overline{WE} = 0$ indicate read and write operations for the RAM chip, respectively. Use a decoder block identical to the 74138.

4.13 Repeat Problem 4.12 to obtain the memory map shown in Table P4.13 using a decoder block identical to the 74138.

4.14 What is meant by *foldback* in linear decoding?

4.15 Define the two types of I/O. Identify each as either CPU-initiated or device-initiated.

4.16 What is the basic difference between standard I/O and memory-mapped I/O? Identify the programmed I/O technique used by the PIC18F.

4.17 What is the difference between memory map in a microcontroller and memory-mapped I/O?

4.18 Discuss the basic difference between polled I/O and interrupt I/O.

4.19 What is the difference between subroutine and interrupt I/O?

4.20 What is an interrupt address vector?

4.21 Summarize the basic difference between maskable and nonmaskable interrupts. Describe how power failure interrupt is normally handled.

4.22 Discuss the basic difference between internal and external interrupts.

5

PIC18F ARCHITECTURE AND ADDRESSING MODES

In this chapter we describe the PIC18F microcontroller architecture and addressing modes. Topics include an introduction to the PIC18F, memory maps, pipelining, register architecture, and addressing modes.

5.1 Introduction

The PIC18F is Microchip's 8-bit RISC-based microcontroller. Since the PIC18F CPU uses the Harvard architecture, program and data memory units use separate memory spaces along with their own buses. This allows the PIC18F to access both programs and data simultaneously. The PIC18F uses flash memory to store program memory and SRAM to contain data memory. Note that F in PIC18F indicates that the chip contains flash memory. In order to illustrate the basic features of microcontrollers, one of the PIC18F family members such as the PIC18F4321 is used in this book for developing the programming examples and illustrating interfacing techniques with the PIC18F.

TABLE 5.1 Basic differences among some of the PIC18F family members (F in PIC18F indicates on-chip flash memory)

On-chip features	PIC18F2221	PIC18F2321	PIC18F4221	PIC18F4321
Flash memory (program memory in bytes)	4K	8K	4K	8K
EEPROM (bytes)	256	256	256	256
SRAM data memory (bytes)	512	512	512	512
Operating frequency	40 MHz	40 MHz	40 MHz	40 MHz
I/O ports	A,B,C, E*	A,B,C, E*	A through E	A through E
Timers	4	4	4	4
Capture/compare/PWM (CCP) module	0	0	1	1
Serial communication interface	Yes	Yes	Yes	Yes
10-bit A/D converter	10 Channels	10 Channels	13 Channels	13 Channels
Instruction set	75 instructions; 83 with extended set enabled	75 instructions; 83 with extended set enabled	75 instructions; 83 with extended set enabled	75 instructions; 83 with extended set enabled
Number of pins	28	28	40, 44	40, 44

* Port E for PIC18F2221 and PIC18F2321 is available under special configuration.

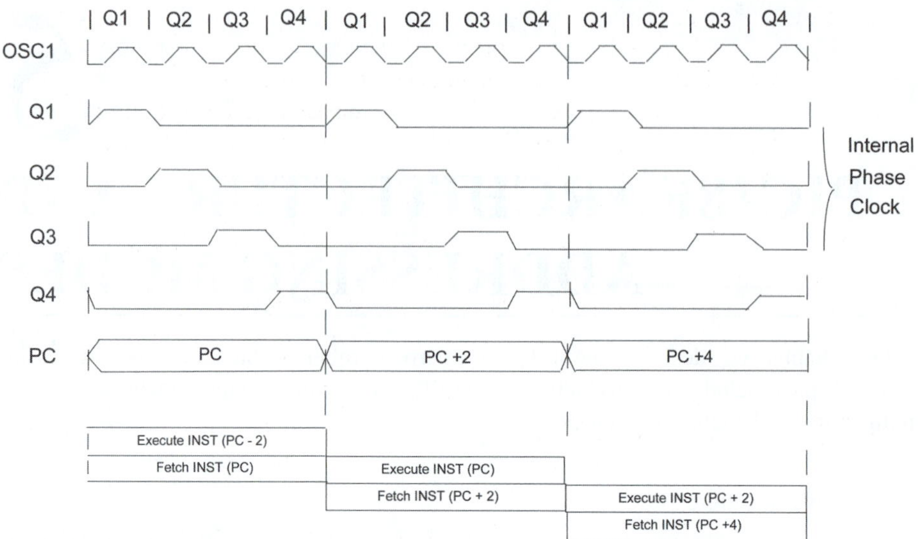

FIGURE 5.1 Clock/instruction cycle

Some versions of the PIC family contain one-time programmable ROM for program memory; this is in addition to data RAM. The PIC16C432 is an example of such a chip. The letter C in PIC16C432 indicates that the chip contains one-time programmable ROM for program memory. The PIC18F is normally used for product development. Once developed, the PIC18C is used for mass production of the product.

The program counter (PC) of the PIC18F is 21 bits wide. Hence, the PIC18F can directly address a maximum of two megabytes (2^{21}) of program memory space. The data memory address, on the other hand, is 12 bits wide. Hence, the PIC18F family members can directly address data memory of up to 4 Kbytes (2^{12}).

There are several versions of the PIC18F microcontroller. The sizes of program and data memories, number of input/output (I/O) ports, and the clock frequency vary from one version to another. For example, the PIC18F4321 contains 8 Kbytes of flash memory, 512 bytes of SRAM, and 256 bytes of EEPROM, and runs at a maximum clock frequency of 40 MHz. The PIC18F8620, on the other hand, includes 64 Kbytes of flash memory, 3840 bytes of SRAM, and 1024 bytes of EEPROM, and runs at a maximum clock frequency of 25 MHz.

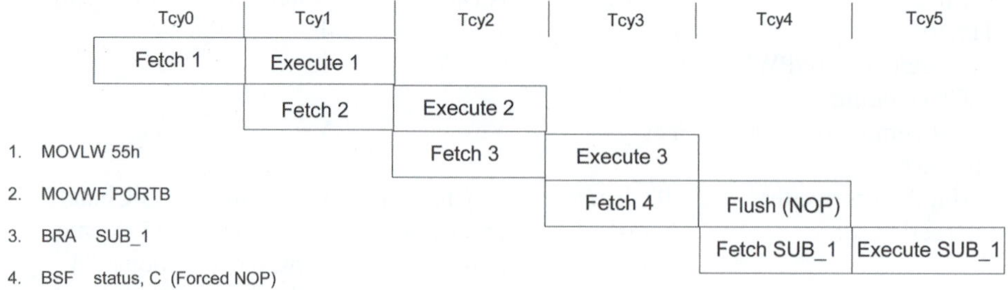

FIGURE 5.2 Instruction pipeline flow

Typical PIC18F family members include PIC18F2221, PIC18F2321, PIC18F4221, and PIC18F4321. Table 5.1 summarizes the basic differences among them. These PIC18F CPUs can be operated from a maximum internal clock frequency of 40 MHz.

Most of the features in Table 5.1 are self-explanatory. However, the purpose of some of the PIC18F on-chip features such as timers, CCP module, A/D converter, and serial communication interface will be provided in the following.

The PIC18F can perform functions such as capture, compare, and pulse width modulation (PWM) using the timers and CCP (capture / compare / PWM) module. The PIC18F can compute the period of an incoming signal using the capture module. The PIC18F can produce a periodic waveform or time delays using the compare module. The PIC18F's on-chip PWM can be used to obtain pulse waveforms with a particular period and duty cycle which are ideal for applications such as motor control.

The PIC18F serial communication interface can be used to facilitate data transmission for serial peripheral devices (telephone systems) which can transmit or receive data one bit at a time.

The on-chip 10-bit A/D converter of the PIC18F can convert an analog signal into 10-bit binary equivalent. This is very convenient since, in practice, physical variables such as temperature, flow, and pressure are analog in nature, and must first be converted into analog electrical signals using transducers. Since the microcontrollers only understand binary numbers, the on-chip A/D converter of the PIC18F then converts the electrical signal to 10-bit binary value before processing.

The PIC18F clock input is internally divided by four to generate four non-overlapping quadrature clocks (Q1, Q2, Q3, and Q4). Internally, the program counter is incremented by 2 (since the PIC18F instruction size is normally 16-bit) on every Q1; the

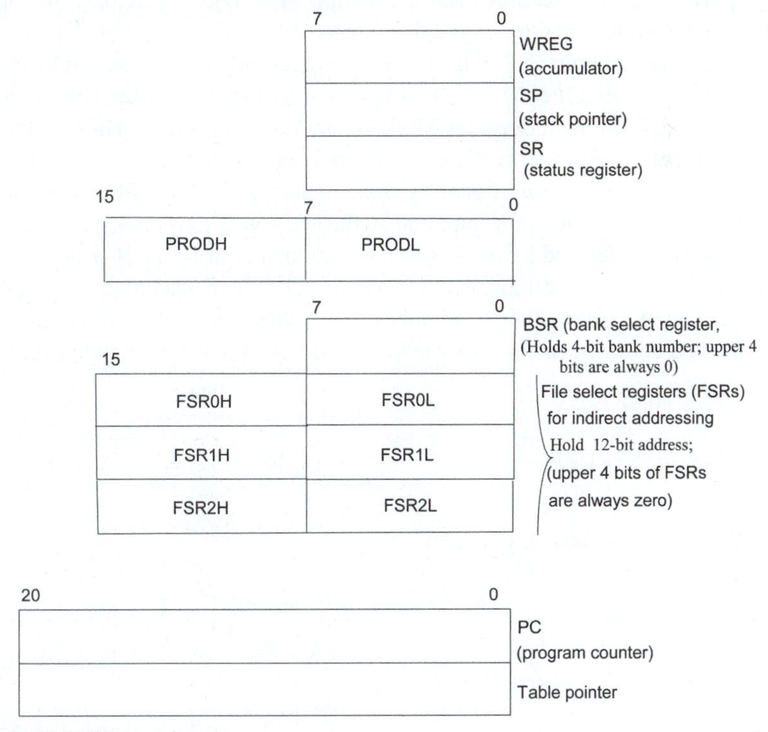

FIGURE 5.3 PIC18F registers

instruction is fetched from the program memory and stored into the instruction register (IR) during Q4. The instruction is decoded and executed during the following Q1 through Q4. The clocks and instruction execution flow are shown in Figure 5.1.

As mentioned before, the PIC18F uses a two-stage pipeline. This means that execution of the previous instruction is overlapped with fetching of the current instruction. This speeds up the program execution by the CPU. The basic concepts associated with PIC18F pipelining will be described in the following.

Figure 5.2 shows the PIC18F instruction pipeline flow. Four PIC18F instructions (MOVLW, MOVWF, BRA, and BSF) are used to illustrate the PIC18F pipelining. These instructions are chosen arbitrarily. The meaning of these four instructions are provided below:

MOVLW 0x55 moves 55 hex (8-bit immediate data) into WREG (accumulator).

MOVWF PORTB moves the contents of WREG (accumulator) into PORTB.

BRA SUB_1 unconditionally jumps to address SUB_1.

BSF STATUS, C sets carry flag in the status register to 1.

In PIC18F, an "instruction cycle" consists of four Q cycles: Q1 through Q4. The instruction fetch and execute are pipelined in such a manner that a fetch takes one instruction cycle, while the decode and execute take another instruction cycle. However, due to the pipelining, each instruction effectively is executed in one cycle. If an instruction causes the program counter to change (e.g., GOTO), then two cycles are required to complete the instruction. A fetch cycle begins with the program counter (PC) incrementing in Q1. In the execution cycle, the fetched instruction is stored into the IR in cycle Q1. This instruction is then decoded and executed during the Q2, Q3, and Q4 cycles. Data memory is read during Q2 (operand read) and written during Q4 (destination write).

The PIC18F executes each instruction in a single cycle, except for any branch instructions. This is explained in the following.

Consider Figure 5.2. The PIC18F fetches MOVLW instruction into IR during Tcy0. The PIC18F executes MOVLW in Tcy1 and also fetches the next instruction MOVWF in Tcy1. The PIC18F then executes MOVWF in Tcy2 and fetches BRA instruction into IR in Tcy2. The PIC18F executes BRA into IR in Tcy3, and also fetches BSF into IR in Tcy3.

The PIC18F unconditionally jumps to address SUB_1, and executes the instruction at address SUB_1. Hence, the pipeline is flushed (NOP) in Tcy4, and the instruction at address SUB_1 is fetched in Tcy4. The instruction at address SUB_1 is executed in Tcy5. Hence, each of the instructions, MOVLW and MOVWF, is executed in a single cycle. The BRA instruction, on the other hand, takes two cycles since the instruction BSF is flushed from the pipeline while the new instruction at address SUB_1 is fetched and then executed.

7	6	5	4	3	2	1	0
STKFUL	STKUNF	—	SP4	SP3	SP2	SP1	SP0

FIGURE 5.4 SP (stack pointer)

7	6	5	4	3	2	1	0
--	--	--	N	OV	Z	DC	C

FIGURE 5.5 SR (status register)

5.2 PIC18F Register Architecture

In order to program the PIC18F in assembly language, one must be familiar with the registers of the PIC18F. Hence, a description of these registers is provided in this section. Figure 5.3 shows the PIC18F CPU registers. All registers in the PIC18F are mapped in the data memory. Hence, each register is assigned with a unique 12-bit memory address. We now briefly describe the functions of these registers in the following.

WREG The WREG (working register) is 8 bits wide. This is basically an accumulator, and has its usual meaning. Most arithmetic and logic operations are performed using the WREG. The address for WREG is 0xFE8.

SP The SP (stack pointer) register is 8 bits wide. The PIC18F stack is a group of 31 21-bit registers to hold memory addresses. The low five bits of the SP are used to address the stack. Figure 5.4 shows the details of the SP. The SP is called STKPTR in the PIC18F.

The PIC18F maps the SP as a special function register with address 0xFFC. The 31 stack registers are neither part of program memory nor data memory. As shown in Figure 5.4, the low five bits of the SP address the stack. The stack overflow bit (STKFUL, Bit 7) is set to one if more than 31 registers are attempted for pushing addresses onto the stack by the programmer; otherwise, the stack overflow bit is cleared to zero. The stack underflow bit (STKUNF, bit 6), on the other hand, is set to one if more addresses than are stored in the stack are attempted to be popped by the programmer; otherwise, the stack underflow bit is cleared to zero. Bit 5 is not implemented and is read as 0.

PC The PC (program counter) is 21 bits wide. The PC normally points to the next instruction. As mentioned before, the 21-bit PC provides the PIC18F with direct addressing capability of a maximum of 2 MB (2^{21}) of program memory. Upon hardware reset, the PC is loaded with zero so that the PIC18F CPU fetches the first instruction from address 0.

The PC is comprised of three 8-bit registers namely, PCL (PC low byte), PCLATH (PC latch high byte), and PCLATU (PC latch upper 5 bits). The 21-bit PC is stored in these registers as follows: bits 0 through 7 in PCL, bits 8 through 15 in PCLATH, and bits 16 through 20 in low five bits of PCATU. Registers PCL, PLATCH, and PCLATU are mapped as special function registers in the data SRAM by the PIC18F as 0xFF9, 0xFFA, and 0xFFB.

Table Pointer The PIC18F uses the 21-bit table pointer register as pointer to a table in program memory for copying bytes between program memory and data memory. This register is mapped by the PIC18F as three 8-bit special function registers in the data SRAM with memory addresses 0xFF6, 0xFF7, and 0xFF8 as follows: bits 0 through 7 in 0xFF6, bits 8 through 15 in 0xFF7, and bits 16 through 20 in low five bits of 0xFF8.

BSR The BSR (bank select register) is 8 bits wide. The lower four bits are used to provide the bank address from 0 to F_{16}; the upper four bits of BSR are zero. The BSR provides the upper four bits of a 12-bit address of data memory. BSR is used for directly addressing the data SRAM. The address for BSR is 0xFE0.

FSR The FSR (file select register) consists of three 16-bit registers (FSR0, FSR1, and FSR2); the upper four bits of each FSR are zero. The lower 12 bits of FSR0, FSR1, or

FSR2 are used to hold the 12-bit memory address of the data SRAM. These registers are used for handling arrays and pointer-based data accessing. The PIC18F indirectly uses these registers to access data in data SRAM. Each of these three registers is divided into two 8-bit registers as follows: FSR0H (high byte of FSR0) and FSR0L (low byte of FSR0), FSR1H (high byte of FSR1) and FSR1L (low byte of FSR1), FSR2H (high byte of FSR2) and FSR2L (low byte of FSR2). The PIC18F maps these three registers as special function registers with the following memory addresses: FSR0 as 0xFE9 (FSR0L) and 0xFEA (FSR0H), FSR1 as 0xFE1 (FSR1L) and 0xFE2 (FSR1H), and FSR2 as 0xFD9 (FSR2L) and 0xFDA (FSR2H).

PRODH / PRODL Each of the PRODH and PRODL registers is 8 bits wide. The PIC18F has two 8-bit X 8-bit unsigned multiplication instructions providing an 16-bit product. The upper byte of the product is stored in the PRODH register while the lower byte of the product is placed in the PRODL register. The PIC18F maps PRODL and PRODH as addresses 0xFF3 and 0xFF4 in the data SRAM.

SR The SR (status register) is 8 bits wide. The address for the SR is 0xFD8. Figure 5.5 shows the PIC18F status register which contains the flags. The meaning of these flags will be explained in the following:

- **C** (carry flag) is set to 1 if there is a carry from addition or a borrow from subtraction; otherwise, C = 0.
- **DC** (digit carry flag) is set to 1 if there is a carry due to addition of the low 4 bits into the high 4 bits or a borrow due to the subtraction of the low 4 bits from the high 4 bits of a number; otherwise, DC = 0. This flag is used by BCD arithmetic instructions.
- **Z** (zero flag) is set to 1 if the result is zero; Z = 0 for a nonzero result.
- **OV** (overflow flag) is set to 1 if there is an arithmetic overflow (i.e., if the size of the result exceeds the capacity of the destination location); otherwise, OV = 0. Note that overflow $OV = C_f \oplus C_p$ where C_f is the final carry and C_p is the previous carry. Overflow is used for signed arithmetic (two's complement).
- **N** (negative flag) is set to 1 if the most significant bit of the result is 1, indicating a negative number; N = 0 if the most significant of the result is 0, indicating a positive number.
- Bits 5 through 7 are not implemented, and are read as zero.

In order to provide a clear understanding of how status flags are affected by arithmetic instructions, numerical examples will be provided in the following.
Consider adding 06_{16} with 14_{16} as follows:

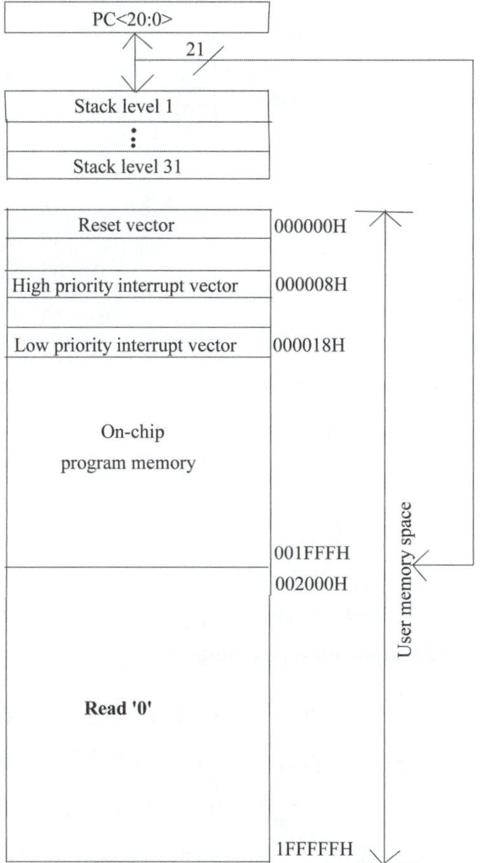

FIGURE 5.6 PIC18F4321 program memory map

```
          0 0 0 0 0 1 1 0        06₁₆
          0 0 0 1 0 1 0 0       +14₁₆
        ─────────────────       ─────
       0  0 0 0 1 1 0 1 0        1A₁₆
```

$C_f = 0$, $C_p = 0$, $DC = 0$

In the above, C = C_f = 0, DC = 0 (no carry from bit 3 to bit 4), Z = 0 (nonzero result), OV = $C_f \oplus C_p = 0 \oplus 0 = 0$ (meaning correct result), and N = 0 (most significant bit of the result is 0 indicating positive number). Next, consider subtracting 06_{16} from 68_{16} using two's complement. The result will be 62_{16}.

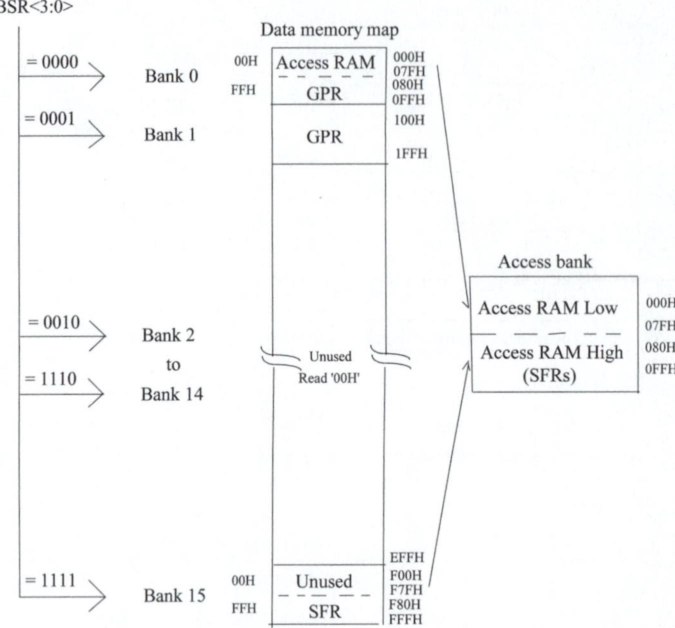

FIGURE 5.7 PIC18F4321 data memory map

68_{16} = 0 1 1 0 1 0 0 0 $\quad 68_{16}$

Add 2's complement of 06_{16} = 1 1 1 1 1 0 1 0 $\quad -62_{16}$

$C_f = 1$ → 1 0 1 1 0 0 0 1 0 $\quad 62_{16}$

Cp=1 DC = 0

In the above, C (borrow) = one's complement of $C_f = 0$, DC = 0 (no carry from bit 3 to bit 4), Z = 0 (nonzero result), OV = $C_f \oplus C_p = 1 \oplus 1 = 0$ (meaning correct result), and N = 0 (most significant bit of the result is 0 indicating positive number). Note that while obtaining two's complement subtraction using paper and pencil, the correct borrow is always the one's complement of the borrow obtained analytically. Hence, microcontrollers perform one's complement operation on the borrow in order to reflect the correct borrow which will be useful in multiprecision subtraction.

5.3 PIC18F Memory Organization

Two types of memories are normally utilized in the PIC18F. They are flash memory and SRAM. The flash memory is used to store programs. The SRAM, on the other hand, contains data. Some versions of the PIC18F family contain EEPROM along with SRAM to hold data. Note that, SRAM is a volatile read/write memory. The EEPROM, on the other hand, is a nonvolatile memory.

The EEPROM is separate from the data SRAM and program flash memory. The EEPROM is used for long-term storage of critical data. The EEPROM is normally used as a read-mostly memory since its read time is faster than write times. It is not directly mapped in either the register file or program memory space but is indirectly addressed through the special function registers (SFRs). One of the main advantage of including EEPROM in the PIC18F is that all critical data stored in the EEPROM can be protected from reading or writing by other users. This can be accomplished by programming appropriate bits in the corresponding SFR. Note that the PIC18F4321 contains 256 bytes of EEPROM.

The data memory in PIC18F devices is implemented as static RAM. Each register in the data memory has a 12-bit address, allowing up to 4096 bytes (2^{12}) of data memory. The memory space is divided into as many as 16 banks that contain 256 bytes each.

5.3.1 PIC18F Program Memory Map

Figure 5.6 shows the program memory map for the PIC18F4321. Program memory is implemented in flash memory in the PIC18F.

The PIC18F4321 contains 8 Kbytes of on-chip flash memory, and can store up to 4096 single 16-bit word instructions. Note that most PIC18F instructions are 16 bits wide.

As mentioned before, the program counter (PC) contains the address of the instruction to be fetched for execution. The PC is 21 bits wide. The PC addresses bytes in the program memory. To prevent the PC from becoming misaligned with 16- or 32-bit-wide instructions, the least significant bit of PC is fixed to a value of '0'. This is because the address is an even number for 16-bit or 32-bit instructions. The PC increments by 2 or 4 to address sequential 16- or 32-bit-wide instructions in the program memory.

The stack operates as a 31-word by 21-bit RAM and a 5-bit stack pointer, STKPTR. The stack space is not part of either program or data space. The stack pointer is readable and writeable. The address on the top of the stack is readable and writeable through the top-of-stack special function registers. Data can also be pushed to, or popped from, the stack using these registers.

The reset vector address is located at 000000H, where H stands for hex. There are two interrupts. These are high-priority interrupt and low-priority interrupt. The starting address for the high priority service routine is 000008H. There are 16 bytes available to the user for writing the high-priority service routine. The starting address for the low priority service routine is 000018H. There is no specific size for the low-priority service routine. Reset and interrupts will be discussed in more detail later in this book. In the PIC18F4321, the user program should be written after the low-priority service routine to a maximum allowable address of 001FFFH. Addresses 002000H through 1FFFFFH are not implemented, and are read as zeros.

5.3.2 PIC18F Data Memory Map

Figure 5.7 shows the data memory organization for the PIC18F4321. As mentioned before, the PIC18F data memory is implemented in SRAM. The PIC18F can have a data memory of up to 4096 (2^{12}) bytes; 12-bit address is needed to address each location. However, the PIC18F4321 implements two banks with a total of 512 bytes of data SRAM.

The data memory contains SFRs and general purpose registers (GPRs). The GPRs are typically used for storing data and as scratch pad registers during programming. The SFRs, on the other hand, are dedicated registers. These registers are used for control and status of the controller and peripheral functions such as registers associated with I/O ports and interrupts, timers, ADC (analog-to-digital converter), and serial I/O. An unimplemented location will be read as 0's. The instruction set and architecture allow operations across all banks. The entire data memory may be accessed by direct or indirect addressing modes.

TABLE 5.2 Selected special function registers (SFRs)

Address	Name	Description
0xFFF	TOSU	Top of stack (upper 5 bits)
0xFFE	TOSH	Top of stack (high byte)
0xFFD	TOSL	Top of stack (low byte)
0xFFC	STKPTR	Stack Pointer
0xFFB	PCLATU	Program counter latch (upper 5 bits)
0xFFA	PCLATH	Program counter latch (upper byte)
0xFF9	PCL	Program counter latch (lower byte)
0xFF8	TBLPTRU	Table pointer (upper 5 bits)
0xFF7	TBLPTRH	Table pointer (high byte)
0xFF6	TBLPTRL	Table pointer (low byte)
0xFF5	TABLAT	Table latch
0xFF4	PRODH	Product register (high byte)
0xFF3	PRODL	Product register (low byte)
0xFEF	INDF0**(1)**	Indirect file register 0; associated with FSR0
0xFEE	POSTINC0**(1)**	Postincrement pointer 0; uses FSR0
0xFED	POSTDEC0**(1)**	Postdecrement pointer 0; uses FSR0
0xFEC	PREINC0**(1)**	Predecrement pointer 0 ; uses FSR0
0xFEB	PLUSW0**(1)**	Add FSR0 to WREG and uses as pointer for data registers
0xFEA	FSR0H	File select register 0 (high byte)
0xFE9	FSR0L	File select register 0 (low byte)
0xFE8	WREG	Working register (accumulator)
0xFE7	INDF1**(1)**	Indirect file register 1; associated with FSR1
0xFE6	POSTINC1**(1)**	Postincrement pointer 1; uses FSR1
0xFE5	POSTDEC1**(1)**	Postdecrement pointer 1; uses FSR1
0xFE4	PREINC1**(1)**	Predecrement pointer 1 ; uses FSR1
0xFE3	PLUSW1**(1)**	Add FSR1 to WREG and uses as pointer for data registers
0xFE2	FSR1H	File select register 1 (high byte)
0xFE1	FSR1L	File select register 1 (low byte)
0xFE0	BSR	Branch select register
0xFDA	FSR2H	File select register 2 (high byte)
0xFD9	FSR2L	File select register 2 (low byte)
0xFD8	SR	Status register

Note 1: This is not a physical register.

Addressing modes are discussed in the next section. Note that a location in the data SRAM is called a "file register." This means that the file registers contain GPRs and SFRs. The file registers are also called "data registers" or simply "registers."

Large areas of data memory require an efficient addressing scheme to make rapid access to any address possible. Ideally, this means that an entire address does not need to be provided for each read or write operation. For PIC18F, this is accomplished with a RAM banking scheme. This divides the memory space into 16 contiguous banks of 256 bytes. Depending on the instruction, each location can be addressed directly by its full 12-bit address, or an 8-bit low-order address and a 4-bit bank pointer.

Most instructions in the PIC18F instruction set make use of the bank pointer, known as the bank select register (BSR). The BSR holds the four Most Significant bits of a location's address; the PIC18F instruction contains the eight least significant bits. Only the four lower bits of the BSR are implemented (BSR3:BSR0). The value of the BSR indicates the bank in data memory; the eight bits in the instruction show the location in the bank and can be thought of as an offset from the bank's lower boundary. The relationship between the BSR's value and the bank division in data memory is shown in Figure 5.7.

In order to access a memory location from one bank to a memory location in a different bank, bank switching is required. For example, to access address F56H in bank F (specified by the upper four bits of the address) from address 150H in bank 1, the programmer must change the bank number from 1 to F. This can be accomplished using the MOVLB K instruction where K is an 8-bit number. The low four bits of K are used to specify the bank, and the upper four bits are always cleared to 0's. For example, in order to switch from bank 1 (assuming active bank) to bank F, the instruction MOVLB 0x0F can be used. This instruction will load 0FH into BSR, and will select bank number F. All data registers in bank F will now become active.

However, the need for bank switching sometimes creates a major problem for the programmer. Obviously, programs will not work if the programmer forgets about bank switching. To facilitate access for the most commonly used data memory locations, the data memory is configured with an "access bank", which allows users to access a mapped block of memory without bank switching. The "access bank" consists of the first 128 bytes of memory (00H-7FH) in Bank 0 and the last 128 bytes of memory (80H-FFH) in Bank F. The lower half is known as the "access RAM" and is composed of GPRs. This upper half is also where the device's SFRs are mapped. These two areas are mapped contiguously in the access bank and can be addressed in a linear fashion by an 8-bit address (Figure 5.7). The GPRs and SFRs are called "file registers," "data registers," or simply "registers." It is convenient to use access bank for file registers. The user does not have to worry about bank switching. Hence, one should use access bank whenever possible. Note that upon power-up, the PIC18F uses the access bank of the file registers as the default bank. In the core PIC18F instruction set, only the MOVFF instruction fully specifies the 12-bit address of the source and target registers. The size of this instruction is two words. This instruction ignores the BSR completely when it executes. All other instructions include only the low-order address as an operand and must use either the BSR or the access bank to locate their target registers.

The SFRs are dedicated registers used by the CPU and peripheral modules for controlling the desired operation of the device. The SFRs can be classified into two sets: those associated with the "core" device functionality (ALU, resets, and interrupts) and those related to the peripheral functions. The SFRs are typically distributed among the peripherals whose functions they control. Unused SFR locations are unimplemented and read as '0's. A list of some of these registers is given in Table 5.2. A complete list of all PIC18F SFRs along with addresses can be found in the Appendix E.

5.4 PIC18F Addressing Modes

Most instructions contain one or more operands. Some instructions have no operands. The manner in which a microcontroller specifies location(s) of operand(s) and destination addresses is called the "addressing mode." Note that an operand may be immediate data (literal), or data stored in a register or in data memory.

The PIC18F provides six addressing modes:

1. Literal or immediate addressing mode
2. Inherent or implied addressing mode
3. Direct or absolute addressing mode
4. Indirect addressing mode
5. Relative addressing mode
6. Bit addressing mode

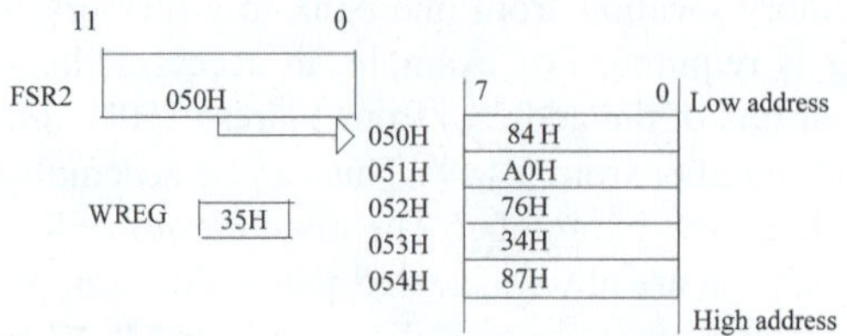

(a) Memory contents after execution of MOVLW and LFSR insructions in the example for indirect addressing mode

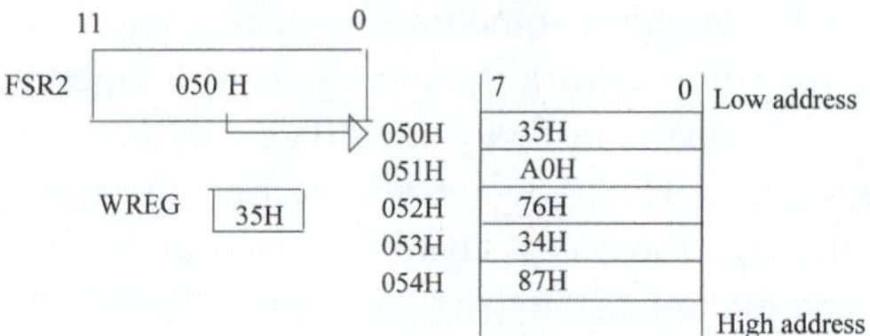

(b) Memory contents after execution of MOVWF in the example for indirect addressing mode

FIGURE 5.8 Illustration of the indirect addressing mode

An additional addressing mode, indexed literal offset, is available when the extended instruction set is enabled. However, this mode will not be described here.

5.4.1 Literal or Immediate Addressing Mode

In the *literal* or *immediate mode*, the operand data are literal or constant data. Immediate data are part of the instruction. This means that the data follow the opcode after assembling an instruction with immediate addressing mode. Constant data can be moved into the WREG or any other specified register such as BSR. However, constant data cannot be moved into file registers using the literal or immediate mode. This mode can also be used by PIC18F arithmetic and logic instructions to be covered in Chapter 6.

An example is the MOVLW 0x2A instruction in which MOVLW uses immediate mode, and moves 8-bit data 2AH into the WREG register.

5.4.2 Inherent or Implied Addressing Mode

In the *inherent* or *implied mode*, instructions do not require operands. These instructions are also called *no-operand instructions*. An example of an instruction with inherent or implied mode is the DAW instruction. The DAW is a no-operand instruction. It adjusts the sum in the WREG register stored after addition of two 8-bit packed BCD numbers. Note that the DAW instruction implicitly or inherently uses the WREG register.

5.4.3 Direct or Absolute Addressing Mode

In the *direct or absolute addressing mode*, the address is included as part of the instruction. This address specifies either a register address in one of the banks of data SRAM or a location in the access bank as the data source for the instruction. Direct addressing mode specifies all or part of the source and/or destination address of the operation within the opcode itself.

An example of the direct addressing mode is MOVWF 0x50. Note that the letter F in the instruction MOVWF indicates the address of the file register in direct addressing mode, and "0x" means hexadecimal number.

The MOVWF 0x50 instruction moves the contents of the WREG register into a file

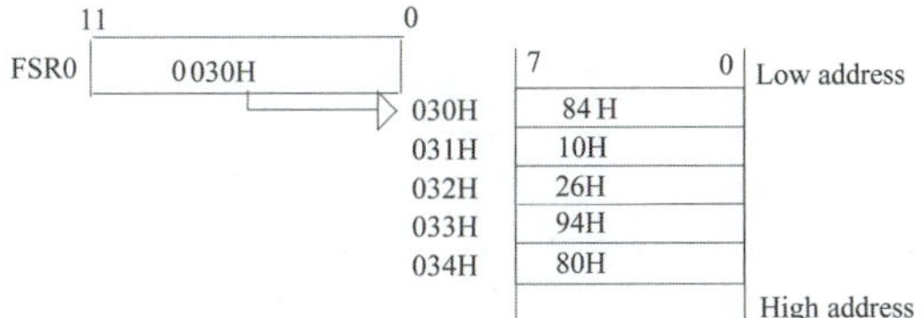

(a) Memory contents before execution of CLRF POSTINC0.

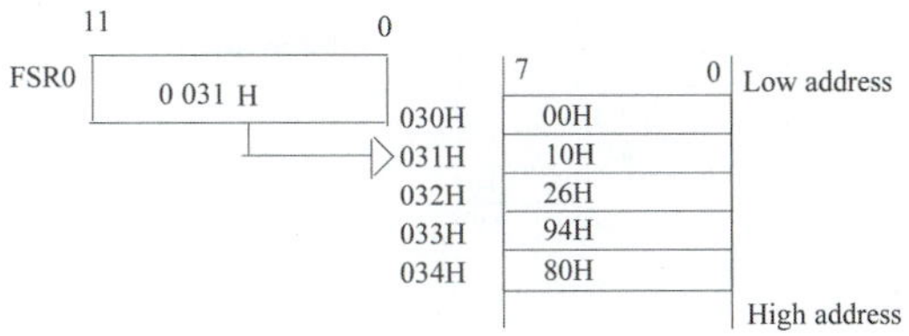

(b) Memory contents after execution of CLRF POSTINC0.

FIGURE 5.9 Illustration of indirect with postincrement mode

register in the data SRAM whose address is 0x50. The contents of WREG are unchanged. MOVWF 0x50 uses direct address mode since address 0x50 is directly specified in the MOVWF instruction.

Note that the 12-bit address 0x050 in the access bank is specified by an 8-bit number 0x50 in the MOVWF instruction since the upper four bits (0H) of the 12-bit address 0x050 specify the bank number 0 (access bank) in this case. The 8-bit address (0x50) is included in the MOVWF (16-bit-wide) instruction. As mentioned before, in the PIC18F, addresses are specified as 8-bit numbers while the bank number is specified by the access bank or BSR.

5.4.4 Indirect Addressing Mode

In *indirect addressing mode*, a register is used as a pointer to an address in the data memory. In the PIC18F, three registers, namely, FSR0, FSR1, and FSR2, are used for this purpose. Note that "FSR" stands for file select register. The PIC18F CPU contains these registers. As mentioned before, each of these 16-bit registers is divided into low byte and high byte as follows: FSR0 as FSR0H and FSR0L, FSR1 as FSR1H and FSR1L, and FSR2 as FSR2H and FSR2L. Each of these registers holds a 12-bit memory address to point to a data memory location. Since FSR's are 16 bits wide, the 12-bit address is stored in the low 12 bits (bits 0 through 11) of the FSRs with the upper four bits (bits 12 through 15) as 0's.

Each FSR is assciated with an INDF (indirect file) register as follows: FSR0 is associated with INDF0, FSR1 is associated with INDF1, and FSR2 is associated with

```
         MOVLW    D'20'       ; Move 20 decimal into WREG
         MOVWF    0x10        ; Initialize counter 0x10 with 20 decimal
         LFSR     0,0x0030    ; Initialize pointer FSR0 with starting address 0x030
REPEAT   CLRF     POSTINC0    ; Clear a location to 0 and increment FSR0 by 1
         DECF     0x10,F      ; Decrement counter by 1
         BNZ      REPEAT      ; Branch to REPEAT if Zero flag = 0; otherwise,
                              ; go to the next instruction
```

FIGURE 5.10 Instruction sequence for illustrating the postincrement mode

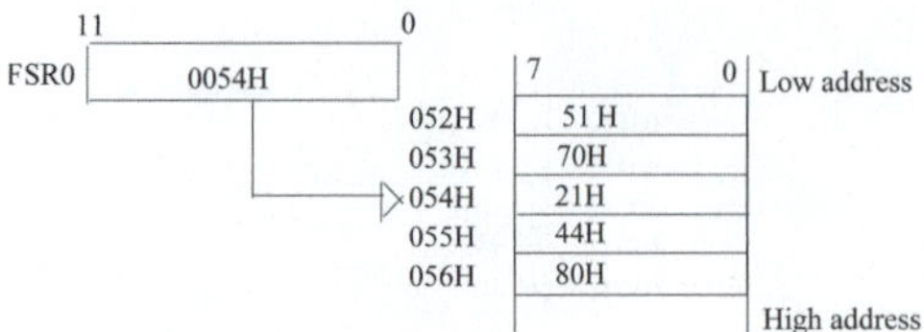

(a) Memory contents before execution of CLRF POSTDEC0.

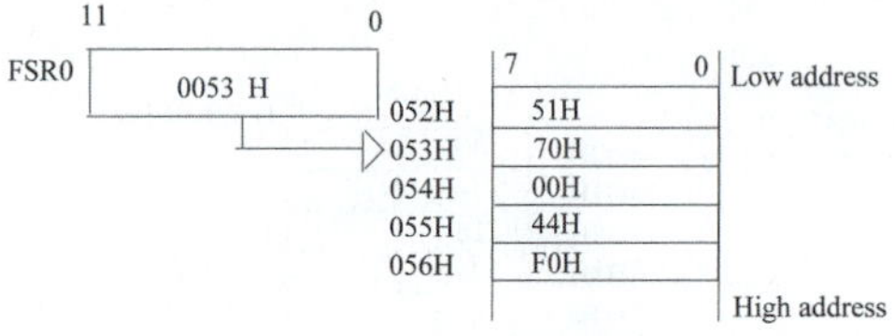

(b) Memory contents after execution of CLRF POSTDEC0.

FIGURE 5.11 Illustration of postdecrement mode

INDF2. These registers can be initialized using the LFSR (load FSR) instruction. The following examples illustrate this:

```
LFSR   0, 0x0010        ; Load 0010H into FSR0
LFSR   1, 0x0040        ; Load 0040H into FSR1
LFSR   2, 0x0080        ; Load 0080H into FSR2
```

After one of the FSR's is initialized, data can be moved into a RAM location indirectly using the associated INDF register. For example, in order to move the contents of the WREG register into a 12-bit data RAM location 050H using the FSR2 register indirectly, the following PIC18F instruction sequence can be used:

```
MOVLW      0x35         ; Move 35H into WREG
LFSR       2,0x0050     ; Initialize FSR2 with the RAM location 050H
MOVWF      INDF2        ; Move contents of WREG (35H) into a data
                        ; RAM address pointed
                        ; to by FSR2 (address 050H) since INDF2 is
                        ; associated with FSR2
```

The above instruction sequence loads 8-bit data 0x35 into a 12-bit data memory location 0x050 via the WREG register. This is depicted in Figure 5.8.

The PIC18F provides the indirect addressing mode with four submodes as follows:

1. Indirect with postincrement mode
2. Indirect with postdecrement mode
3. Indirect with preincrement mode
4. Indirect with 8-bit indexed mode

These four submodes are described next. As mentioned earlier, the PIC18F includes three file select registers (FSR0 through FSR2). Each FSR is comprised of two 8-bit registers: FSRH and FSRL. Also, each FSR has a corresponding INDFx register (INDF0-INDF2) used for indirect addressing. In addition to these registers, the four sub-modes utilize four SFRs namely: POSTINC, POSTDEC, PREINC, and PLUSW.

```
          MOVLW    D'100'      ; Move 100 decimal into WREG
          MOVWF    0x20        ; Initialize counter reg (0x20) with 100 decimal
          LFSR     0,0x0044    ; Initialize pointer FSR0 with starting address 044H
REPEAT    CLRF     POSTDEC0    ; Clear a location to 0 and  decrement FSR0 by 1
          DECF     0x20,F      ; Decrement counter by 1
          BNZ      REPEAT      ; Branch to  REPEAT if  Zero flag = 0; otherwise,
                               ; go to the next instruction
```

FIGURE 5.12 Instruction sequence for illustrating posdecrement mode

These SFRs use the contents of FSR0 through FSR2 to achieve the four submodes. The submodes can be used with various PIC18F instructions. The SFRs are utilized by the submodes as follows:

- Indirect with postincrement mode uses POSTINC0 through POSTINC2 registers. POSTINC0 is associated with FSR0, POSTINC1 with FSR1 and, POSTINC2 with FSR2.
- Indirect with postdecrement mode uses POSTDEC0 through POSTDEC2. POSDEC0 is associated with FSR0, POSDEC1 with FSR1, and POSDEC2 with FSR2.
- Indirect with preincrement mode uses PREINC0 through PREINC2 registers. PREINC0 is associated with FSR0, PREINC1 with FSR1, and PREINC2 with FSR2.
- Indirect with 8-bit indexed mode uses PLUSW0 through PLUSW2. PLUSW0 is associated with FSR0, PLUSW1 with FSR1, and PLUSW2 with FSR2.

Indirect with postincrement mode reads the contents of the FSR specified in the instruction, useing the low 12-bit value as the address for the operation to be performed. The specified FSR is then incremented by 1 to point to the next address. The special function register POSTINC is used for this purpose.

As an example, consider CLRF POSTINC0. Prior to execution of this instruction, suppose that the 16-bit contents of FSR0 are 0030H, and the 8-bit contents of the data memory location addressed by 12-bit address 030H are 84H. After execution of the instruction CLRF POSTINC0, the contents of address 030H will be cleared to 00H, and the contents of FSR0 will be incremented by 1 to hold 0031H. This may be used as a pointer to the next data. This is depicted in Figure 5.9. Note that all addresses and data are chosen arbitrarily.

The postincrement mode is typically used with memory arrays stored from LOW to HIGH memory locations. For example, to clear 20 bytes starting at data memory address 030H and above, the instruction sequence in Figure 5.10 can be used. In Figure 5.10, MOVLW D'20' moves 20 decimal into WREG while MOVWF 0x10 moves the contents of WREG (20 decimal) into address 010H. This will initialize the counter register with 20 decimal. The LFSR 0,0x0030 loads FSR0 with 0030H; 030H is the address of the first byte in the array to be cleared to 0. The CLRF POSTINC0 clears the contents of the data memory addressed by FSR0 to 0 and increments FS0 by 1 to hold 031H. This is because POSTINC0 is associated with FSR0. Since the 16-bit contents of FSR0 are 0030H, contents addressed by the the low 12 bits (030H) of FSR0 are cleared to 0. The DECF 0x10,F decrements the contents of data register 010H by one and then places the result in the data register 010H. After the first pass, data register 010H will contain 19 decimal.

The BNZ REPEAT instruction checks if Z flag in the flag register is 0. Note that Z = 0 since the contents of counter are nonzero (19) after execution of DECF. The program branches to label REPEAT, and the loop will be performed 20 times clearing 20 bytes of the array to 0's.

Indirect with postdecrement mode reads the contents of the FSR specified in the instruction, using the low 12-bit value as the address for the operation to be performed. The specified FSR is then decremented by 1. The special function register POSTDEC is used for this purpose.

As an example, consider CLRF POSTDEC0. Prior to execution of this instruction, suppose that the 16-bit contents of FSR0 are 0054H, and the 8-bit contents of the data register addressed by 12-bit address 054H are 21H. After execution of the instruction CLRF POSTDEC0, the contents of address 054H will be cleared to 00H, and the contents of FSR0 will be decremented by 1 to hold 0053H. This is depicted in Figure 5.11. Note that all addresses and data are chosen arbitrarily.

The postdecrement mode can be used with arrays stored from HIGH to LOW addresses. For example, to clear 100 bytes starting at address 044H and below, the instruction sequence in Figure 5.12 can be used. Note that all instructions are self-explanatory from the example of Figure 5.10.

Indirect with preincrement mode reads the contents of the FSR specified in the instruction. First, the contents of the FSR are incremented by 1 to contain the next address. The low 12 bits of the FSR are then used as the address for the operation to be performed. The special function register PREINC is used for this purpose.

As an example, consider CLRF PREINC0. Prior to execution of this instruction, suppose that the 16-bit contents of FSR0 are 0030H, and the 8-bit contents of the data register addressed by 12-bit address 031H are 84H. After execution of the instruction CLRF PREINC0, the contents of FSR0 will be incremented by one to hold 0031H. The contents of data register with address 031H will be cleared to 00H.

Indirect with 8-bit indexed mode adds the contents of the FSR specified in the instruction with the 8-bit contents of the WREG register. The sum is used as an address of a data register in the RAM. The instruction is then executed using these data. The contents of the specified FSR and WREG are unchanged. As an example, consider CLRF PLUSW2. Prior to execution of this instruction, suppose that the 16-bit contents of FSR2 are 0020H, the 8-bit contents of WREG are 04H, and the 8-bit contents of address 024H in data RAM are 37H. After execution of CLRF PLUSW2, the contents of the data register 024H will be cleared to 00H. The contents of FSR2 and WREG are 0020H (unchanged) and 04H (unchanged), respectively.

The indirect with 8-bit indexed mode can be used for code conversion. Two 8-bit ports (Port C and Port D) of the PIC18F will be used to illustrate this example. Assume that a PIC18F is interfaced to an ASCII keyboard via Port C and to an EBCDIC printer via Port D. Suppose that it is desired to enter numerical data via the ASCII keyboard and then print them on the EBCDIC printer.

Note that numerical data entered into the PIC18F via the keyboard will be in ASCII code. Since the printer only understands EBCDIC code, an ASCII-to-EBCDIC code conversion program is required. As discussed in Chapter 1, the ASCII codes for numbers 0 through 9 are 30H through 39H, while the EBCDIC codes for numbers 0 to 9 are F0H to F9H.

An array can be stored in the access bank starting at address 030H EBCDIC code F0H (decimal 0) at address 030H, EBCDIC code F1H (decimal 1) at address 031H, and so on. Now, suppose that '1' is pushed on the ASCII keyboard. The PIC18F can input these data via PORT C into WREG as 31H (ASCII for 1). The EBCDIC printer will print '1' if the PIC18F outputs F1H to Port D. This can be accomplished by initializing one of the FSRs (assume FSR2 in this example) with 0000H and then execute the MOVFF instruction using indirect with 8-bit indexed mode as follows:

```
LFSR      2, 0x0000        ; Load 0000H into FSR2
MOVFF     PLUSW2,PORTD     ; Add WREG to FSR2, move
                           ; the byte content  of that address to
                           ; Port D
```

In the above instruction sequence, since the content of WREG, in this example, is 31H, the instruction MOVFF PLUSW2,PORTD will output the contents of address 031H (F1H), to the EBCDIC printer. The printer will then print '1'.

Note that the PIC18F TBLRD* and TBLWT* instructions use indirect, postincrement, postdecrement, and preincrement modes. In the PIC18F, the address and data sizes of the program memory and data memory are not compatible. The 16-bit contents of the program memory are addressed by 21-bit addresses while the 8-bit contents of the data memory (data registers) are addressed by 12-bit addresses. In order to transfer data between program and data memories, the PIC18F is provided with two SFRs namely, 21-bit TBLPTR (table pointer) and 8-bit TABLAT (table latch). The TABLPTR is used as a pointer for program memory. The TABLAT, on the other hand, is used to hold a byte to be transferred.

Two instructions, namely, TBLRD* and TBLWT*, are provided with the PIC18F in order to perform data transfers between program and data memories. The TABLRD* instruction reads a byte from the program memory into the TABLAT register that can be moved into a data register. The TBLWT*, on the other hand, writes a data byte from the TABLAT register (already moved from a data register) into the program memory. The TBLRD* and TBLWT* instructions can use register indirect, postincrement, postdecrement, and preincrement modes. These instructions along with the addressing modes can be used to transfer a block of data between program memory and data memory. The concept of table pointers for transferring data between program memory and data memory will be covered in detail in Chapter 6.

5.4.5 Relative Addressing Mode

All conditional and one unconditional branch (BRA) instructions in the PIC18F use relative addressing mode.

The conditional branch instructions in the PIC18F are based on four flags namely, C, Z, OV, and N. Each conditional branch instruction specifies an 8-bit offset. This offset is a two's complement signed number. This means that for forward branching, the range of the offset value is from 00H to 7FH. For backward branching, this range varies from 80H to FFH. Since conditional branch instructions are 16 bits wide in the PIC18F, the PC (program counter) is incremented by 2 to point to the next instruction while executing the conditional branch instruction. The offset is multiplied by 2 and then added to PC+2 to find the branch address if the condition is true. Note that the offset is multiplied by 2 since the contents of the PC must always be an even number for 16- and 32-bit instruction lengths.

As an example, consider BNC 0x03. Note that BNC stands for "Branch if no carry." If the C (carry flag) in the status register is 0, then the PC is loaded with the (PC + 2 + 03H x 2). When the PIC18F executes the BNC instruction, the PC points to the next instruction. This means that if BNC is located at address 0050H in program memory, the

PC will contain 0052H (PC + 2) when the PIC18F executes BNC. Hence, if C = 0, then after execution of the BNC 0x03 instruction, the PC will be loaded with address 0058H (0052H + 03H x 2). Hence, the program will branch to address 0058H which is six steps forward relative to the current contents of PC. This is called "relative addressing mode". Note that the relative mode is useful for developing position independent code.

The unconditional branch instruction BRA (branch always) also uses the relative addressing mode. However, the BRA d instruction unconditionally branches to (PC + 2 + d x 2) where offset 'd' is a signed 11-bit number specifying a range from -1024_{10} to $+1023_{10}$ with 0 being positive. For example, consider BRA 0x05 is stored at location 0040H in the program memory. This means that the PC will contain 0042H (PC + 2) when the PIC18F executes BRA. Hence, after execution of the BRA 0x05 instruction, the PC will be loaded with address 004CH (0042H + 05H x 2). Hence, the program will branch to address 004CH, which is 10 (A_{16}) steps forward relative to the current contents of PC. Next, consider the following instruction sequence:

```
      ORG   0x100
HERE  BRA   HERE
```

The machine code for the above instruction is 1101011111111111_2 ($D7FF_{16}$). Note that 0xD7 is the opcode and 0xFF (- 1_{10}) is the offset. During execution of the BRA instruction, the PC points to 0x102. The target branch address = (PC + 2 + d x 2) = 0x102 + (-1) x 2 = 0x100. The instruction HERE BRA HERE unconditionally branches to address 0x100. This is equivalent to HALT instruction in other processors.

To illustrate the concept of relative branching, consider the following PIC18F disassembled instruction sequence along with the machine code (all numbers in hex):

```
                              1:          #INCLUDE<P18F4321.INC>
                              2:          ORG       0x00
0000  0E02  MOVLW 0x2         3:  BACK    MOVLW     0x02
0002  0802  SUBLW 0x2         4:          SUBLW     0x02
0004  E001  BZ      0x8       5:          BZ        DOWN
0006  0E04  MOVLW 0x4         6:          MOVLW     0x04
0008  0804  SUBLW  0x4        7:  DOWN    SUBLW     0x04
000A  EOFA  BZ 0              8:          BZ        BACK
000C  0003  SLEEP             9:          SLEEP
```

Note that all instructions, addresses, and data are chosen arbitrarily. The first branch instruction, BZ DOWN (line 5) at address 0x0004, has a machine code 0xE001. Upon execution of the instruction BZ (branch if Z-flag = 1), the PIC18F branches to label DOWN if Z = 1; the PIC18F executes the next instruction if Z = 0. The BZ instruction uses the relative addressing mode. This means that DOWN is a positive number (the number of steps forward relative to the current program counter) indicating a forward branch. The machine code 0xE001 means that the opcode for BZ is 0xE0 and the relative 8-bit signed offset value is 0x01 (+1). This is a positive value indicating a forward branch. Note that while executing BZ DOWN at address 0x0004, the PC points to address 0x0006 since the program counter is incremented by 2. This means that the program counter contains 0x0008. The offset 0x01 is multiplied by 2 and added to address 0x0006 to find the target branch address where the program will jump if Z = 1.

The branch address can be calculated as follows:

```
  0x0006 = 0000 0000  0000 0110
+ 0x0002 = 0000 0000  0000 0010  (0x01 is multiplied by 2, and sign-extended to 16 bits)
----------------------------------------------------------------------
           0000 0000  0000 1000 = 0x0008
```

Hence, the PIC18F branches to address 0x0008 if Z = 1. This can be verified in the instruction sequence above.

Next, consider the second branch instruction, BZ BACK (line 8). Upon execution of BZ BACK, the PIC18F branches to label BACK if Z = 1; otherwise, the PIC18F executes the next instruction. The machine code this instruction at address 0x000A is 0xE0FA, where 0xE0 is the opcode and 0xFA is the signed 8-bit offset value. The offset is represented as an 8-bit two's complement number. Since 0xFA is a negative number (-6_{10}), this is a backward jump. Note that while executing BZ BACK at address 0x000A, the PC points to address 0x000C since the program counter is incremented by 2. This means that the program counter contains 0x000C. The offset -6 is multiplied by 2, and then added to 0x000C to find the address value where the program will branch if Z = 1. The branch address is calculated as follows:

```
  0x000C  =  0000 0000  0000 1100
+ 0xFFF4  =  1111 1111  1111 0100  (0xFA is multiplied by 2, and then
                                    sign-extended to 16 bits)
----------------------------------------------------------------------
         ↗1  0000 0000  0000 0000  = 0x0000
     Ignore final carry
```

The branch address is 0x0000, which can be verified in the instruction sequence above. As mentioned in Chapter 1, in order to add a 16-bit signed number with an 8-bit signed number, the 8-bit signed number must first be sign-extended to 16 bits. The two 16-bit numbers can then be added. Any carry resulting from the addition must be discarded. This will provide the correct answer.

5.4.6 Bit Addressing Mode

The instructions using the bit addressing mode directly specify a single bit to be operated on. For example, BCF 0x10,3 will clear bit 3 to 0 in the data register addressed by 0x10. Also for example, if the contents of the data register 0x10 are 01111000_2 (78H), then after execution of BCF 0x10,3, the contents of data register 0x10 are 01110000_2 (70H). Note that, in this addressing mode, bit 3 to be cleared to zero is directly specified in the instruction BCF 0x10,3. Three bits in the instructions using bit addressing mode are used to specify the bit number from 0 to 7. These instructions will be covered in Chapter 6.

Questions and Problems

5.1 What is the size of the program counter in the PIC18F? What is the maximum size of the PIC18F program memory? Justify your answer.

5.2 What type of memory is used for program memory in the PIC18F?

5.3 What is the maximum size of the data memory in the PIC18F? How many bits are needed to address data memory? Justify your answer.

5.4 What type of memory is used for data memory in the PIC18F?

5.5 What are the maximum sizes of the program and data memories in the PIC18F4321? What is the main purpose of the EEPROM?

5.6 What is the maximum clock frequency of the PIC18F4321?

5.7 What is the purpose of the CCP module in the PIC18F?

5.8 Summarize the basic features of the PIC18F pipeline.

5.9 Assume a PIC18F microcontroller. What is the basic difference between

(a) program counter (PC) and function select registers (FSRs)?

(b) working register (WREG) and instruction register (IR)?

5.10 Assume that the PIC18F is currently executing a 16-bit instruction addressed by 4000H. What are the current contents of the program counter?

5.11 What is the PIC18F instruction for switching from bank 1 to bank F in the data memory?

5.12 What are the addresses of the PIC18F status and stack pointer registers?

5.13 Is the stack in the PIC18F implemented in the data SRAM? Discuss briefly how the hardware stack is implemented in the PIC18F.

5.14 Find the sign, carry, zero, and overflow flags in the PIC18F for the following arithmetic operation: 6AH - 6AH.

5.15 What is the advantage of incorporating the "access bank" in the PIC18F?

5.16 What is meant by addressing mode?

5.17 Identify the addressing modes for the following instructions:

(a) NOP
(b) MOVLW 0x2A
(c) CLRF PREINC2

5.18 Using the PIC18F instructions described in Section 5.4.4, write an instruction sequence using indirect with preincrement mode to clear 50 bytes in the data memory starting at address 010H and above. Use register addresses of your choice.

6

ASSEMBLY LANGUAGE PROGRAMMING WITH THE PIC18F: PART 1

In this chapter we provide the first part of the PIC18F's instruction set. We also cover the fundamental concepts associated with assembly language programming relating to the PIC18F microcontroller. Topics include PIC18F instruction format, instruction set, and assembly language programming.

6.1 Basic Concepts

As mentioned in Chapter 3, the assembly language program is translated into binary via a program called an *assembler*. In this section, we will review the concepts described in Chapter 3, relating them to the PIC18F MPLAB assembler. The assembler reads each instruction of an assembly language program as ASCII characters and translates them into the respective binary opcodes. For example, the PIC18F assembler translates the SLEEP (places the PIC18F in sleep mode; same as HALT instruction in other processors) instruction into its 16-bit binary opcode as 0000 0000 0000 0011 (0003 in hex), as depicted in Table 6.1.

A notable advantage of the assembler is its ability to handle address computation. Most programs use addresses within the program as data storage or as targets for jumps or calls. When programming in machine language, these addresses must be calculated by hand. The assembler solves this problem by allowing the programmer to assign a symbol to an address. The programmer may then reference that address elsewhere by using the symbol. The assembler computes the actual address for the programmer and fills it in automatically.

One can obtain hands-on experience with a typical assembler for a microcontroller by downloading it from the Internet. The MPLAB assembler/simulator is used to assemble

TABLE 6.1 Conversion of SLEEP instruction into its binary opcode

Assembly code	Binary form of ASCII codes as seen by the assembler	Binary opcode created by the MPLAB PIC18F assembler
S	0101 0011	
L	0100 1000	0000 0000 0000 0011
E	0100 0101	
E	0100 0101	
P	0101 0000	

and debug all PIC18F assembly language programs in this book. It can be downloaded free of charge from the Web site: www.microchip.com.

As mentioned in Chapter 4, each line in an assembly language program includes four fields:

- label field
- mnemonic or opcode field
- operand field
- comment field

The rules associated with the MPLAB will be used for explaining the meaning of these fields. The assembler ignores the comment field but translates the other fields. The label field must start with an alphabetic character, and can have a maximum length of 32 characters. The label field ends in a colon or nothing.

The assembler must know where one field starts and another ends. The MPLAB, like most assemblers, allows the programmer to use a special symbol or delimiter to indicate the beginning or end of each field. Typical delimiters used are spaces, commas, semicolons, and colons:

- Spaces are used between fields.
- Commas (,) are used between addresses in an operand field.
- A semicolon (;) is used before a comment.
- A colon (:) or none is used after a label.

To handle numbers, most assemblers consider all numbers as decimal numbers unless specified otherwise. The MPLAB assembler, however, recognizes a number without any prescript or postscript as a hexadecimal number. Most assemblers, including the MPLAB assembler, will also allow other number systems, including hexadecimal. For example, with the MPLAB assembler, the user can define a hexadecimal number usually in three ways using:

- 0x before the number
- H after the number
- default (without any prescript or postscript)

This means that 60 in hexadecimal can be represented as 0x60, 60H, or 60. A '0' is used by most assemblers if the first digit of the hexadecimal number is a letter; otherwise, the assembler will generate an error. This is done to distinguish between numbers and labels. However, the MPLAB assembler does not require the first digit of the hexadecimal number to be a letter.

The MPLAB uses D before a 'number' to specify a decimal number. For example, decimal number 60 can be represented as D'60'. A binary number is specified by the MPLAB using B before the 'number'. For example, 8-bit binary number 01011100 can be represented by the MPLAB as B'01011100'.

ASCII characters are represented using MPLAB by the symbols ' ' or A''. For example, PIC18F is represented as 'PIC18F' or A'PIC18F' to be recognized as ASCII characters in MPLAB.

Assemblers use pseudoinstructions or directives to make the formatting of the edited text easier. These directives are not translated directly into machine language instructions. Typical assembler directives are discussed in the following.

ORIGIN (ORG) The directive ORG specifies the starting address (must be an even number for MPLAB assembler) of a program or data. For example, after assembling the following statements, the MPLAB will place the assembled code for MOVLW 0x50 starting at address 0x100:

```
        ORG         0x100
        MOVLW       0x50
```

Equate (EQU) The EQU assigns a value in its operand field to an address in its label field. This allows the user to assign a numerical value to a symbolic name. The user can then use the symbolic name in the program instead of its numerical value. A typical example of EQU is START EQU 0x20, which assigns the value 20 in hexadecimal to the label START. As mentioned before, some assemblers require hexadecimal numbers to start with a digit when the EQU directive is used. A 0 is used if the first digit of the hexadecimal number is a letter; otherwise, an error will be generated by the assembler. For example, TEST EQU 0A5H will assign A5 in hex to the label TEST. However, the MPLAB PIC18F does not have this restriction. The statement TEST EQU A5H in the MPLAB will assign A5 in hex to the label TEST.

To illustrate the EQU directive, consider the following instruction sequence:

```
TEST    EQU         0x20
        MOVLW       TEST
        MOVWF       TEST
```

The instruction, MOVLW TEST in the above moves the constant 0x20 into WREG. The instruction MOVWF TEST, on the other hand, moves the contents of WREG into the data register with address 0x20.

Define Byte (DB) The directive DB is generally used to set a memory location to a certain byte value. For example,

```
START       DB          0x45
```

will store the data value 45 hex to the address START. The DB directive can be used to generate a table of data as follows:

```
        ORG     0x70
TABLE   DB      0x20,0x30,0x40,0x50
```

In this case, 20 hex is the first data of the memory location 70 hex; 30 hex, 40 hex, and 50 hex occupy the next three memory locations. Therefore, the data in memory will look like this:

```
70      20
71      30
72      40
73      50
```

Define Word (DW) The directive DW is typically used to assign a 16-bit value to two memory locations. For example,

```
        ORG     50
START   DW      0x4AC2
```

will assign C2 to location 50 hex and 4A to location 51 hex. It is assumed that the assembler will assign the low byte first (C2) and then the high byte (4A). The DW directive can be used to generate a table of 16-bit data as follows:

```
            ORG     0x80
POINTER     DW      0x5000,0x6000, 0x7000
```

In this case, the three 16-bit values 5000 hex, 6000 hex, and 7000 hex are assigned to memory locations starting at the address 80 hex. That is, the array would look like this:

80	00
81	50
82	00
83	60
84	00
85	70

INCLUDE The directive INCLUDE or #INCLUDE includes source code or a file from the MPLAB library for a specific device. This will allow the MPLAB to assemble a program for that device. For example, using INCLUDE <P18F4321.INC> at the beginning of a program will add appropriate files from the MPLAB library required to assemble the program. Note that the MPLAB C compiler only accepts the statement #INCLUDE <P18F4321.INC>; because the PIC18F C compiler will not accept INCLUDE <P18F4321.INC>, the # sign must be placed before the statement.

END The END directive indicates the end of an assembly language program.

Let us now explain how one can write a typical PIC18F assembly language program. A simple program for adding two 8-bit numbers written in PIC18F assembly language is provided below:

Label field	Mnemonic field	Operand field	Comment field
	INCLUDE	<P18F4321.INC>	
SUM	EQU	0x40	
	ORG	0x100	; STARTING ADDRESS
	MOVLW	0x02	; MOVING 2 INTO WREG
	ADDLW	0x05	; ADDING WREG and 5
	MOVWF	SUM	; STORE RESULT IN SUM
	SLEEP		; HALT
	END		; END OF PROGRAM

Let us explain the above program. INCLUDE <P18F4321.INC> at the beginning of the program will add appropriate files from the MPLAB library required to assemble the program. Note that # INCLUDE <P18F4321.INC> could have been used instead of INCLUDE <P18F4321.INC>. SUM EQU 0x08 assigns 08 (hex) to label SUM. ORG 0x100 assembles the program at address 100 (hex). MOVLW 0x02 moves 02 (hex) into WREG. ADDLW 0x05 adds the contents of WREG with 05 (hex), and stores the 8-bit result 07 (hex) in WREG. The instruction MOVWF SUM stores the 8-bit contents of WREG (07 hex) in file register SUM with address 40 (hex). Also, the PIC18F has a SLEEP instruction which is the same as the HALT instruction in other processors. The PIC18F instruction for unconditionally jumping to the same location such as “FINISH GOTO FINISH” can be used instead of the “SLEEP” instruction. Both “SLEEP” and “FINISH GOTO FINISH” are equivalent to the "HALT" instruction in other processors. The assembler directive END indicates the end of assembly language program.

The assembly language program described above, called a *source file*, contains all of the instructions required to execute a program. It should be mentioned that the source code for the MPLAB assembler is not case sensitive. However, it is a good idea to type the source program in all uppercase or all lower-case letters. Also, it is a good practice to use the TAB key between two fields while typing the source code. This will enhance the visual appearance of the program.

The assembler converts the source file into an object file containing the binary codes or machine codes that the PIC18F will understand. In typical assemblers including the MPLAB (Microchip's PIC18F assembler), the source file must be stored with a file extension called .ASM. Suppose that the programmer stores the source file as SUM.ASM. To assemble the program, the source file SUM.ASM is presented as input to the assembler. The assembler typically generates two files: SUM.OBJ (object file) and SUM.LST (list file).

SUM.OBJ is an *object file*, a binary file containing the machine code and data that correspond to the assembly language program in the source file (SUM.ASM). The object file, which includes additional information about relocation and external references, is not normally ready for execution.

SUM.LST is a *list file* that shows how the assembler interprets the source file SUM.ASM. The list file may be displayed on the screen. The source file SUM.ASM is assembled using the MPLAB assembler. The SUM.LST file is as follows:

```
                          1:         INCLUDE <P18F4321.INC>
                          2:  SUM    EQU      0x40
                          3:         ORG      0x100  ; STARTING ADDRESS
0100 0E02 MOVLW 0x2       4:         MOVLW    0x02   ; MOVING 2 INTO WREG
0102 0F05 ADDLW 0x5       5:         ADDLW    0x05   ; ADDING WREG and 5
0104 6E40 MOVWF 0x40      6:         MOVWF    SUM    ; STORE RESULT
0106 0003 SLEEP           7:         SLEEP           ; HALT
```

Note that the assembled code shown on the left above is in hex. ORG 0x100 (line 3) generates the starting address 0100 in hex. The machine code (0E02 hex) for the first instruction, MOVLW 0x02, is stored at the address 0100 (hex). Since this instruction takes 16-bit (two bytes), the machine code for the next instruction, ADDLW 0x05, starts at address 0102 (hex). Similarly, the machine codes for MOVWF SUM and SLEEP start at addresses 0104 (hex) and 0106 (hex), respectively. Note that the comment fields in the SUM.ASM file are not translated by the MPLAB assembler.

When a large program is being developed by a group of programmers, each programmer may write only a portion of the whole program. The individual program parts must be tested and assembled to ensure their proper operation. When all portions of the program are verified for correct operation, their object files must be combined into a single object program using a *Linker*, a program that checks each object file and finds certain characteristics such as the size in bytes and its proper location in the single object program. The linker also resolves any problems with regard to cross-references to labels. A *library* of object files is typically used to reduce the size of the source file. The library files may contain frequently used subroutines and/or sections of codes. Rather than these codes being written repeatedly in the source file, a special pseudoinstruction is used to tell the assembler that the code must be inserted by the linker at linking time. When linking is completed, the final object file is called an *executable* (.EXE) *file*. Finally, a program called a *Loader* can be used to load the .EXE file in memory for execution.

Appendix F provides a tutorial showing the step-by-step procedure of assembling and debugging a PIC18F assembly language program using Microchip MPLAB PIC18F assembler/debugger.

6.2 PIC18F Instruction Format

The instruction format is specified by the operation performed. The instruction set is grouped into four basic categories based on the operations:

- byte-oriented operations
- bit-oriented operations
- literal operations
- control operations

Table 6.2 lists these operations.

Most byte-oriented instructions have three operands:

1. The file register (specified by 'f')
2. The destination of the result (specified by 'd')
3. The accessed memory (specified by 'a')

The file register designator 'f' specifies which file register is to be used by the instruction. The destination designator 'd' specifies where the result of the operation is to be placed. If 'd' is zero, the result is placed in the WREG register. If 'd' is one, the result is placed in the file register specified in the instruction. If a = 0, the file register is the access bank. On the other hand, if a = 1, the BSR specifies the data memory bank. For example, consider ADDWF 0x04, 0, 0. This instruction adds the contents of WREG register with the contents of file register 0x04 and stores the result in WREG. By inspecting ADDWF 0x04, 0, 0 and comparing with the byte-oriented format of Table 6.2, d = 0, a = 0, and f = 0x04. Since the 6-bit opcode for ADDWF is 001001, the binary 16-bit code for ADDWF 0x04, 0, 0 is 0010010000000100 (2404H).

All bit-oriented instructions have three operands:

1. The file register (specified by 'f')
2. The bit in the file register (specified by 'b')
3. The accessed memory (specified by 'a')

The bit field designator 'b' selects the number of the bit affected by the operation, while the file register designator 'f' represents the file address in which the bit is located. If a = 0, the file register is access bank. On the other hand, if a = 1, the BSR specifies the data memory bank. For example, consider BSF 0x27, 5, 0. This instruction sets bit 5 to one in file register 0x27 in the access bank, and comparing with the bit-oriented format of Table 6.2, b = 101, a = 0, and f = 0x27. Since the 4-bit opcode for BSF is 1000, the binary 16-bit code for BSF 0x27, 5, 0 is 1000101000100111 (8A27H).

The literal instructions may use some of the following operands:

- a literal or constant value to be loaded into a file register (specified by 'k')
- the desired FSR register to load the literal value into (specified by 'f')
- no operand required (specified by '—')

As an example, consider MOVLW 0x2A. This instruction moves 0x2A into

TABLE 6.2 General format for instructions

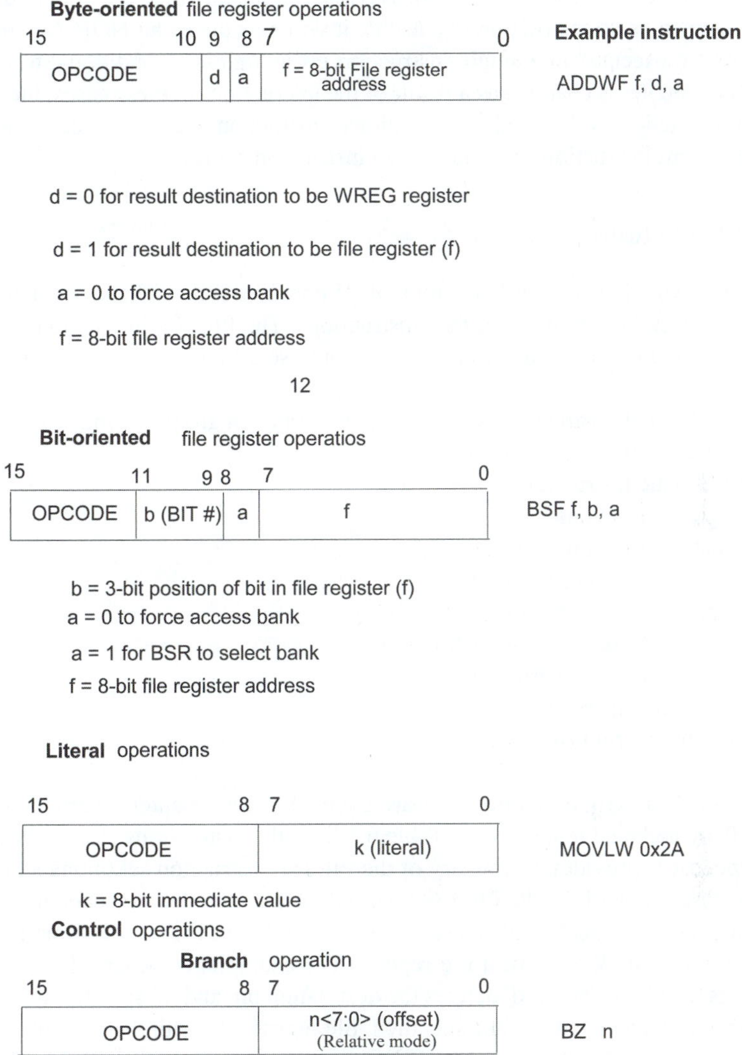

the WREG register. When compared with the literal-oriented format of Table 6.2, k = 00101010 (0x2A). Since the 8-bit opcode for MOVLW is 00001110, the binary 16-bit code for MOVLW 0x2A is 0000111000101010 (0E2AH).

An example of the control instructions includes conditional branch instructions with the following operand:

- a signed 8-bit offset (specified by '*n*')

As an example, consider BZ 0x04 where 04 (hex) is the offset (*n*). This instruction branches to an address (PC+2+ 2 x 4) if Z = 1; otherwise, the next instruction is executed. Since the 8-bit opcode for BZ is 11100000_2 and *n* = 00000100_2 (0x04), the binary 16-bit code for BZ 0x04 is 1110000000000100_2 (E004H).

Most PIC18F instructions are a single word; only four instructions are double-word instructions. These instructions were made double-word to contain the required information in 32 bits. In the second word, the four MSBs (most significant bits) are '1's. If this second word is executed as an instruction (by itself), it will execute as an NOP. All single-word instructions are executed in a single instruction cycle, unless a conditional test is true or the program counter is changed as a result of the instruction. In these cases, the execution takes two instruction cycles, with the additional instruction cycle(s) executed as an NOP. The double-word instructions execute in two instruction cycles.

6.3 PIC18F Instruction Set

The PIC18F instruction set contains a total of 75 core instructions. In addition, the PIC18F provides an extended set of eight new instructions. The PIC18F instruction set is highly orthogonal. This means that most instructions can use all addressing modes with data of any type.

The PIC18F instructions can be classified into ten groups as follows:

1. Data movement instructions
2. Arithmetic instructions
3. Logic instructions
4. Rotate instructions
5. Bit manipulation instructions
6. Jump/branch instructions
7. Test , compare, and skip instructions
8. Table read/write instructions
9. Subroutine instructions
10. System control instructions

Instruction groups 1 through 5 are covered in this chapter. Instruction groups 6 through 10 are included in Chapter 7. Table 6.3 lists the instructions along with the status flags. Appendix C provides a summary of the PIC18F instruction set (alphabetical order). A detailed description of the the PIC18F instructions are included in Appendix D.

It should be mentioned that there are several PIC18F instructions that specify the destination as the WREG or a file register. For example, consider ADDWF F, d, a, where 'F ' is the file register , 'd' defines the destination bit, and 'a' specifies the bank. This instruction adds the contents of the specified file register 'F' with the contents of WREG. If d = 0, then the result is stored in the WREG register. However, if d = 1, then the result is placed in the file register 'F' specified in the instruction. Note that a = 0 means that the data register is located in the access bank while a = 1 means that the contents of BSR specify the address of the bank. As mentioned before, the file (data) register can be one of the general purpose register (GPRs) or one of the special function registers (SFRs) in the access bank.

For example, ADDWF 0x18, 0, 0 will add the contents of file register 0x18 with the contents of WREG. The result will be placed in WREG since d = 0. The instruction ADDWF 0x18, 1, 0 will also add the contents of file register 0x18 with the contents of WREG. But the result will be stored in memory location 0x18 since d = 1.

Note that for instructions such as ADDWF F, d, a, the MPLAB assembler allows the PIC18F programmer to use 'W' instead of '0', and 'F' instead of '1', as far as 'd' is concerned. This means that the programmer, in this case, can use ADDWF 0x18, W, 0 instead of ADDWF 0x18, 0, 0, and ADDWF 0x18, F, 0 instead of ADDWF 0x18, 1, 0. Also, as mentioned in Chapter 5, upon power-up, the PIC18F uses the access bank (a =

0) of the file (data) registers as the default bank (addresses 0x00 through 0x7F and 0xF80 through 0xFFF). Hence, we will not specify 'a' in the instruction (assuming the access bank as the default bank upon power-up). Also, for better clarity, the MPLAB assembler uses W or F instead of 0 or 1 for 'd'. This means that ADDWF 0x18, W will be used instead of ADDWF 0x18, 0, 0, and ADDWF 0x18, F will be used instead of ADDWF 0x18,1,0. No bank switching will be used in all of the programming examples.

In the following, the brackets [ ] are used to indicate the contents of a register or a data memory location. For example, [WREG] will mean the contents of WREG. Also, all numbers are chosen arbitrarily in order to illustrate the instructions using numerical examples. In the PIC18F, the term "file register" can be one of the GPRs in the low access bank (addresses 0x00 to 0x7F in bank 0) or one of the SFRs in the high access bank (addresses 0x80 to 0xFF in bank F).

TABLE 6.3 PIC18F instructions and the status flags

Instruction	N	OV	Z	DC	C
ADDLW, ADDWF, ADDWFC	✓	✓	✓	✓	✓
ANDLW, ANDWF	✓	–	✓	–	–
BC, BNC, BZ, BNZ, BN, BNN	–	–	–	–	–
BOV, BNOV, BRA	–	–	–	–	–
BCF, BSF, BTG	–	–	–	–	–
BTFSC, BTFSS	–	–	–	–	–
CALL	–	–	–	–	–
CLRF	–	–	✓	–	–
CLRWDT	–	–	–	–	–
COMF	✓	–	✓	–	–
CPFSEQ, CPFSGT, CPFSLT	–	–	–	–	–
DAW	–	–	–	–	–
DECF	✓	✓	✓	✓	✓
DECFSZ, DECFSNZ	–	–	–	–	–
GOTO	–	–	–	–	–
INCF	✓	✓	✓	✓	✓
INCFSZ, INCFSNZ	–	–	–	–	–
IORLW, IORWF	✓	–	✓	–	–
LFSR	–	–	–	–	–
MOVF	✓	–	✓	–	–
MOVFF, MOVWF, MULWF	–	–	–	–	–
MOVLB, MOVLW	–	–	–	–	–
NEGF	✓	✓	✓	✓	✓
NOP	–	–	–	–	–
POP, PUSH	–	–	–	–	–
RCALL, RETFIE,	–	–	–	–	–
RESET	✓	✓	✓	✓	✓
RETLW, RETURN,	–	–	–	–	–
RLCF, RRCF	✓	–	✓	–	✓
RLNCF, RRNCF	✓	–	✓	–	–
SETF	–	–	–	–	–
SLEEP, SWAPF	–	–	–	–	–
SUBFWB, SUBLW	✓	✓	✓	✓	✓
SUBWF, SUBWFB,	✓	✓	✓	✓	✓
XORLW, XORWF	✓	–	✓	–	–

✓ Affected, – Not affected

Note: TBLRD*, TBLRD*+, TBLRD*-, TBLRD+*, TBLWT*, TBLWT*+, TBLWT*- and, TBLWT+* do not affect any flags.

TABLE 6.4 PIC18F data movement instructions

Instruction	Comment
CLRF F, a	Clear data register F to zero. F is located in access bank if a = 0, and F is located in the bank specified by BSR if a =1.
LFSR F, K	Load low 12 bits of K into the specified file select register (F can be 0 or 1 or 2).
MOVLB K	Move 8-bit value K to BSR; low 4 bits of K are moved into low 4 bits of BSR and upper 4 bits of BSR are always cleared to zero regardless of upper 4 bits of K.
MOVLW data 8	Move 8-bit immediate data into WREG.
MOVWF F, a	Move data from WREG into data register F. F is located in access bank if a = 0 and in the bank specified by BSR if a =1.
MOVFF Fs, Fd	Move data from Fs (source data register) to Fd (destination data reg).
MOVF F, d, a	Move the contents of the file register into WREG (d = 0) or into the same file register (d = 1). F is located in access bank if a = 0 and in the bank specified by BSR if a =1. This is the only MOVE instruction that affects N and Z status flags.
SETF F, a	Set all eight bits in the specified data register F to ones. See note for 'a'.
SWAPF F, d, a	Swaps low-order 4 bits with the high-order 4 bits of the file register F; see note for 'a' and 'd'.

- All instructions in the above table except MOVFF and LFSR are executed in one cycle; MOVFF and LFSR are executed in two cycles.
- a = 0 means that the data register is located in the access bank while a = 1 means that the contents of BSR specify the address of the bank.
- For destination, d = 0 means that the destination is WREG while d = 1 means that the destination is file register.
- The size of each instruction except LFSR and MOVFF is one word; the sizes of LFSR and MOVFF are two words.

6.3.1 Data Movement Instructions

The PIC18F data movement instructions are given in Table 6.4. Next, we explain the data movement instructions using the access bank and specifying F or W in place of 'd'.

CLRF F, a clears the contents of the specified data register F to zero. If a = 0, then F is located in the access bank . On the other hand, if a = 1, then F is located in the bank specified by BSR. For example, CLRF 0x20 will clear the 8-bit contents of data register with address 0x20 to zero. After clearing, the Z-flag is set to one. No other flags are affected.

The CLRF instruction can be used to configure an I/O port. For example, in the PIC18F, the TRISC register is used to configure Port C. Writing 0's to all 8 bits of TRISC will configure Port C as an output port. This can be accomplished using the CLRF TRISC instruction. This topic will be discussed later.

LFSR F, K loads low 12 bits of K into one of three FSRs (file select registers). The specified FSR can then be used to point to a data register. For example, LFSR 2, 0x0020 will load 00H into FSR2H and 20H into FSR2L. The low 12-bit contents of FSR2 (0x020) can then be used as a pointer to data memory. Note that since FSRs are 16 bits wide registers, the 12-bit address is stored in low 12 bits (bits 0 through 11) of the FSRs with the upper four bits (bits 12 through 15) as 0's.

MOVLB K moves 8-bit value K to BSR; low 4 bits of K are moved into low 4 bits of BSR and the upper 4 bits of BSR are always cleared to zero regardless of the upper 4 bits of K. For example, MOVLB 0x01 will move 01H into BSR. This instruction is useful for bank switching.

MOVLW data8 moves an 8-bit literal (constant data) into WREG. For example, MOVLW 25H will move 25H into the WREG register. The previous contents of WREG are lost.

MOVWF F, a moves data from WREG into data register F. F is located in access bank if a = 0 and in the bank specified by BSR if a =1. As an example, consider MOVWF 0x40.

Prior to execution of MOVWF 0x40, [0x40] = F1H, and [WREG] = 53H.

After execution of MOVWF 0x40, [0x40] = 53H, and [WREG] = 53H (unchanged).

MOVFF Fs, Fd moves data from source data register Fs to destination data register Fd. WREG can be used as either Fs or Fd. Also, Fs and Fd can be any data memory location from 000H to FFFH. This is a two-word (32 bits) instruction. As an example, consider MOVFF 0x04, 0x03.

Prior to execution of MOVFF 0x04, 0x03, [0x03] = 2FH, and [0x04] = 57H.

After execution of MOVFF 0x04, 0x03, [0x03] = 57H, and [0x04] = 57H (unchanged).

The MOVFF instruction is useful in transferring the contents of a data register to an I/O port.

MOVF F, d, a moves the contents of the data register F into WREG (d = 0) or into the same data register F (d = 1). F is located in access bank if a = 0 and in the bank specified by BSR if a =1. As an example, consider MOVF 0x30, W.

Prior to execution of MOVF 0x30, W, [WREG] = 70H, and [0x30] = 2AH.

After execution of MOVF 0x30, W, [WREG] = 2AH, and [0x30] = 2AH (unchanged).

SETF F, a sets the contents of the specified data register (F) to FFH. If a = 0, then the data register is located in the access bank while a = 1 means that the contents of BSR specify the address of the bank. As an example, consider SETF 0x20.

Prior to execution of SETF 0x20, [0x20] = 24H.
After execution of SETF 0x20, [0x20] = FFH.

The SETF instruction can be used to configure an I/O port. For example, the TRISB register in the PIC18F is used to configure Port B. Writing 1's to all 8 bits of TRISB will configure Port B as an input port. This can be accomplished using the SETF TRISB instruction. This topic will be discussed later.

SWAPF F, d, a exchanges low-order 4 bits with the high-order 4 bits of the file register F. F is located in access bank if a = 0 and in the bank specified by BSR if a =1. d = 0 means that the destination is WREG while d = 1 means that the destination is file register.

As an example, consider SWAPF 0x60, W.

Prior to execution of the instruction, SWAPF 0x60, W, [0x60] = 48H, and [WREG] = 50H.

After execution of the instruction, SWAPF 0x60, W, [0x60] = 48H, and [WREG] = 84H.

Note that, in the above, since d = 0 in the instruction, SWAPF 0x60, W stores the result 84H in the WREG.

Example 6.1 Determine the effect of each of the following PIC18F instructions:

- CLRF PREINC1
- MOVWF INDF1
- MOVFF 0x40, 0x81
- MOVF 0x40, F; also find the flags that are affected
- SWAPF 0x45, W

Assume the following initial configuration before each instruction is executed; also assume that all numbers are in hex:

[FSR0] = 0044, [FSR1] = 0075
[043] = 66, [076] = FF
[075] = 24, [WREG] = 33
[040] = 78, [045] = 61
[081] = 55

Solution See Table 6.5

Example 6.2 Find the affected FSR, WREG, and data register contents for the following PIC18F instruction sequence:

```
LFSR      2,0x0044
MOVLW     D'20'
MOVWF     0x40
MOVFF     PLUSW2, 0x40
CLRF      POSTINC2
SETF      0x40
```

Assume [0x58] = 1AH prior to execution of the instruction sequence.

TABLE 6.5 Results for Example 6.1 (all numbers in hex)

Instruction	Affected register	Net effect (hex)
CLRF PREINC1	Data register address = 076	[076]= 00
MOVWF INDF1	Data register address = 075	[075] = 33
MOVFF 0x40, 0x81	Data register address = 081	[0x81] = 78
MOVF 0x40, F	Data register address = 040	[0x40] = 78,N = 0, Z=0
SWAPF 0x45, W	WREG	[WREG] = 16

Solution

After execution of LFSR 2,0x0044, the file select register FSR2 is loaded with 0044H. MOVLW D'20' moves immediate decimal data 20 (14H) into WREG. The instruction MOVWF 0x40 moves [WREG] into data register 0x40. Hence, [0x40] = 14H.

MOVFF PLUSW2, 0x40 adds [WREG] with [FSR2] and then moves the byte content of that address into data register 0x40. This means that the contents of data register 0x58 are moved to data register 0x40. Hence, [0x40] = 1AH.

CLRF POSTINC2 clears the contents of data register 0x44 to zero, and then increments FSR2 by 1. Hence, [0x44] = 00H, and [FSR2] = 0x45.

SETF 0x40 sets all bits in data register 0x40 to ones. Hence, [0x40] = FFH.

Example 6.3 It is desired to clear 10 consecutive bytes to zero from LOW to HIGH data register addresses starting at data register 0x40 pointed to by FSR1.
(a) Flowchart the problem.
(b) Convert the flowchart to a PIC18F assembly language program starting at address 0x100.

Solution

(a) The flowchart is provided below:

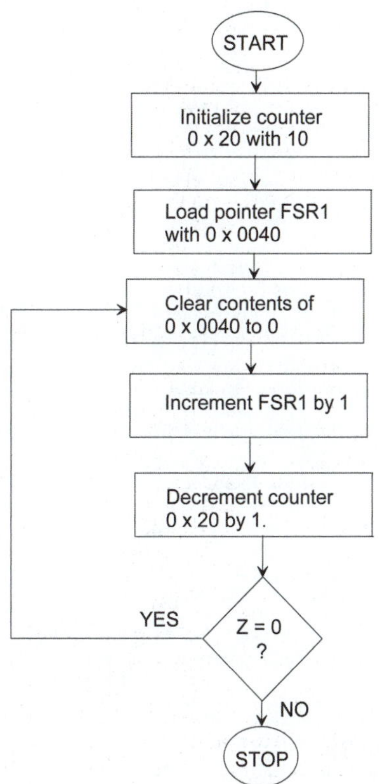

(b) The flowchart is converted to PIC18F assembly language program as follows:

```
        INCLUDE <P18F4321.INC>
        ORG     0x100
        MOVLW   D'10'      ; Move 10 decimal into WREG
        MOVWF   0x20       ; Initialize counter reg (20H) with 10 decimal
        LFSR    1, 0x0040  ; Initialize FSR1 with 040H as starting
                           ; address since postincrement mode to be used
REPEAT  CLRF    POSTINC1   ; then clear a location to 0 and increment FSR1 by 1
        DECF    0x20, F    ; Decrement counter by 1
        BNZ     REPEAT     ; Branch to  REPEAT if  Zero flag = 0; otherwise,
                           ; go to the next instruction
        SLEEP              ; HALT
        END
```

Example 6.4 It is desired to move a block of 8-bit data of length 10 from the source block (from HIGH to LOW address) starting at data register address 0x55 to the destination block (from LOW to HIGH addresses) starting at data register address 0x30. That is, [0x55] will be moved to [0x30], [0x54] to [0x31], and so on. Assume that data for the source block and the destination block are already stored in data memory addresses.

(a) Flowchart the problem

(b) Convert the flowchart to PIC18F assembly language program starting at address 0x100.

Solution

(a) The flowchart along with data memory layout is provided below:

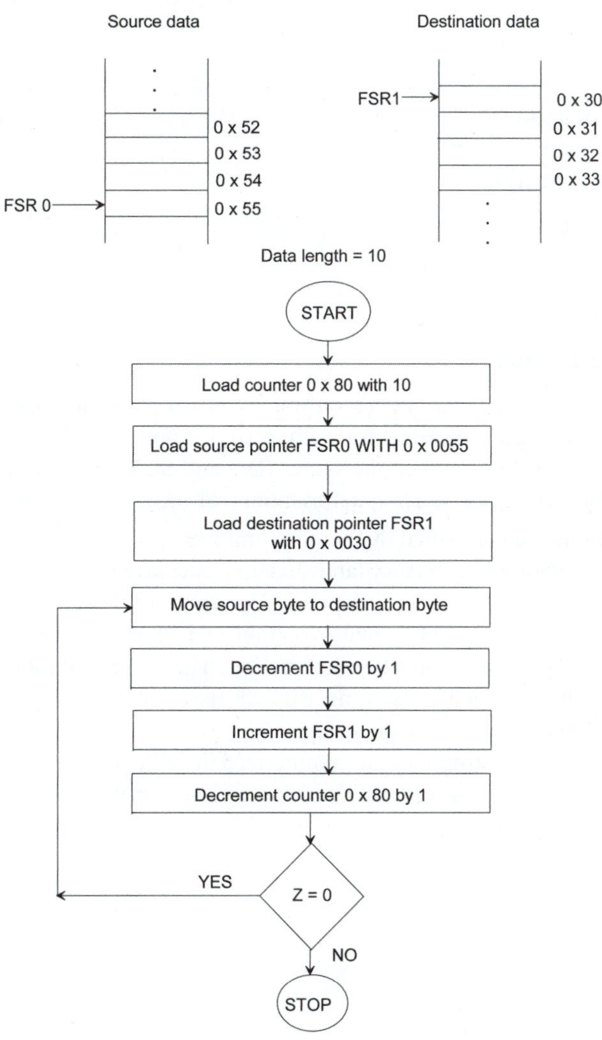

(b) The flowchart can be converted to PIC18F assembly language program as follows:

```
        INCLUDE   <P18F4321.INC>
COUNTER EQU       0x80
        ORG       0x100
        MOVLW     D'10'           ; Move 10 decimal into WREG
        MOVWF     COUNTER         ; Initialize counter reg (0x80) with 10
                                  ; decimal
        LFSR      0, 0x0055       ; Initialize FSR0 with source starting
                                  ; address
        LFSR      1, 0x0030       ; Initialize FSR1 with destination
                                    starting address
```

```
BACK    MOVFF   POSTDEC0, POSTINC1 ; Move source data to destination
        DECF    COUNTER, F         ; Decrement counter by 1
        BNZ     BACK               ; Branch to BACK if Zero flag = 0,
                                   ; otherwise, go to the next instruction
        SLEEP                      ; HALT
        END
```

6.3.2 Arithmetic Instructions

The PIC18F arithmetic instructions allow

- 8-bit additions and subtractions
- 8-bit by 8-bit unsigned multiplication
- negate instruction
- decrement and increment instructions
- BCD adjust (BCD correction)

As mentioned before, typical microcontrollers utilize common hardware to perform addition and subtraction operations for both unsigned and signed numbers. The instruction set of microcontrollers include the same ADD and SUBTRACT instructions for both unsigned and signed numbers. The interpretations of unsigned and signed ADD and SUBTRACT operations are performed by the programmer.

Unsigned and signed multiplication and division operations can be performed using various algorithms. Typical 32-bit microprocessors such as the Pentium contain separate instructions for performing these multiplication and division operations. The PIC18F provides only unsigned multiplication instruction. The other multiplication and division instructions can be obtained by writing PIC18F assembly language programs using appropriate algorithms. This is shown in Chapter 7.

The PIC18F arithmetic instructions are summarized in Table 6.6. Next, we explain the arithmetic instructions using the access bank and specifying F or W in place of 'd'.

TABLE 6.6 PIC18F arithmetic instructions

Instruction	Operation
	Addition and subtraction instructions
ADDLW data8	[WREG] + [8-bit data] → [WREG]
ADDWF F, d, a	[WREG] + [F] → destination; see note for 'a' and 'destination'.
ADDWFC F, d, a	[WREG] + [F] + carry→ destination; see note for 'a' and 'destination'.
SUBLW data8	[8-bit data] – [WREG] → [WREG]
SUBWF F, d, a	[F] – [WREG] → destination; see note for 'a' and 'destination'
SUBWFB F, d, a	[F] – [WREG] – carry → destination; see note for 'a' and 'destination'.
SUBFWB F, d, a	[WREG] – [F] – carry → destination; see note for 'a' and 'destination'.
	Unsigned multiplication instructions
MULLW data8	[WREG] x [8-bit data] → [PRODH]: [PRODL] (unsigned multiplication)
MULWF F, a	[WREG]x [F] → [PRODH]: [PRODL] (unsigned multiplication)
	Negate instruction
NEGF F, a	0 – [F] →[F]; see note for 'a'.
	Decrement and increment instructions
DECF F, d, a	[F] – 1→ destination; see note for 'a' and 'destination'.
INCF F, d, a	[F] + 1 →destination; see note for 'a' and 'destination'.
	BCD Adjust (BCD correction) *instruction*
DAW	Decimal adjust [WREG]

- All instructions in the above table are executed in one cycle.
- The size of each instruction is one word.
- a = 0 means that the data register is located in the access bank while a = 1 means that the contents of BSR specify the address of the bank.
- For destination d = 0 means that the destination is WREG while d = 1 means that the destination is file register. As mentioned before, W or F instead of 0 or 1 will be used in this book for better clarity.
- The PIC18F does not provide any multiplication (signed) and division (signed and unsigned) instructions.

Addition and Subtraction Instructions The PIC18F addition and subtraction instructions are illustrated by means of numerical examples in the following. All flags are affected. Assume signed numbers.

- ADDLW data8 instruction adds the 8-bit contents of WREG with 8-bit immediate data, and stores the result in WREG. For example, consider ADDLW 0x02.

  Prior to execution of ADDLW 0x02, [WREG] = 12H.

  After execution of ADDLW 0x02, [WREG] = 12H + 02H = 14H.

  The flags are affected based on the result as follows:

```
             Previous carry → 0000 010 ← Intermediate carries
            [WREG]= 12H = 0001 0010
  Add immediate data, 02H = 0000 0010
                          ---------------------
              Final carry → 0  0001 0100 = 14H
```

N = 0 (most significant bit of the result is 0), OV = 0 (no overflow since the previous carry and the final carry are the same), Z = 0 (nonzero result), DC = 0 (no carry from bit 3 to bit 4), and C = 0 (no carry).

- ADDWF F, d, a instruction adds the contents of WREG with the contents of the specified data register (F). The result is stored in WREG if d = 0 or in the data register if d = 1. The data register is in the access bank if a = 0 or specified by BSR if a = 1

As an example, consider ADDWF 0x50, W.

Prior to execution of ADDWF 0x50, W, [WREG] = 73H, [0x50] = 2AH.

After execution of ADDWF 0x50,W, [WREG]= 73H + 2AH = 9DH, and [0x50] = unchanged = 2AH. The flags are affected based on the result as follows:

```
  Previous carry → 1100 010 ← Intermediate carries
  [WREG] = 73H =  0111 0011
  Add [0x50] = 2AH = 0010 1010
                  ---------------------
     Final carry → 0  1001 1101 = 9DH
```

N = 1 (most significant bit of the result is 0), OV = $C_f \oplus C_p = 0 \oplus 1 = 1$ (overflow indicating wrong result; addition of two positive numbers generated a negative result), Z = 0 (nonzero result), DC = 0 (no carry from bit 3 to bit 4), and C = 0 (no carry). Note that correct result can be rectified by increasing the number of bits for the two signed numbers to be added (73H and 2AH).

- ADDWFC F, d, a adds the contents of WREG with the contents of the specified data register and the carry flag. The result is stored in WREG if d = 0 or in the data register if d = 1. The data register is in the access bank if a = 0 or specified by BSR if a = 1.

  As an example, consider ADDWFC 0x60, W.

  Prior to instruction execution, carry bit = 1, [0x60] = 03H, and [WREG] = 07H.

  After instruction execution, [0x60] = 03H (unchanged), [WREG] = 0BH, N = 0, OV = 0, Z = 0, DC = 0, and C = 0.

- SUBLW data8 subtracts [WREG] from 8-bit immediate data. The result is placed in WREG. As an example, consider SUBLW 07H.

  Prior to instruction execution, [WREG] = 03H.

  After Instruction execution, [WREG] = 07H – 03H = 04H.

  The flags are affected as follows:

  Using two's complement subtraction, 1111 111 ← Intermediate carries

```
            Immediate data = 07H  =  0000 0111
  Add two's complement of 03 = 1111 1101
                               ---------------------
            Final carry → 1  0000 0100 (04H)
```

The final carry is one's complemented after subtraction to reflect the correct borrow. Hence, C = 0. Also, N = 0 (most significant bit of the result is zero), OV = $C_f \oplus C_p = 1 \oplus 1 = 0$, Z = 0 (nonzero result), and DC = 1. Note that, in the above, the final carry is 1, indicating a borrow while performing the operation (07H - 03H). The correct result should have been 04H without any borrow. But, in the above, the result is 04H with a borrow. When two's complement subtraction is performed analytically using pencil and paper, the borrow is always one's complement of the true borrow. Hence, the PIC18F complements the carry to reflect the true borrow.

- The SUBWF F, d, a instruction subtracts the contents of WREG from the contents of the specified data register. The result is stored in WREG if d = 0 or in the data register if d = 1. The data register is in access bank if a = 0 or specified by BSR if a = 1. An example is SUBWF 0x30, F. Flags are affected in the same way as in the SUBLW instruction. Note that the carry is one's complemented to reflect the correct borrow.
- The SUBWFB F, d, a instruction subtracts the contents of WREG and the carry from the contents of the specified data register. The result is stored in WREG if d = 0 or in the data register if d = 1. The data register is in access bank if a = 0 or specified by BSR if a = 1. An example is SUBWFB 0x50, F. Flags are affected in the same way as in the SUBLW instruction. Note that the carry is one's complemented to reflect the correct borrow.
- The SUBFWB F, d, a instruction subtracts the contents of the specified data register and the carry from the contents of the WREG. That is, the SUBWFB performs the following operation using two's complement: [dest] ← [F] - [WREG + carry]. The result is stored in WREG if d = 0 or in the data register if d = 1. The data register is in access bank if a = 0 or specified by BSR if a = 1 . An example is SUBFWB 0x70, W. Flags are affected in the same way as in the SUBLW instruction. Note that the carry is one's complemented to reflect the correct borrow.

Multiplication Instructions The PIC18F includes an 8 x 8 hardware multiplier as part of the ALU. The multiplier performs an unsigned operation and provides a 16-bit result that is stored in the product register pair PRODH:PRODL. Because of hardware implementation, the multiplier executes the multiplication operation in a single instruction cycle. This has the advantages of higher computational throughput and reduced code size for multiplication algorithms, and allows the PIC18F to be used in many applications previously reserved for digital signal processors (DSPs). The PIC18F includes two instructions for performing 8-bit x 8-bit unsigned multiplication providing 16-bit result in the product register PRODH:PRODL. None of the status flags are affected. Note that neither overflow nor carry is possible in this operation. A zero result is possible but not detected. The PIC18F provides neither signed multiplication instruction nor any division instruction (signed and unsigned). The PIC18F unsigned multiplication instructions are described in the following.

- The MULLW data8 instruction performs an unsigned multiplication between the 8-bit contents of WREG and 8-bit immediate data. The 16-bit result is placed in the PRODH:PRODL register pair. PRODH contains the high byte, and PRODL contains the low byte. The contents of WREG are unchanged. As an example, consider MULLW 0x03.

  Prior to instruction execution, [WREG] = 02H.

  After Instruction execution, [PRODH] = 00H, [PRODL] = 06H, and [WREG] = 02H = unchanged.

- The MULWF F, a instruction performs an unsigned multiplication between the 8-bit contents of WREG and 8-bit contents of the specified data register. The 16-bit result is placed in the PRODH:PRODL register pair. PRODH contains the high byte, and PRODL contains the low byte. The contents of the WREG and the data register are unchanged. The data register is in the access bank if a = 0 or specified by BSR if a = 1. An example is MULWF 0x50.

Negate Instruction The PIC18F negate instruction is illustrated by means of numerical examples in the following.

- The NEGF F, a instruction negates the contents of the specified data register using two's complement. The result is stored in the data register. The data register is in the access bank if a = 0 or specified by BSR if a = 1. An example is NEGF 0x70.

  Prior to instruction execution, [0x70] = 02H.
  After instruction execution, [0x70] = FEH = -2_{10}.

Decrement and Increment Instructions The PIC18F decrement and increment instructions are illustrated in the following by means of numerical examples.

- The DECF F, d, a decrements the contents of the specified data register by 1. The result is stored in WREG if d = 0 or in the data register if d = 1. The data register is in the access bank if a = 0 or specified by BSR if a = 1. All flags are affected. An example is DECF 0x50, F.

  Prior to instruction execution, [0x50] = 01H.

  After Instruction execution, [0x50] = 00H.

- The INCF F, d, a increments the contents of the specified data register by 1. The result is stored in WREG if d = 0 or in the data register if d = 1. The data register is in the access bank if a = 0 or specified by BSR if a = 1. An example is INCF 0x50, F.

Prior to instruction execution, [0x50] = FFH = -1_{10}.

After instruction execution, [0x50] = 00H.

BCD Adjust (BCD Correction) Instruction The PIC18F contains a BCD adjust instruction which will be illustrated in the following by means of a numerical example.

- The DAW instruction adjusts the 8-bit result in WREG after adding two packed BCD numbers using ADDLW or ADDWF or ADDWFC to provide the correct packed BCD result.

  If, after the addition, the low 4 bits of the result in WREG are greater than 9 (or if DC = 1), the DAW adds 6 to the low 4 bits of WREG. On the other hand, if the high 4 bits of the result in WREG are greater than 9 (or if C = 1), DAW adds 6 to the high 4 bits in WREG. Consider the following instruction sequence:

```
MOVLW   0x29   ; Move 29H into WREG
ADDLW   0x54   ; Add 29H with 54H and store the result in WREG
DAW            ; Decimal adjust WREG to provide the correct packed
               ;   BCD result
```

The details of the result obtained by the instruction sequence above are provided in the following:

```
[WREG] = 29H = 0010 1001 (Packed BCD 29, same as 29H)
   Add   54H = 0101 0100  (Packed BCD 54, same as 54H)
               ----------
      [WREG] = 0111 1101
                    0110   Add 6 (BCD correction by DAW since low 4 bits
               ---------   of the sum  in WREG are  greater than 9)
               1000 0011 = 83H correct packed BCD result since 29 + 54 = 83
```

Note that packed BCD is covered in Section 1.2.3 of Chapter 1.

Example 6.5 Write a PIC18F assembly language instruction sequence to implement the following C segment;

```
if (x<0)
y++;
else
y--;
```

Solution

```
        MOVF  X,F    ; Move [X] into [F]; MOVF affects N flag
        BNN   ELSE   ; Branch to ELSE if N = 0  (if [X] is positive)
        INCF  Y      ; Else, if N = 1 ( [X] is negative), increment [Y] by 1
        BRA   NEXT
ELSE    DECF  Y      ; Decrement [Y] by 1 if N = 0
NEXT    GOTO  NEXT   ; Halt
```

Example 6.6 Write a PIC18F assembly language instruction sequence to add four numbers 1, 2, 3, 4, and then store the result in a data register 0x40 as follows:
(a) without using a loop (b) using a loop

Solution

(a) without using a loop

```
SUM     EQU         0x40
        MOVLW       D'1'     ; Move 1 into WREG
        ADDLW       D'2'     ; Add 2 with [WREG] and store 3 in WREG
        ADDLW       D'3'     ; Add 3 with [WREG] and store 6 in WREG
        ADDLW       D'4'     ; Add 4 with [WREG] and store 10 in WREG
        MOVWF       SUM      ; Store [WREG] in SUM
```

(b) using a loop

```
SUM       EQU      0x40
COUNTER   EQU      0x50
          MOVLW    D'4'          ;  Move 4 into WREG
          MOVWF    COUNTER       ;  Initialize COUNTER with 4
          MOVLW    0             ;  Clear [WREG] to 0
LOOP      ADDWF    COUNTER, W    ;  Add [COUNTER] with [WREG]; store
                                    result in W
          DECF     COUNTER       ;  Decrement [COUNTER] by 1
          BNZ      LOOP          ;  Branch to loop if Z not equal to 0
          MOVWF    SUM           ;  Move result (10) from WREG into SUM
```

Example 6.7 Write a PIC18F instruction sequence to implement the following C statement:

c = a + b;

Assume data registers 0x50, 0x60, and 0x70 store a, b, and c, respectively.

Solution

```
A       EQU         0x50
B       EQU         0x60
C       EQU         0x70
        MOVF        A, W     ; Move [A] to [WREG]
        ADDWF       B, W     ; Add [B] with [WREG]; store result in WREG
        MOVWF       C        ; Store result in WREG in C
```

Example 6.8 Write a PIC18F instruction sequence to implement the following C statement:

c = 2a + b;

Assume data registers 0x24, 0x31, and 0x50 store a, b, and c, respectively.

Solution

```
A       EQU      0x24
B       EQU      0x31
C       EQU      0x50
        MOVF     A, W     ; Move [A] to [WREG]
        ADDWF    A, W     ; Add [A] with [WREG]; result 2 x [A] in WREG
        ADDWF    B, W     ; Add 2x[A] with [B]; store result in WREG
        MOVWF    C        ; Store result in WREG in C
```

Example 6.9 Write a PIC18F assembly language program at address 0x100 that implements the following C language program segment:

```
        sum = 0;
for (i = 0; i <= 9; i = i + 1)
sum = sum + a[i];
```

where sum is the address of the 8-bit result of addition. Assume sum as 0x50 and the address of the first element of the array a[0] is stored in data register 0x20, the second element a[1] in data register 0x21, and so on. Assume that the array is already stored in data memory. Also, assume addition of two consecutive will generate no carry.

Solution

Assume that register FSR0 holds the address of the first element of the array. The assembly language program is listed below:

```
        INCLUDE <P18F4321.INC>
        ORG      0x100
SUM      EQU     0x50           ; Initialize SUM to 0x50 for result
        LFSR     0, 0x0020      ; Point FS0 to a[0]
        CLRF     SUM            ; Clear [SUM] to zero
        MOVLW    D'10'          ; Move  WREG with 10
        MOVWF    0x70           ; Initialize 0x70 with loop count (10)
LOOP    MOVF     SUM, W         ; Move [SUM] into WREG
        ADDWF    POSTINC0, W    ; Add and store result in WREG
        MOVWF    SUM            ; Save [WREG] IN SUM
        DECF     0x70, F        ; Decrement counter 0x70 by 1
        BNZ      LOOP           ; Branch to LOOP if  Z not equal 0
        SLEEP                   ; Halt
         END
```

Example 6.10 Write a PIC18F assembly language program at address 0x100 to add two 16-bit numbers as follows:

```
         [0x51] [0x50]
    PLUS [0x61] [0x60]
      ----------------------
         [0x61] [0x60]
```

Assume data registers 0x50 and 0x51 contain low and high bytes of the first 16-bit numbers while data registers 0x60 and 0x61 contain the second 16-bit number. Also, assume that data are already loaded into the data registers.

Solution

```
        INCLUDE  <P18F4321.INC>
        ORG      0x100
        MOVF     0x50, W        ; Move low byte of first data into WREG
        ADDWF    0x60, F        ; Add low 8 bits; store result in 0x60
        MOVF     0x51, W        ; Move high byte of first data into WREG
        ADDWFC   0x61, F        ; Add high bytes with carry; store result in 0x61
        SLEEP                   ; Halt
        END
```

EXAMPLE 6.11 Write a PIC18F assembly language program at address 0x100 to add four 8-bit numbers stored in consecutive data registers from high to low addresses starting at data register 0x78. Store the 8-bit result in WREG. Assume that no carry is generated, due to the addition of two consecutive 8-bit numbers. Assume that data are already loaded into data memory addresses 0x75 through 0x78 with the first data byte at 0x78 and the last data byte at 0x75.

Solution

```
        INCLUDE <P18F4321.INC>
        ORG     0x100
        MOVLW   D'4'            ; Move WREG with 4
        MOVWF   0x50            ; Initialize 0x50 with loop count (4)
        MOVLW   0x00            ; Clear WREG to store sum
        LFSR    0, 0x0078       ; Initialize pointer FSR0 with 0x0078
START   ADDWF   POSTDEC0, W     ; Add two bytes and store sum in WREG
        DECF    0x50, F         ; Decrement counter 0x50 by 1
        BNZ     START           ; Branch to START if Z ≠ 0
        SLEEP                   ; Halt
        END
```

Example 6.12 Write a PIC18F assembly language program at 0x100 to subtract two 32-bit numbers as follows:

```
        [0x43] [0x42] [0x41] [0x40]
MINUS   [0x75] [0x74] [0x73] [0x72]
     ----------------------------------------
        [0x43] [0x42] [0x41] [0x40]
```

Assume data registers 0x40 through 0x43 contain the first 32-bit number while data registers 0x72 through 0x75 contain the second 32-bit number. Also, assume that data are already loaded into the data registers. Store 32-bit result in data registers 0x40 (lowest byte) through 0x43 (highest byte).

Solution

```
        INCLUDE<P18F4321.INC>
        ORG     0x100
        MOVLW   D'4'          ; Move  WREG with 4
        MOVWF   0x50          ; Initialize 0x50 with loop count (4)
        LFSR    0, 0x0072     ; Initialize pointer FSR0 with 0x0072
        LFSR    1, 0x0040     ; Initialize pointer FSR1 with 0x0040
        ADDLW   0x00          ; Clear carry flag
START   MOVF    POSTINC0, W   ; Move byte into WREG and update pointer
        SUBWFB  POSTINC1, F   ; Subtract [WREG] and carry  from byte; store result
                              ; in data register
        DECF    0x50, F       ; Decrement counter 0x50 by 1
        BNZ     START         ; Branch to START  if  Z ≠ 0
        SLEEP                 ; Halt
        END
```

Example 6.13 Write a PIC18F assembly language program at address 0x100 to compute ($X^2 + Y^2$) where X and Y are two 8-bit unsigned numbers stored in data registers 0x40 and 0x41, respectively. Store the 16-bit result in data registers 0x61 (high byte) and 0x60 (low byte). Assume X and Y are already loaded into data registers 0x40 and 0x41.

Solution

```
        INCLUDE   <P18F4321.INC>
        ORG       0x100
        MOVF      0x40, W       ;  Move X into WREG
        MULWF     0x40          ;  Multiply X by X; result in PRODH:PRODL
        MOVFF     PRODL, 0x50   ;  Save low byte of result in data reg 0x50
        MOVFF     PRODH, 0x51   ;  Save high byte of result in data reg 0x51
        MOVF      0x41, W       ;  Move Y into WREG
        MULWF     0x41          ;  Multiply Y by Y; result in PRODH:PRODL
        MOVFF     PRODL, 0x60   ;  Save low byte of result in data reg 0x60
        MOVFF     PRODH, 0x61   ;  Save high byte of result in data reg 0x61
        MOVF      0x50, W       ;  Move low byte of  X times X  into WREG
        ADDWF     0x60, F       ;  Add and store low byte of result in 0x60
        MOVF      0x51, W       ;  Move high byte of  X times X  into WREG
        ADDWFC    0x61, F       ;  Add high bytes with carry. Store result in 0x61
        SLEEP                   ;  Halt
        END
```

Example 6.14 Write a PIC18F assembly language program at address 0x100 to add two packed BCD bytes stored in data registers 0x20 and 0x21. Store the correct packed BCD result in data register 0x40. Load packed BCD bytes 0x72 and 0x45 into data registers 0x20 and 0x21, respectively, using PIC18F instructions. Note that data are arbitrarily chosen.

Solution

```
INCLUDE     <P18F4321.INC>
ORG         0x100
MOVLW       0x72          ; Load first packed BCD data
MOVWF       0x20          ; into 0x20
MOVLW       0x45          ; Load second packed BCD data
MOVWF       0x21          ; into 0x21
MOVF        0x20, W       ; Move 72H into WREG
ADDWF       0x21, W       ; Add and store binary result in WREG
DAW                       ; Convert WREG to correct packed BCD byte
SLEEP                     ; Halt
END
```

6.3.3 Logic Instructions

The PIC18F logic instructions include logic AND, NOT (one's complement) OR, and exclusive-OR operations. Table 6.7 lists PIC18F logic instructions. Next, we explain the logic instructions using the access bank and specifying F or W in place of 'd'.

TABLE 6.7 PIC18F logic instructions

Instruction	Comment
ANDLW data8	[WREG] AND 8-bit data → [WREG]
ANDWF F, d, a	[WREG] AND [F] → destination; see note for 'a' and 'destination'.
COMF F, d, a	NOT [F] → destination; see note for 'a' and 'destination'.
IORLW data8	[WREG] OR 8-bit data → [WREG]; Performs Inclusive OR or simply OR operation.
IORWF F, d, a	[WREG] AND [F] → destination; see note for 'a' and 'destination'. Performs inclusive OR or simply OR operation.
XORLW data8	[WREG] ⊕ [8-bit data] → [WREG]
XORWF F, d, a	[WREG] ⊕ [F] → [destination] ; see note for 'a' and 'destination'.

- All instructions in the above are executed in one cycle.
- The size of each instruction is one word.
- All instructions affect N and Z flags; other flags are not affected.
- a = 0 means that the data register is located in the access bank while a = 1 means that the contents of BSR specify the address of the bank.
- For destination, d = 0 means that the destination is WREG while d = 1 means that the destination is file register F.

Let us now explain these instructions.

- The ANDLW data8 instruction ANDs the contents of WREG with the 8-bit literal (immediate data, data8). The result is placed in WREG. As an example, consider ANDLW 0x8F.

```
Prior to instruction execution: [WREG] = 0x72
After instruction execution:
        [WREG] = 0x72 =    0111 0010
           AND  0x8F =    1000 1111
                          ---------------
                 [WREG] = 0000 0010
```

N and Z flags are affected. Z = 0 (result is nonzero) and N = 0 (most significant bit of the result is 0). The status flags are affected in the same way after execution of other logic instructions such as OR, XOR, and NOT.

- The ANDWF F, d, a instruction ANDs the contents of WREG with register 'F'. If 'd' is '0', the result is stored in W. If 'd' is '1', the result is stored back in register 'F'. If 'a' is '0', the access bank is selected. If 'a' is '1', the BSR is used to select the bank .

As an example, consider ANDWF 0x60, F.

Prior to instruction execution, [0x60] = 0xFF, and [WREG] = 0x01.
After Instruction execution, [0x60] = 0x01, [WREG] = 0x01 (unchanged), Z = 0, and N = 0.

The AND instruction can be used to perform masking operation. If the bit value in a particular bit position is desired in a data byte, the data can be logically ANDed with appropriate data to accomplish this. For example, the bit value at bit 2 of an 8-bit number 0100 1Y10 (where the unknown bit value of Y is to be determined) can be obtained as follows:

```
        0 1 0 0 1 Y 1 0 -- 8-bit number
AND     0 0 0 0 0 1 0 0 -- Masking data
        ---------------
        0 0 0 0 0 Y 0 0 -- Result
```

If the bit value Y at bit 2 is 1, then the result is nonzero (flag Z = 0); otherwise, the result is zero (Z = 1). The Z flag can be tested using typical conditional branch instructions such as BZ (branch if Z = 1) or BNZ (branch if Z = 0) to determine whether Y is 0 or 1. This is called masking operation. The AND instruction can also be used to determine whether a binary number is ODD or EVEN by checking the Least Significant bit (LSB) of the number (LSB = 0 for even and LSB = 1 for odd).

- The COMF F, d, a instruction complements (one's) the contents of register 'F'. If 'd' is '0', the result is stored in WREG. If 'd' is '1', the result is stored back in register 'F'. If 'a' is '0', the access bank is selected. If 'a' is '1', the BSR is used to select the bank.

  For example, consider COMF 0x50, W.

  Prior to instruction execution, [0x50] = 0x01, and [WREG] = 0x57.

  After Instruction execution, [WREG] = 0xFE, [0x50] = 0x01 (unchanged), Z = 0, and N = 1; no other flags are affected.

- The IORLW data8 instruction ORes the contents of WREG with the 8-bit literal (immediate data, data8). The result is placed in WREG. As an example, consider IORLW 0x7F.

  Prior to instruction execution, [WREG] = 0x01.

  After Instruction execution, [WREG] = 0x7F, Z = 0, and N = 0; other flags are not affected.

  The OR instruction can typically be used to insert a 1 in a particular bit position of a binary number without changing the values of the other bits. For example, a 1 can be inserted using the OR instruction at bit number 3 of the 8-bit binary number 0 1 1 1 0 0 1 1 without changing the values of the other bits as follows:

```
      0 1 1 1 0 0 1 1 -- 8-bit number
OR    0 0 0 0 1 0 0 0 -- Data for inserting a 1 at bit number 3
      ---------------
      0 1 1 1 1 0 1 1 -- Result
```

- The IORWF F, d, a instruction ORes [WREG] with register 'F'. If 'd' is '0', the result is placed in WREG. If 'd' is '1',the result is placed back in register 'F'. If 'a' is '0', the access bank is selected. If 'a' is '1', the BSR is used to select the bank.

  As an example, consider IORWF 0x50, W.

  Prior to instruction execution, [WREG] = 0xA2, and [0x50] = 0x5D

  After Instruction execution, [WREG] = 0xFF, [0x50] = 0x5D (unchanged), Z = 0, and N = 1; other flags are not affected.

- The XORLW data8 instruction exclusive-ORes the contents of WREG with 8-bit literal (immediate data, data8). The result is placed in WREG.

  As an example, consider XORLW 0x02.

  Prior to instruction execution, [WREG] = 0x42.

  After Instruction execution, [WREG] = 0x40, Z = 0, and N = 0 ; no other flags are affected.

  The exclusive-OR instruction can be used to find the one's complement of a binary number by XORing the number with all 1's as follows:

```
        0 1 0 1 1 1 0 0 - -  8-bit number
XOR     1 1 1 1 1 1 1 1 - -  Data
        --------------------------
        1 0 10 0 0 1 1  - - Result ( one's complement of the 8-bit number
                             0 1 0 1 1 1 0 0 )
```

- The XORWF F, d, a instruction exclusive-ORes the contents of WREG with register 'F'. If 'd' is '0', the result is stored in WREG. If 'd' is '1', the result is stored back in the register 'F'. If 'a' is '0', the access bank is selected. If 'a' is '1', the BSR is used to select the bank .

  As an example, consider XORWF 0x42, W.

  Prior to instruction execution, [WREG] = 0xFF, and [0x42] = 0xFF.

  After Instruction execution, [WREG] = 0x00, Z = 1, and N = 0; no other flags are affected.

6.3.4 Rotate Instructions

The PIC18F rotate instructions are listed in Table 6.8. Next, we explain the rotate instructions using the access bank and specifying F or W in place of 'd'.

TABLE 6.8 PIC18F rotate instructions

Instruction	Operation
RLCF F, d, a	Rotate register F one bit to the left through carry. See notes for 'd' and 'a'.
RLNCF F, d, a 	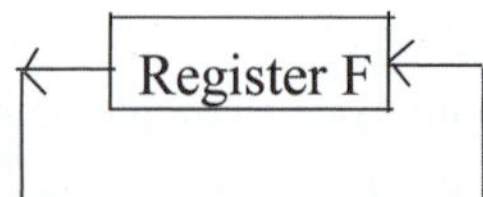 Rotate register F one bit to the left without carry. See notes for 'd' and 'a'.
RRCF F, d, a	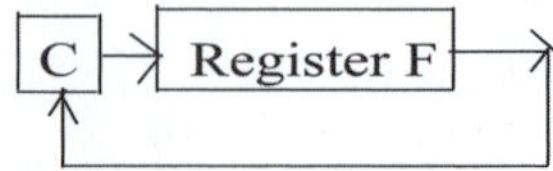 Rotate register F one bit to the right through carry. See notes for 'd' and 'a'.
RRNCF F, d, a	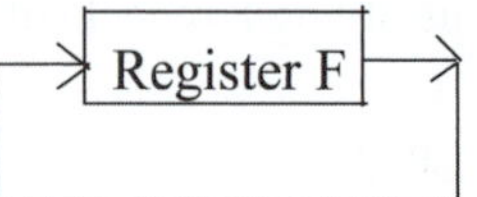 Rotate register F one bit to the right without carry. See notes for 'd' and 'a'.

- All instructions in the above are executed in one cycle.
- The size of each instruction is one word.
- a = 0 means that the data register is located in the access bank while a = 1 means that the contents of BSR specify the address of the bank.
- For destination, d = 0 means that the destination is WREG while d = 1 means that the destination is file register F.
- RLCF and RRCF affect N, Z, and C flags according to the result, whereas RLNCF and RRNCF affect N and Z flags based on the result.
- Note that the PIC 18F does not have any shift instructions.

Let us now explain the instructions in Table 6.8 using numerical examples.

- The RLCF F, d, a instruction rotates the contents of register 'F' one bit to the left through the carry flag. If 'd' is '0', the result is placed in WREG. If 'd' is '1', the result is stored back in register 'F'. If 'a' is '0', the access bank is selected. If 'a' is '1', the BSR is used to select the bank.

  As an example, consider RLCF 0x40, W.

  Prior to instruction execution, [WREG] = 0xFF, [0x40] = 0xAF, and C = 0.

  After instruction execution, [WREG] = 0x5E, C = 1, Z = 0 (result in WREG after rotating is nonzero), and N = 0 (most significant bit of result 0x5E is 0); no other flags are affected.

- The RLNCF F, d, a instruction rotates the contents of register 'F' one bit to the left. If 'd' is '0', the result is placed in WREG. If 'd' is '1', the result is stored back in register 'F'. If 'a' is '0', the access bank is selected. If 'a' is '1', the BSR is used to select the bank.

  As an example, consider RLNCF 0x70, F.

  Prior to instruction execution, [WREG] = 0x89, and [0x70] = 0x32.

  After instruction execution, [0x70] = 0x64, [WREG] = 0x89 (unchanged), Z = 0 (result in 0x70 after rotating is nonzero), and N = 0 (most significant bit of result 0x64 is 0); no other flags are affected.

- The RLCF instruction can be used after clearing the carry flag to 0, to multiply an unsigned number by 2^n by shifting the number *n* times to the left using a loop as long as a '1' is not shifted out of the most significant bit. In the above example, after shifting [0x70] once, the contents 0x32 of data register 0x70 are multiplied by 2. Hence, [0x70] = 0x64 after shifting.

- The RRCF F, d, a instruction rotates the contents of register 'F' one bit to the right through the carry flag. If 'd' is '0', the result is placed in WREG. If 'd' is '1', the result is stored back in register 'F'. If 'a' is '0', the access bank is selected. If 'a' is '1', the BSR is used to select the bank.

  As an example, consider RRCF 0x30, F.

  Prior to instruction execution, [WREG] = 0x91, [0x30] = 0x27, and C = 0.

  After Instruction execution, [0x30] = 0x13, [WREG] = 0x91 (unchanged), C = 1, Z = 0 (result 0x13 is nonzero), and N = 0 (most significant bit of result 0x13 is 0) ; no other flags are affected.

- The RRNCF F, d, a instruction rotates the contents of register 'F' one bit to the right. If 'd' is '0', the result is placed in WREG. If 'd' is '1', the result is stored back in register 'F'. If 'a' is '0', the access bank is selected. If 'a' is '1', the BSR is used to select the bank.

  As an example, consider RRNCF 0x60, F.

  Prior to instruction execution, [WREG] = 0xB3, and [0x60] = 0x28

  After instruction execution, [0x60]= 0x14, [WREG] = 0xB3 (unchanged), Z = 0 (result 0x14 after rotating is nonzero), and N = 0 (most significant bit of result 0x14 is 0); no other flags are affected.

- The RRCF instruction can be used after clearing the carry flag to 0, to divide an unsigned number by 2^n by shifting the number n times to the right using a loop as long as a '1' is not shifted out of the least significant bit. This means that the remainder is discarded. Since the PIC18F does not have unsigned division instruction, the RRCF can be used for this purpose. In the above example, after shifting [0x60] once, the contents 0x28 of data register 0x60 are divided by 2. Hence, [0x60] = 0x14 after shifting.

Example 6.15 Write a PIC18F logic instruction to convert a 4-bit unsigned number in low 4 bits of WREG into an 8-bit unsigned number in WREG. Assume the 4-bit unsigned number is already loaded into WREG.

Solution

```
ANDLW 0x0F
```

Note that the ANDLW 0x0F instruction logically ANDs 8-bit data in WREG with 0x0F. This operation will clear the upper 4 bits to zero, and retain the lower 4 bits. Thus, the 4-bit unsigned number will be converted to an 8-bit unsigned number.

Example 6.16 Write a PIC18F assembly language program at address 0x100 to convert a 4-bit signed number stored in the low 4 bits of WREG into an 8-bit signed number in WREG. That is, if the 4-bit signed number is positive, zero-extend to 8 bits. On the other hand, if the 4-bit signed number is negative, sign-extend to 8 bits. Assume the 4-bit signed number is already loaded into WREG. Store result in data register 0x50. Do not use any ROTATE instructions.

(a) Flowchart the problem.

(b) Convert the flowchart to PIC18F assembly language program starting at address 0x100.

Solution

(a) The flowchart is provided below:

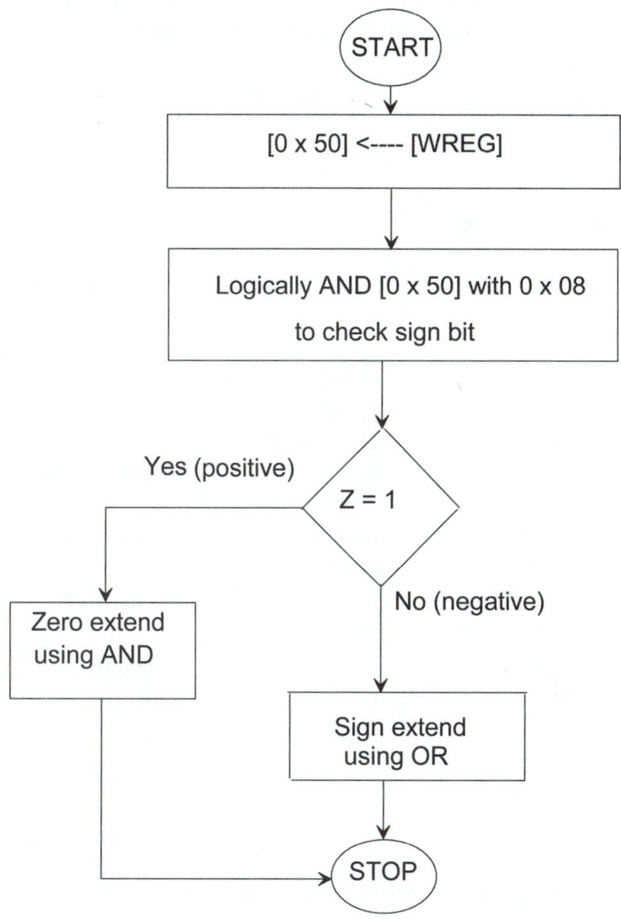

(b) The PIC18F assembly language program is provided below:

```
         INCLUDE   <P18F4321.INC>
         ORG     0x100
         MOVWF   0x50       ; Copy [WREG] into data register 0x50
         ANDLW   0x08       ; Check whether bit 3 it is 0 (positive) or 1 (negative)
         BZ      POSITIVE   ; If Z = 1, the number is positive
         MOVF    0x50, W    ; Else, the number is negative. Move [0x50] into W
         IORLW   0xF0       ; Sign extend by inserting 1's in upper 4 bits of W
         SLEEP              ; Halt. Result in WREG
POSITIVE MOVF    0x50, W    ; The number is positive. Move [0x50] into WREG
         ANDLW   0x0F       ; The number is positive. Hence, zero extend upper 4
                            ; bits
         SLEEP
         END
```

Example 6.17 Write a PIC18F assembly language program at address 0x100 to multiply an 8-bit unsigned number in data register 0x50 by 16. Store the 8-bit result in WREG. Do not use any multiplication instructions. Use ROTATE instruction. Assume that a '1' is not shifted out of the most significant bit each time after rotating to the left. Also, assume that the 8-bit unsigned number is already loaded into data register 0x40.
Write the program
(a) without using a loop (b) using a loop

Solution

(a) without using a loop

The following program will multiply [0x50] by 16 by shifting [0x50] four times to the left:

```
            INCLUDE     <P18F4321.INC>
            ORG         0x100
            BCF         STATUS, C
            RLCF        0x50, F         ; Rotate [0x50] once to left
            BCF         STATUS, C
            RLCF        0x50, F         ; Rotate [0x50] once to left
            BCF         STATUS, C
            RLCF        0x50, F         ; Rotate [0x50] once to left
            BCF         STATUS, C
            RLCF        0x50, F         ; Rotate [0x50] once to left
            MOVF        0x50, W         ; Save result in WREG
FOREVER     GOTO        FOREVER         ; Stop
            END
```

(b) using a loop

The following program will multiply [0x50] by 16 by shifting [0x50] four times to the left in a loop:

```
            INCLUDE     <P18F4321.INC>
            ORG         0x100
COUNTER     EQU         0x70
            MOVLW       4               ; Initialize COUNTER with 4
            MOVWF       COUNTER
BACK        BCF         STATUS, C
            RLCF        0x50, F         ; Rotate [0x50] four times
            DECF        COUNTER, F      ; to left to multiply [0x50] by 16
            BNZ         BACK            ; Branch to BACK if Z = 0
            MOVF        0x50, W         ; Move result to WREG
FOREVER     GOTO        FOREVER         ; Stop
            END
```

Example 6.18 Write a PIC18F assembly language program at address 0x100 that will multiply an 8-bit unsigned number in data register 0x50 by 4 to provide an 8-bit product, and then perform the following operations on the contents of data register 0x50:

- Set bits 0 and 3 to one without changing other bits in data register 0x50.
- Clear bit 5 to zero without changing other bits in data register 0x50.
- One's complement bit 7 without changing other bits in data register 0x50.

Use only "logic", and "rotate" instructions. Do not use any multiplication instruction. Assume data are already in data register 0x50. Store result in WREG. Assume that a '1' is not shifted out of the most significant bit each time after rotating to the left.

Solution

```
        INCLUDE <P18F4321.INC>
        ORG      0x100
        BCF      STATUS, C
        RLCF     0x50, W    ;  Unsigned multiply [0x50] by 2
        BCF      STATUS, C
        RLCF     0x50, W    ;  Unsigned multiply [0x50] by 4; result in W
        IORLW    0x09       ;  Set bits 0 and 3 in WREG to one's
        ANDLW    0xDF       ;  Clear bit 5  in WREG to zero
        XORLW    0x80, W    ;  One's complement  bit 7 in WREG
FINISH  GOTO     FINISH     ;  Stop
        END
```

As mentioned before, FINISH GOTO FINISH (unconditionally jumping to the same location) and the instruction SLEEP are equivalent to HALT instruction in other processors. Either can be used in the PIC18F as HALT in the assembly language program.

Example 6.19 Write a PIC18F assembly language program at address 0x100 to check whether an 8-bit signed number (x) is positive or negative. If the number is positive, then compute 16-bit value $y1 = x^2$ and store result in PRODH:PRODL. If the number is negative, then compute the 8-bit value $y2 = 2x$. Store result in WREG. Do not use any logic instructions. Assume that the 8-bit number x is already loaded in data register 0x60.

Solution

```
          INCLUDE  <P18F4321.INC>
          ORG      0x100
          MOVF     0x60, W        ; Move  x into WREG
          MOVFF    0x60, 0x70     ; Save x in 0x70
          RLCF     0x60, F        ; Rotate sign bit to carry to check whether 0 or 1
          BN       NEGATIVE       ; Branch if N = 1
          MULWF    0x60           ; Compute y1 and store in PRODH:PRODL
          GOTO     FINISH         ; Jump to FINISH
NEGATIVE  ADDWF    0x70, W        ; Compute y2 by adding x to itself
FINISH    GOTO     FINISH
          END
```

TABLE 6.9 Bit manipulation instructions

Instruction	Comment
BCF F, b, a	Clear bit number 'b' to 0 in file register F
BSF F, b, a	Set bit number 'b' to 1 in file register F
BTG F, b, a	Toggle (one's complement) bit number 'b' in file register F

- All instructions in the above are executed in one cycle.
- The size of each instruction is one word.
- No flags are affected.
- 'b' can be from 0 to 7.
- a = 0 means that the data register is located in the access bank while a = 1 means that the contents of BSR specify the address of the bank.

6.3.5 Bit Manipulation Instructions

The PIC18F has three bit manipulation instructions, and these are listed in Table 6.9. Next, we explain the bit manipulation instructions using the access bank and specifying F or W in place of 'd'.

- The BCF F, b, a instruction clears the specified bit 'b' in register 'F' to zero. If 'a' is '0', the access bank is selected. If 'a' is '1', the BSR is used to select the bank. As an example, consider BCF 0x50, 2.

  Prior to instruction execution, [0x50] = 0x36 = 00110110_2.

  After Instruction execution, bit 2 is cleared to 0. Hence, [0x50] = 00110010_2 = 0x32.

- The BCF instruction can be used to clear the carry flag to 0. For example, BCF STATUS, C will clear the carry flag in the status register to 0.
- The BSF F, b, a instruction sets the specified bit 'b' in register 'F' to one. If 'a' is '0', the access bank is selected. If 'a' is '1', the BSR is used to select the bank . As an example, consider BSF 0x30, 5.

  Prior to instruction execution, [0x30] = 0x0F = 00001111_2.

  After Instruction execution, bit 5 is set to 1. Hence, [0x30] = 00101111_2 = 0x2F.

- The BSF instruction can be used to set the carry flag to 1. For example, BSF STATUS, C will set the carry flag in the status register to 1.
- The BTG F, b, a instruction one's complements (toggles) the specified bit 'b' in register 'F'. If 'a' is '0', the access bank is selected. If 'a' is '1', the BSR is used to select the bank. As an example, consider BTG 0x40, 2.

  Prior to instruction execution, [0x40] = 0x16 = 00010110_2.

  After instruction execution, bit 2 is one's complemented from 1 to 0. Hence,

  [0x40] = 00010010_2 = 0x12.

  Note that the BTG instruction can be used to toggle a specific bit in an I/O port. For example, BTG PORTB, 1 will toggle bit 1 of Port B. This may be useful sometimes in some I/O applications.

Example 6.20 Write a PIC18F assembly language program at address 0x100 that will multiply an 8-bit unsigned number in data register 0x50 by 4 to provide an 8-bit product, and then perform the following operations on the contents of data register 0x50 :

- Set bits 0 and 3 to one without changing other bits in data register 0x50.
- Clear bit 5 to zero without changing other bits in data register 0x50.
- One's complement bit 7 without changing other bits in data register 0x50.

Use only "rotate" and "bit manipulation" instructions. Do not use any multiplication instruction. Assume data are already in data register 0x50. Store result in WREG. Assume that a '1' is not shifted out of the most significant bit each time after rotating to the left.

This example is a repeat of Example 6.18, but uses "bit manipulation instructions" instead of "logic instructions."

Solution

```
            INCLUDE     <P18F4321.INC>
            ORG         0x100
            BCF         STATUS, C
            RLCF        0x50, F  ; Unsigned multiply [0x50] by 2
            BCF         STATUS, C
            RLCF        0x50, F  ; Unsigned multiply [0x50] by 4; result in F
            BSF         0x50, 0  ; Set bit 0 in [0x50] to one
            BSF         0x50, 3  ; Set bit 3 in [0x50] to one
            BCF         0x50, 5  ; Clear bit 5 in [0x50] to zero
            BTG         0x50,7   ; One's complement bit 7 in D0
            MOVF        0x50, W ; Store result in WREG
FINISH      GOTO        FINISH ; Stop
            END
```

Example 6.21 Write a PIC18F assembly language program at address 0x100 that will perform $5 \times X + 6 \times Y + [Y/2] \rightarrow$ [0x71][0x70], where X is an unsigned 8-bit number stored in data register 0x40 and Y is a 4-bit unsigned number stored in the upper 4 bits of data register 0x50. Discard the remainder of $Y/2$. Save the 16-bit result in 0x71 (upper byte) and in 0x70 (lower byte).

(a) Flowchart the problem.

(b) Convert the flowchart to PIC18F assembly language program starting at address 0x100.

Solution

(a) The flowchart is provided below:

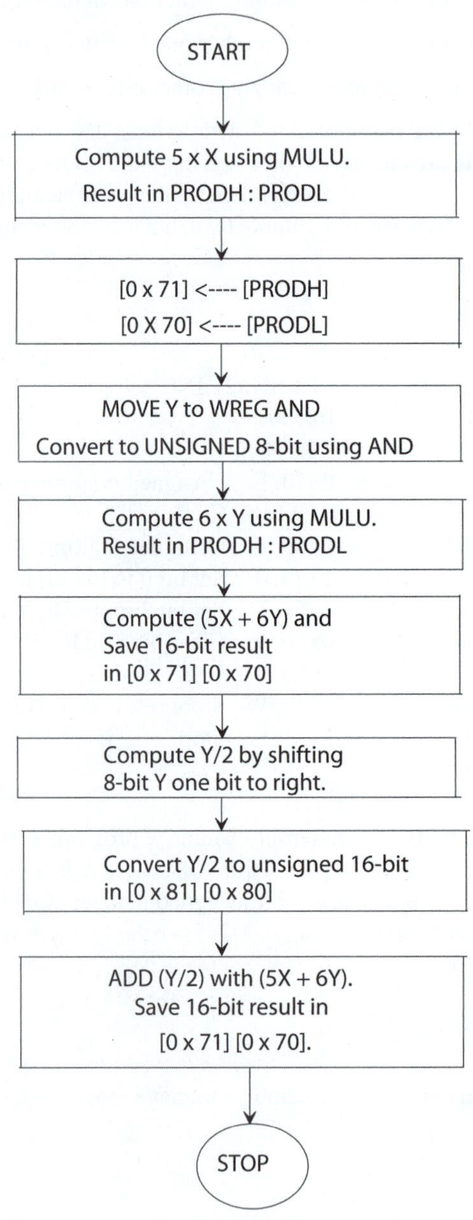

(b) The PIC18F assembly language program is provided below:

```
INCLUDE <P18F4321.INC>
ORG      0x100
MOVF     0x40, W      ;   MOVE X TO WREG
MULLW    5            ;   COMPUTE UNSIGNED 16-BIT 5xX IN
```

```
                                ; PRODH:PRODL
        MOVFF   PRODH, 0x71     ; SAVE UPPER BYTE  OF 5xX  IN 0x71
        MOVFF   PRODL, 0x70     ; SAVE LOWER BYTE OF 5xX  IN 0x70
        SWAPF   0x50, W         ; MOVE Y TO LOW 4 BITS IN  WREG
        ANDLW   0x0F            ; CONVERT Y TO UNSIGNED 8-BIT IN WREG
        MOVWF   0x80            ; SAVE Y  FROM WREG TO 0x80
        MULLW   6               ; COMPUTE UNSIGNED 16-bit  6 x Y IN
                                ; PRODH:PRODL
        MOVFF   PRODL, W        ; MOVE  PRODL  INTO WREG
        ADDWF   0x70, F         ; ADD  LOW BYTES OF 5*X WITH 6*Y, SAVE
                                ; IN 0x70
        MOVFF   PRODH, W        ; MOVE  PRODH  INTO WREG
        ADDWFC  0x71, F         ; ADD HIGH BYTES OF 5*X WITH 6*Y
                                ; WITH CARRY, AND SAVE IN 0x71
        BCF     STATUS, C
        RRCF    0x80, F         ; COMPUTE  Y/2 , 8-BIT RESULT IN  0x80
        CLRF    0x81            ; CONVERT  Y/2 TO UNSIGNED 16-BIT IN
                                ; [0x81][0x80]
        MOVFF   0x80, W         ; MOVE LOW BYTE OF  Y/2 INTO WREG
        ADDWF   0x70, F         ; PERFORM 5 × X + 6 × Y + [Y/2] FOR LOW
                                ; BYTES, RESULT IN 0x70
        MOVFF   0x81, W         ; MOVE HIGH BYTE OF  Y/2 INTO WREG
        ADDWFC  0x71, F         ; PERFORM 5 × X + 6 × Y + [Y/2] FOR HIGH
                                ; BYTES PLUS CARRY , RESULT IN 0x71
FINISH  GOTO    FINISH          ; HALT
        END
```

QUESTIONS AND PROBLEMS

6.1 Write a PIC18F instruction sequence to implement the following C statement: c = a + b; assume data registers 0x30, 0x40, and 0x50 store a, b, and c, respectively.

6.2 Write a PIC18F instruction sequence to implement the following C statement: e = a + b+c-d; assume data registers 0x30, 0x40, 0x50, 0x60, and 0x70 store a, b, c, d, and e, respectively.

6.3 (a) Find the contents of data register 0x20 after execution of the MOVWF 0x20. Assume [WREG] = FFH prior to execution of this PIC18F MOVWF instruction.

(b) If [FSR0] = 0x0070, [FSR1] = 0x0060, [FSR2] = 0x0024,{WREG] = 0x2A, [0x24] = 0x0B, [0x60] = 0x43, and [0x70] = 0x57, what happens after execution of the PIC18F instruction: MOVWF INDF1?

6.4 Determine the contents of registers / memory locations affected by each of the following PIC18F instructions:

(a) MOVWF POSTDEC2

Assume the following data prior to execution of this MOVWF:

[FSR0] = 0x0050 [0x50] = 0x51
[FSR1] = 0x0025 [0x25] = $52
[FSR2] = 0x0075 [0x75] = 0x7F
[WREG] = 0xFE

(b) MOVFF PLUSW1, 0x70

Assume the following data prior to execution of this MOVFF:
[FSR0] = 0x0020, [FSR1] = 0x0025, [FSR2] = 0x0028, [WREG] = 0x05, [0x20] = 0x05, [0x25] = 0x0A, [0x28] = 0xC2, [0x25] = 0x07, [0x2A] = 0x09, [0x2D] = 0x19

6.5 Find the machine code for the following PIC18F instruction sequence:

```
MOVLW    0x00
MOVWF    TRISB
NOP
MOVLW    0xAA
MOVWF    PORTB
SLEEP
```

6.6 Rewrite the following PIC18F instruction sequence with fewer instructions:

```
MOVLW    0x00
MOVWF    0x20
MOVLW    0xFF
MOVWF    0x22
```

6.7 Write PIC18F instruction sequence that is equivalent to the following C code: P- = Q;

6.8 Write PIC18F instruction sequence that is equivalent to the following C code:

```
if (p <= q)
   p = p + 5;
else
   p = 10;
```

6.9 What is the content of WREG after execution of the following PIC18F instruction sequence?

```
MOVLW  0x33
ADDLW  0x77
DAW
```

6.10 Find two ways to clear [WREG] to 0 using
(a) a single PIC18F instruction
(b) two PIC18F instructions

6.11 Using a single PIC18F instruction, clear the carry flag without changing the contents of any data registers, WREG, or other status flags.

6.12 Write the machine code for the following PIC18F instruction sequence:

```
       ORG   0x200
HERE   BRA   HERE
```

6.13 Write a PIC18F assembly language program at address 0x100 to add two 8-bit numbers (N1 and N2). Data register 0x20 contains N1. The low four bits of N2 are stored in the upper nibble of data register 0x21 while the high four bits of N2 are stored in the lower nibble of data register 0x21. Store result in data register 0x30.

6.14 Write a PIC18F assembly language program at address 0x100 to add two 24-bit data items in memory, as shown in Figure P6.14. Store the result pointed to by 0x50. The operation with sample data is given by

```
        F1 91 B5
PLUS    07 A2 04
--------------------------------
        F8 03 B9
```

Assume that the data pointers and the data are already initialized.

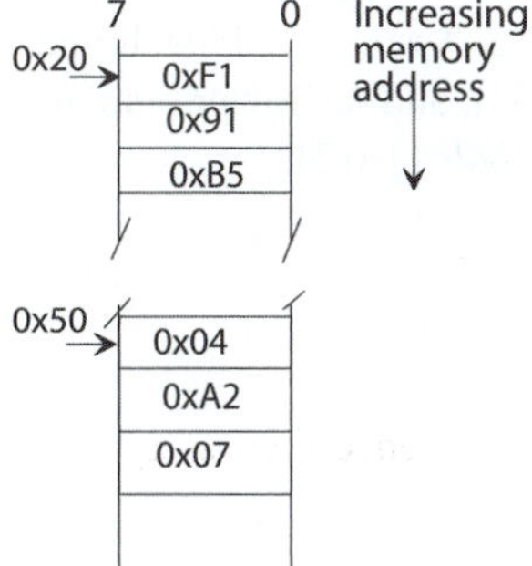

FIGURE P6.14

6.15 Write a PIC18F assembly language program to subtract two 16-bit numbers as follows:

```
         [0x40] [0x50]
MINUS    [0x20] [0x25]
       ---------------------
         [0x40] [0x50]
```

6.16 Write a PIC18F assembly language program at address 0x 0x100 to compute $\sum_{i=1}^{N} X_i Y_i$, where the X_i's and Y_i's are unsigned 8-bit numbers and $N = 10$. Store the 16-bit result in data registers 0x40 (low byte) and 0x41 (high byte). Assume that the X_i's are already stored in data registers 0x50 through 0x59 while Y_i's are already stored in data registers 0x70 through 0x79.

6.17 Write a PIC18F assembly program at address 0x50 to compute the following: $I = 6 \times J + (K/8)$, where the data registers 0x32 and 0x33 contain the 8-bit unsigned integers J and K. Store the 16-bit result into data registers 0x50:0x51. Discard the remainder of $K/8$.

6.18 Write a PIC18F assembly language program at address 0x50 that will check whether the 8-bit unsigned number in WREG is odd or even. If the number is even, the program will clear all bits in data register 0x40 to 0's. On the other hand, if the number is odd, the program will change all bits in data register 0x40 to 1's. Assume that the 8-bit numbers are already loaded into WREG.

6.19 Write a PIC18F assembly language program at address 0x70 to insert a '1' at bit 2 of WREG without changing the other bits if WREG contains a negative number. On the other hand, insert a '0' at bit 2 of WREG without changing the other bits if WREG contains a positive number. Assume that the 8-bit signed number is already loaded into WREG.

6.20 Write a PIC18F assembly language program at address 0x100 to check the parity of an 8-bit number in data register 0x70. If the parity is even, the program will store EE (hex) in data register 0x50. On the other hand, if the parity is odd, the program will store DD (hex) in data register 0x50.

6.21 Write a PIC18F assembly language at address 0x100 program to divide an unsigned 16-bit number by 2. Assume that the higher byte of the 16-bit number is stored in data register [0x20], and the lower byte in data register [0x21]. Discard remainder. Store result in [0x20] [0x21].

7

ASSEMBLY LANGUAGE PROGRAMMING WITH THE PIC18F: PART 2

In this chapter we provide the second part of the PIC18F's instruction set. Topics include jump/branch, test/compare/skip, table read/write, subroutine, and system control instructions. Several assembly language programming examples using most of these instructions are provided. Finally, delay routines using PIC18F's instructions are covered.

7.1 PIC18F Jump/Branch Instructions

These instructions include jumps and branches, as listed in Table 7.1.

There is one unconditional JUMP such as GOTO k instruction, where 'k' is an address. Hence, the GOTO instruction uses the "direct" or "absolute" addressing mode. There is also an unconditional branch such as BRA d instruction, where 'd' is the signed 11-bit offset. Hence, this instruction uses the "relative" addressing mode.

There are eight conditional branch conditions. They use the "relative" addressing mode. For example, consider Bcc d instruction where 'd' is an 8-bit signed offset. Note that the cc (condition code) in Bcc can be replaced by eight conditions providing eight instructions: BC, BNC, BZ, BNZ, BN, BNN, BOV, and BNOV. It should be mentioned

TABLE 7.1 PIC18F jump/branch instructions

Instruction	Operation
GOTO k	Unconditionally jumps to an address defined by the k. Uses direct or absolute mode.
Bcc d	If the condition cc is true, then (PC+2) + 2 x d → PC. The PC value is current instruction location plus 2. Displacement d is an 8-bit signed number. There are eight conditions such as BC (branch if carry = 1), BCC (branch if carry clear), BZ (branch if result equal to zero, i.e., Z = 1), BNZ (branch if not equal to zero, i.e., Z = 0), BN (branch if negative, i.e., N = 1), BNN (Branch if not negative i.e. N = 0), BOV (branch if overflow, i.e., OV = 1), and BNOV (branch if no overflow, i.e., OV = 0).
BRA d	Branch always to (PC+2) + 2 x d, where PC value is the current instruction location plus 2. d is a signed 11-bit number. This is an unconditional branch instruction with relative mode.

- All instructions in the above except GOTO and BRA are executed in one cycle; GOTO and BRA are executed in two cycles. The size of each instruction except GOTO is one word; the size of GOTO is two words.

that these instructions are applicable to signed numbers.

The instructions in Table 7.1 will now be discussed using numerical examples.

- The GOTO k instruction unconditionally jumps to a 21-bit address; 20-bit 'k' is loaded into the PC (bit 1 through bit 20) with the least significant bit (bit 0 of the PC) as 0. This will make the target address an even number. Note that the PIC18F instruction sizes are even multiple(s) of a byte (one or two words). The target address for the GOTO instruction must be an even number. Hence, the least significant bit of the PC is automatically fixed at 0. The 21-bit address with the GOTO instruction will allow the PIC18F to unconditionally jump to anywhere in the two megabytes (2^{21}) of program memory. The GOTO instruction is used to unconditionally jump to any location in the program memory.
- The Bcc d instruction (discussed in Chapter 5) will branch if the condition cc is true. The two's complement number '2 x d' is added to the PC. Since the PC will be incremented to fetch the next instruction, the new address will be PC + 2 + 2d. This instruction is then a two-cycle instruction. If the condition is false, then the next instruction is executed. Note that displacement 'd' is an 8-bit signed number.

In order to illustrate the concept of relative branching, the following example will be repeated from Chapter 5 for convenience. Hence, consider the PIC18F disassembled instruction sequence along with the machine code (all numbers in hex) provided below:

```
                                1:              INCLUDE<P18F4321.INC>
                                2:              ORG        0x00
0000   0E02   MOVLW  0x2        3:   BACK       MOVLW      0x02
0002   0802   SUBLW  0x2        4:              SUBLW      0x02
0004   E001   BZ       0x8      5:              BZ         DOWN
0006   0E04   MOVLW  0x4        6:              MOVLW      0x04
0008   0804   SUBLW  0x4        7:   DOWN       SUBLW      0x04
000A   E0FA   BZ 0              8:              BZ         BACK
000C   0003   SLEEP             9:              SLEEP
```

Note that all instructions, addresses, and data are chosen arbitrarily. The first branch instruction, BZ DOWN (line 5) at address 0x0004, has a machine code 0xE001. Upon execution of the instruction BZ (branch if Z-flag = 1), the PIC18F branches to label DOWN if Z = 1; the PIC18F executes the next instruction if Z = 0. The BZ instruction uses the relative addressing mode. This means that DOWN is a positive number (the number of steps forward relative to the current program counter) indicating a forward branch. The machine code 0xE001 means that the opcode for BZ is 0xE0 and the relative 8-bit signed offset value is 0x01 (+1). This is a positive value indicating a forward branch. Note that while executing BZ DOWN at address 0x0004, the PC points to address 0x0006 since the program counter is incremented by 2. This means that the program counter contains 0x0008. The offset 0x01 is multiplied by 2 and added to address 0x0006 to find the target branch address where the program will jump if Z = 1. The branch address can be calculated as follows:

```
 0x0006  = 0000 0000 0000 0110
+0x0002  = 0000 0000 0000 0010  (0x01 is multiplied by 2 , and sign-extended to 16 bits)
           0000 0000 0000 1000 = 0x0008
```

Hence, the PIC18F branches to address 0x0008 if Z = 1. This can be verified in the instruction sequence above.

Next, consider the second branch instruction, BZ BACK (line 8). Upon execution of BZ BACK, the PIC18F branches to label BACK if Z = 1; otherwise, the PIC18F executes the next instruction. The machine code for this instruction at address 0x000A is 0xE0FA, where 0xE0 is the opcode and 0xFA is the signed 8-bit offset value. The offset is represented as an 8-bit two's complement number. Since 0xFA is a negative number (-6_{10}), this is a backward jump. Note that while executing BZ BACK at address 0x000A, the PC points to address 0x000C since the program counter is incremented by 2. This means that the program counter contains 0x000C. The offset -6 is multiplied by 2, and then added to 0x000C to find the address value where the program will branch if Z = 1. The branch address is calculated as follows:

```
  0x000C = 0000 0000 0000 1100
+ 0xFFF4 = 1111 1111  1111 0100 (or 0xFA multiplied by 2 and then sign-extended
                                 to 16 bits)
     ↗ 1 0000 0000  0000 0000  = 0x0000
  Ignore final carry
```

The branch address is 0x0000, which can be verified in the instruction sequence above. As mentioned in Chapter 1, in order to add a 16-bit signed number with an 8-bit signed number, the 8-bit signed number must first be sign-extended to 16 bits. The two 16-bit numbers can then be added. Any carry resulting from the addition must be discarded. This will provide the correct answer.

- The BRA (branch always) instruction uses the relative addressing mode. As mentioned in Chapter 5, the BRA d instruction unconditionally branches to (PC + 2 + 2 x d), where offset 'd' is a signed 11-bit number specifying a range from -1024 to + 1023, with 0 being positive. For example, consider BRA 0x05 is stored at location 0040H in the program memory. This means that the PC will contain 0042H (PC + 2) when the PIC18F executes BRA. Hence, after execution of the BRA 0x05 instruction, the PC will be loaded with address 004CH (0042H + 05H x 2). Thus, the program will branch to address 004CH, which is 10 (A_{16}) steps forward relative to the current contents of PC. Next, consider the following instruction sequence:

```
        ORG   0x100
HERE    BRA   HERE
```

The machine code for the above instruction is 1101011111111111_2 ($D7FF_{16}$). Note that 0xD7 is the opcode and 0xFF (-1_{10}) is the offset. During execution of the BRA instruction, the PC points to 0x102. The target branch address = (PC + 2 + d x 2) = 0x102 + (-1) x 2 = 0x100. The instruction HERE BRA HERE unconditionally branches to address 0x100. This is equivalent to HALT instruction in other processors.

7.2 PIC18F Test, Compare, and Skip Instructions

Table 7.2 lists these instructions. Next, we explain these instructions using the access bank and specifying F or W in place of 'd'.

TABLE 7.2 PIC18F test, compare, and skip instructions

Instruction	Operation
BTFSC F, b, a	Bit test file register, skip if clear. If bit 'b' in register 'F' is '0', then the next instruction is skipped. If bit 'b' is '1', then the next instruction is executed.
BTFSS F, b, a	Bit test file, skip if set. If bit 'b' in register 'F' is '1', then the next instruction is skipped. If bit 'b' is '0', then the next instruction is executed.
CPFSEQ F, a	Compare F with W; skip if [F] = [W]. Compares the contents of data memory location 'F' to the contents of W by performing an unsigned subtraction. If [F] = [W], then the next instruction is skipped; else, the next instruction is executed.
CPFSGT F, a	Compare F with W; skip if [F] > [W]. Compares the contents of data memory location 'F' to the contents of the W by performing an unsigned subtraction. If the contents of F are greater than the contents of W, then the next instruction is skipped; else, the next instruction is executed.
CPFSLT F, a	Compare F with W; skip if [F] < [W]. Compares the contents of data memory location 'F' to the contents of W by performing an unsigned subtraction. If the contents of 'F' are less than the contents of W, then the next instruction is skipped; else, the next instruction is executed.
DECFSNZ F, d, a	Decrement F; skip if not 0. The contents of register 'F' are decremented by 1. If the result is not '0', then the next instruction is skipped; else, the next instruction is executed.
DECFSZ F, d, a	Decrement F; skip if 0. The contents of register 'F' are decremented by 1. If the result is '0', then the next instruction is skipped; else, the next instruction is executed.
INCFSNZ F, d, a	Increment F; skip if not 0. The contents of register 'F' are incremented by 1. If the result is not '0', then the next instruction is skipped; else, the next instruction is executed.
INCFSZ F, d, a	Increment F; skip if 0. The contents of register 'F' are incremented by 1. If the result is '0', then the next instruction is skipped; else, the next instruction is executed.
TSTFSZ F, a	Test F; skip if 0. If 'F' = 0, then the next instruction is skipped; else, the next instruction is executed.

- All instructions in the above are executed in one to two cycles. No flags are affected. The size of each instruction is one word.
- a = 0 means that the data register is located in the access bank while a = 1 means that the contents of BSR specify the address of the bank.
- For destination, d = 0 means that the destination is WREG while d = 1 means that the destination is file register.

- The BTFSC F, b, a instruction tests the specified bit 'b' in the file register 'F', and skips the next instruction if the bit 'b' is 0. On the other hand, if bit 'b' is 1, the

PIC18F executes the next instruction. Hence, GOTO or BRA instruction is typically used after the BTFSC instruction. The BTFSC instruction is useful for conditional (polled) I/O. This topic is discussed later.

Next, as an example, consider BTFSC 0x40, 5.

Prior to execution of BTFSC 0x40, 5, [0x40] = F1H.

After execution of BTFSC 0x40, 5, since bit 5 of [0x40] is 1, the BTFSC executes the next instruction.

The BTFSC instruction can be used to write the PIC18F assembly language instruction sequence for the following C segment:

```
if (x<0)
        y++ ;
else
        y--;
```

The PIC18F assembly language instuction sequence can be written as follows:

```
        BTFSC   X, 7      ; Check sign bit (bit 7) of [X]. If negative, increment [Y]
        BRA     NEG       ; If [X] is positive, decrement [Y]
        DECF    Y         ; Increment [Y] if [X] is negative
        BRA     NEXT
NEG     INCF    Y         ; Decrement [Y] if [X] is positive
NEXT    ----------        ; Next instruction
```

- The BTFSS F, b, a instruction tests the specified bit 'b' in the file register 'F', and skips the next instruction if the bit 'b' is '1'. On the other hand, if bit 'b' is '0', the PIC18F executes the next instruction. Like the BTFSC instruction, the GOTO or BRA instruction is typically used after the BTFSS instruction. The BTFSS instruction can be used for conditional (polled) I/O. This topic is discussed later.

  Next, as an example, consider BTFSS 0x70, 0.

  Prior to execution of BTFSS 0x70, 0, [0x70] = 0xF1.

  After execution of BTFSS 0x70, 0, since bit 0 of [0x70] is 1, the BTFSS skips the next instruction.

The same example for BTFSC, described in the last section, can be used to illustrate the BTFSS instruction. The PIC18F assembly language instruction sequence is written using the BTFSS instruction written for the following C segment:

```
if (x<0)
y++ ;
else
y--;
```

The PIC18F assembly language instruction sequence using the BTFSS is provided below:

```
BTFSS   X, 7      ; Check sign bit (bit 7) of [X]. If negative, increment [Y]
BRA     POS       ; If [X] is positive, decrement [Y]
INCF    Y         ; Increment [Y] if [X] is negative
```

```
        BRA     NEXT
POS     DECF    Y           ; Decrement [Y] if [X] is positive
NEXT    ---------           ; Next instruction
```

- The CPFSEQ F, a instruction compares [F] with [WREG] by performing an unsigned subtraction, and skips the next instruction if [F] = [WREG]; if [F] ≠ [WREG], the following instruction is executed. The CPFSGT F, a, on the other hand, compares [F] with [WREG] by performing an unsigned subtraction, and skips the next instruction if [F] > [WREG]; if [F] ≤ [WREG], the next instruction is executed. The CPFSLT F, a instruction compares [F] with [WREG] by performing an unsigned subtraction, and skips the next instruction if [F] < [WREG]; if [F] ≥ [WREG], the next instruction is executed. Note that in all three cases, [F] and [WREG] are considered as 8-bit unsigned (positive) numbers. The GOTO or BRA instruction is typically used after each of these COMPARE instructions. These instructions do not provide any result of subtraction, and also, they do not affect any status flags.

  In order to illustrate the use of one of the PIC18F COMPARE instructions, consider the following example. Suppose it is desired to find the number of matches for an 8-bit unsigned number in data register 0x80 with a data array (stored from low to high memory) of 50 bytes in memory pointed to by 0x50. Assume that data are already stored in memory. The following PIC18F instruction sequence with CPFSEQ can be used :

```
            CLRF    0x40        ; Clear 0x40 to 0. Register 0x40 will hold
                                ; the number of matches
            MOVLW   D'50'       ; Move 50 into WREG
            MOVWF   0x20        ; Initialize 0x20 with the array count 50
            LFSR    0, 0x50     ; Initialize indirect pointer FSR0 with 0x50
            MOVF    0x80, W     ; Move [0x80] to WREG
BACK        CPFSEQ  POSTINC0    ; Compare the number to be matched with [WREG]
            BRA     NOMATCH
            INCF    0x40        ; If there is a match, increment [0x40] by 1
NOMATCH     DECF    0x20        ; Decrement [0x20] by 1
            BNZ     BACK        ; Go to BACK if Z is not 0
            ------------        ; Next instruction
```

  Note that, in the above, CPFSEQ rather than SUBWF is used. This is because we are not interested in the subtraction result. Rather, we are interested in the number of matches. If SUBWF is used, one needs to load the number to be matched or the data byte from the array after each SUBWF; the subtraction result would erase the data. Hence, CPFSEQ, instead of SUBWF, is ideal for the above example.

- The DECFSNZ F, d, a instruction decrements [F] by 1, and if the result is not '0', skips the instruction; else, the next instruction is executed. The DECFSZ , on the other hand, decrements [F] by 1, and if the result is '1', skips the instruction; else, the next instruction is executed. The GOTO or BRA instruction is typically used after each of these COMPARE instructions.

  Both instructions can be used to execute a certain loop '*n*' times, where '*n*' is an 8-bit number. This is another way of executing a loop without using the conditional branch instructions.

For example, consider executing a loop to obtain the 8-bit SUM (10 x A) by repeated addition, assuming that the 8-bit unsigned number 'A' is stored in register 0x70, and the sum will be stored in 0x50. The following PIC18F instruction sequence using DECFSNZ will accomplish this:

```
        CLRF     0x50      ; Clear register 0x50 to 0 for SUM
        MOVLW    D'10'     ; Move 10 to WREG
        MOVWF    0x60      ; Initialize register 0x60 with 10
        MOVF     0x70,W    ; Move 'A' into WREG
REPEAT  ADDWF    0x50, F   ; Add 'A' 10 times; store result in 0x50
        DECFSNZ  0x60, F   ; Decrement counter; skip if reg 0x60 is not 0
        GOTO     NEXT
        GOTO     REPEAT    ; Repeat addition until counter 0x60 is 0
NEXT    ------------       ; Next instruction
```

Using the DECFSZ, the above program to compute (10 x A) can be written as follows:

```
        CLRF     0x50      ; Clear register 0x50 to 0 for SUM
        MOVLW    D'10'     ; Move 10 to WREG
        MOVWF    0x60      ; Initialize register 0x60 with 10
        MOVF     0x70,W    ; Move 'A' into WREG
REPEAT  ADDWF    0x50, F   ; Add 'A' 10 times; store result in 0x50
        DECFSZ   0x60, F   ; Decrement counter; skip if 0
        GOTO     REPEAT    ; Repeat addition until counter 0x60 is 0
        ---------          ; Next instruction
```

- The INCFSNZ F, d, a increments the contents of register 'F' by one, and skips the next instruction if the result is not 0; else, the next instruction is executed. The INCFSZ F, d, a , on the other hand, increments the contents of register 'F' by one, and skips the next instruction if the result is 0; else, the next instruction is executed. The GOTO or BRA instruction is typically used after each of these instructions.

Both instructions can be used to execute a certain loop '*n*' times, where '*n*' is an 8-bit number. This is another way of executing a loop without using the conditional branch instructions.

Consider the same examples of the last section using the DECFSNZ and DECFSZ instructions. As before, a loop will be executed to obtain the 8-bit sum (10 x A) by repeated addition, assuming that the 8-bit unsigned number 'A' is stored in register 0x70, and the sum will be stored in 0x50. The INCFSNZ and INFSZ will be used this time.

The following PIC18F instruction sequence using INCFSNZ will accomplish this:

```
        CLRF     0x50      ; Clear register 0x50 to 0 for SUM
        MOVLW    0xF6      ; Move -10 to WREG
        MOVWF    0x60      ; Initialize register 0x60 with -10
```

```
        MOVF      0x70,W    ; Move 'A' into WREG
REPEAT  ADDWF     0x50, F   ; Add 'A' 10 times; store result in 0x50
        INCFSNZ   0x60, F   ; Increment counter; skip if reg 0x60 not 0
        GOTO      NEXT
        GOTO      REPEAT    ; Repeat addition until counter 0x60 is 0
NEXT    ------------        ; Next instruction
```

Using the INCFSZ , the above program to compute (10 x A) can be written as follows:

```
        CLRF      0x50      ; Clear register 0x50 to 0 for SUM
        MOVLW     0xF6      ; Move -10 to WREG
        MOVWF     0x60      ; Initialize register 0x60 with -10
        MOVF      0x70,W    ; Move 'A' into WREG
REPEAT  ADDWF     0x50, F   ; Add 'A' 10 times; store result in 0x50
        INCFSZ    0x60, F   ; Increment counter; skip if counter is 0
        GOTO      REPEAT    ; Repeat addition until counter 0x60 is 0
        ---------           ; Next instruction
```

- The TSTFSZ F, a checks if [F] = 0, and skips the next instruction if it is zero; otherwise, the next instruction is executed. The TSTFSZ instruction can be used to check the contents of a register for 0 without using the conditional branch instruction. Note that a typical decrementing counter can be implemented using the conditional branch instruction such as BNZ, as follows:

```
COUNTER  EQU      0x40
         MOVLW    D'50'      ; Initialize loop counter with 50
         MOVWF    COUNTER
LOOP     DECF     COUNTER    ; Decrement COUNTER by 1
         BNZ      LOOP       ; Branch if [COUNTER] is not 0
         -------             ; Next instruction
```

The above loop can be implemented using the TSTFSZ instruction as follows:

```
COUNTER  EQU      0x40
         MOVLW    D'50'     ; Initialize loop counter with 50
         MOVWF    COUNTER
LOOP     DECF     COUNTER   ; Decrement COUNTER by 1
         TSTFSZ   COUNTER   ; Test COUNTER for 0, and if not 0,
         GOTO     LOOP      ; go to LOOP. If [COUNTER] is 0, skip.
         -------            ; Next instruction
```

7.3 PIC18F Table Read/Write Instructions

As mentioned before, the PIC18F program memory is 16 bits wide, while the PIC18F data memory space is 8 bits wide. Programs are stored in program memory with the data register contents defined using the assembler's DB directive. In order to execute a program requiring

TABLE 7.3 PIC18F table read/write instructions

TBLRD*	Move 8-bit data from program memory addressed by 21-bit TBLPTR into the 8-bit register TABLAT.
TABLRD*+	Move 8-bit data from program memory addressed by 21-bit TBLPTR into the 8-bit register TABLAT, and then increment TBLPTR by 1.
TBLRD*-	Move 8-bit data from program memory addressed by 21-bit TBLPTR into the 8-bit register TABLAT, and then decrement TBLPTR by 1.
TBLRD+*	Increment TBLPTR by 1, and then move 8-bit data from program memory addressed by 21-bit TBLPTR into the 8-bit register TABLAT.
TBLWT*	Move 8-bit data from 8-bit register TABLAT into program memory addressed by 21-bit TBLPTR .
TBLWT*+	Move 8-bit data from 8-bit register TABLAT into program memory addressed by 21-bit TBLPTR , and then increment TBLPTR by 1.
TBLWT*-	Move 8-bit data from 8-bit register TABLAT into program memory addressed by 21-bit TBLPTR , and then decrement TBLPTR by 1.
TBLWT+*	Increment TBLPTR by 1, and then move 8-bit data from 8-bit register TABLAT into program memory addressed by 21-bit TBLPTR .

- All TBLRD and TBLWT instructions are executed in two cycles.
- The size of each instruction is one word.

data, the data bytes stored in program memory using DB directive must be transferred to the specified data registers in data memory. Since the sizes of program memory and data memory are different, it would be difficult to accomplish this data transfer. However, four table read and four table write instructions provided in the PIC18F facilitate transferring data between these two memory spaces through an 8-bit TABLAT (register called Table latch) and a 21-bit TBLPTR (pointer register called the Table pointer). The TBLPTR includes three registers, namely, TABLPTRU (bits 20 through 16), TBLPTRH (bits 15 through 8), and TBLPTRL (bits 7 through 0).

Two operations that allow the PIC18F to move bytes between the program memory and the data memory are

- Table read (TBLRD)
- Table write (TBLWT)

The table read operation retrieves data bytes from program memory and places them into the data memory. Figure 7.1 shows the operation of a table read with program memory and data memory. The table read operation reads data from program memory onto TABLAT using the TBLRD instruction with four addressing modes (register indirect, postincrement, postdecrement, and preincrement).

The table write operation stores data from the data memory space into holding registers in program memory. Figure 7.2 shows the operation of a table write with program memory and data memory. The table write operation writes the contents of TABLAT into program memory using the TBLWT instruction with four addressing modes (register indirect, postincrement, postdecrement, and preincrement). Table operations work with data bytes . A table block containing data, rather than program instructions, is not required to be word aligned. Therefore, a table block can start and end at any byte address.

Table 7.3 lists the TBLRD and TBLWT instructions.

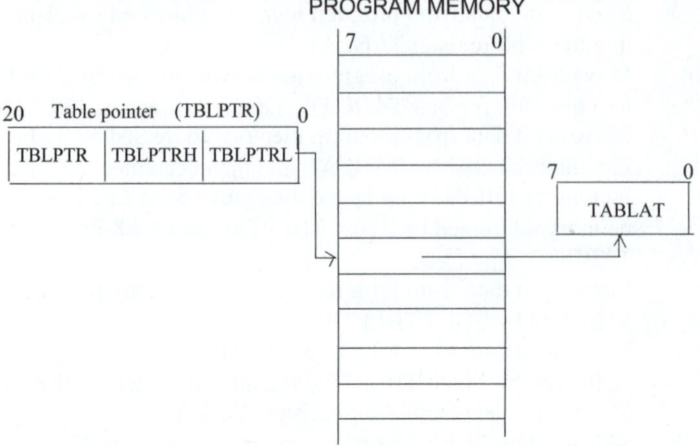

FIGURE 7.1 Table read operation (instruction TBLRD*)

The table read/write instructions will now be explained in the following using numerical examples with similar data for each instruction:

- Consider TBLRD* instruction.

  Prior to execution of TBLRD*, [TBLPTR] = 0x02318, [TABLAT] = 0x24, and [0x002318] = 0xF2.

  After execution of TBLRD*, [TABLAT] = 0xF2, [TBLPTR] = 0x002318 (unchanged), and [0x002318] = 0xF2 (unchanged).

- Consider TBLRD*+ instruction.

  Prior to execution of TBLRD* +, [TBLPTR] = 0x002318, [TABLAT] = 0x24, and [0x002318] = 0xF2.

  After execution of TBLRD*+, [TABLAT] = 0xF2, [TBLPTR] = 0x002319, and [0x002318] = 0xF2 (unchanged).

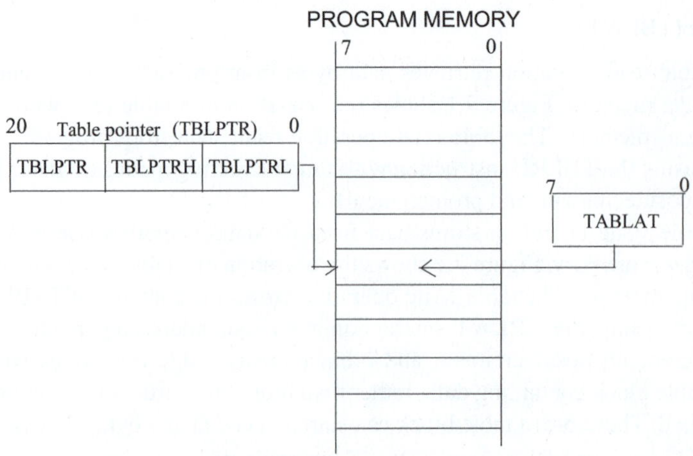

FIGURE 7.2 Table write operation (instruction TBLWT*)

- Consider TBLRD*- instruction.

  Prior to execution of TBLRD*-, [TBLPTR] = 0x002318, [TABLAT] = 0x24, and [0x002318] = 0xF2.

  After execution of TBLRD*-, [TABLAT] = 0xF2, [TBLPTR] = 0x002317, and [0x002318] = 0xF2 (unchanged).

- Consider TBLRD+* instruction.

  Prior to execution of TBLRD+*, [TBLPTR] = 0x002318, [TABLAT] = 0x24, and [0x002319] = 0xF2.

  After execution of TBLRD+*, [TABLAT] = 0xF2, [TBLPTR] = 0x002319, and [0x002319] = 0xF2 (unchanged).

- Consider TBLWT* instruction.

  Prior to execution of TBLWT*, [TBLPTR] = 0x002318, [TABLAT] = 0x24, and [0x002318] = 0xF2.

  After execution of TBLWT*, [0x002318] = 0x24, [TABLAT] = 0x24 (unchanged), and [TBLPTR] = 0x002318 (unchanged).

- Consider TBLWT*+ instruction.

  Prior to execution of TBLWT*+, [TBLPTR] = 0x002318, [TABLAT] = 0x24, and [0x002318] = 0xF2.

  After execution of TBLWT*+, [0x002318] = 0x24, [TBLPTR] = 0x002319, and [TABLAT] = 0x24 (unchanged).

- Consider TBLWT*- instruction.

  Prior to execution of TBLWT*-, [TBLPTR] = 0x002318, [TABLAT] = 0x24, and [0x002318] = 0xF2.

  After execution of TBLWT*-, [0x002318] = 0x24, [TBLPTR] = 0x002317, and [TABLAT] = 0x24 (unchanged).

- Consider TBLWT+* instruction.

  Prior to execution of TBLWT+*, [TBLPTR] = 0x002318, [TABLAT] = 0x24, and [0x002319] = 0xF2.

  After execution of TBLWT+*, [0x002319] = 0x24, [TBLPTR] = 0x002319, and [TABLAT] = 0x24 (unchanged).

Example 7.1 Write a PIC18F assembly language program at address 0x100 to move the ASCII codes (30H through 39H) for BCD numbers 0 through 9 from program memory starting at address 0x200 (30H at address 0x200, 31H at 0x201, and so on) into data memory starting at address 0x40 (30H to be stored at address 0x40, 31H at 0x41, and so on).

Solution

```
        INCLUDE  <P18F4321.INC>
        ORG      0x100          ;   #1 Starting address of program
```

```
COUNTER  EQU     0x20
         MOVLW   UPPER ADDR  ;  #2 Move upper 5 bits (00H) of address
         MOVWF   TBLPTRU     ;  #3 to TBLPTRU
         MOVLW   HIGH ADDR   ;  #4 Move bits 15-8 (02H) of address
         MOVWF   TBLPTRH     ;  #5 to  TBLPTRH
         MOVLW   LOW ADDR    ;  #6 Move bits 7-0 (00H) of address
         MOVWF   TBLPTRL     ;  #7 to TBLPTRL
         LFSR    0, 0x40     ;  #8 Initialize FSR0 to 0x40 to be used as
                             ;  destination pointer in data memory
         MOVLW   D'10'       ;  #9 Initialize COUNTER with 10
         MOVWF   COUNTER     ;  #10 Move [WREG] into COUNTER
LOOP     TBLRD*+             ;  #11 Read data from program memory
                             ;  into TABLAT; increment TBLPTR by 1
         MOVF    TABLAT, W   ;  #12 Move [TABLAT] into WREG
         MOVWF   POSTINC0    ;  #13 Move W into data memory pointed
                             ;  to by FSR0, and then increment FSR0
                             ;  by 1
         DECF    COUNTER, F  ;  #14 Decrement COUNTER BY 1
         BNZ     LOOP        ;  #15 Branch  if  Z = 0; else stop
FINISH   BRA     FINISH      ;  Stop
         ORG     0x200
         ADDR    DB             30H, 31H, 32H, 33H, 34H, 35H, 36H, 37H
                 DB             38H, 39H
         END
```

In the above program, the # sign along with the line number is placed before the comment in order to identify the specific line for explanation. Line #1 specifies the starting address of the program at 0x000200. Note that the programmer does not have to define the address 0xFF5 for TABLAT. This is a predefined address (special function register) by Microchip. The MPLAB assembler determines this internally. As mentioned before, the TBLPTR is divided into three registers as TBLPTRU (predefined address 0xFF8), TBLPTRH (predefined address 0xFF7), and TBLPTRL (predefined address 0xFF6). These are also special function registers. The MPLAB assembler determines these addresses internally. Also, the MPLAB assembler identifies TBLPTRU as UPPER, TBLPTRH as HIGH, and TBLPTRL as LOW. Line #'s 2 through 7 initialize TBLPTR with the 21-bit address 0x000200.

Line #8 initializes data memory pointer FSR0 to 0x40. Line #'s 9 and 10 initialize COUNTER with 10. Line #11 reads a byte from program memory addressed by TBLPTR into TABLAT, and then increments TBLPTR by 1. Line #12 moves the contents of TABLAT into WREG. Line #13 moves [WREG] into the destination data memory address pointed to by FSR0, and then increments FSR0 by 1. Line #14 decrements [COUNTER] by 1. BNZ at line #15 checks the Z flag, and if Z = 0, the LOOP is executed 10 times, and thus the ASCII numbers 30H through 39H are transferred from program memory to data memory.

7.4 PIC18F Subroutine Instructions

Table 7.4 lists PIC18F subroutine instructions. These include PUSH/POP and subroutine CALL/RETURN instructions. The subroutine instructions automatically use the "hardware stack" implemented by the manufacturer. The programmer can create a "software stack"

TABLE 7.4 PIC18F subroutine instructions

Instruction	Operation
CALL k, s	Call the subroutine at address k within the two megabytes of program memory. First, return address (PC + 4) is pushed onto the return stack. If 's' = 1, the WREG, STATUS, and BSR registers are also pushed into their respective shadow registers (internal to the CPU), WS, STATUSS and BSRS. If 's' = 0, these registers are unaffected (default). Then, the value 'k' is loaded into PC.
POP	Discards top of stack pointed to by SP and decrements PC by 1.
PUSH	PUSHes or writes the PC onto the stack, and increments PC by 1.
RCALL n	Subroutine CALL with relative mode.
RETFIE	Returns from interrupt. Stack is popped and top-of-stack (TOS) is loaded into the PC. Interrupts are enabled by setting either the high or low priority global interrupt enable bit. This instruction is normally used at the end of an interrupt service routine.
RETLW k	WREG is loaded with the eight-bit literal 'k'. The program counter is loaded from the top of the hardware stack (the return address).
RETURN s	Returns from subroutine. The stack is popped and the top of the stack (TOS) is loaded into the program counter. If 's'= 1, the contents of the shadow registers, WS, STATUSS, and BSRS, are loaded into their corresponding registers, WREG, STATUS, and BSR. If 's' = 0, these registers are not affected (default).

- CALL and RETURN instructions are executed in two cycles.
- POP and PUSH are executed in one cycle.
- The size of each instruction except CALL is one word; the size of the CALL instruction is two words.

if needed for storing local variables. The "hardware stack" and "software stack" will be discussed in the next section in more detail.

- POP instruction reads (pops) the TOS (top of stack) value from the return stack and discards it; [STKPTR] is decremented by 1. The TOS value becomes the previous value that was pushed onto the return stack.

  As an example, consider the POP instruction with numerical data in the following:

  Prior to execution of the POP, [STKPTR] = 0x65, [0x65] = TOS (top of stack) = 0x000502, stack (1 level down), and [0x64] = 0x002418.

  After execution of the POP, [STKPTR] = 0x64, TOS, and [0x64] = 0x002418.

  Note that previous TOS (0x000502) is discarded, and previous stack (1 level down) is the current TOS. Figures 7.3 (a) and (b) depict this.

- PUSH writes (pushes) PC+2 onto the top of the return stack; [STKPTR] is incremented

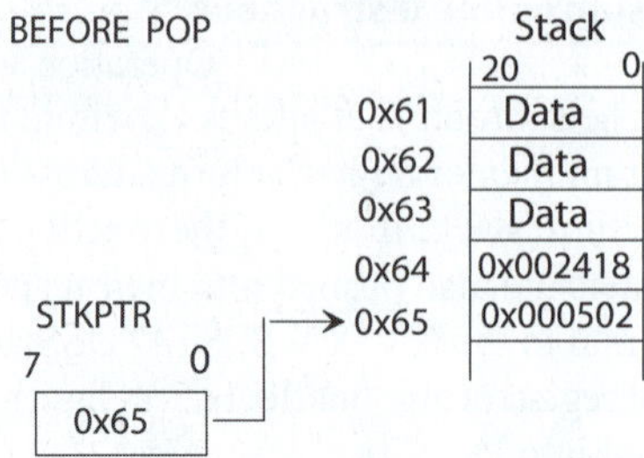

FIGURE 7.3 (a) PIC18F hardware stack with arbitrary data before execution of POP instruction

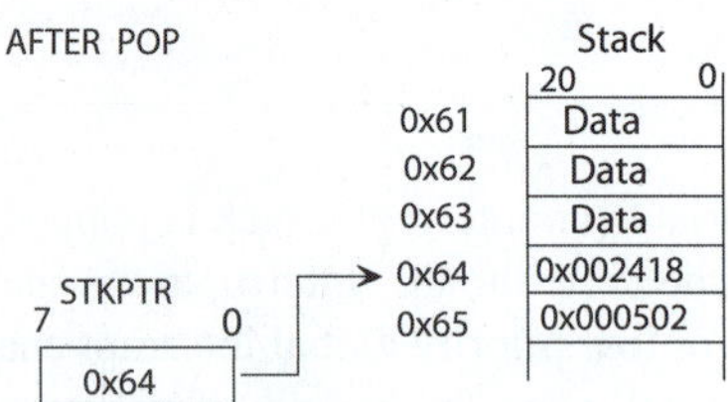

FIGURE 7.3 (b) PIC18F hardware stack with arbitrary data after execution of POP instruction; address 0x65 is assumed to be free

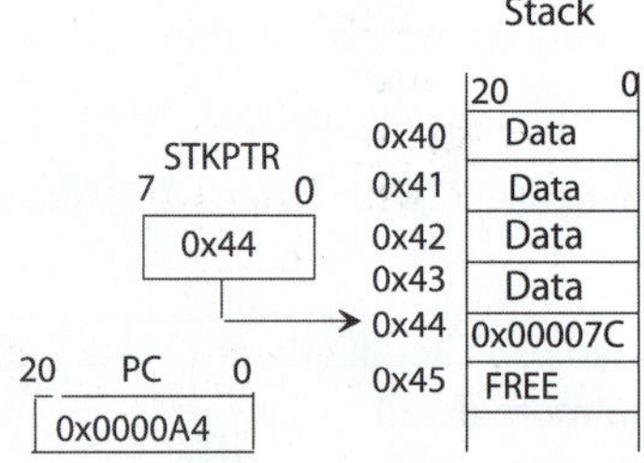

FIGURE 7.4 (a) PIC18F hardware stack with arbitrary data before execution of PUSH instruction

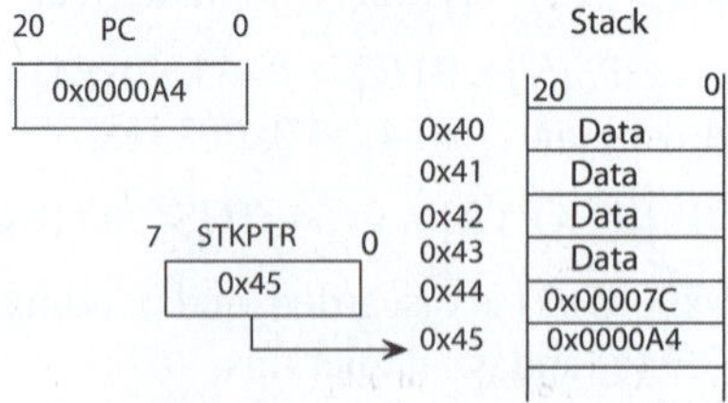

FIGURE 7.4 (b) PIC18F hardware stack with arbitrary data after execution of PUSH instruction

by 1. The previous TOS value is pushed down on the stack.

As an example, consider the PUSH instruction with numerical data in the following: Prior to execution of PUSH, [STKPTR] = 0x44, [0x44] = TOS (top of stack) = 0x00007C, and [PC+ 2] = 0x0000A4; that is, the PUSH instruction is stored at address 0x0000A2. After execution of the PUSH, [STKPTR] = 0x45, [0x45] = TOS = 0x0000A4, previous TOS (one level down), and [0x44] = 0x00007C.

Figures 7.4 (a) and (b) depict this.

- The "CALL k, s" with s = 0 (or CALL k) instruction is the simplest way of CALLing a subroutine; s = 0 is the default case. As an example, the CALL START instruction automatically pushes the current contents of the PC onto the stack, and loads PC with the label called START. Note that address START contains the starting address of the subroutine. The "RETURN s" instruction with s = 0 (or RETURN since s = 0 is default) pops the return address (PC pushed onto the stack by the CALL START instruction) from TOS, and loads PC with this address. Thus, control is returned to the main program, and program execution continues with the instruction next to the CALL START.

Consider the following PIC18F program segment:

Main Program	*Subroutine*		
—	SUB	—	; First instruction of subroutine
—		—	
—		—	
START CALL SUB		—	
—		—	
—		—	
—		RETURN	; Last instruction of the subroutine

Here, the CALL SUB instruction in the main program calls the subroutine SUB. In response to the CALL instruction, the PIC18F pushes the current PC contents (START in this case) onto the stack and loads the starting address SUB of the subroutine into PC. After the subroutine is executed, the RETURN instruction at the end of the subroutine pops the address START from the stack into PC, and program control is then returned to the main program.

7.5 PIC18F System Control Instructions

The system control instructions are associated with the operation of the PIC18F. Table 7.5 lists these instructions.

7.6 PIC18F Hardware vs. Software Stack

As mentioned in Chapter 5, the PIC18F stack is a group of thirty-one 21-bit registers to hold memory addresses. This stack (also called the "hardware stack") is part of neither data memory nor program memory. Note that the size of the stack (21-bit) is the same as the size of the PC (21-bit). The SP (stack pointer) is 5 bits wide in order to address 31 registers.

TABLE 7.5 PIC18F system control instructions

Instruction	Operation
CLRWDT	Clears watchdog timer to 0.
RESET	Resets all registers and flags to their "hardware reset" values. The hardware RESET is performed upon activation of the PIC18F $\overline{\text{MCLR}}$ input pin. The RESET instruction provides software reset.
SLEEP	The PIC18F is put into sleep mode with the oscillator stopped.
NOP	No Operation

- All instructions in the above are executed in one cycle.
- The size of each instruction is one word.

In the PIC18F, after a POP (stack read), the SP is decremented by one while the SP is incremented by one after a PUSH (stack write). Also, the SP points to the last-used address. Although with a 5-bit SP, 32 (2^5) registers are available, the PIC18F stack provides 31 registers with addresses 00001_2 through 11111_2.

In the PIC18F, these 31 registers are typically used to store return addresses after execution of the subroutine CALL instructions. Sometimes it may be necessary to save local variables before executing a subroutine CALL instruction. The 31 registers provided in the hardware stack may not be adequate. In that case, the user can create a "software stack" in the PIC18F using one or more of the three file select registers (FSRs) as the SP along with the registers in data memory.

The PIC18F software stack can be implemented by the programmer using data registers and FSRs namely, FSR0-FSR2. The PIC18F uses one of the FSRs (referred to as FSRn with n = 0 to 2) as the software stack pointer, and supports the software stack with the register indirect postincrement and predecrement addressing modes. In addition to the software stack pointers (FSRn), any bank of data registers can be used for the

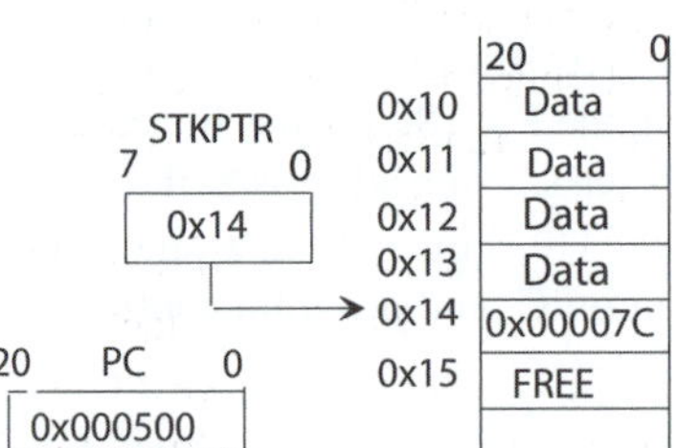

FIGURE 7.5 (a) PIC18F hardware stack with arbitrary data before execution of CALL 0x000200 instruction

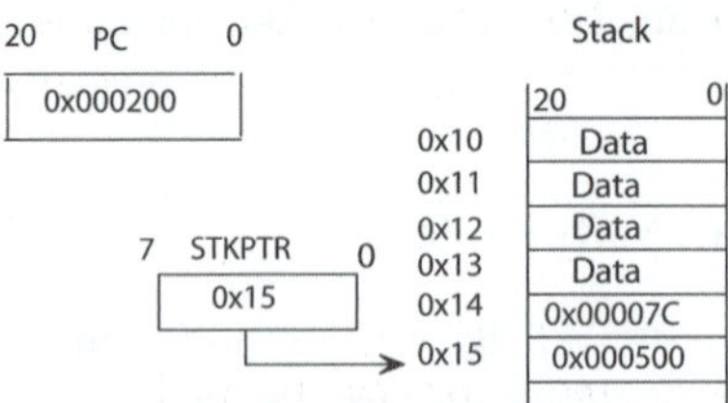

FIGURE 7.5 (b) PIC18F hardware stack with arbitrary data after execution of CALL 0x000200 instruction

software stack. Subroutine CALLs and interrupts automatically use the hardware stack pointer (STKPTR). As mentioned before, subroutine CALLs push the current PC onto the hardware stack; RETURN pops the PC from the hardware stack.

The PIC18F accesses the system stack from the top for operations such as subroutine calls or interrupts. This means that stack operations such as subroutine calls or interrupts access the hardware stack automatically from HIGH to LOW memory. As mentioned before, the low five bits of the STKPTR are used as the stack pointer for the hardware stack. Note that the STKPTR can be initialized using PIC18F MOVE instructions. For example, in order to load 0x20 into the STKPTR, the following PIC18F instruction sequence can be used:

```
MOVLW      0x20        ; Load 0x20 into WREG
MOVWF      STKPTR      ; Load [WREG] into STKPTR
```

Also, the STKPTR is incremented by 1 after a push and decremented by one after a pop. As an example, suppose that a PIC18F CALL instruction such as CALL 0x000200 is executed when [PC] = 0x000500; then, after execution of the subroutine call, the PIC18F will push the current contents of PC (0x000500) onto the hardware stack, and then load PC with 0x000200 (starting address of the subroutine specified in the CALL 0x000200 instruction). This is shown in Figures 7.5 (a) and (b). The RETURN instruction at the end of the subroutine will pop 0x000500 from the hardware stack into the PC and return control to the main program. All data are arbitrarily chosen.

In the PIC18F, the software stack can be created using appropriate addressing modes. Typical PIC18F memory instructions such as the MOVFF instruction can be used to access the stack. Also, by using one of the three FSRns (FSR0–FSR2) as software stack pointers, stacks can be filled from either HIGH to LOW memory or vice versa:

Filling a stack from HIGH to LOW memory (top of the stack) is implemented with postdecrement mode for push and preincrement mode for pop.

Filling a stack from LOW to HIGH (bottom of the stack) memory is implemented with preincrement for push and postdecrement for pop.

The programmer can create software stack growing from HIGH to LOW memory addresses using FSRn as the stack pointer. To push the contents of a data register onto the software stack, the MOVFF instruction with appropriate addressing modes can be used. For example, to push contents of a data register 0x30 using FSR0 as the stack pointer, the following PIC18F instruction sequence can be used:

```
LFSR    0, 0x0070          ; Initialize FSR0 with 0x70 to be used as the SP
MOVFF   0x30, POSTDEC0     ; Push [0x30] to stack, decrement SP (FSR0) by 1
```

This is shown in Figures 7.6 (a) and (b). Figure 7.6 (a) shows the software stack with arbitrary data prior to execution of the above instructions. Figure 7.6 (b) shows the software stack with arbitrary data after execution of the above instructions. Note that the stack pointer FSR0 in this case is decremented by 1 after PUSH.

The 8-bit data 0xF2 can be popped from the stack into another data register 0x20, for example, using the MOVFF PREINC0, 0x20 instruction. Note that the stack pointer FSR1 in this case is incremented by 1 after POP.

Next, consider the stack growing from LOW to HIGH memory addresses in which the programmer also utilizes FSRn as the stack pointer.

To push the 8-bit contents of a data register onto the software stack, the MOVFF instruction with appropriate addressing modes can be used. For example, to push contents of a data register 0x20 using FSR1 as the stack pointer, the following PIC18F instruction sequence can be used:

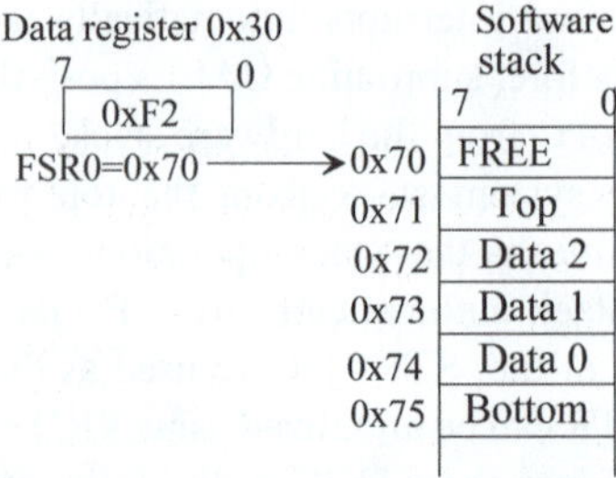

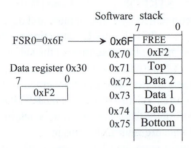

FIGURE 7.6 (a) **PIC18F software stack with arbitrary data growing from HIGH to LOW memory prior to PUSH**

Software stack
7 0
FSR0=0x6F → 0x6F FREE
0x70 0xF2
Data register 0x30 0x71 Top
7 0 0x72 Data 2
0xF2 0x73 Data 1
0x74 Data 0
0x75 Bottom

FIGURE 7.6 (b) **PIC18F software stack with arbitrary data growing from HIGH to LOW memory after PUSH**

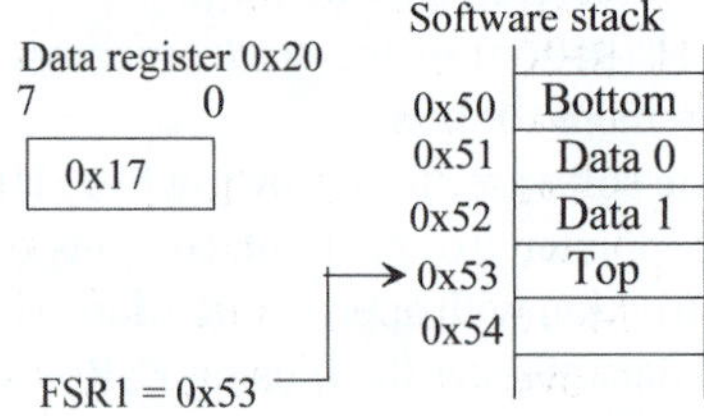

FIGURE 7.7 (a) **PIC18F software stack with arbitrary data growing from HIGH to LOW memory before PUSH**

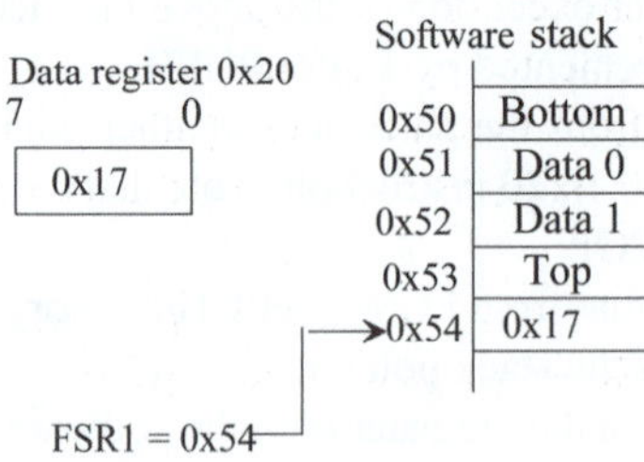

FIGURE 7.7 (b) **PIC18F software stack with arbitrary data growing from HIGH to LOW memory after PUSH**

```
        LFSR    1, 0x0053       ; Initialize FSR1 to 0x53 to be used as the SP
        MOVFF 0x20, PREINC0     ; Increment SP (FSR1) by 1, and Push [0x20] to stack.
```

This is shown in Figures 7.7 (a) and (b). Figure 7.7 (a) shows the software stack with arbitrary data prior to execution of the above instructions. Figure 7.7 (b) shows the software stack with arbitrary data after execution of the above instructions. Note that the stack pointer FSR1 in this case is incremented by 1 after PUSH.

The 8-bit data 0x17 can be popped from the stack into another data register 0x26, for example, using the MOVFF POSTDEC1, 0x26 instruction. Note that the stack pointer FSR1 in this case is decremented by 1 after POP.

Example 7.2 Write a PIC18F subroutine at address 0x100 to compute $Y = \sum_{i=1}^{N} Xi^2$. Assume the X_i's are 8-bit unsigned integers already stored in data memory, and $N = 4$. The numbers are stored in consecutive locations. Assume data register 0x40 points to the first element of the array for X_i's. The array elements are stored from LOW to HIGH memory addresses. The subroutine will store the 16-bit result *(Y)* in data memory registers 0x21 (high byte) and 0x20 (low byte), Also, write the main program at address 0x50 that will load data, initialize STKPTR to 0x05, FSR0 to 0x0040, call the subroutine, compute *(Y*/4) by discarding the remainder, and then stop.

Verify the correct operation of the programs using the MPLAB. Show screen shots as necessary.

Solution

```
        INCLUDE<P18F4321.INC>
        ORG     0x50            ; Starting address of the main program
;LOAD FOUR ARBITRARILY CHOSEN DATA INTO DATA MEM ADDR .
                                ; 0x40 TO 0x43
        MOVLW   0x7E            ; Move 0x7E into WREG
        MOVWF   0x40            ; Move 0x7E into file register 0x40
        MOVLW   0x08            ; Move 0x08 into WREG
        MOVWF   0x41            ; Move 0x08 into file register 0x41
        MOVLW   0x23            ; Move 0x23 into WREG
        MOVWF   0x42            ; Move 0x23 into file register 0x42
        MOVLW   0x30            ; Move 0x30 into WREG
        MOVWF   0x43            ; Move 0x43 into file register 0x40
; INITIALIZE STKPTR, CALL SUBROUTINE, AND DIVIDE BY 4 BY
                                ; RIGHT SHIFT TWICE
        MOVLW   0x05            ; Move 0x05 into WREG
        MOVWF   STKPTR          ; Load 0x05 into STKPTR
        LFSR    0,0x0040        ; Load file select register 0 with register 0x0040
        CALL    SQR             ; Call the function SQR
        BCF     STATUS, C       ; Clear the carry flag
        RRCF    0x21,F          ; Rotate right, or divide by 2
        RRCF    0x20,F
        BCF     STATUS, C       ; Clear the carry flag
        RRCF    0x21.F          ; Divide by 4
        RRCF    0x20,F
```

```
FINISH  GOTO     FINISH       ; Halt
        ORG      0x100        ; Starting address of the subroutine
SQR     MOVLW  0x00
        MOVWF  0x21           ; Clear register 0x21
        MOVWF 0x20            ; Clear register 0x20
        MOVLW  0x04
        MOVWF 0x60            ; Move 0x04 into register 0x60
BACK    MOVFF   INDF0, 0x50   ; Move the value in memory pointed to by FSR0
                              ; into register 0x50.
                              ; 0x50 is used as a holding register in data memory
                              ; It should not be confused with the starting address
                              ; 0x50 of the main program which is in program
                              ; memory of the PIC18F
        MOVF     POSTINC0, W; Move value pointed to by FSR0  into WREG, and
                              ; then increment FSR0 by 1
        MULWF   0x50          ; Multiply WREG by 0x50, or X squared
        MOVF     PRODL, W     ; Move low byte of answer to WREG
        ADDWF   0x20, F       ; Sum with value in 0x20
        MOVF     PRODH, W     ; Move high byte of product to WREG
        ADDWFC 0x21, F        ; Sum with carry with value in 0x21
        DECFSZ   0x60, F      ; Decrement register 0x60 by one, and skip next
                              ; step if 0
        GOTO     BACK         ; Start over
        RETURN                ; Return to main code
        END
```

Verification of the programs using MPLAB:

The following sample data are used:

[0x40] = 0x7E = 126 (decimal)
[0x41] = 0x08 = 8 (decimal)
[0x42] = 0x23 = 35 (decimal)
[0x43] = 0x30 = 48 (decimal)

$$\sum_{i=1}^{N} X_i^2 = (126)^2 + (8)^2 + (35)^2 + (48)^2 = 15876 + 64 + 1225 + 2304 = 19469 \text{ (decimal)}$$

Hence, result = (19469)/4 = 4867.25, which is approximately 4867 (decimal) or 1303 (hex).

The following example will also demonstrate how the hardware stack on the PIC18F changes with the execution of the CALL and RETURN instructions.

The “PIC18F disassembly” function can be displayed from the “Disassembly Listing” option in the “View” menu as follows:

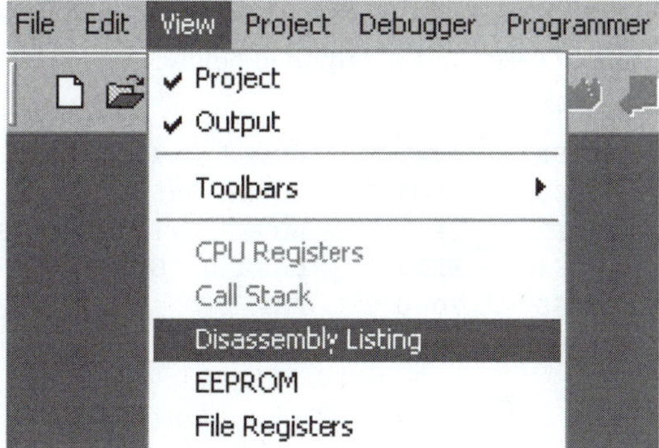

The "PIC18F hardware stack" can be displayed by selecting the "Hardware Stack" from the "View" menu as follows:

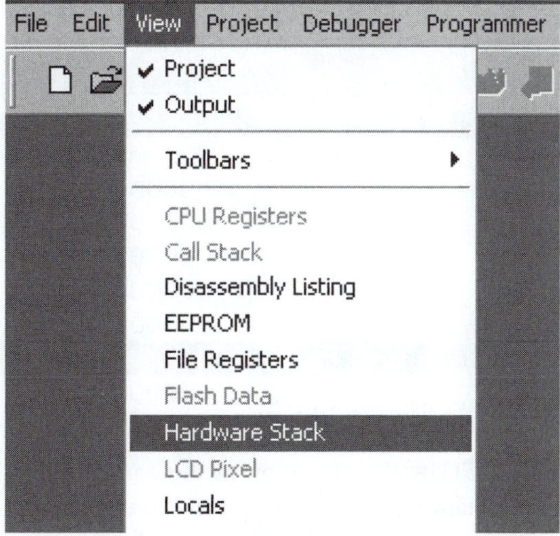

Next, each one of the instructions of the PIC18F assembly language main program is executed using the MPLAB SIM debugger. After execution of the instruction "MOVWF STKPTR" in the main program, the value 0x05 is loaded into the STKPTR. The following screen shot displaying the contents of the hardware stack verifies this:

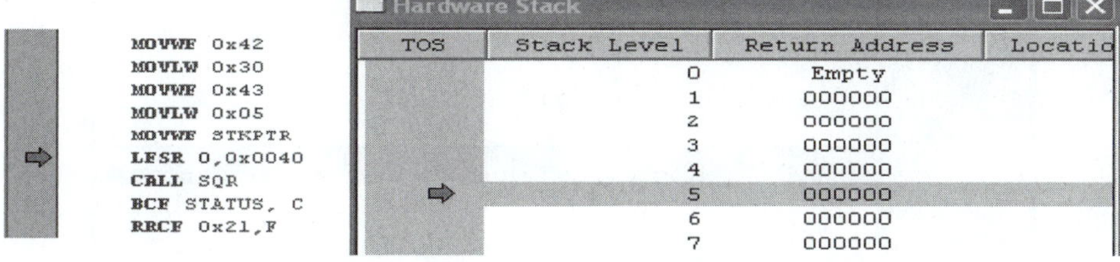

The next screen shot shows that the "CALL SQR" instruction is located at address 0x0068 in the program memory. Since the CALL instruction is four bytes (two words) wide, the program counter will contain the address of the next instruction which is (0x0068 + 4 = 0x006C). The screen shot to verify this is provided below:

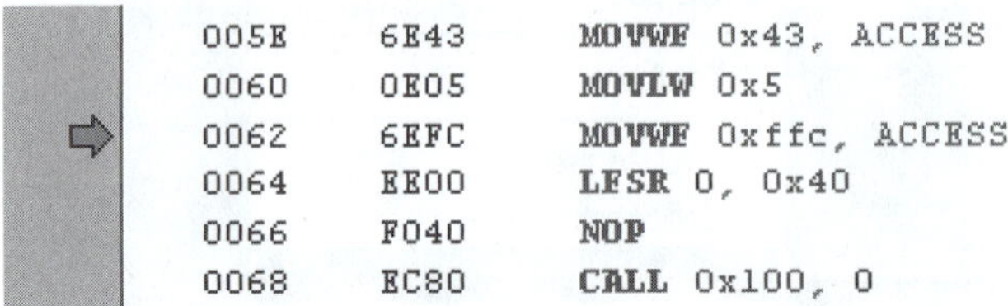

```
005E   6E43   MOVWF 0x43, ACCESS
0060   0E05   MOVLW 0x5
0062   6EFC   MOVWF 0xffc, ACCESS
0064   EE00   LFSR 0, 0x40
0066   F040   NOP
0068   EC80   CALL 0x100, 0
```

When the "CALL SQR" is executed, the following screen shot shows that the return address 00006C is PUSHed onto the hardware stack, and the STKPTR is incremented by 1 to contain 6 as follows:

Hardware Stack

TOS	Stack Level	Return Address	Locatio
	0	Empty	
	1	000000	
	2	000000	
	3	000000	
	4	000000	
	5	000000	
⇨	6	00006C	.file
	7	000000	

After execution of the "RETURN" instruction at the end of the subroutine, the return address is popped from the hardware stack, and is placed in the program counter so that the program goes back to the main program. The STKPTR is then decremented by 1 to contain 5 as follows:

Hardware Stack

TOS	Stack Level	Return Address	Locatio
	0	Empty	
	1	000000	
	2	000000	
	3	000000	
	4	000000	
⇨	5	000000	
	6	00006C	.file
	7	000000	

After execution of the main program and the subroutine, the final answer (1303 hex) is stored in the file registers 0x21 (high byte) and 0x20 (low byte) as follows:

File Registers

Address	00	01	02	03	04	05	06
000	00	00	00	00	00	00	00
010	00	00	00	00	00	00	00
020	03	13	00	00	00	00	00
030	00	00	00	00	00	00	00
040	7E	08	23	30	00	00	00
050	30	00	00	00	00	00	00
060	00	00	00	00	00	00	00

7.7 Multiplication and Division Algorithms

As mentioned in Chapter 1, an *unsigned binary number* has no arithmetic sign, and therefore, is always positive. Typical examples are your age or a memory address, which are always positive numbers. An 8-bit unsigned binary integer represents all numbers from 00_{16} through FF_{10} (0_{10} through 255_{10}).

A *signed binary number*, on the other hand, includes both positive and negative numbers. It is represented in the microcontroller in two's complement form. For example, the decimal number +15 is represented in 8-bit two's complement form as 00001111 (binary) or 0F (hexadecimal). The decimal number -15 can be represented in 8-bit two's complement form as 11110001 (binary) or F1 (hexadecimal). Also, the most significant bit (MSB) of a signed number represents the sign of the number. For example, bit 7 of an 8-bit number represents the signs of the respective numbers. A "0" at the MSB represents a positive number; a "1" at the MSB represents a negative number. Note that the 8-bit binary number 11111111 is 255_{10} when represented as an unsigned number. On the other hand, 11111111_2 is -1_{10} when represented as a signed number.

As mentioned before, the PIC18F includes only unsigned multiplication instruction. The PIC18F instruction set does not provide any instructions for signed multiplication, or unsigned and signed division instructions. These algorithms are covered in detail in Section 4.3.6 of Chapter 4. A summary of the algorithms is provided in this section for convenience. The PIC18F assembly language programs using these algorithms are written in this section.

7.7.1 Signed Multiplication Algorithm

Signed multiplication can be performed using various algorithms. A simple algorithm follows. Assume that *M* (multiplicand) and *Q* (multiplier) are in two's complement form. Assume that M_n and Q_n are the most significant bits (sign bits) of the multiplicand (*M*) and the multiplier (*Q*), respectively. To perform signed multiplication, proceed as follows:

1. If $M_n = 1$, compute the two's complement of *M*; else, keep *M* unchanged.
2. If $Q_n = 1$, compute the two' s complement of *Q*; else, keep *Q* unchanged.
3. Multiply the $n - 1$ bits of the multiplier and the multiplicand using unsigned multiplication.
4. The sign of the result $S_n = M_n \oplus Q_n$.

5. If $S_n = 1$, compute the two's complement of the result obtained in step 3; else, keep result unchanged.

Next, consider a numerical example. Assume that M and Q are two's complement numbers. Suppose that $M = 1100_2$ and $Q = 0111_2$. Because $M_n = 1$, take the two's complement of $M = 0100_2$; because $Q_n = 0$, do not change Q. Multiply 0111_2 and 0100_2 using the unsigned multiplication method discussed before. The product is 00011100_2. The sign of the product $S_n = M_n \oplus Q_n = 1 \oplus 0 = 1$. Hence, take the two's complement of the product 00011100_2 to obtain 11100100_2, which is the final answer: -28_{10}.

Example 7.3 Using the signed multiplication algorithm just described, multiply two 8-bit signed numbers stored in data registers 0x15 and 0x17. Save 16-bit result in PRODH:PRODL.

(a) Flowchart the problem.

(b) Convert the flowchart to PIC18F assembly language program starting at address 0x100.

Solution

(a)

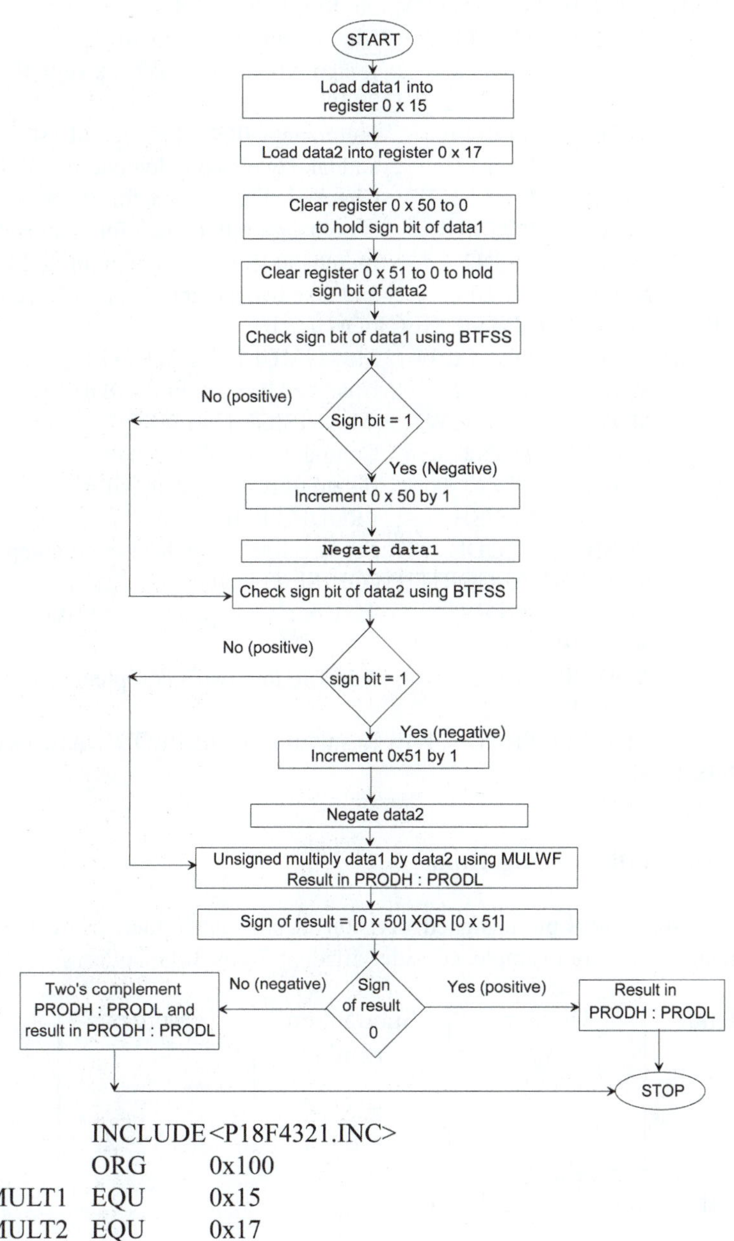

(b)

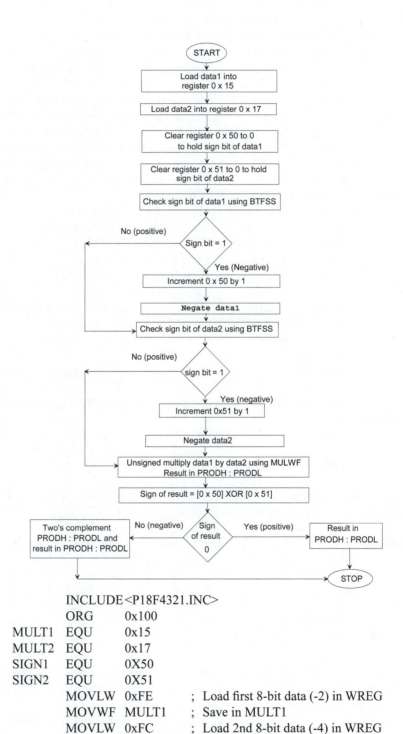

```
        INCLUDE<P18F4321.INC>
        ORG     0x100
MULT1   EQU     0x15
MULT2   EQU     0x17
SIGN1   EQU     0X50
SIGN2   EQU     0X51
        MOVLW   0xFE        ; Load first 8-bit data (-2) in WREG
        MOVWF   MULT1       ; Save in MULT1
        MOVLW   0xFC        ; Load 2nd 8-bit data (-4) in WREG
```

```
          MOVWF  MULT2     ; Save in MULT2
          CLRF   SIGN1     ; Clear [SIGN1] to 0
          CLRF   SIGN2     ; Clear [SIGN2] to 0
; STEPS 1 AND 2 OF THE ALGORITHM OF  SECTION 7.7.1
          BTFSS  MULT1, 7  ; Check sign bit 7  for 1 for 1st #
          BRA    NEG       ; If sign = 0, branch to check sign of
                           ; 2nd #
          INCF   SIGN1     ; Increment [SIGN1] if  sign of 1st # = 1
          NEGF   MULT1     ; and take two's complement of [MULT1]
NEG       BTFSS  MULT2, 7  ; Check sign bit 7  for 1 for 2nd #
          BRA    POSMUL    ; If both sign = 0, branch for unsigned mul
          INCF   SIGN2     ; Increment [SIGN2] if  sign of 2nd # = 1
          NEGF   MULT2     ; and take two's complement of [MULT2]
; STEP 3 OF THE ALGORITHM OF  SECTION 7.7.1
POSMULMOVF       MULT1, W  ; Move [MULT1] to WREG
          MULWF  MULT2     ; Unsigned product in PRODH:PRODL
          MOVF   SIGN1, W  ;  Move [SIGN1] to WREG
          XORWF  SIGN2     ; Compute sign of the result
          BTFSS  SIGN2, 0  ; If sign of result is 0, result in
          BRA    FINISH    ; PRODH:PRODL and stop
          COMF   PRODL     ; For negative result, take one's comp of PROD
; STEPS  4  AND 5  OF THE ALGORITHM OF  SECTION 7.7.1
          COMF   PRODH     ; Take one's complement of PRODH
          MOVLW  1
          ADDWF  PRODL     ; Add 1 to find two's complement
          MOVLW  0
          ADDWFC PRODH, F  ; Result in PRODH:PRODL in two's comp
FINISH    SLEEP
          END
```

7.7.2 Unsigned Division Algorithm

The 8-bit by 8-bit unsigned division can be performed using the repeated subtraction algorithm. For example, consider dividing 7_{10} by 3_{10} as follows:

Dividend	Divisor	Subtraction result	Counter
7_{10}	3_{10}	$7 - 3 = 4$	1
		$4 - 3 = 1$	$1 + 1 = 2$

Quotient = counter value = 2
Remainder = subtraction result = 1

Here, one is added to a counter whenever the subtraction result is greater than the divisor. The result is obtained as soon as the subtraction result is smaller than the divisor. The unsigned division algorithm can be summarized as follows:

First, load dividend and divisor into data registers, and initialize a counter to 0 to hold quotient (number of times divisor can be subtracted until subtraction result is less than the divisor). Data register storing the dividend will eventually contain the quotient (result of subtraction). The algorithm can be verified using numerical data.

1. Compare the dividend with the divisor for equality.
2. If equal, increment the counter by 1, and then perform (dividend - divisor). Store subtraction result in the data register holding the dividend. If not equal, go to step 3.
3. Compare if dividend > divisor. If greater, increment the counter by 1 and then perform (dividend - divisor). Store subtraction result in the data register holding the dividend.
4. Go to Step 1, and repeat steps 1 through 3 until subtraction result is less than or equal to 0.
5. When the subtraction result in the dividend is less than or equal to the divisor, go to halt. The counter will contain the quotient (number of times divisor can be subtracted until subtraction result is less than the divisor). Data register holding the dividend will contain the remainder (result of subtraction).

Example 7.4 Using the unsigned division algorithm just described, divide an 8-bit unsigned number (dividend) stored in data register 0x20 by another 8-bit unsigned number (divisor) stored in data register 0x30. Save 16-bit result in data registers 0x20 (remainder) and 0x30 (quotient).

(a) Flowchart the problem.

(b) Convert the flowchart to PIC18F assembly language program starting at address 0x100.

Solution

(a)

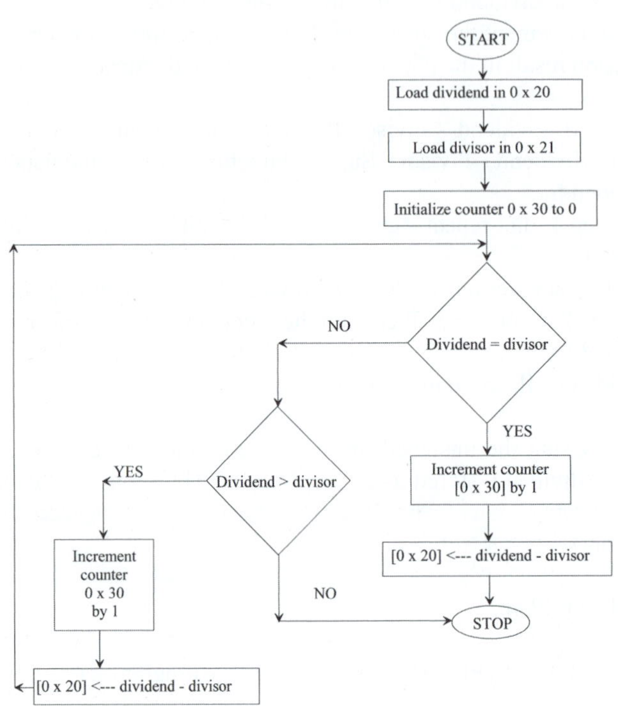

(b)

```
(b)        INCLUDE <P18F4321.INC>
           ORG      0x100
DIVIDEND   EQU      0x20
DIVISOR    EQU      0x21
COUNTER    EQU      0x30
           MOVLW    16          ; Dividend in WREG
           MOVWF    DIVIDEND    ; Store dividend in 0x20
           MOVLW    4           ; Divisor in WREG
           MOVWF    DIVISOR     ; Store divisor in 0x21
           CLRF     COUNTER     ; Clear Counter to 0
; STEPS 1 AND 2 OF THE ALGORITHM OF SECTION 7.7.2
BACK       CPFSEQ   DIVIDEND    ; If  dividend equals divisor, skip next instr.
           BRA      RESULT      ; If not equal, branch to RESULT
           INCF     COUNTER, F  ; Increment  [0x30] by 1
           SUBWF    DIVIDEND, F ; Subtract divisor from dividend,
                                ; remainder in 0x20
           BRA      FOREVER     ; Go to halt
; STEPS 3 , 4 AND 5 OF THE ALGORITHM OF SECTION 7.7.2
RESULT     CPFSGT   DIVIDEND    ; If  dividend greater than divisor, skip next inst.
           BRA      FOREVER     ; Quotient in 0x30, remainder in 0x20, halt
           INCF     COUNTER, F  ; Increment [0x20] by 1
           SUBWF    DIVIDEND, F ; Subtract divisor from dividend, result in 0x20
```

```
          BRA      BACK       ; Repeat
FOREVER   GOTO     FOREVER    ; Halt
          END
```

7.7.3 Signed Division Algorithm

The 8-bit by 8-bit signed division algorithm uses the equation for division: Dividend = quotient x divisor + remainder.

Signed division can be performed using various algorithms. A simple algorithm follows. Assume that DV (dividend) and DR (divisor) are in two's complement form. For the first case, perform unsigned division using repeated subtraction of the magnitudes without the sign bits. The sign bit of the quotient is determined as $DV_n \oplus DR_n$, where DV_n and DR_n are the most significant bits (sign bits) of the dividend (DV) and the divisor (DR), respectively. To perform signed division, proceed as follows:

1. If DVn = 1, compute the two's complement of DV; else, keep DV unchanged.
2. If DRn = 1, compute the two's complement of DR; else, keep DR unchanged.
3. Divide the n - 1 bits of the dividend by the divisor using unsigned division algorithm (repeated subtraction).
4. The sign of the Quotient, $Q_n = DV_n \oplus DR_n$. The sign of the remainder is the same as the sign of the dividend unless the remainder is zero.
5. If $Q_n = 1$, compute the two's complement of the quotient obtained in step 3; else, keep the quotient unchanged.

Example 7. 5 Write a PIC18F assembly language program at address 0x100 to divide an 8-bit signed number (dividend) in register 0x30 by another 8-bit signed number (divisor) in register 0x40. Use the signed division algorithm described in Section 7.7.3.

Solution

```
          INCLUDE<P18F4321.INC>
          ORG      0x100
COUNTER   EQU      0x20
DIVIDEND  EQU      0x30
DIVISOR   EQU      0x40
SIGN1     EQU      0X50
SIGN2     EQU      0X51
          MOVLW    4           ; Load 8-bit data (+4) in WREG
          MOVWF    DIVIDEND    ; Save in DIVIDEND
          MOVLW    0xFE        ; Load 8-bit data (-2) in WREG
          MOVWF    DIVISOR     ; Save in DIVISOR
          CLRF     SIGN1       ; Clear [SIGN1] to 0
          CLRF     SIGN2       ; Clear [SIGN2] to 0
; STEPS 1 AND 2 OF THE ALGORITHM OF SECTION 7.7.3
          BTFSS    DIVIDEND, 7 ; Check sign bit 7 for 1 for 1st #
          BRA      NEG         ; If sign= 0, branch to check sign of 2nd #
          INCF     SIGN1       ; Increment [SIGN1] if sign of 1st # = 1
          NEGF     DIVIDEND    ; Take two's complement of [DIVIDEND]
NEG       BTFSS    DIVISOR, 7  ; Check sign bit 7 for 1 for [DIVISOR]
          BRA      POSDIV      ; If both sign = 0, branch for unsigned division
```

```
            INCF      SIGN2          ; Increment [SIGN2] if sign of 2nd # = 1
            NEGF      DIVISOR        ; and take two's complement of [DIVISOR]
; STEP 3 OF THE ALGORITHM OF SECTION 7.7.3
POSDIV      MOVF      DIVISOR, W     ; Load divisor into WREG
            CLRF      COUNTER        ; Clear Counter to 0
BACK        CPFSEQ    DIVIDEND       ; If dividend equals divisor, skip next instr.
            BRA       RESULT1        ; If not equal, branch to RESULT
            INCF      COUNTER, F     ; Increment [0x20] by 1
            SUBWF     DIVIDEND, F    ; Subtract divisor from dividend, result in 0x20
RESULT1     CPFSGT    DIVIDEND       ; If dividend greater than divisor, skip next inst.
            BRA       RESULT         ; Quotient in 0x20, remainder in 0x30
            INCF      COUNTER, F     ; Increment [0x20] by 1
            SUBWF     DIVIDEND, F    ; Subtract divisor from dividend, result in 0x30
            BRA       BACK           ; Repeat
; STEPS 4 AND 5 OF THE ALGORITHM OF SECTION 7.7.3
RESULT      MOVF      SIGN1, W       ; Move [SIGN1] to WREG
            XORWF     SIGN2          ; Compute sign of the result
            BTFSS     SIGN2, 0       ; If sign of the quotient is 0, result in
            BRA       FINISH         ; 0x70 and stop
            NEGF      0x20           ; For negative result, take two's comp of [0x20]
            BTFSS     SIGN1, 0       ; Check sign of DIVIDEND
            BRA       FINISH         ; If plus, positive remainder in 0x30
            NEGF      0x30           ; If negative, negate remainder in 0x30
FINISH      SLEEP
            END
```

7.8 Advanced Programming Examples

In this section, more challenging assembly language programming examples using the PIC18F instruction set will be provided.

Example 7. 6 Write a PIC18F assembly language program at address 0x100 for copying a string in the program memory starting at address 0x500 into the data memory starting at address 0x40. Assume that the string is null terminated. The assembly language program is basically equivalent to the following string library function in C:

```
void strcpy (char t[ ]), char s[ ]){   // copy the string s into t and advance pointer
   while ( *t++ = *s++)
             ;
}
```

Solution

```
      INCLUDE   <P18F4321.INC>
      ORG       0x500                    ; Starting address of the source string
      DB        "CAL POLY POMONA", 0     ; A null terminated string
      ORG       0x100
```

```
        MOVLW    0
        MOVWF    TBLPTRL          ; Initialize  8-bit TBLPTRL with 0
        MOVWF    TBLPTRU          ; Initialize 5-bit TBLPTRU with 0
        MOVLW    0x05             ; Initialize 8-bit TBLPTRH with 0x05
        MOVWF    TBLPTRH
        LFSR     2, 0x40          ; Initialize FSR2 with destination address
LOOP    TBLRD*+                   ; Read a character from program memory
        MOVF     TABLAT, W        ; Save in WREG
        BZ       EXIT             ; If it is a null, then EXIT
        MOVWF    POSTINC2         ; Copy to destination and increment pointer
        BRA      LOOP
EXIT    MOVWF    INDF2            ; Copy the null character into the destination
        SLEEP                     ; Halt
        END
```

Example 7.7 Write a subroutine in PIC18F assembly language program at address 0x60 to find the nth number (for example, $n = 0$ to 6) of the Fibonacci sequence. The subroutine will obtain the desired Fibonacci number using a lookup table stored starting at an address 0x200 in the program memory. Also, write the main program at address 0x100 that will transfer Fibonacci array from program memory stored at address starting at 0x200 to data memory stored starting at address 0x40, initialize STKPTR to 0x15, store a number (0 to 6) in WREG, initialize data pointer FSR1 to 0x40, call the sbroutine, and stop. The Fibonacci sequence for $n = 0$ to 6 is provided below:

n	Fib(n)
0	1
1	1
2	2
3	3
4	5
5	8
6	13

Solution

```
; MAIN  PROGRAM
                INCLUDE  <P18F4321.INC>
                ORG      0x100            ; Starting address of  the main program
       COUNTER  EQU      0x20
; FIBONACCI ARRAY TRANSFER FROM PROGRAM MEMORY TO DATA MEMORY
                MOVLW  UPPER ADDR  ; Move upper 5 bits (00H) of address
                MOVWF  TBLPTRU     ; to TBLPTRU
                MOVLW  HIGH ADDR   ; Move bits 15-8 (02H) of address
                MOVWF  TBLPTRH     ; to  TBLPTRH
                MOVLW  LOW ADDR    ; Move bits 7-0 (00H) of address
                MOVWF  TBLPTRL     ; to TBLPTRL
                LFSR   0, 0x40     ; Initialize FSR0 to 0x40 to be used as
                                   ; destination pointer in data memory
                MOVLW  D'7'        ; Initialize COUNTER with 7
```

```
            MOVWF  COUNTER      ; Move [WREG] into COUNTER
LOOP        TBLRD*+             ; Read data from program memory into
                                ; TABLAT, increment TBLPTR by 1
            MOVF   TABLAT, W    ; Move [TABLAT] into WREG
            MOVWF  POSTINC0     ; Move W into data memory pointed to
                                ; by FSR0, and then increment FSR0 by 1
                                ; memory address  0x000200
            DECF   COUNTER, F   ; Decrement COUNTER BY 1
            BNZ    LOOP         ; Branch  if  Z = 0
; INITIALIZE STKPTR, LOAD n, INITIALIZE DATA POINTER, CALL SUBROUTINE
            MOVLW  0x15         ; Initialize STKPTR to 0x15
            MOVWF  STKPTR
            MOVLW  4            ; Move n into WREG
            LFSR   1, 0x40      ; Load 0x40 into FSR0 to be used as pointer
            CALL   FIBNUM
FINISH      BRA    FINISH
; READ THE FIBONACCI NUMBER FOR n FROM DATA MEMORY INTO 'W' USING
; MOVF WITH INDEXED ADDRESSING MODE
      END
; SUBROUTINE
            ORG    0x60
FIBNUM      MOVF   PLUSW1, W    ; Result in WREG
                   RETURN       ; Return to FINISH in main
            ORG    0x200
ADDR        DB     1, 1, 2, 3, 5, 8, 13 ;     Fibonacci numbers
                   END
```

Example 7.8 Without using a lookup table and the MOVFF with indexed addressing mode as in Example 7.7, write a subroutine in PIC18F assembly language at address 0x50 to find the nth number (0 to 6) of the Fibonacci sequence. The subroutine will return the desired Fibonacci number in WREG based on 'n' stored by the main program. Also, write the main program at address 0x100 that will store the nth number (0 to 6) in WREG, call the subroutine, and stop. The Fibonacci sequence for $n = 0$ to 6 is provided below:

n	Fib(n)
0	1
1	1
2	2
3	3
4	5
5	8
6	13

Solution

This program can be written with the RETLW instruction that is ideal for returning the desired value using an operation alternate to using a table lookup with indexed addressing mode shown in Example 7.7. Note that, the RETLW k loads the 8-bit immediate data k into WREG, and returns to the main program by loading the program counter with the address from the top of the hardware stack. The assembly language program is provided below:

```
        INCLUDE <P18F4321.INC>
        ORG     0x100       ; Main program
        MOVLW   0x10        ; Initialize STKPTR with  0x10
        MOVWF   STKPTR
        MOVLW   5           ; Load n into WREG
        CALL    FIBNUM      ; Call subroutine FIBNUM to find Fibonacci #
HERE    BRA     HERE        ; Halt
        ORG     0x50        ; Subroutine
FIBNUM  MULLW   2           ; PRODH:PRODL ← 2 x n, offset of RETLW table.
                            ; 'n' is  multiplied by 2 since the  instruction size is
                            ; word
        MOVFF   PRODL, W    ; Save low order 8 bits of the product in WREG
        ADDWF   PCL         ; PCL = PCL + 2 x n
        ;Fibonacci number table follows
        RETLW   0x00
        RETLW   0x01
        RETLW   0x02
        RETLW   0x03
        RETLW   0x05
        RETLW   0x08
        RETLW   0x0D        ; 13 in decimal
        END
```

Example 7.9 Write a PIC18F assembly language program at address 0x200 to add two 16-bit numbers (N1 and N2), each containing two ASCII digits. The first 16-bit number (N1) is stored in two consecutive locations (from LOW to HIGH) in data memory with the low byte pointed to by address 0x40, and the high byte pointed to by address 0x41. The second 16-bit number (N2) is also stored in two consecutive locations (from LOW to HIGH) in data memory with low byte pointed to by 0x50, and the high byte pointed to by 0x51. Store the packed BCD result in WREG.

Solution

Note that ASCII codes for decimal numbers 0 through 9 are 30H through 39H (see Chapter 1).

Numerical example: Assume [N1] = 3439H and [N2] = 3231H. The procedure for adding the two 16-bit ASCII numbers (N1 and N2) will be as follows:

1. Convert N1 and N2 to unpacked BCD numbers by retaining the low four bits using ANDWF instruction. This means that N1 = 0409H and N2 = 0201H.

2. Logically shift the high byte of N1 four times to the left so that the high byte will be converted from 04H to 40H. This is equivalent to swapping the low four bits (nibble) with the high four bits (nibble) using the SWAPF instruction. Logically OR this with the low byte of N1. Hence, N1 will be converted from unpacked BCD (0409H) to packed BCD 49H. Similarly, convert N2 from unpacked BCD (0201H) to packed BCD (21H).

3. Add (binary addition) the two packed BCD numbers (49H, 21H) using ADDWF instruction to obtain the following result:

```
First packed BCD byte    = 49H = 0100 1001
Second packed BCD byte = 21H = 0010 0001
                                   -----------------
Result after binary addition       0110  1010 (6AH)
```

4. Perform BCD correction on the binary result 0110 1010

```
Add 6 using DAW instruction                  0110
                                        -----------------
                                        0111 0000 = 70H (Correct packed BCD result)
```

The PIC18F assembly language is provided below:

```
          INCLUDE <P18F4321.INC>
          ORG     0x200
COUNTER   EQU     0x45
          MOVLW   0x39          ; #1 LOAD LOW BYTE OF N1 INTO 0x40
          MOVWF   0x40
          MOVLW   0x34          ; #2 LOAD HIGH BYTE OF N1 INTO 0x41
          MOVWF   0x41
          MOVLW   0x31          ; #3 LOAD LOW BYTE OF N2 INTO 0x50
          MOVWF   0x50
          MOVLW   0x32          ; #4 LOAD HIGH BYTE OF  N2 INTO 0x51
          MOVWF   0x51
          MOVLW   2             ; #5 INITIALIZE  COUNTER
          MOVWF   COUNTER
          LFSR    0, 0x40       ; #6  INITIALIZE FSR0 TO 0x40
          LFSR    1, 0x50       ; #7 INITIALIZE FSR1 TO 0x50
; STEP1: CONVERT N1 AND N2 TO UNPACKED BCD.
START     MOVLW   0x0F
          ANDWF   POSTINC0, F   ; #8 CONVERT  N1 TO UNPACKED BCD
          ANDWF   POSTINC1, F   ; #9 CONVERT  N2 TO UNPACKED BCD
          DECF    COUNTER, F    ; #10 DECREMENT  COUNTER BY 1
          BNZ     START         ; #11 BRANCH  IF NOT  ZERO
; UNPACKED BCD RESULT 0x41 (N1 HIGH BYTE), 0x40 (N1 LOW BYTE), 0x51(N2 HIGH
; BYTE), 0x50 (N2 LOW  BYTE)
; STEP2: CONVERT N1 AND N2 FROM  UNPACKED TO PACKED BCD
          SWAPF   0x41, W       ; #12 SWAP LOW NIBBLE OF 0x41 WITH
                                ; HIGH NIBBLE AND STORE  IN WREG
          IORWF   0x40, F       ; #13 OR [W] WITH [0x40], PACKED BCD N1
          SWAPF   0x51, W       ; #14 SWAP LOW NIBBLE OF 0x51 WITH
                                ; HIGH NIBBLE AND STORE IN WREG
          IORWF   0x50, W       ; #15 OR [WREG] WITH [0x50], PACKED BCD
                                ; N2 IN WREG
; STEP3: PERFORM BINARY ADDITION BETWEEN N1 (PACKED BCD) in WITH N2
; (PACKED BCD) AND STORE RESULT IN WREG
          ADDWF    0x40, W      ; #16  BINARY RESULT IN WREG
```

```
; STEP4: ADJUST (BCD CORRECTION) [WREG] TO CONTAIN CORRECT PACKED BCD
            DAW                       ; #17 ADJUST THE RESULT TO CONTAIN
                                      ; CORRECT PACKED BCD
FINISH      BRA       FINISH          ; HALT
            END
```

Note: The above program will be explained in the following. Note that the # sign along with the line number is placed before each comment in order to explain the program. ASCII data to be added are assumed to be 3439H and 3231H. The purpose of the program is to convert the first number, ASCII 3439H to unpacked BCD 0409H, and then to packed BCD 49H, and similarly, the second number, ASCII 3231H, to unpacked BCD 0201H, and then to packed BCD 21H. Finally, the two packed BCD numbers will be added in binary using PIC18F's ADDWF instruction, and then the result in WREG will be converted to correct packed BCD using DAW.

Line #'s 1 through 4 initialize N1 and N2 so that [0x40] = 39H, [0x41] = 34H, [0x50] = 31H, and [0x51] = 32H. Line #5 initializes COUNTER with loop count of 2 for converting the numbers from ASCII to unpacked BCD. Line #'s 6 and 7 initialize FSR0 and FSR1 with 0x40 and 0x50, respectively. Line #'s 8 through 11 convert the two bytes of ASCII codes in 0x41 (high byte) and 0x40 (low byte) into unpacked BCD in 0x41 (high byte) and 0x40 (low byte). Also, Line #'s 8 through 11 convert the ASCII numbers, N1 and N2 into their corresponding unpacked BCD bytes.

Line #'s 12 through 15 convert the unpacked BCD numbers (N1 and N2) into packed BCD bytes. This is done by swapping high unpacked bytes of N1 and N2 , and then ORing with the corresponding low unpacked bytes. Line #16 performs binary addition of the two packed BCD bytes (N1 and N2), and stores the binary result in WREG. The DAW instruction at Line #17 adjusts the contents of WREG to provide the correct packed BCD result.

7.9 PIC18F Delay Routine

Typical PIC18F software delay routines can be written by loading a "counter" with a value equivalent to the desired delay time, and then decrementing the "counter" in a loop, using typically MOVE, DECREMENT, and conditional BRANCH instructions. For example, the following PIC18F instruction sequence can be used for a delay loop:

```
          MOVLW   COUNT
          MOWF    0x20
DELAY     DECF    0X20, F
          BNZ     DELAY
```

Note that DECF in the above decrements the register 0x20 by one, and if [0x20] ≠0, branches to DELAY; if [0x20] = 0, the PIC18F executes the next instruction. The initial loop counter value of "COUNT" can be calculated using the machine cycles (Appendix D) required to execute the following PIC18F instructions:

```
MOVLW     (1 cycle)
MOVWF     (1 cycle)
DECF      (1 cycle)
BNZ       (2/1 cycles)
```

Note that the BNZ instruction requires two different execution times. BNZ requires two cycles when the PIC18F branches if Z = 0. However, the PIC18F goes to the next instruction and does not branch when Z = 1. This means that the DELAY loop will require two cycles for “COUNT” times, and the last iteration will take one cycle. The desired delay time can be obtained by loading register 0x20 with the appropriate COUNT value.

Assuming 1 MHz default crystal frequency, the PIC18F’s clock period will be 1 μsec. Note that the PIC18F divides the crystal frequency by 4. This is equivalent to multiplying the clock period by 4. Hence, each instruction cycle will be 4 microseconds. For a 100-microsecond delay, total cycles = $\frac{\text{100 micro sec}}{\text{4 micro sec}}$ = 25. The BNZ in the loop will require two cycles for (COUNT - 1) times when Z = 0 and the last iteration will take 1 cycle when no branch is taken (Z = 1). Thus, total cycles including the MOVLW = 1 + 1 + 1 + 2 × (COUNT - 1) + 1 = 25. Hence, COUNT = 11.5. Therefore, register 0x20 should be loaded with an integer value of 12 for an approximate delay of 100 microseconds.

Now, in order to obtain delay of one millisecond, the above DELAY loop of 100 miicroseconds can be used with an external counter. Counter value = $\frac{\text{1 milli sec}}{\text{100 micro sec}}$ = 10. The following instruction sequence will provide an approximate delay of one millisecond:

```
        MOVLW  D'10'
        MOVWF  0x30      ; Initialize counter 0x30 for one-millisecond delay
BACK    MOVLW  D'12'
        MOVWF  0x20      ; Initialize counter 0x20 for 100-microsecond delay
DELAY   DECF   0X20, F   ; 100-microsec delay
        BNZ    DELAY
        DECF   0X30, F
        BNZ    BACK
```

Next, the delay time provided by the above instruction sequence can be calculated.

As before, assuming 1 MHz crystal, each instruction cycle is 4 microseconds. Total delay in seconds from the above instruction sequence

= Execution time for MOVLW + Execution time for MOVWF +
10 x (100-microsecond delay) + Execution time for DECF +
Execution time for BNZ (Z = 1) + Execution time for DECF +
(10-1) x Execution time for BNZ (Z = 0) + Execution time for BNZ (Z = 1)

= 1 x (4 microsec) + 1 x (4 microsec) + (1000 microseconds)
+ 1x (4 microsec) + 9 x (2 x 4 microsec) + 1 x (4 microsec)

= 1.088 milliseconds.

This is approximately equivalent to the desired 1-millisecond delay. In other words, the delay is 1.088 milliseconds rather than 1 millisecond. This is because the execution times of MOVLW D’10’, MOVWF 0x30, DECF 0x30, F, and BNZ DELAY are discarded.

Example 7.10 Assume 1 MHz PIC18F. Consider the following subroutine:

```
DELAY   MOVLW   D'100'
        MOVWF   0x20
DLOOP   DECFSZ  0x20, F
        BRA     DLOOP
        RETURN
```

(a) Calculate the time delay provided by the above subroutine.

(b) Calculate the counter value to be loaded into data register 0x20 for 1 msec delay.

Solution

(a) Each instruction in the above subroutine is executed in one cycle except the DECFSZ instruction. DECFSZ is executed in one cycle if it does not skip, and two cycles if it skips.

Hence, total instruction cycles = Cycle for MOVLW + Cycle for MOVWF + (100-1) (Cycles for DECFSZ if it does not skip and BRA instructions) + (Cycle for DECFSZ if it skips) + Cycle for RETURN
= 1 + 1 + 99 (1 + 1) + 2 + 1
= 203

Since for the PIC18F, one instruction cycle = 4 clock cycles, total delay = (203) x 4 = 812 clock cycles. Also, for 1 MHz clock, each clock cycle is 1 μsec. Hence, total time delay = 812 μsec .

(b) Let n be the counter value. Hence, (2 + 2 x (n-1) + 3) x 4 = 1000 μsec.
Note that 1 msec = 1000 μsec. Therefore, n = 123.5, and data register 0x20 should be loaded with 124 for an approximate delay of 1 msec.

Questions and Problems

7.1 Write a PIC18F assembly language program at address 0x150 to subtract two 16-bit numbers as follows: [0x21][0x20] - [0x31][0x30] → [0x40][0x41], if [0x50] is odd. If [0x50] is even, store 0's in [0x41][0x40].

7.2 Write a PIC18F assembly program at address 0x200 to multiply a 4-bit unsigned number in the low nibble of 0x30 by another 4-bit unsigned number in the high nibble of 0x30. Store the result in 0x31.

7.3 Write a PIC18F assembly language program at address 0x100 to multiply a 4-bit unsigned number stored in the high nibble of data register 0x30 by a 4-bit signed number stored in the low nibble of data register 0x30. Store the 8 bit result in 0x30.

7.4 Write a PIC18F assembly language program at address 0x150 to convert temperature from Fahrenheit to Celsius using the equation: $C = [(F - 32)/9] \times 5$; assume that the temperature in Fahrenheit is 8 bits wide to be loaded into data register 0x20. Assume that the temperature is always positive. Store the 8-bit result in data register 0x21.

7.5 Write a PIC18F assembly language program at address 0x100 to find $X^2/128_{10}$, where X is an 8-bit unsigned number stored in data register 0x40. Store the 8-bit result in data register 0x50. Discard the remainder of the division.

7.6 Write a subroutine in PIC18F assembly language at address 0x200 to perform $(X^2 + Y^2)$, where X is a signed 8-bit number and Y is an unsigned 8-bit number. Use subroutine for signed multiplication in PIC18F assembly as needed. Also, write the main program at address 0x100 in PIC18F assembly language that will initialize FSR0 to 0x0070, X and Y to arbitrary data, initialize STKPTR to 0x10, call the subroutines to compute $(X^2 + Y^2)$, and then push 8-bit result onto the software stack pointed to by FSR0.

7.7 Write a PIC18F assembly program at address 0x100 that is equivalent to the following C language segment:

```
sum = 0;
for (i = 0; i <= 9; i = i + 1)
sum = sum + x[i] * y[i];
```

Assume that the arrays x[i] and y[i] contain unsigned 8-bit numbers already stored in memory starting at data memory addresses 0x20 and 0x30, respectively. Store the 8-bit result at address 0x50.

7.8 Write a PIC18F assembly program at address 0x150 to compare two strings of 10 ASCII characters. The first string is stored starting at 0x30. The second string is stored at location 0x50. The ASCII character in location 0x30 of string 1 will be compared with the ASCII character in location 0x50 of string 2, [0x31] will be compared with [0x51], and so on. Each time there is a match, store 0xEE onto the software stack; otherwise, store 0x00 onto the software stack. Initialize software stack pointer FSR0 to 0x60.

7.9. Write a PIC18F assembly program at address 0x100 to divide a 9-bit unsigned number in the high 9 bits (bits 8-1 in bits 7-0 of register 0x30 and bit 0 in bit 7 of register 0x31) by 8_{10}. Do not use any division instruction. Store the result in register 0x50. Discard the remainder.

7.10 Write a PIC18F assembly language program at address 0x200 that will check whether the 16-bit signed number in registers [0x31][0x30] is positive or negative. If the number is positive, the program will multiply the 16-bit unsigned number (bits 12 through 15 as 0's) in [0x21][0x20] by 16, and provide a 16-bit result; otherwise, the program will set the byte in register 0x40 to all ones. Use only data movement, shift, bit manipulation, and program control instructions. Assume the 16-bit signed and unsigned numbers are already loaded into the data registers.

7.11 Assume that several 8-bit packed BCD numbers are stored in data memory locations from 0x10 through 0x2D. Write a PIC18F assembly language program at address 0x100 to find how many of these numbers are divisible by 5, and save the result in data memory location 0x40.

7.12 Write a program at address 0x100 in PIC18F assembly language to add two 32-bit packed BCD numbers. BCD number 1 is stored in data registers starting from 0x20 through 0x23, with the least significant digit at register 0x23 and the most significant digit at 0x20. BCD number 2 is stored in data registers starting from 0x30 through 0x33, with the least significant digit at 0x33 and the most significant digit at 0x30. Store the result as packed BCD digits in 0x20 through 0x23.

7.13 Write a subroutine at address 0x100 in PIC18F assembly language program to find the square of a BCD digit (0 to 9) using a lookup table. The subroutine will store the desired result in WREG based on the BCD digit stored by the main program. The lookup table will store the square of the BCD numbers starting at program memory address 0x300. Also, write the main program at address 0x200 that will initialize STKPTR to 0x30, store the BCD digit (0 to 9) in WREG, call the subroutine, and stop. Use indexed addressing mode.

7.14 Write a subroutine in PIC18F assembly language program at address 0x100 to find the square of a BCD digit (0 to 9) and store it in WREG. The subroutine will return the desired result based on the BCD digit stored by the main program. Also, write the main program at address 0x200 that will initialize STKPTR to 0x20, store the BCD digit (0 to 9) in WREG, call the subroutine, and stop. Do not use indexed addressing mode.

7.15 Write a subroutine at address 0x150 in PIC18F assembly language to convert a 3-digit unpacked BCD number to binary using unsigned multiplication by 10, and additions. The most significant digit is stored in a memory location starting at register 0x30, the next digit is stored at 0x31, and so on. Store the 8-bit binary result (*N*) in register 0x50. Note that arithmetic operations for obtaining *N* will provide binary result. Use the value of the 3-digit BCD number,

$$N = N2 \times 10^2 + N1 \times 10^1 + N0$$
$$= ((10 \times N2)+N1) \times 10+N0$$

7.16 Write a subroutine in PIC18F assembly language at 0x100 to compute

$$Z = \sum_{i=1}^{8} X_i$$

Assume the X_i's are unsigned 8-bit and stored in consecutive locations starting at 0x30 and Z is 8-bit. Also, assume that FSR1 points to the X_i's. Write the main program in PIC18F assembly language at 0x150 to perform all initializations (FSR1 to 0x30, STKPTR to 0x20), call the subroutine, and then compute *Z*/8. Discard remainder of Z/8. Assume data are already loaded in data registers.

7.17 Write a PIC18F assembly language program to estimate the square root of an 8-bit integer number P using the algorithm provided in the following:
The sum of odd integers is always a perfect square. For example, $1 = 1^2$, $1+3 = 2^2$, $1+3+5 = 3^2$, and so on. Specifically, $\sum_{i=1}^{k}(2i - 1) = k^2$. This property is useful in approximating the square root of an 8-bit unsigned number P. For example, if P = 17, the square root of P can be estimate as follows:
Subtract 1 from P so that P becomes 16; since the subtraction went through, add 1 to a counter. Hence, counter value is 1.
Subtract 3 from P so that P becomes 13; since the subtraction went through, add 1 to a counter. Hence, counter value is 2.
Subtract 5 from P so that P becomes 8; since the subtraction went through, add 1 to a counter. Hence, counter value is 3.
Subtract 7 from P so that P becomes 1; since the subtraction went through, add 1 to a counter. Hence, counter value is 4. Now, when 9 is subtracted from the existing value of P (P = 1), the result becomes negative (-8) meaning that the subtraction did not go through. The process terminates, and the integer square root approximation for 17 is 4.

7.18 Consider the following PIC18F DELAY subroutine:

```
DELAY    MOVLW    Q
         MOVWF    Q
LOOP1    MOVLW    100
         MOVWF    P
LOOP2    DECF     P, F
         BNZ      LOOP2
         DECF     Q, F
         BNZ      LOOP1
         RETURN
```

Assuming 1 MHz PIC18F, determine the value of Q such that when this subroutine is called, a delay of 145.940 msec will be generated.

7.19 Consider the following loop statement in C language;

for (p = 80; p > 0; p--); // a dummy loop with no statement

The above C loop can be represented using PIC18F assembly language in two ways:

Method 1: Using conventional way

```
            MOVWLF    0x50
            MOVWF     P
LOOP        DECF      P, F
            BNZ       LOOP
```

Method 2: Using SKIP instruction

```
            MOVWLF    0x50
            MOVWF     P
LOOP        DECFSZ    P, F
            BRA       LOOP
DONE        -------
```

Assume 1 MHz PIC18F clock. Calculate the execution time of each delay loop (Method 1 and Method 2). Which delay loop will be executed faster?

8

PIC18F HARDWARE AND INTERFACING: PART 1

In this chapter we describe the first part of hardware aspects of the PIC18F4321. Topics include PIC18F4321 pins and signals, clock and reset circuits, programmed and interrupt I/O, seven-segment and LCD displays, and hexadecimal keyboard interfacing techniques.

8.1 PIC18F Pins and Signals

The PIC18F4321 is contained in three types of packaging as follows:

- 40-pin plastic dual in-line package (PDIP)
- 44-pin quad flat no-lead plastic package (QFN)
- 44-pin thin plastic quad flat pack package (TQFP)

Figure 8.1 shows the PIC18F4321 pin diagram for a PDIP. A brief description of all pins and signals for the PIC18F4321 contained in the 40-pin PDIP is provided in Table 8.1.

40-Pin PDIP

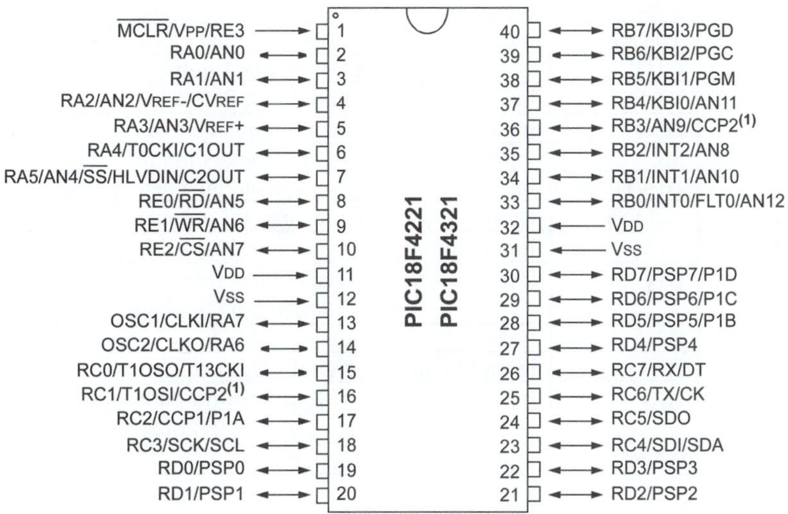

FIGURE 8.1 PIC18F4321 pins and signals

There are two VDD (Vcc) pins and two VSS (ground) pins which are not shared (multiplexed) with other pins. The range of voltages for the VDD pins are from + 4.2 V to +5.5 V. However, the VDD pins are normally connected to +5 V. The VSS pins are connected to ground. The maximum power dissipation for the PIC18F4321 is one watt. Note that multiple pins for power and ground are used in order to distribute the power and reduce noise problems at high frequencies.

All other 36 pins are multiplexed (shared) with other signals. There are 36 pins assigned to five I/O ports, namely, Port A (8-bit, RA0-RA7), Port B (8-bit, RB0-RB7), Port C (8-bit, RC0-RC7), Port D (8-bit, RD0-RD7), and Port E (4-bit, RE0-RE3). These pins are multiplexed with other signals such as the clock/oscillator, reset, external interrupt, analog inputs, and CCP (Capture/Compare/Pulse Width Modulation).
Note 1: RB3 is the alternate pin for CCP2 multiplexing.

TABLE 8.1 PIC18F4321 pinout description

Pin number	Pin name	Pin type	Description
1	$\overline{MCLR}$	Input	Master clear reset ; active low input
	Vpp	Input	Programming voltage input for the flash memory
	RE3	Input	Digital input; Port E bit 3
2	RA0	Input/Output	Digital I/O; Port A bit 0
	AN0	Input	Analog input 0
3	RA1	Input/Output	Digital I/O; Port A bit 1
	AN1	Input	Analog input 1
4	RA2	Input/Output	Digital I/O; Port A bit 2
	AN2	Input	Analog input 2
	VREF-	Input	A/D reference voltage (low) input
	CVREF	Output	Comparator reference voltage output
5	RA3	Input/Output	Digital I/O; Port A bit 3
	AN3	Input	Analog input 3
	VREF+	Input	A/D reference voltage (high) input
6	RA4	Input/Output	Digital I/O; Port A bit 4
	TOCKI	Input	Timer0 external clock input
	C1OUT	Output	Comparator 1 output
7	RA5	Input/Output	Digital I/O; Port A bit 5
	AN4	Input	Analog input 4
	SS	Input	SPI slave select input
	HLVDIN	Input	High/low-voltage detect input
	C2OUT	Output	Comparator 2 output
8	RE0	Input/Output	Digital I/O; Port E bit 0
	$\overline{RD}$	Input	Read control for parallel slave port (see also $\overline{WR}$ and $\overline{CS}$ pins).
	AN5	Input	Analog input 5
9	RE1	Input/Output	Digital I/O; Port E bit 1
	$\overline{WR}$	Input	Write control for parallel slave port (see $\overline{CS}$ and $\overline{RD}$ pins).
	AN6	Input	Analog input 6

TABLE 8.1 PIC18F4321 Pinout description (continued)

Pin number	Pin name	Pin type	Description
10	RE2	Input/Output	Digital I/O; Port E bit 2
	$\overline{CS}$	Input	Chip Select control for parallel slave port (see related $\overline{RD}$ and $\overline{WR}$).
	AN7	Input	Analog input 7
11	VDD	Power	Positive supply for logic and I/O pins.
12	VSS	Power	Ground reference for logic and I/O pins.
13	OSC1	Input	Oscillator crystal input or external clock source input.
	CLKI	Input	External clock source input. Always associated with pin function OSC1. (See related OSC1/CLKI, OSC2/CLKO pins.)
	RA7	Input/Output	Digital I/O; Port A bit 7
14	OSC2	Output	Oscillator crystal output. Connects to crystal or resonator in crystal oscillator mode. In RC, EC, and INTIO modes, OSC2 pin outputs.
	CLKO	Output	CLKO which has one-fourth the frequency of OSC1 and denotes the instruction cycle rate.
	RA6	Input/Output	Digital I/O; Port A bit 6
15	RC0	Input/Output	Digital I/O; Port C bit 0
	T1OSO	Output	Timer1 oscillator analog output
	T13CKI	Input	Timer1/Timer3 external clock input
16	RC1	Input/Output	Digital I/O; Port C bit 1
	T1OSI	Input	Timer1 oscillator analog input
	CCP2	Input/Output	Capture 2 input/Compare 2 output/PWM 2 output; default assignment for CCP2 when Configuration bit, CCP2MX, is set.
17	RC2	Input/Output	Digital I/O; Port C bit 2
	CCP1	Input/Output	Capture 1 input/Compare 1 output/PWM 1 output
	P1A	Output	Enhanced CCP1 output
18	RC3	Input/Output	Digital I/O; Port C bit 3
	SCK	Input/Output	Synchronous serial clock input/output for SPI mode.
	SCL	Input/Output	Synchronous serial clock input/output for I^{2}C™ mode.
19	RD0	Input/Output	Digital I/O; Port D bit 0
	PSP0	Input/Output	Parallel Slave Port data
20	RD1	Input/Output	Digital I/O; Port D bit 1
	PSP1	Input/Output	Parallel slave port data
21	RD2	Input/Output	Digital I/O; Port D bit2
	PSP2	Input/Output	Parallel slave port data
22	RD3	Input/Output	Digital I/O; Port D bit 3
	PSP3	Input/Output	Parallel slave port data

TABLE 8.1 PIC18F4321 Pinout description (continued)

Pin number	Pin name	Pin type	Description
23	RC4	Input/Output	Digital I/O
	SDI	Input	SPI data in
	SDA	Input/Output	I^2C data I/O
24	RC5	Input/Output	Digital I/O; Port C bit 5
	SDO	Output	SPI data out
25	RC6	Input/Output	Digital I/O; Port C bit 6
	TX	Output	EUSART asynchronous transmit
	CK	Input/Output	EUSART synchronous clock (see related RX/DT).
26	RC7	Input/Output	Digital I/O; Port C bit 7
	RX	Input	EUSART asynchronous receive
	DT	Input/Output	EUSART synchronous data (see related TX/CK)
27	RD4	Input/Output	Digital I/O; Port D bit 4
	PSP4	Input/Output	Parallel slave port data
28	RD5	Input/Output	Digital I/O; port D bit 5
	PSP5	Input/Output	Parallel slave port data
	P1B	Output	Enhanced CCP1 output
29	RD6	Input/Output	Digital I/O; port D bit 6
	PSP6	Input/Output	Parallel slave port data
	P1C	Output	Enhanced CCP1 output
30	RD7	Input/Output	Digital I/O; Port D bit 7
	PSP7	Input/Output	Parallel slave port data
	P1D	Output	Enhanced CCP1 output
31	VSS	Power	Ground reference for logic and I/O pins
32	VDD	Power	Positive supply for logic and I/O pins
33	RB0	Input/Output	Digital I/O; Port B bit 0
	INT0	Input	External interrupt 0
	FLT0	Input	PWM fault input for enhanced CCP1
	AN12	Input	Analog input 12
34	RB1	Input/Output	Digital I/O; Port B bit 1
	INT1	Input	External interrupt 1
	AN10	Input	Analog input 10
35	RB2	Input/Output	Digital I/O; Port B bit 2
	INT2	Input	External interrupt 2
	AN8	Input	Analog input 8
36	RB3	Input/Output	Digital I/O; Port B bit 3
	AN9	Input	Analog input 9
	CCP2	Input/Output	Capture 2 input/compare 2 output/ PWM 2 output; alternate assignment for CCP2 when configuration bit CCP2MX, is cleared.
37	RB4	Input/Output	Digital I/O; Port B bit 7
	KBI0	Input	Interrupt-on-change pin
	AN11	Input	Analog input 11
38	RB5	Input/Output	Digital I/O; Port B bit 5
	KBI1	Input	Interrupt-on-change pin

TABLE 8.1 PIC18F4321 Pinout description (continued)

Pin number	Pin name	Pin type	Description
	PGM	Input/Output	Low-voltage programming enable pin
39	RB6	Input/Output	Digital I/O; Port B bit 6
	KBI2	Input	Interrupt-on-change pin
	PGC	Input/Output	In-circuit debugger and programming clock pin
40	RB7	Input/Output	Digital I/O; Port B bit 7
	KBI3	Input	Interrupt-on-change pin
	PGD	Input/Output	In-circuit debugger and ICSP programming data pin.

The PIC18F pins associated with clock, reset, and I/O will be discussed in the following.

8.1.1 Clock

Upon hardware reset, the PIC18F4321 operates at an internal clock frequency of 1 MHz (default). This means that with no crystal oscillator circuit connected to the PIC18F4321, the microcontroller operates from an internal clock of 1 MHz.

The PIC18F4321 can also be operated in ten different oscillator modes. The user can program the Configuration bits, FOSC3:FOSC0, in Configuration Register to select one of these ten modes. Note that this configuration regisgter is mapped as address 300001H in program memory, and can be accessed using TBLRD and TBLWT instructions. These modes can be classified into following groups:

1. By connecting a crystal oscillator or ceramic resonator at OSC1 and OSC2 pins.
2. By connecting an external clock source at the OSC1 pin. The oscillator frequency divided by 4 is output at the OSC2 pin.
3. By connecting an RC oscillator circuit at the OSC1 pin. The oscillator frequency divided by 4 is output at the OSC2 pin.
4. Using a frequency multiplier for a crystal oscillator to produce an internal clock frequency of up to 40MHz. The "frequency multiplier mode" is only available to the crystal oscillator when the FOSC3:FOSC0 configuration bits are programmed for this mode. This will be useful for applications requiring higher clock speed.
5. Using an internal oscillator block which generates two different clock signals; either can be used as the microcontroller's clock source. This may eliminate the need for external oscillator circuits on the OSC1 and/or OSC2 pins.
6. By switching the PIC18F4321 clock source from the main oscillator to an alternate clock source. When an alternate clock source is enabled, the various power-managed operating modes are available.

In the following, typical oscillator circuits using a crystal and an RC circuit will be provided. Figure 8.2 shows a typical quartz crystal oscillator circuit for the PIC18F. The crystal frequency can vary from 4MHz to 25 MHz.

Figure 8.3 shows how the PIC18F clock is generated at the OSC2 pin by connecting an RC oscillator circuit at the OSC1 pin. The oscillator frequency at the OSC1 pin is divided by 4 by the PIC18F and then generated on the OSC2 output pin. The OSC2 clock may be used for test purposes to synchronize other logic.

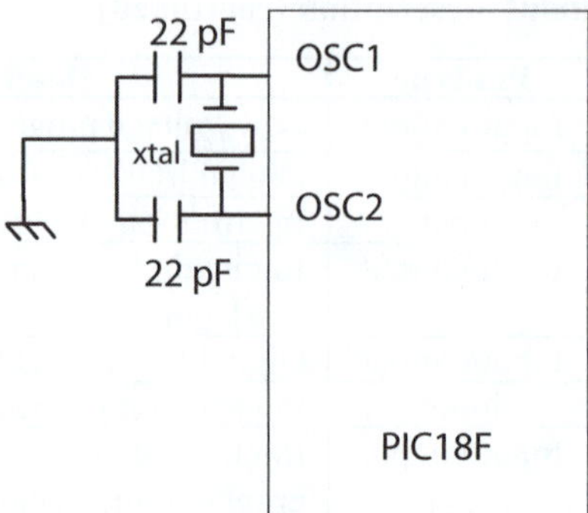

FIGURE 8.2 Typical crystal oscillator circuit

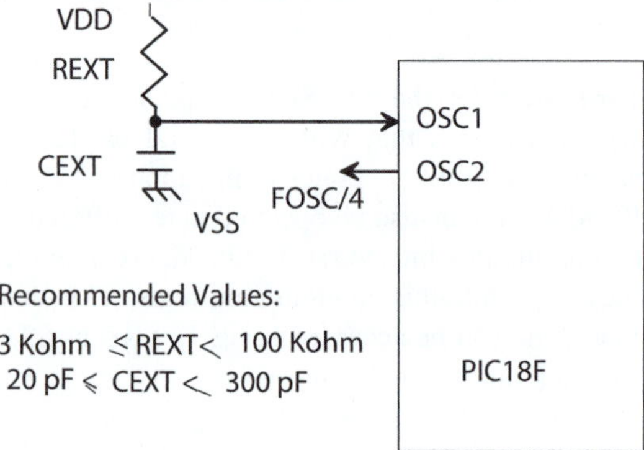

FIGURE 8.3 RC oscillator

8.1.2 PIC18F Reset

Upon activating the reset, the PIC18F loads '0' into program counter. Thus the PIC18F reads the first instruction from the contents of address 0 in the program memory. Most registers are unaffected by a Reset. Their statuses are unknown on POR (Power On Reset) and unchanged by all other resets. The other registers are forced to a "Reset state" depending on the type of Reset that occurred.

The PIC18F4321 can be reset in several different ways. For simplicity, the two most commonly used RESET techniques are Power-on and Manual resets. These two resets will be discussed in this following. A summary of some of the other resets will then be provided.

Power-On Reset (POR) A power-on reset pulse is generated on-chip upon power-up whenever VDD rises above a certain threshold. This allows the device to start in the initialized state when VDD is adequate for operation. The reset circuit in Figure 8.4 provides a simple power-on reset circuit with a pushbutton (manual) switch. When the power is turned ON, the resistors in Figure 8.4 with the switch open will provide power-on reset. When the PIC18F exits the reset condition, and starts normal operation, a program can be executed by pressing the pushbutton, and program execution can be restarted upon activation of the pushbutton.

Power-on reset events are captured by the $\overline{\text{POR}}$ bit (bit 1 of RCON, Figure 8.5). The state of the bit is set to '0' whenever a POR occurs; $\overline{\text{POR}}$ bit = 1 indicates that a POR has not occured.

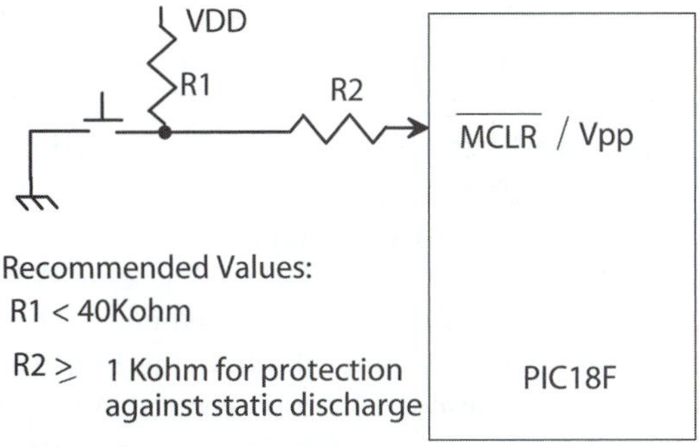

VDD
R1
R2
$\overline{\text{MCLR}}$ / Vpp

Recommended Values:
R1 < 40Kohm
R2 ≥ 1 Kohm for protection against static discharge

PIC18F

FIGURE 8.4 PIC18F manual reset circuit

7	6	5	4	3	2	1	0
IPEN	SBOREN	-----------	$\overline{\text{RI}}$	$\overline{\text{TO}}$	$\overline{\text{PD}}$	$\overline{\text{POR}}$	$\overline{\text{BOR}}$

Bit 7 = IPEN (interrupt pending, 1 = enable, 0 = disable; to be discussed later) **Bit 6** = SBOREN (BOR software enable bit, 1 = enable, 0 = disable) **Bit 5** = unimplemented (read as '0') **Bit 4** = $\overline{\text{RI}}$ (RESET instruction bit, 1 = RESET instruction executed, 0 = RESET instruction not executed) **Bit 3** = $\overline{\text{TO}}$ (watchdog timeout bit, 1 = upon power-up or execution of CLRWDT or SLEEP instruction, 0 = a watchdog timer timeout occurred) **Bit 2** = $\overline{\text{PD}}$ (power- down detection bit, 1 = upon power-up or execution of CLRWDT instruction, 0 = execution of SLEEP instruction. **Bit 1** = A $\overline{\text{POR}}$ (power-on reset status bit, 1 = A Power-on reset has not occurred, 0 = A power-on reset has occurred) **Bit 0 =** $\overline{\text{BOR}}$ (brown-out reset status bit, 1 = A brown-out reset has not occurred, 0 = A brown-out reset has occurred)

FIGURE 8.5 RCON (RESET CONTROL) register

Manual Reset Figure 8.4 shows a typical circuit for manual reset. The $\overline{\text{MCLR}}$/Vpp pin is normally HIGH. Upon activating the push button, the $\overline{\text{MCLR}}$/Vpp pin is driven from HIGH to LOW. The internal on-chip circuitry connected to the $\overline{\text{MCLR}}$/Vpp pin ensures that the pin must be LOW for at least 2 μsec (minimum requirement for reset). Note that the PIC18F can be reset manually by the circuit of Figure 8.4 since it can be shown that the minimum timing requirement is satisfied by this circuit.

Other Resets Note that device reset events are tracked through the RCON (reset control) register (Figure 8.5). The lower five bits of the register indicate that a specific reset event has occurred. In most cases, these bits can be cleared only by the event and must be set by the application after the event. The state of these flag bits, taken together, can be read to indicate the type of reset that just occurred. The RCON register also has control bits for setting interrupt priority (IPEN) and software control of the BOR (brown-out reset). Some of the other resets include:

(a) Brown-out reset (BOR)
(b) Watchdog timer (WDT) reset (during program execution)
(c) RESET instruction

(a) Brown-out reset. The PIC18F implements a BOR circuit that resets the PIC18F when the power drops below a specified voltage. The BOR is controlled by the specified bits in the configuration register. The BOR threshold is set by the BORV1:BORV0 bits in the configuration register. The BOR can be enabled or disabled by the user in software. This is done with the control bit, SBOREN (bit 6 in RCON). Setting SBOREN enables the BOR. Clearing SBOREN disables the BOR entirely.

If BOR is enabled, any drop of VDD (5 V) below VBOR (brown-out reset voltage, 4.59 V typical) for greater than TBOR (brown-out reset pulse width, 200 μsec) will automatically reset the device; the PIC18F clears the $\overline{\text{BOR}}$ (bit 0) bit in the RCON register to 0 to indicate that a brown-out reset has occurred. A reset may or may not occur if VDD falls below VBOR for less than TBOR. The chip will remain in brown-out reset until VDD rises above VBOR.

(b) Watchdog timer (WDT) reset (during program execution). The PIC18F WDT is driven by an internal clock source. When the WDT is enabled via software using WDTCON (watchdog timer control register), the internal clock source is also enabled. The time delay provided by the WDT varies from 4 ms to 131.072 seconds. If the timer associated with the WDT is enabled, and then times out after a specific amount of time during program execution, the PIC18F resets itself automatically. Also, the $\overline{\text{TO}}$ bit in the RCON register is cleared to 0. This "time out" may happen in certain situations such as if the program is caught in a loop or if the program takes unexpectedly longer execution time. The WDT can be useful for debugging programs.

(c) RESET instruction. This is a software RESET. Upon execution of the RESET instruction, the PIC18F resets all registers and flags that are affected by a MCLR reset, and the $\overline{\text{RI}}$ bit in the RCON register is cleared to 0.

8.1.3 A Simplified Setup for the PIC18F4321

Figure 8.6 shows a simplified setup for the 4321 microcontroller using the default clock of 1 MHz. Appendix H shows the hardware and software aspects of how to interface the PIC18F4321 to a personal computer or a laptop using PicKit3. This setup can be used for performing inexpensive meaningful experiments in laboratories using a breadboard.

There are two pairs of pins on the PIC18F4321 that must be connected to power and ground; pins 11 (VDD) and 32 (VDD) are normally connected directly to +5 V and pins 12 (VSS) and 31 (VSS) are connected directly to ground. Note that the operating voltage for VDD is between 4.2 and 5.5 V.

The manual reset circuit is connected to pin 1 of the PIC18F4321 chip. When the push button is activated, the PIC18F4321 is reset. This also allows for an easy way to restart a program in the PIC18F4321.

8.2 PIC18F4321 I/O Ports

The PIC18F4321 contains five ports namely Port A (8-bit), Port B (8-bit), Port C (8-bit), Port D (8-bit), and Port E (4-bit). All pins of the PIC18F4321 I/O ports are multiplexed with an alternate functions from the peripheral features on the device. In general, when a peripheral is enabled, that pin may not be used as a general purpose I/O pin.
For simple I/O operation, three latches are associated with each I/O port bit. They are

1. TRIS (Tristate) latch

2. Input latch
3. Output latch

Writing a '1' in the TRIS latch will configure the corresponding bit in the port as an input. Writing a '0' at a particular bit in the TRIS latch will configure the corresponding bit in the port as an output.

A simplified model for a single pin of a generic I/O port is shown in Figure 8.7. This circuit does not include the peripheral functions that may be multiplexed to the I/O pin. The circuit in Figure 8.7 is a simplified version of the internal circuitry associated with an I/O port bit. For simplicity, other components connected to this circuit are not shown. This circuit will provide a basic understanding of how a port bit is configured as an input or an output. Note that Figure 8.7 shows three D latches, namely, input latch, TRIS (Tristate) latch, and output latch.

Writing a '1' in the TRIS latch using the PIC18F instruction will make its Q output HIGH. This will enable the input buffer buffer, and disable the output buffer. Thus the selected I/O pin of the I/O port will be configured as an input bit. Upon execution of an input instruction, the $\overline{RD}$ line will be LOW. The inverter at the bottom of Figure 8.7 will make the EN line of the input latch HIGH. Hence, the I/O pin connected at the D input of the Input latch via the enabled input buffer will be transferred to the selected data bus pin of the PIC18F.

Writing a '0' in the TRIS latch using the PIC18F instruction will make its Q output LOW. This will enable the output buffer, and disable the input buffer. Thus the selected I/O pin of the I/O port will be configured as an output bit. Upon execution of an output instruction, the pin connected to the clock of the output latch will transfer the selected data

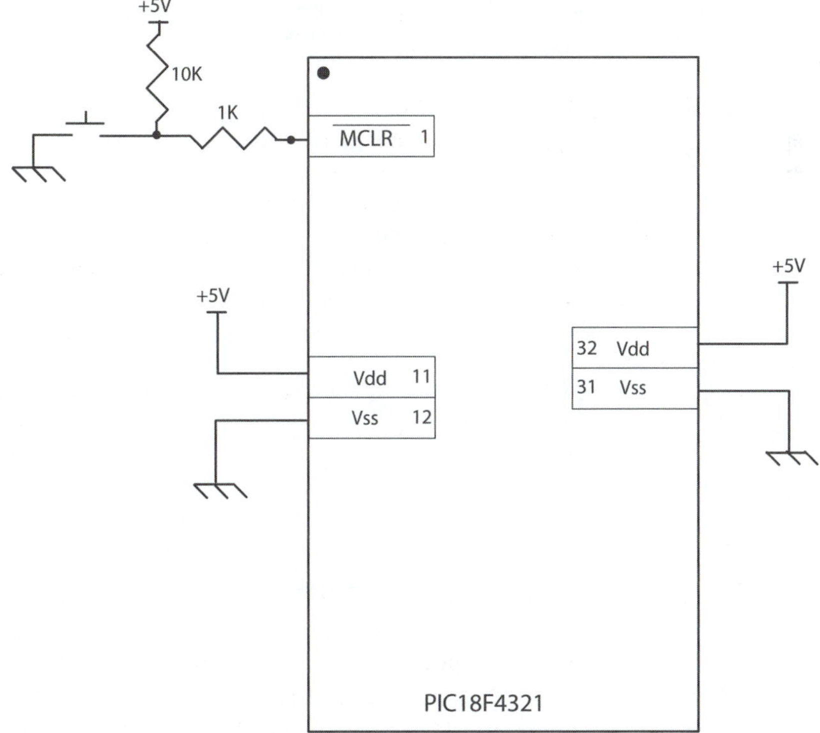

FIGURE 8.6 Simplified PIC18F4321 setup

bus pin of the PIC18F to the I/O pin via the enabled Output buffer.

Table 8.2 shows a list of the PIC18F4321 I/O ports along with the associated TRISx registers. Note that these ports and registers are mapped as Special Function Registers (SFR's) in the PIC18F4321 data memory. Hence, these addresses are also included in the table for convenience.

Note that all bits of Ports A through D are available for general I/O operation. On the other hand, only three bits of Port E (bits 0, 1, 2) are available for general I/O. The fourth pin (bit 3) of PORT E (MCLR/VPP/RE3) is an input only pin. Its operation is controlled by the MCLRE configuration bit in a special register called CONFIG3H (Configuration Register 3 High) . When selected as a port pin (MCLRE = 0), it functions as a digital input only pin; as such, it does not have TRIS or LAT bits associated with its operation. Otherwise, it functions as the device's master clear input. In either configuration, RE3 also functions as the programming voltage input (VPP) during programming the on-chip memory.

8.2.1 PIC18F I/O Instructions

The PIC18F does not provide IN or OUT instructions for inputting from or outputting to ports. The PIC18F uses memory mapped I/O. Therefore, typical memory-oriented instructions such as MOVWF, MOVF, and MOVFF can be used for

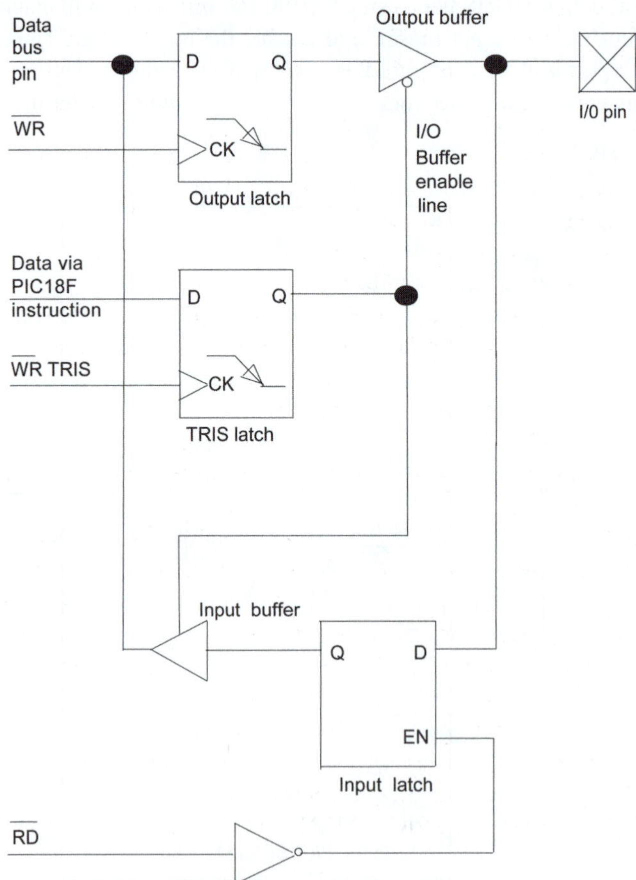

FIGURE 8.7 Generic I/O port operation (simplified)

TABLE 8.2 PIC18F4321 I/O PORTS, TRISx REGISTERS, ALONG WITH ADDRESSES

(Upon RESET, all ports are configured as inputs)

Port Name	Size	Mapped SFR address	Comment
Port A	8-bit	0xF80	Port A
TRISA	8-bit	0xF92	Data Direction Register for Port A
Port B	8-bit	0xF81	Port B
TRISB	8-bit	0xF93	Data Direction Register for Port B
Port C	8-bit	0xF82	Port C
TRISC	8-bit	0xF94	Data Direction Register for Port C
Port D	8-bit	0xF83	Port D
TRISD	8-bit	0xF95	Data Direction Register for Port D
Port E	4-bit	0xF84	Port E
TRISE	4-bit	0xF96	Data Direction Register for Port E

inputting from or outputting to ports. As an example, consider the PIC18F instruction, MOVF PORTD, W will input the contents of PORTD into WREG.

The MOVWF PORTC instruction, on the other hand, will output the contents of WREG into PORTC. Data can also be output from one port to another. For example, the MOVFF PORTC, PORTD instruction will output the contents of PORTC to PORTD.

The PIC18F bit-oriented instructions such as BSF and BCF can be used to output a '1' or '0' to a specific bit of an I/O port. For example, the instruction BSF PORTD, 6 will set bit 6 of Port D; in other words, the PIC18F will output a '1' to bit 6 of Port D. The instruction BCF PORTC, 3 , on the other hand, will clear bit 3 of Port A to zero; in other words, the PIC18F4321 will output a '0' to bit 3 of PORTC.

8.2.2 Configuring PIC18F4321 I/O Ports

As mentioned before, writing a '1' at a particular bit position in the TRISx register will make the corresponding bit in the associated port as an input. On the other hand, writing a '0' at a particular bit position in the TRISx register will make the corresponding bit in the associated port as an output. Upon reset all TRIS registers are automatically loaded with 1's, and hence, all ports will be configured as inputs.

Next, in order to illustrate how PIC 18F4321 ports are configured using the associated TRISx registers, consider the following PIC18F instruction sequence:

```
MOVLW     0x34    ; Move 0x34 into WREG
MOVWF     TRISD   ; Configure PORT D
```

In the above instruction sequence, MOVLW loads WREG with 34 (hex), and then moves these data into TRISD (8-bit data direction register for PORTD) which then contains 34(Hex); the corresponding port is defined as shown in Figure 8.8. In this example, because 34H (0011 0100) is written into TRISD, bits 0, 1, 3, 6, and 7 of the port are set up as outputs, and bits 2, 4, and 5 of the port are defined as inputs. The microcontroller can then send output to external devices, such as LEDs, connected to bits 0, 1, 3, 6, and 7 through a proper interface. Similarly, the PIC18F4321 can input the status of external devices, such as switches, through bits 2, 4, and 5. To input data from the input switches, the PIC18F4321 inputs the complete byte, including the bits to which output devices such as LEDs are connected. While receiving input data from an I/O port, however, the PIC18F4321 places a value, probably 0, at the bits configured as outputs and the program must interpret them as "don't cares." At the same time, the PIC18F4321's outputs to bits configured as inputs are disabled.

The PIC18F instructions such as SETF and CLRF can be used to configure I/O

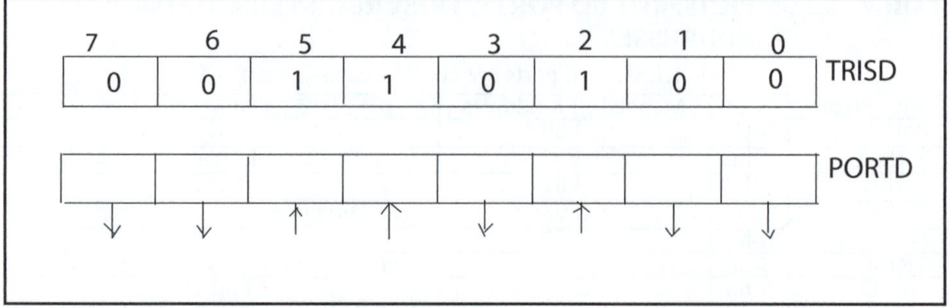

FIGURE 8.8 PORT D along with TRISD

ports. For example, to configure all bits in Port C as inputs, and Port D as outputs, SETF or CLRF instructions can be used as follows:

```
SETF    TRISC  ; Set all bits in TRISC to 1's and configure
               ; configure Port C as an input port.
CLRF    TRISD  ; Clear all bits in TRISD to 0's and configure
               ; configure Port D as an output port
```

Also, a specific bit in a port can be configured as an input or as an output using PIC18F bit-oriented instructions such as BSF and BCF. For example, the instruction BSF TRISD, 7 will make bit 7 of Port D as an input bit. On the other hand, BCF TRISC, 1 will make bit 1 of Port C as an output.

Note that configuring Port A, Port B and Port E is different than configuring Port C and Port D. This is because, certain bits of Port A, Port B, and Port E are multiplexed with analog inputs. For example, bits 0-3 and bit 5 of Port A are multiplexed with analog inputs AN0 -AN4, bits 0-4 of Port B are multiplexed with analog inputs AN8-AN12, and bits 0-2 of PORT E are multiplexed with analog inputs AN5-AN7 (Figure 8.1). When a port bit is multiplexed with an analog input, bits 0-3 of a special function register (SFR) called ADCON1 (A/D Control Register 1) must be used to configure the port bit as input. The other bits in ADCON1 are associated with the A/D converter.

Figure 8.9 shows the ADCON1 register along with the associated bits for digital I/O. When bits 0 through 3 of the ADCON register are loaded with 1111, the analog inputs (AN0- AN12) multiplexed with the associated bits of Port A, Port B, and Port E are configured as digital inputs. This will also make these port bits as inputs automatically; the corresponding TRISx registers are not required to configure the ports. However, for configuring these ports as outputs, the corresponding TRISx bits must be loaded with 0's; the ADCON1 register is not required for configuring these port bits as outputs. The following examples will illustrate this.

For example, the following instruction sequence will configure all 13 port bits multiplexed with AN0 - AN12 as inputs:

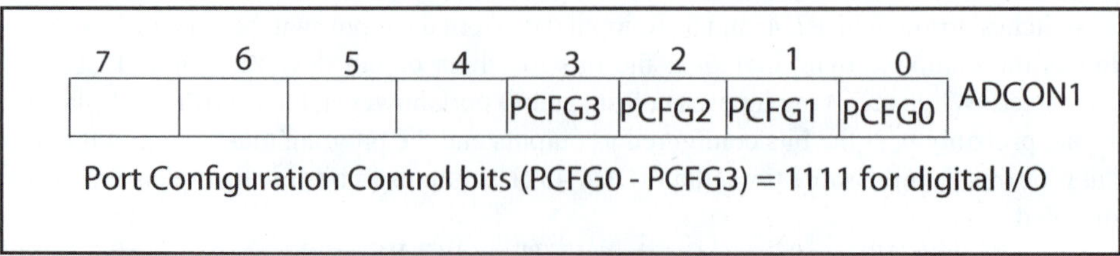

FIGURE 8.9 ADCON1 register for digital I/O

```
MOVLW     0x0F       ; Move 0xF into WREG
MOVWF     ADCON1     ; Move WREG into ADCON1
```

Next, in order to configure bit 1 of Port A, and bits 2 and 4 of Port B as outputs, the following instruction sequence can be used:

```
BCF       TRISA, 1   ; Configure bit 1 of Port A as output
BCF       TRISB, 2   ; Configure bit 2 of Port B as output
BCF       TRISB, 4   ; Configure bit 4 of Port B as output
```

It should be mentioned that if a bit of an I/O port in the PIC18F family of microcontrollers such as the PIC18F4321 and PIC18F4520 is multiplexed with an analog input, the bit must be configured as an input using the ADCON1 register; the same bit can be configured as an output using the corresponding bit in the associated TRISx register. However, if a port bit is not multiplexed with an analog input, it can be configured as an input or an output using the associated TRISx register.

For simplicity, Port C, Port D, and multiplexed bits of Port A, Port B, and Port E with analog inputs will be used to illustrate the concept of programmed I/O associated with the PIC18F4321 in this book.

8.2.3 Interfacing LEDs (Light Emitting Diodes) and Seven-segment Displays

The PIC18F sources and sinks adequate currents so that LEDs and seven-segment displays can be interfaced to the PIC18F without buffers (current amplifiers) such as 74HC244. An LED can be connected in two ways. Figure 8.10 (a) and (b) shows these configurations.

In Figure 8.10 (a), the PIC18F will output a HIGH to turn the LED ON; the PIC18F will output a 'LOW' will turn it OFF. In Figure 8.10 (b), the PIC18F will output a LOW to turn the LED ON; the PIC18F will output a 'HIGH' will turn it OFF. Also, when an LED is turned on, a typical current of 10 mA flows through the LED with a voltage drop of 1.7 V. Hence,

$$R = \frac{5 - 1.7}{10 \text{ mA}} = 330\ \Omega$$

As discussed in Chapter 3, a seven-segment display can be used to display, for example, decimal numbers from 0 to 9. The name "seven segment" is based on the fact that there are seven LEDs — one in each segment of the display. Figure 8.11 shows a typical seven-segment display.

Figure 8.12 shows two different seven-segment display configurations, namely, common cathode and common anode. Note that Figures 8.11 and 8.12 are redrawn from Chapter 3 for convenience. In Figure 8.12, each segment contains an LED. All decimal

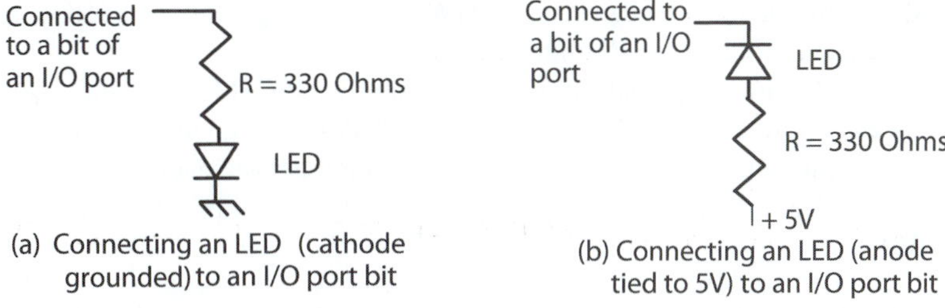

FIGURE 8.10 Interfacing LED to PIC18F

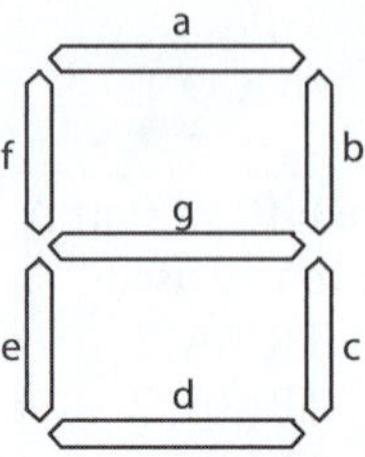

FIGURE 8.11 A seven-segment display

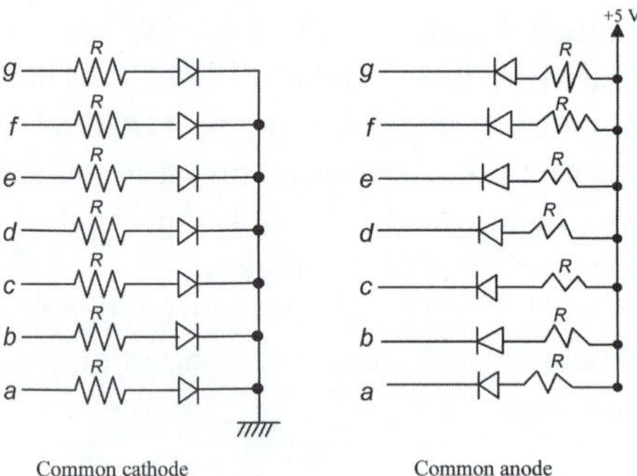

FIGURE 8.12 Seven-segment display configurations

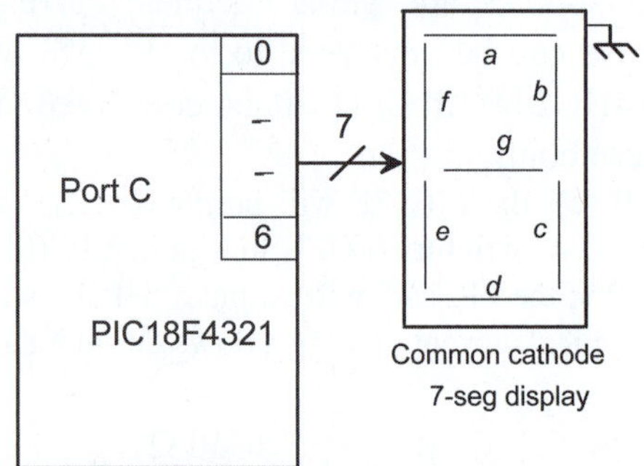

FIGURE 8.13 PIC18F4321 interface to a common cathode seven-segment display via PORT C

numbers from 0 to 9 can be displayed by turning the appropriate segment "ON" or "OFF". For example, a zero can be displayed by turning the LED in segment *g* "OFF" and turning the other six LEDs in segments *a* through *f* "ON." There are two types of seven-segment displays. These are common cathode and common anode.

In a common cathode arrangement, the microcotroller can be programmed to send a HIGH to light a segment and a LOW to turn it off. In a common anode configuration, on the other hand, the microcontroller can send a LOW to light a segment and a HIGH to turn it off. In both configurations, R = 330 ohms can be used.

Figure 8.13 shows a typical interface between the PIC18F4321 and a common cathode seven-segment display via Port C. Each bit of Port C is connected to a segment of the seven-segment display via 330 ohm resistor. Note that the seven resistors are not shown in the figure. A common anode seven-segment display can similarly be interfaced to the PIC18F4321.

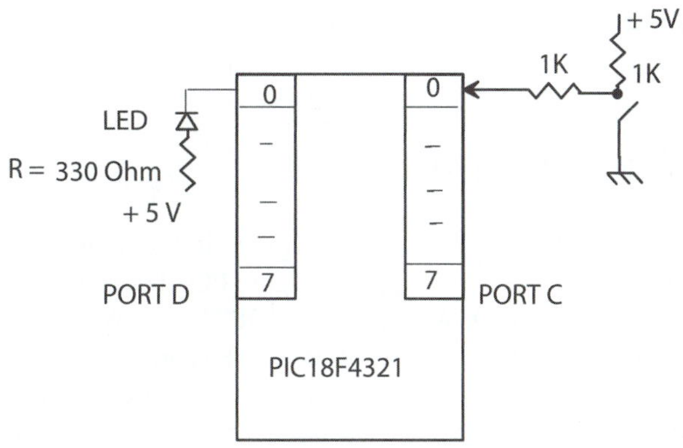

FIGURE 8.14 Figure for Example 8.1

Example 8.1 Write a PIC18F assembly language program at 0x100 to drive an LED connected to bit 0 of Port D based on a switch input at bit 0 of Port C, as shown in Figure 8.14. If the switch is opened, turn the LED OFF; turn the LED ON if the switch is closed.

Solution

From Figure 8.14, since the cathode of the LED is connected to bit 0 of Port D, a '0' output from the PIC18F4321 will turn the LED ON, and a '1' will turn it OFF. The PIC18F sinks (for LOW output) adequate current to turn an LED OFF or ON without any buffer such as 74HC244; only a current limiting resistor R = 330 ohm is required.

The PIC18F assembly language program is provided below:

```
        INCLUDE   <P18F4321.INC>
        ORG       0x100
        BSF       TRISC, 0        ; Configure bit 0 of Port C as an input
        BCF       TRISD, 0        ; Configure bit 0 of Port D as an output
START   MOVFF     PORTC, PORTD    ; Output switch input to LED
        BRA       START           ; Repeat
        END
```

In the above, since the switch and the LED data are aligned (both connected at bit 0 of the respective ports), the MOVFF PORTC, PORTD instruction directly inputs the switch input from bit 0 of PORTC and outputs to bit 0 of PORTD. Also, the infinite loop using BRA START will make the LOOP continuous. This means that after execution of the above program once, the LED will be turned ON and OFF automatically as soon as the switch is pressed. However, if SLEEP or FINISH BRA FINISH is placed at the end of the program , the program will be executed once upon activation of the manual reset. Each time the switch pressed in Figure 8.14, the above program must be re-executed by activating a manual reset.

Example 8.2 A PIC18F4321 microcontroller is required to drive an LED connected to bit 7 of Port C based on two switch inputs connected to bits 6 and 7 of Port D. If both switches are equal (either HIGH or LOW), turn the LED ON; otherwise turn it OFF.

Assume that a HIGH will turn the LED ON and a LOW will turn it OFF. Write a PIC18F assembly program at 0x200 to accomplish this.

Solution

```
        INCLUDE  <P18F4321.INC>
        ORG      0x200
        BCF      TRISC, 7     ; Configure bit 7 of PORTC as an output
        SETF     TRISD        ; Configure PORTD as an input port
        MOVF     PORTD, W     ; Input PORTD into WREG
        ANDLW    0xC0         ; Retain bits 6 and 7
        BZ       LEDON        ; If both switches are LOW, turn the LED ON
        SUBLW    0xC0         ; If both switches are HIGH, turn the LED ON
        BZ       LEDON
        BCF      PORTC, 7     ; Turn LED OFF
        BRA      FINISH
LEDON   BSF      PORTC, 7     ; Turn LED ON
FINISH  BRA      FINISH
        END
```

Example 8.3 The PIC18F4321 microcontroller shown in Figure 8.15 is required to output a BCD digit (0 to 9) to a common-anode seven-segment display connected to bits 0 through 6 of Port D. The PIC18F4321 inputs the BCD number via four switches connected to bits 0 through 3 of PortT C. Write a PIC18F main program at address 0x40 that will initialize Port C and Port D, initialize STKPTR to 0x10, CALL a subroutine to obtain the seven-segment for the BCD input, and then output to the display via Port D. Write the PIC18F subroutine at address 0x70 that will return the seven-segment code using the RETLW instruction based on the BCD input digit.

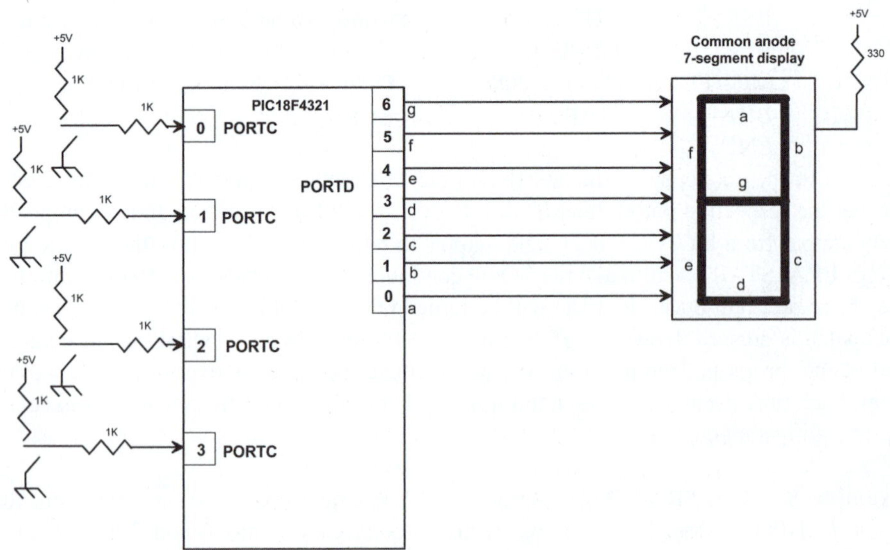

FIGURE 8.15 Figure for Example 8.3

Solution

To find the proper values for the display, a table containing the seven-segment code for each BCD digit can be obtained from Figure 8.15 as follows:

	g	f	e	d	c	b	a		Hex:
0:	1	0	0	0	0	0	0	=	0x40
1:	1	1	1	1	0	0	1	=	0x79
2:	0	1	0	0	1	0	0	=	0x24
3:	0	1	1	0	0	0	0	=	0x30
4:	0	0	1	1	0	0	1	=	0x19
5:	0	0	1	0	0	1	0	=	0x12
6:	0	0	0	0	0	1	1	=	0x03
7:	1	1	1	1	0	0	0	=	0x78
8:	0	0	0	0	0	0	0	=	0x00
9:	0	0	1	1	0	0	0	=	0x18

Note that for a common-anode seven-segment display, a ‘0’ will turn a segment ON and a ‘1’ will turn it OFF. Also, bit 7 of Port D is assumed to be ‘0’.

The subroutine returns the seven-segment code for a specific BCD digit using the RETLW k instruction. As mentioned before, the RETLW k loads the 8-bit immediate data k into WREG, and returns to the main program by loading the program counter with the address from the top of the hardware stack.

The following PIC18F main and assembly language program are provided in the following:

```
            INCLUDE <P18F4321.INC>
; MAIN PROGRAM
            ORG       0x40
            SETF      TRISC        ; Configure PORTC is input
            CLRF      TRISD        ; Configure PORTD is output
            MOVLW     0x10
            MOVWF     STKPTR       ; Initialize STKPTR to 0x10
LOOP        MOVF      PORTC, W     ; Move switch data to the WREG
            ANDLW     B'00001111'  ; Mask the lower 4-bits
            CALL      LOOKUP       ; Call the subroutine LOOKUP
            MOVWF     PORTD        ; Move WREG to PORTD
            BRA       LOOP         ; Loop
; SUBROUTINE
            ORG       0x70
LOOKUP      MULLW     2            ; Double the WREG value
            MOVF      PRODL,W      ; Place the answer back into WREG
            ADDWF     PCL          ; Use PCL to find the location on the table
; PCL  CONTAINS THE STARTING ADDRESS OF THE  TABLE
            RETLW     0x40         ; Value for 0 display
            RETLW     0x79         ; Value for 1 display
            RETLW     0x24         ; Value for 2 display
            RETLW     0x30         ; Value for 3 display
```

```
RETLW   0x19    ; Value for 4 display
RETLW   0x12    ; Value for 5 display
RETLW   0x03    ; Value for 6 display
RETLW   0x78    ; Value for 7 display
RETLW   0x00    ; Value for 8 display
RETLW   0x18    ; Value for 9 display
END
```

In the above, the main program at address 0x40 configures Port C and Port D, initializes STKPTR at 0x10, and then inputs the BCD digit via switches into WREG. The low four bits (BCD digit) are retained via masking using the ANDLW instruction. The subroutine is then called using the "CALL LOOKUP" instruction. The CALL will push the current PC contents (address of the next instruction MOVWF PORTD) onto the hardware stack, and then jump to subroutine.

The subroutine "LOOKUP" at address 0x70 multiplies the BCD digit in WREG by 2 since each instruction is 2 bytes. The low byte of PRODH:PRODL will contain the 16-bit product. Since the maximum value of this product will be 0x0012 (9 x 2 = 18 decimal, maximum value of the BCD digit is 9), PRODL will contain the product. The PCL (program counter LOW) can be added with PRODL to find the appropriate seven-segment code for the BCD digit included with the RETLW instruction.

For example, suppose that the switch inputs are 0010_2(BCD 2). After multiplying by 2, WREG will contain 4. During execution of "ADDWF PCL", the PCL will contain 0x76. After execution of "ADDWF PCL", the contents of PCL will be 0x7A. Hence, the instruction, RETLW 0x24 will be executed. This instruction will load WREG with 0x24 (seven segment code for 2), pops the address of "MOVWF PORTD" (previously pushed during execution of THE CALL) and returns to the main program. The "MOVWF PORTD" will output 0x24 (contents of WREG; seven segment code for 2) to PORTD to display 2, and then "BRA LOOP" will go back for new data input from the switches.

8.3 PIC18F Interrupts

The concept of interrupt is discussed in detail in Chapter 4. Certain interrupt topics will be repeated for convenience. As mentioned before, interrupts are basically divided into two types, namely, external and internal interrupts. Figure 8.16 shows a typical interrupt structure.

External interrupts are initiated through a microcontroller's interrupt pins by external devices. External interrupts can be divided further into two types: maskable and nonmaskable. Nonmaskable interrupt cannot be enabled or disabled by instructions, whereas a microcontroller's instruction set contains instructions to enable or disable maskable interrupt. A nonmaskable interrupt has a higher priority than a maskable interrupt. If maskable and nonmaskable interrupts are activated at the same time, the processor will service the nonmaskable interrupt first.

A nonmaskable interrupt is typically used as a power failure interrupt. Microcontrollers normally use +5 V dc, which is transformed from 110 V ac. If the power falls below 90 V ac, the DC voltage of +5 V cannot be maintained. However, it will take a few milliseconds before the ac power drops below 90 V ac. In these few milliseconds, the power-failure-sensing circuitry can interrupt the processor. The interrupt service routine

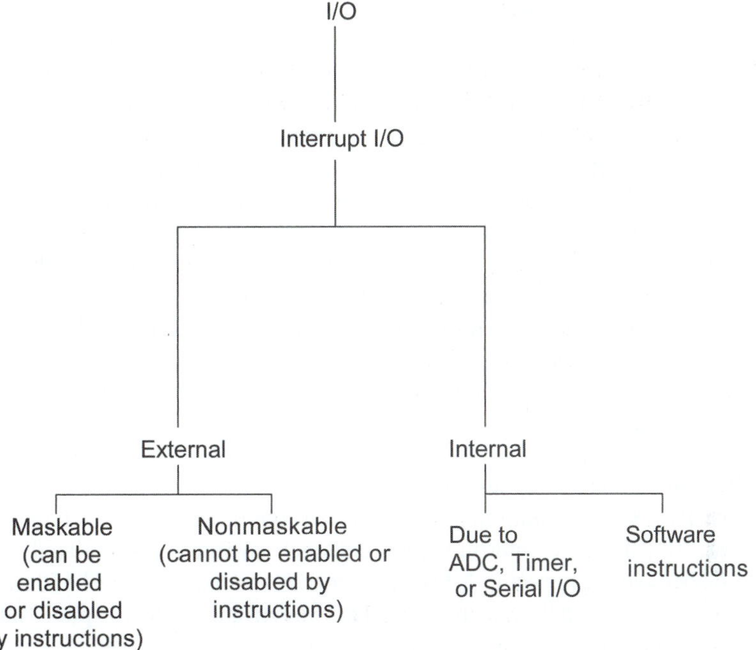

FIGURE 8.16 Typical interrupt structure

can be written to store critical data in nonvolatile memory such as battery-backed CMOS RAM, and the interrupted program can continue without any loss of data when the power returns. *Internal interrupts* are usually activated internally by conditions such as interrupts due to completion of conversion of ADC (Analog-to-Digital Conversion), Timer, and Serial I/O. Interrupts are handled in the same manner as external interrupts.

Some microcontrollers include software interrupt instructions. When one of these instructions is executed, the microcontroller is interrupted and serviced similarly to external or internal interrupts.

Some microcontrollers such as the Motorola/Freescale HC11/HC12 provide both external (maskable and nonmaskable) and internal (exceptional conditions and software instructions). The PIC18F provides external maskable interrupts only. The PIC18F does not have any external nonmaskable interrupts. However, the PIC18F provides internal interrupts. The internal interrupts are activated internally by conditions such as timer interrupts, completion of ADC, and serial I/O. Internal interrupts are handled in the same manner as external interrupts. The user writes a service routine for each of these interrupts.

8.3.1 Interrupt Procedure

Upon reset, the PIC18F operates in default mode. In this mode, interrupt priorities cannot be assigned. For simplicity, default interrupts will first be discussed. A detailed coverage of the PIC18F interrupts will then be explained.

When an interrupt is recognized after reset, the PIC18F completes execution of the current instruction, pushes the current contents of the program counter (address of the next instruction) onto the hardware stack automatically, and loads the program counter with an address (predefined by Microchip Technology) called the "Interrupt address vector." The programmer writes a program called the Interrupt Service Routine (ISR) at this address.

The RETFIE instruction is normally used at the end of the service routine. The PIC18F RETFIE 0 or simply RETFIE instruction pops the contents of the program counter previously pushed before going to the service routine, enables all interrupts by clearing the GIE flag to 0, and returns control to the appropriate place in the main program.

The interrupt procedure is similar in concept to the procedure associated with subroutine CALL and RETURN instructions. The subroutine CALL /RETURN includes a main program and a subroutine while the interrupt contains a main program and a service routine. The subroutine CALL instruction pushes the current contents of the program counter onto the stack. The RETURN instruction placed at the end of the subroutine pops the previously pushed program counter, and returns control to the main program. The RETURN instruction does not clear the GIE (Global Interrupt Enable) flag. Note that the interrupt is initiated externally via hardware or internally via occurrence of events such as completion of ADC. Once the interrupt is recognized, a similar procedure associated with subroutine CALL/RETURN is followed by the PIC18F.

8.3.2 PIC18F Interrupt Types

The PIC18F4321 interrupts can be of two types. These are external interrupts initiated via PIC18F4321's interrupt pins, and internal interrupts initiated by internal peripheral devices such as on-chip A/D converter and on-chip timers. The PIC18F4321 is provided with three external maskable pins. These interrupts are INT0 (pin 33), INT1 (pin 34), and INT2 (pin 35).

These interrupts can be programmed to be activated by either leading edge or trailing edge pulses (to be discussed later). Also, signals KB10 (pin 37), KB11 (pin 38), and KB12 (pin 39) can be used as external interrupts. Each of these signals is recognized as an interrupt if there is a change in logic level on the pin.

The PIC18F 4321 on-chip peripherals can generate internal interrupts. These interrupts are also maskable. These peripheral interrupts will be covered later.

8.3.3 PIC18F External Interrupts in Default Mode

The concept of PIC18F external interrupts (default mode) described in this section can be used to perform simple and meaningful experiments in laboratories.

Upon power-on reset, the PIC18F handles the three external interrupts (INT0, INT1, INT2) in default mode. The interrupt address vector for all three interrupts is 0x000008 in the program memory in this mode. The INT0 interrupt has an individual interrupt enable bit along with a corresponding flag bit located in a SFR called the INTCON register. The mapped 12-bit data memory address for the INTCON is 0xFF2.

Each of the two other external interrupts (INT1 and INT2), on the other hand, has an individual interrupt enable bit along with the corresponding flag bit located in the SFRs called the INTCON3 register. The mapped 12-bit data memory address for the INTCON3 is 0xFF1. Figure 8.17 shows the INTCON and INTCON3 registers along with the associated interrupt bits. The other bits in these registers are specified for other functions such as timers, and will be discussed as these topics are covered.

All PIC18F interrupts are disabled upon reset. However, these interrupts can be enabled via software by setting the GIE bit (bit 7 in INTCON register) to one. For example, the PIC18F instruction, BSF INTCON, GIE will set the GIE bit (bit 7) in the INTCON register to 1; this will enable all interrupts. Next, the respective Interrupt Enable (IE) for each one of the three external interrupts must be set to 1 in order to enable a particular interrupt. For example, INT0 can be enabled by the instruction, BSF INTCON,

7	6	5	4	3	2	1	0	
GIE/GIEH	PEIE/GIEL	TMR0IE	INT0 IE	RBIE	TMR0 IF	INT0 IF	RBIF	INTCON

(a) INTCON Register

bit 7 **GIE/GIEH:** Global Interrupt Enable bit

When IPEN = 0: 1 = Enables all unmasked interrupts 0 = Disables all interrupts

When IPEN = 1: 1 = Enables all high priority interrupts 0 = Disables all interrupts

bit 6 **PEIE/GIEL:** Peripheral Interrupt Enable bit

When IPEN = 0: 1 = Enables all unmasked peripheral interrupts 0 = Disables all peripheral interrupts

When IPEN = 1: 1 = Enables all low priority peripheral interrupts 0 = Disables all low priority peripheral interrupts

bit 5 **TMR0IE:** TMR0 Overflow Interrupt Enable bit,

1 = Enables the TMR0 overflow interrupt 0 = Disables the TMR0 overflow interrupt

bit 4 **INT0IE:** INT0 External Interrupt Enable bit

1 = Enables the INT0 external interrupt, 0 = Disables the INT0 external interrupt

bit 3 **RBIE:** RB Port Change Interrupt Enable bit

1 = Enables the RB port change interrupt, 0 = Disables the RB port change interrupt

bit 2 **TMR0IF:** TMR0 Overflow Interrupt Flag bit

1 = TMR0 register has overflowed (must be cleared in software), 0 = TMR0 register did not overflow

bit 1 **INT0IF:** INT0 External Interrupt Flag bit

1 = The INT0 external interrupt occurred (must be cleared in software), 0 = The INT0 external interrupt did not occur

bit 0 **RBIF:** RB Port Change Interrupt Flag bit

1 = At least one of the RB7:RB4 pins changed state (must be cleared in software)

0 = None of the RB7:RB4 pins have changed state

(b) INTCON3 Register

bit 7 **INT2IP:** INT2 External Interrupt Priority bit, 1 = High priority 0 = Low priority

bit 6 **INT1IP:** INT1 External Interrupt Priority bit, 1 = High priority 0 = Low priority

bit 5 **Unimplemented:** Read as '0'

bit 4 **INT2IE:** INT2 External Interrupt Enable bit, 1 = Enables the INT2 interrupt 0 = Disables the INT2 interrupt

bit 3 **INT1IE:** INT1 External Interrupt Enable bit, 1 = Enables the INT1 interrupt 0 = Disables the INT1 interrupt

bit 2 **Unimplemented:** Read as '0'

bit 1 **INT2IF:** INT2 External Interrupt Flag bit, 1 = The INT2 interrupt occurred (must be cleared in software)
0 = The INT2 external interrupt did not occur

bit 0 **INT1IF:** INT1 External Interrupt Flag bit, 1 = The INT1 external interrupt occurred (must be cleared in software)
0 = The INT1 external interrupt did not occur

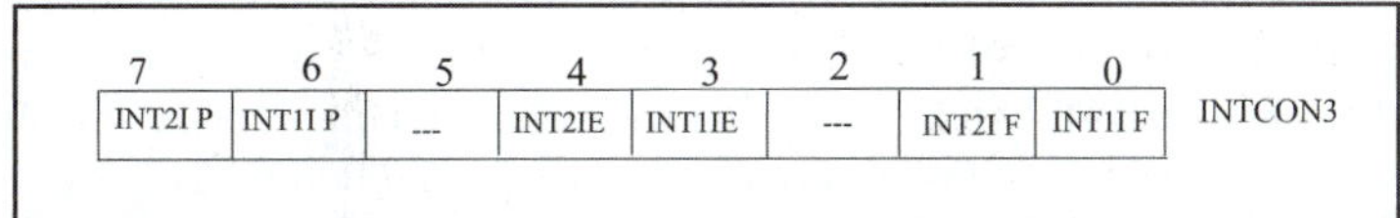

7	6	5	4	3	2	1	0	
INT2I P	INT1I P	---	INT2IE	INT1IE	---	INT2I F	INT1I F	INTCON3

FIGURE 8.17 INTCON (interrupt control) and INTCON3 (interrupt control 3)

INT0IE which will set the INT0IE (bit 4) in the INTCON register to 1. Finally, since the external interrupts (INT2, INT1, and INT0) are multiplexed with analog inputs (AN8, AN10, and AN12), the ADCON1 register must be configured as digital input. The PIC18F will now recognize any interrupt via the INT0-INT2 pins.

The following PIC18F instruction sequence will enable all external interrupts:

```
BSF      INTCON, INT0IE     ; Enable INT0
BSF      INTCON3, INT1IE    ; Enable INT1
BSF      INTCON3, INT2IE    ; Enable INT2
MOVLW    0X0F               ; Configure INT0-INT2
MOVWF    ADCON1             ; as inputs
BSF      INTCON, GIE        ; Enable all interrupts
```

Once an interrupt is recognized by the PIC18F, the corresponding flag bits

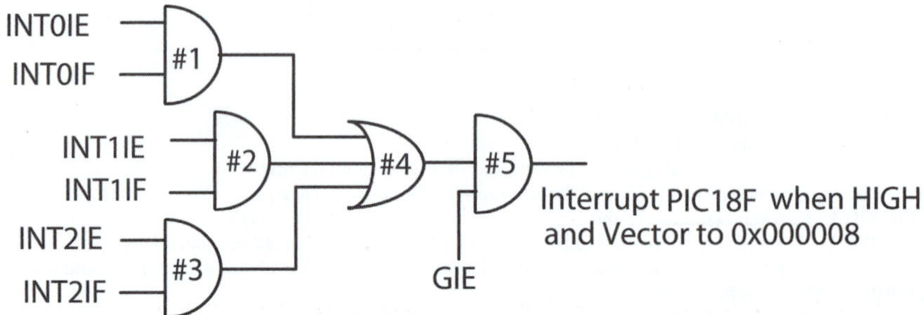

FIGURE 8.18 Simplified schematic for the PIC18F external interrupts for power-on reset

(INT0IF for INT0 in INTCON, INT1IF for INT1 in INTCON3, and INT2IF for INT2 in INTCON3) are set to one.

Figure 8.18 shows a simplified schematic for the PIC18F external interrupts for power-on reset. The PIC18F external interrupts must first be enabled using instructions by setting the GIE and INTxIE bits to one. The PIC18F will automatically set the corresponding interrupt flag bit to one after occurrence of each interrupt (INT0 through INT2).

As an example, consider INT0. In order for the PIC18F to recognize INT0, the user must first enable this interrupt using the following instruction sequence:

```
BSF    INTCON, INT0IE      ; Enable INT0
BSF    INTCON, GIE         ; Enable all interrupts
```

In Figure 8.18, the above instruction sequence will make the GIE input of AND gate #5 to one and also the INT0IE input of AND gate #1 to one.

Now, as soon as the interrupting device connected to the PIC18F4321 interrupts the microcontroller, the PIC18F4321 pushes the current program counter contents onto the hardware stack, and sets the INT0IF flag to 1, indicating that the INT0 interrupt has occurred. This will make the INT0IF input of AND gate #1 to one. Hence, the output of AND gate #5 in Figure 8.18 will be one. This will enable the appropriate hardware inside the PIC18F4321, and will load the program counter with 0x000008. The user can write a service routine at this address.

Note that before executing the RETFIE instruction (the last instruction of the interrupt service routine), the user must clear the INT0IF bit (in this case) using the BCF INTCON, INT0IF instruction to make sure that the INT0 interrupt is serviced only once, and not a multiple number of times, The RETFIE instruction at the end of the service routine will set the GIE bit (bit 7 in INTCON) to one, enabling all interrupts, and will load the program counter from the hardware stack with the previously pushed value. Thus, the control is returned to the main program. Note that external interrupts INT1 and INT2 can be explained in a similar manner.

Example 8.4 Assume that the PIC18F4321 microcontroller shown in Figure 8.19 is required to perform the following:

(a) If Vx > Vy , turn the LED ON if the switch is open; otherwise, turn the LED OFF. Write a PIC18F assembly language program starting at address 0x100 to accomplish the above by inputting the comparator output via bit 0 of Port B.

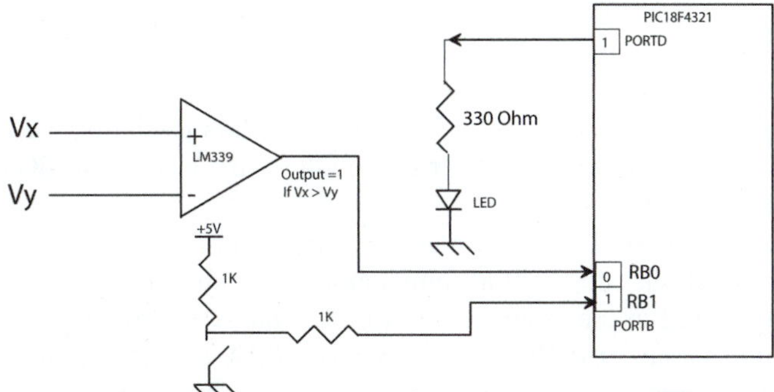

FIGURE 8.19 Figure for Example 8.4 (a) using programmed I/O

(b) Repeat part (a) using INT0 external interrupt. Use Port D for the LED and bit 1 of Port B for the switch as above. Write the main program at address 0x100 in PIC18F assembly language. Connect INT0 to the output of the comparator. The main program will initialize hardware stack pointer (STKPTR) to 0x30, configure Port B and Port D and enable GIE and INT0IE. Write the service routine in PIC18F assembly language which will clear the INT0 flag, input the switch, output to LED, and then return to the main program.

Solution

Example 8.4(a) uses programmed (polled or conditional) I/O while Example 8.4(b) uses interrupt I/O.

(a) In this example, an LM339 comparator is connected to the PIC18F4321 in order to control when the LED will be turned ON or OFF, based on the switch. In Figure 8.19, when Vx > Vy, then the comparator will output a one, and the PIC18F4321 will turn the LED ON or OFF depending on the switch status. If Vx < Vy, then the comparator will output a zero and the LED will be turned OFF. In the program, the ADCON1 register is used to configure RB0 and RB1 as inputs. The TRISD is used to make RD1 of Port D as output.

The PIC18F assembly language program for programmed I/O is provided below:

```
        INCLUDE <P18F4321.INC>
        ORG     0           ; RESET VECTOR
        GOTO    MAIN        ; JUMP TO MAIN
        ORG     0x100
MAIN    BCF     TRISD, 1    ; CONFIGURE BIT 1 OF PORTD AS OUTPUT
        MOVLW   0x0F        ; CONFIGURE BITS 0 AND 1 OF PORTB
        MOVWF   ADCON1      ; AS DIGITAL INPUTS
BEGIN   BCF     PORTD, 0    ; TURN LED OFF
CHECK   BTFSS   PORTB, 0    ; CHECK IF COMPARATOR IS ONE
        BRA     BEGIN       ; WAIT IN LOOP UNTIL ONE
        BTFSS   PORTB, 1    ; CHECK IF SWITCH IS OPEN
        BRA     BEGIN
        BSF     PORTD, 1    ; IF SWITCH OPEN, TURN LED ON
```

```
        BRA     CHECK
        END
```

In the above program, upon reset, the PIC18F starts executing the program at address 0x000000 in program memory. After execution of the instruction GOTO MAIN, the program will jump to address 0x100.

Next, Port B and Port D are configured. The instruction BCF PORTD, 0 turns the LED OFF. In order to check whether the comparator is outputting a one, the instruction BTFSS PORTB, 0 is used in the program. After execution of this instruction, if bit 0 of PORTB (comparator output) is 0, the next instruction, BRA BEGIN, continues looping until the comparator outputs a one. However, if the comparator output is 1, the BTFSS PORTB, 0 will skip the next instruction (BRA BEGIN), and will execute BTFSS PORTB, 1 to see whether the switch is ON or OFF. If it is OFF, the program will branch to BEGIN where it will turn the LED OFF. If the switch is OPEN (bit 1 of PORTB is 1), the program will skip the next instruction (BRA BEGIN), output a 1 to bit 1 of PORTD, and then turn the LED ON. The program will then branch to CHECK to make the loop is continuous.

(b) Figure 8.20 shows the relevant connections of the comparator to the PIC18F4321 using interrupt I/O. Note that the comparator output is connected to bit 0 of Port B to be used as an INT0 pin. In this example, using ADCON1 register, bit 0 of Port B can be configured as digital input to accept interrupt via INT0. The INT0IE bit of the INTCON register must be set to one in order to enable the external interrupt along with GIE to enable global interrupts.

The PIC18F assembly language program using external interrupt INT0 is provided in the following:

```
            INCLUDE <P18F4321.INC>
            ORG      0               ; RESET
            GOTO     MAIN_PROG
; MAIN PROGRAM
            ORG      0x00100         ; MAIN PROGRAM
MAIN_PROG   MOVLW    0x30            ; Initialize STKPTR to 0x30
            MOVWF    STKPTR
            BCF      TRISD,1         ; Configure bit 1 OF PORTD as output
            MOVLW    0x0F            ; Configure bit 0 of PORTB as INT0
            MOVWF    ADCON1          ; and bit 1 as input
```

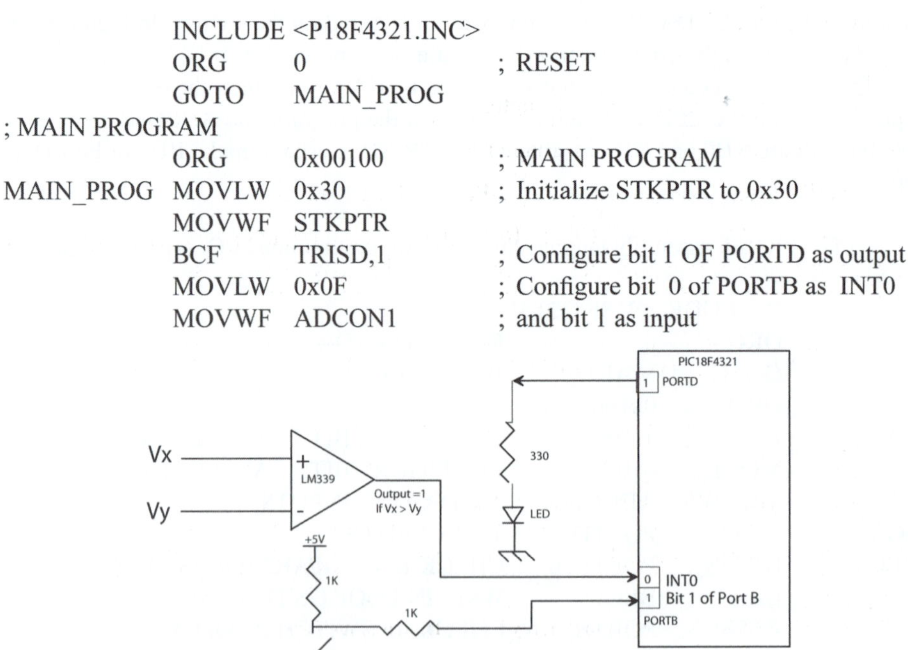

FIGURE 8.20 Figure for Example 8.3(b) using interrupt I/O

```
            BSF     INTCON, INT0IE ; Enable the external interrupt
            BSF     INTCON, GIE    ; Enable global interrupts
OVER        BRA     OVER           ; Wait for interrupt
            GOTO    MAIN_PROG      ; Repeat
; INTERRUPT SERVICE ROUTINE
            ORG     0x000008       ; Interrupt Address Vector
INT_SERV    MOVFF   PORTB, PORTD   ; Output switch status to turn LED ON/OF
            BCF     INTCON, INT0IF ; Clear the external interrupt flag to avoid
                                   ; double interrupt
            RTFIE                  ; Enable interrupt and return
            END
```

In the above, upon recognition of the interrupt, the PIC18F4321 pushes the program counter onto the stack, and automatically jumps to address 0x000008 (interrupt address vector, 0x000008 for INT0). The interrupt service routine is written at address 0x000008. The interrupt flag bit does not need to be checked to determine the source of interrupt for the single interrupt in this example. It will be shown in the next section that for multiple interrupts, the interrupt flag bit for each individual interrupt must be checked in the routine at the interrupt address vector to find the source of interrupt.

8.3.4 Interrupt Registers and Priorities

The PIC18F4321 contains ten registers which are used to control interrupt operation. These registers are

- RCON (Figure 8.5)
- INTCON (Figure 8.17)
- INTCON2 (Figure 8.21)
- INTCON3 (Figure 8.17)
- PIR1, PIR2 (to be discussed in Chapter 9)
- PIE1, PIE2 (to be discussed in Chapter 9)
- IPR1, IPR2 (discussed in Microchip's PIC18F4321 manual)

Registers RCON, INTCON, INTCON2, and INTCON3 are associated with external and port change interrupts. Hence, they will be covered in this section. Registers

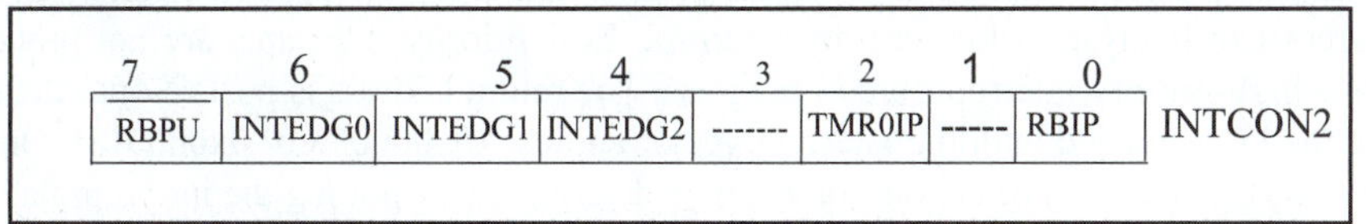

bit 7 **RBPU:** PORTB Pull-up Enable bit
1 = All PORTB pull-ups are disabled
0 = PORTB pull-ups are enabled by individual port latch values
bit 6 **INTEDG0:** External Interrupt 0 Edge Select bit, 1 = Interrupt on rising edge, 0 = Interrupt on falling edge
bit 5 **INTEDG1:** External Interrupt 1 Edge Select bit, 1 = Interrupt on rising edge, 0 = Interrupt on falling edge
bit 4 **INTEDG2:** External Interrupt 2 Edge Select bit, 1 = Interrupt on rising edge, 0 = Interrupt on falling edge
bit 3 **Unimplemented:** Read as '0'
bit 2 **TMR0IP:** TMR0 Overflow Interrupt Priority bit, 1 = High priority, 0 = Low priority
bit 1 **Unimplemented:** Read as '0'
bit 0 **RBIP:** RB Port Change Interrupt Priority bit, 1 = High priority, 0 = Low priority

FIGURE 8.21 INTCON2 register

PIR1, PIR2, PIE1, PIE2, IPR1, and IPR2 are used by peripheral interrupts. Hence, the functions of these registers except IPR1 and IPR2 will be discussed when these topics are covered in Chapter 9.

In general, the operation of each interrupt source can be controlled by three bits. They are

- Flag bit that indicates that an interrupt event has occurred.
- Enable bit that allows program execution to branch to the interrupt vector address when the flag bit is set.
- Priority bit that is used to select high priority or low priority.

The PIC18F4321 interrupts can be classified into two groups: high-priority interrupt levels and low-priority interrupt levels. The high-priority interrupt vector is at address 0x000008 and the low-priority interrupt vector is at address 0x000018 in the program memory. High-priority interrupt events will interrupt any low-priority interrupts that may be in progress.

As mentioned before, upon power-on reset, the interrupt address vector is 0x000008 (default), and no interrupt priorities are available. Upon power-on reset, IPEN is automatically cleared to 0, and the PIC18F operates as a high-priority interrupt (single interrupt) system. Hence, the interrupt vector address is 0x000008. The IPEN bit (bit 7 of the RCON register) of the RCON register in Figure 8.5 can be programmed to assign interrupt priorities. During normal operation, the IPEN bit can be set to one by executing the BSF RCON, IPEN to assign priorities in the system.

When interrupt priority (IP) is enabled (IPEN = 1), there are two bits that enable interrupts globally. Setting the GIEH bit (bit 7 of INTCON register of Figure 8.17) enables all interrupts that have the priority bit set (high-priority). Setting the GIEL bit (bit 6 of INTCON register of Figure 8.17) enables all low-priority interrupts. When the interrupt flag bit, enable bit, and appropriate global interrupt enable bit are set, the interrupt will vector immediately to address 0x000008 or 0x000018, depending on the priority bit setting. Individual interrupts can be disabled through their corresponding enable bits.

In summary, when the IPEN bit is cleared (default state), the interrupt priority feature is disabled. Bit 6 of the INTCON register is the PEIE bit, which enables/disables all peripheral interrupt sources. Bit 7 of the INTCON register is the GIE bit, which enables/disables all interrupt sources. All interrupts branch to address 0x000008 upon power-on reset (default). When an interrupt is responded to, the global interrupt enable bit is cleared to disable further interrupts. If interrupt priority levels are used, high-priority interrupt sources can interrupt a low-priority interrupt. Low-priority interrupts are not processed while high-priority interrupts are in progress. The return address is pushed onto the stack and the PC is loaded with the interrupt vector address (0x000008 or 0x000018). Once in the ISR, the source(s) of the interrupt can be determined by polling the interrupt flag bits. The interrupt flag bits must be cleared in software before re-enabling interrupts to avoid recursive interrupts.

The "return from interrupt" instruction, RETFIE, exits the interrupt routine and sets the GIE bit (GIEH or GIEL if priority levels are used), which re-enables interrupts.

Next, interrupt priorities associated with the PIC18F4321 external interrupts (INT0, INT1, and INT2) will be discussed. Setting up interrupt priorities for peripherals such as the hardware timers will be discussed as these topics are covered.

Table 8.3 shows the three external interrupts of the PIC18F4321 along with the corresponding IP bit. As mentioned in the table, since an IP is not assigned to INT0, it

TABLE 8.3 PIC18F4321 external interrupts along with interrupt priority (IP) bits

Interrupt name	Interrupt priority (IP) bit	Comment
INT0	Unassigned	Since no interrupt priority bit is assigned, INT0 always has the high priority level.
INT1	INT1IP	Bit 6 of INTCON3 register (Figure 8.17); 1 = high priority, 0 = low priority
INT2	INT2IP	Bit 7 of INTCON3 register (Figure 8.17); 1 = high priority, 0 = low priority

always has the high-priority level. However, INT1 and INT2 can be programmed as high- or low-level priorities. For example, in order to program INT1 as a high-priority interrupt and INT2 as a low-priority interrupt, the following instruction sequence can be used:

```
BSF    RCON, IPEN       ; Set IPEN to 1, enable interrupt level
BSF    INTCON, GIEL     ; Set low-priority levels
BSF    INTCON, GIEH     ; Set high-priority levels
BSF    INTCON3, INTIP   ; INT1 has high level
BCF    INTCON3, INT2P   ; INT2 has low level
```

8.3.5 Setting the Triggering Levels of INTn Pin Interrupts

External interrupts on the RB0/INT0, RB1/INT1, and RB2/INT2 pins are edge-triggered. Upon power-on reset, each of these external interrupts (INT0, INT1, INT2) is activated by a rising edge pulse (LOW to HIGH). The PIC18F has the flexibility of changing the triggering levels for these interrupts to falling edge pulses (HIGH to LOW). This can be accomplished by programming bits 4 through 6 of the INTCON2 register (Figure 8.21). For example, the instruction BCF INTCON2, INTEDG0 will change INT0 from positive-triggered to negative-triggered interrupt. Note that the other bits in the INTCON2 are either unimplemented or contain control bits such as RB Port Change Interrupt Priority bit (to be discussed later).

If the corresponding INTEDGx bit in the INTCON2 register is set (= 1), the interrupt is triggered by a rising edge; if the bit is clear, the trigger is on the falling edge. When a valid edge appears on the RBx/INTx pin, the corresponding flag bit, INTxF, is set. This interrupt can be disabled by clearing the corresponding enable bit, INTxE. Flag bit INTxF must be cleared in software in the ISR before re-enabling the interrupt. All external interrupts (INT0, INT1, and INT2) can wake up the processor from Idle or Sleep modes if bit INTxE was set prior to going into those modes. If the GIE bit, is set, the processor will branch to the interrupt vector following wake-up. Interrupt priority for INT1 and INT2 is determined by the value contained in the interrupt priority bits, INT1IP (bit 6 of INTCON3) and INT2IP (bit 7 of INTCON3). There is no priority bit associated with INT0. It is always a high-priority interrupt source.

8.3.6 Return from Interrupt Instruction

The "RETFIE s" instruction is normally used at the end of the service routine. 's' can be 0 or 1. When s = 0, the PIC18F RETFIE 0 or simply RETFIE instruction pops the contents of the program counter previously pushed before going to the service routine, enables the global interrupt enable bit, and returns control to the appropriate place in the main program. The "RETFIE 1" instruction, on the other hand, pops the contents of WREG, BSR, and STATUS registers (previously PUSHed) from shadow registers WS, STATUSS, and BSRS before going to the main program, enables the global interrupt enable bit, and returns control to the appropriate place in the main program.

8.3.7 PORTB Interrupt-on-Change

The PIC18F4321 provides four interrupt-on-change pins (KB10 through KB13). These pins are multiplexed among others with bits 4 through 7 of Port B.

An input change (HIGH to LOW or LOW to HIGH) on one or more of these interrupts sets the flag bit RBIF (bit 0 of INTCON register). Note that a single flag bit is assigned to all four interrupts.

The interrupt can be enabled/disabled by setting/clearing a single enable bit, RBIE (bit 3 of INTCON register). Interrupt priority for PORTB interrupt-on-change is determined by the value contained in the interrupt priority bit RBIP (bit 0 of INTCON2 register). As before, BSF and BCF instructions can be used to set or clear a bit in a register. The PORTB interrupt-on-charge is typically used for interfacing devices such as keyboard.

8.3.8 Context Saving During Interrupts

During interrupts, the return PC address is saved onto the hardware stack. For high-priority interrupts, the PIC18F also saves WREG, STATUS, and BSR registers automatically in the associated shadow registers (internal to the PIC18F) called WS, STATUSS, and BSRS. Note that these three registers are saved internally upon recognition of a high-priority interrupt, and before going to the ISR. The contents of WREG, STATUS, and BSR are normally changed by the instructions in the ISR. Hence, it is desirable for the user to restore the contents of these registers before returning to the main program. This can be accomplished by placing the "RETFIE 1" at the end of the ISR which will restore these registers, enable global interrupt, and return control to the appropriate place in the main program.

For low priority interrupts, only the return PC address is saved onto the hardware stack. Additionally, the user may need to save WREG, STATUS and BSR registers in data memory on entry to the ISR. Depending on the user's application, other registers may also need to be saved. For low-priority interrupts, the user may save the desired registers such as WREG, STATUS, and BSR in data registers, and retrieve them before returning to the main program. The following PIC18F instruction sequence illustrates this:

```
; MAIN PROGRAM
W_TEMP              EQU     0x20
STATUS_TEMP         EQU     0x30
BSR_TEMP            EQU     0x40
                    ---
                    ---
; INTERRUPT SERVICE ROUTINE
; SAVING STATUS, WREG AND BSR REGISTERS IN DATA MEMORY
          MOVWF     W_TEMP                   ; Save WREG in 0x20
          MOVFF     STATUS, STATUS_TEMP      ; Save STATUS in 0x30
          MOVFF     BSR, BSR_TEMP            ; Save BSR in 0x40
          ---
          ---
          MOVFF     BSR_TEMP, BSR            ; Restore BSR
```

```
MOVF     W_TEMP, W                ; Restore WREG
MOVFF    STATUS_TEMP, STATUS      ; Restore STATUS
RETFIE                            ; POP PC from hardware stack,
                                  ; Enable global interrupt, and
                                  ; return to the main program
```

Example 8.5 In Figure 8.22, if Vx > Vy, the PIC18F4321 is interrupted via INT0. On the other hand, opening the switch will interrupt the microcontroller via INT1. Note that in the PIC18F4321, INT0 has higher priority than INT1. Write the main program in PIC18F assembly language at address 0x100 that will perform the following:

- Initialize STKPTR to 0x50.
- Configure PORTB as interrupt inputs.
- Clear interrupt flag bits of INT0 and INT1.
- Set INT1 as low-priority interrupt.
- Enable global HIGH and LOW interrupts.
- Turn both LEDs at PORTD OFF (comparator LED at bit 0 of PORTD and switch LED at bit 1 of PORTD)
- Wait in an infinite loop for one or both interrupts to occur.

Also, write a service routine for the high-priority interrupt (INT0) in PIC18F assembly language at address 0x200 that will perform the following:

- Clear interrupt flag for INT0.
- Check to see if the comparator output is still 1. If it is, turn LED at bit 0 of PORTD ON. If the comparator output is 0, return.

Finally, write a service routine for the low-priority interrupt (INT1) in PIC18F assembly language at address 0x300 that will perform the following:

- Clear interrupt flag for INT1.
- Check to see if the switch is still 1. If it is, turn LED at bit 1 of PORTD ON. If the switch input is 0, return.

Solution

This example will demonstrate the interrupt priority scheme of the PIC18F4321 microcontroller. With interrupt priority, the user has the option to have the interrupts declared as either low or high interrupts. If, at anytime, the low- and high-priority interrupts occur at the same time, the microcontroller will always service the high-priority interrupt.

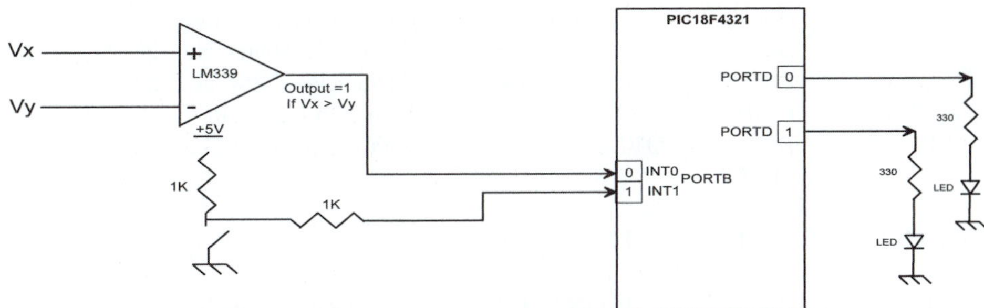

FIGURE 8.22 Figure for Example 8.5

In the above example, the comparator is set as the high-priority and the switch is set as the low-priority, so if both interrupts are triggered simultaneously, then only the LED associated with the comparator will be turned ON.

. Note that the external interrupt INT0 can only be a high-priority interrupt. When implementing a single interrupt, the interrupt service routine is written at address 0x08. On the other hand, when priority interrupts are enabled, the service routine for the high-priority interrupt is written at address 0x08 while the service routine for the low-priority interrupt is written at address 0x018.

In order to enable the second external interrupt INT1, the register INTCON3 is configured. Also, INT1IE must be enabled, and INT1IF must be cleared to 0. Furthermore, the INT1IP bit in INTCON3 register that sets the priority of INT1 in the INTCON3 register must be cleared to 0 for low priority. Next, the IPEN bit in the RCON register that enables the interrupt priority functionality of the PIC18F4321 must be set to one. Finally, the GIEH and GIEL bits in the INTCON register must be set to one in order to enable global high and low interrupts. The following code implements priority interrupts on the PIC18F4321 using assembly language:

```
INCLUDE <P18F4321.INC>
; RESET
                ORG 0                             ; Reset vector
                GOTO      MAIN                    ; Jump to main program
; HIGH PRIORITY INTERRUPT ADDRESS VECTOR
                ORG       0x0008                  ; High-priority interrupt
                BRA       HIGH_INT_ISR            ; Jump to service routine for the comparator
; LOW PRIORITY INTERRUPT ADDRESS VECTOR
                ORG       0x0018                  ; Low-priority interrupt
                BRA       LOW_INT_ISR             ; Jump to service routine for the switch
                                                  ; Main Program
                ORG       0x0100
MAIN            MOVLW     0x50                    ; Initialize STKPTR to 0x50
                MOVWF     STKPTR
                CLRF      TRISD                   ; PORTD is output
                MOVLW     0x0F                    ; Configure ADCON1 to set up
                MOVWF     ADCON1                  ; INT0 and INT1 as digital inputs
                BSF       INTCON, INT0IE          ; Enable the external interrupt INT0
                BSF       INTCON3,INT1IE          ; Enable the external interrupt INT1
                BCF       INTCON,INT0IF           ; Clear the INT0 flag
                BCF       INTCON3,INT1IF          ; Clear the INT1 flag
                BCF       INTCON3, INT1IP         ; Set INT1 as low priority
                BSF       RCON, IPEN              ; Enable interrupt priority
                BSF       INTCON, GIEH            ; Enable global high interrupts
                BSF       INTCON, GIEL            ; Enable global low interrupts
                CLRF      PORTD                   ; Turn both LEDs off
OVER            BRA       OVER                    ; Wait for interrupt
                SLEEP                             ; Halt

; SERVICE ROUTINE FOR HIGH PRIORITY
                ORG       0x200
HIGH_INT_ISR    BCF       INTCON,INT0IF           ; Clear the interrupt flag
CHECK           BTFSS     PORTB,0                 ; Check to see if comparator output is one
```

```
                RETFIE
                MOVLW    0x01              ; Turn on LED at bit 0 of PORTD
                MOVWF    PORTD
                BRA      CHECK
; SERVICE ROUTINE FOR LOW PRIORITY
                ORG      0x300
LOW_INT_ISR     BCF      INTCON3, INT1IF   ; Clear the interrupt flag
CHECK1          BTFSS    PORTB,1           ; Check to see if switch is one
                RETFIE
                MOVLW    0x02              ; Turn on LED at bit 1 of PORTD
                MOVWF    PORTD
                BRA      CHECK1
                END
```

8.4 PIC18F Interface to an LCD (Liquid Crystal Display)

Seven-segment LEDs are easy to use, and can display only numbers and limited characters. LCDs are very useful for displaying numbers, and several ASCII characters along with graphics. Furthermore, LCDs consume low power. Because of inexpensive price of LCDs these days, they have been becoming popular. LCDs are widely used in notebook computers.

Figure 8.23 shows the PIC18F4321's interface to a typical LCD display such as the Optrex DMC16249 LCD with a 2-line x 16-character display screen. In order to illustrate the basic concepts associated with LCDs, the phrase "Switch Value:" along with the numeric BCD value (0 through 9) of the four switch inputs will be displayed.

The Optrex DMC16249 LCD shown in Figure 8.23 contains 14 pins. The VCC pin is connected to +5 V and the VSS pin is connected to ground. The VEE pin is the contrast control for brightness of the display. VEE is connected to a potentiometer with a value between 10k and 20k. The eight data pins (D0-D7) are used to input data and commands to display the desired message on the screen.

The three control pins EN, $R/\overline{W}$, and RS allow the user to let the display know what kind of data is sent. The EN pin latches the data from the D0-D7 pins into the LCD

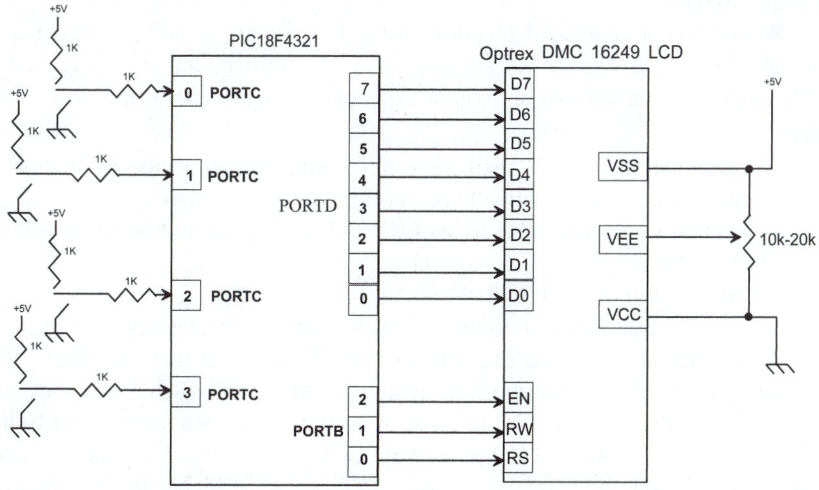

FIGURE 8.23 PIC18F4321 interface to Optrex DMC 16249 LCD

TABLE 8.4 Typical LCD commands along with 8-bit codes in hex

Hex	Command
0x01	Clear the screen
0x02	Return home
0x04	Shift cursor to left
0x05	Shift display to right
0x06	Shift cursor to right
0x07	Shift display to left
0x08	Display off, cursor off
0x0A	Display off, cursor on
0x0C	Display on, cursor off
0x0E	Display on, cursor blinking
0x10	Shift cursor position to the left
0x14	Shift cursor position to the right
0x80	Move cursor to the start of the first line

display. Data on D0-D7 pins will be latched on the trailing edge (high-to-low) of the EN pulse. The EN pulse must be at least 450 ns wide. The $R/\overline{W}$ (read/write) pin allows the user to either write to the LCD or read data from the LCD. In this example, the $R/\overline{W}$ pin will always be zero since only a string of ASCII data is written to the LCD. The $R/\overline{W}$ pin is set to one for reading data from the LCD.

The command or data can be output to the LCD in two ways. One way is to provide time delays of a few milliseconds before outputting the next command or data. The second approach utilizes a busy flag to determine whether the LCD is free for the next data or command. For example, in order to display ASCII characters one at a time, the LCD must be read by outputting a HIGH on the $R/\overline{W}$ pin. The busy flag can be checked to ensure whether the LCD is busy or not before outputting another string of data. Note that the busy flag can thus be used instead of time delays.

Finally, the RS (Register Select) pin is used to determine whether the user is sending command or data. The LCD contains two 8-bit internal registers. They are command register and data register. When RS = 0, the command register is accessed, and typical LCD commands such as shift cursor left (hex code 0x04) can be used. Table 8.4 shows a list of some of the commands. Note that the busy flag is bit 7 of the LCD's command register. The busy bit can be read by outputting 0 to RS pin, 1 to $R/\overline{W}$ pin, and a leading edge (LOW to HIGH) pulse to the EN pin.

When attempting to send data or commands to the LCD, the user must make sure that the values of EN, $R/\overline{W}$, and RS are correct, along with appropriate timing. A PIC18F assembly language program can be written to output appropriate values to these pins via I/O ports.

For example, in order to send the 8-bit command code to the LCD, a PIC18F assembly language program is written to perform the following steps:

- output the command value to the PIC18F4321 I/O port that is connected to the LCD's D0- D7 pins.
- send 0 to RS pin and 0 to $R/\overline{W}$ pin.
- Send a '1' and then a '0' to the EN pin to latch the LCD's D0-D7 code.

As mentioned earlier, the example in Figure 8.23 will display the phrase "Switch Value:" along the BCD value of the four switch inputs. Four switches are connected to bits 0 through 3 of PORTC. The D0-D7 pins of the LCD are connected to bits 0 through 7 of PORTD. The RS, $R/\overline{W}$, and EN pins of the LCD are connected to bits 0, 1, and 2 of PORTB of PIC18F4321.

The PIC18F assembly language program is shown in Figure 8.24. Note that time delay rather than the busy bit is used before outputting the next character to the LCD. Three subroutines are used: one for outputting command code, one for delay, and one for LCD data. Since subroutines are used, the hardware STKPTR is initialized in the main program with an arbitrarily chosen value of 0x40. PORTB and PORTD are configured as output ports, and PORTC is set up as an input port. Also, assume 1-MHz default crystal frequency for the PIC18F4321.

As an example, let us consider the code for outputting a command code such as the command "move cursor to the beginning of the first line" to the LCD. From Table 8.4, the command code for this is 0x80. From Figure 8.24, the code MOVLW 0x80 moves 0x80 into WREG. The CALL CMD calls the subroutine CMD. The CMD subroutine first outputs the command code 0x80 to PORTD using MOVWF PORTD. Since PORTD is connected to LCD's D0-D7 pins, these data will be available to be latched by the LCD. The following few lines of the code of the CMD subroutine are for outputting 0's to RS and $R/\overline{W}$ pins, and a trailing edge (1 to 0) pulse to EN pin along with a delay of 20 msec. Hence, the LCD will latch 0x80, and the cursor will move to the start of the first line. Note that an external counter of 10 loaded into a register 0x21 with a 2 msec inner loop for LOOP2 is used for the 20 msec delay. Typical delays should be 10 to 30 milliseconds. Also, 1 MHz default crystal frequency for the PIC18F4321 is assumed. The program then returns to the main program.

The first few lines of the main program at address MAIN perform initializations. Next, in order to display 'S', the MOVLW D'10' moves 10 (decimal) into WREG, and CALL DELAY provides 20 msec delay using this value in the routine. After executing the DELAY routine, MOVLW A'S' moves the 8-bit ASCII code for S into WREG. The instruction CALL LCDDATA calls the subroutine LCDDATA. The MOVWF PORTD instruction in this subroutine outputs the ASCII code for S into the D0-D7 pins of the LCD via PORTD. The next few instructions in the LCDDATA subroutine outputs 1 to the RS pin (for selecting LCD data register to display data), 0 to the $R/\overline{W}$ pin, and a trailing edge (1 to 0) pulse to EN pin along with delay so that the LCD will latch ASCII code for S, and will display S on the screen.

Similarly, the program logic in Figure 8.24 for outputting other ASCII characters and switch input data can be explained.

The PIC18F assembly language program is provided in Figure 8.24 as follows.

For 1 MHz default crystal frequency, the PIC18F clock period will be 1 μsec. Hence, each instruction cycle will be 4 microseconds. For 2 msec delay, total cycles = (2 msec)/(4 μsec)= 500. The DECFSZ in the loop will require 2 cycles for (COUNT - 1) times when Z = 0 and the last iteration will take 1 cycle when skip is taken (Z = 1). Thus, total cycles including the MOVLW = $1 + 1 + 1 + 2 \times (\text{COUNT} - 1) + 2 = 500$. Hence, COUNT will be approximately 255 (decimal), discarding execution times of certain instructions.. Therefore, register 0x21 should be loaded with an integer value of 255 for an approximate delay of 2 msec.

8.5 Interfacing PIC18F4321 to a Hexadecimal Keyboard and a Seven-segment Display

In this section we describe the basics of interfacing the PIC18F4321 microcontroller to a hexadecimal keyboard and a seven-segment display.

```
        INCLUDE <P18F4321.INC>
        ORG     0x100       ; Start of the MAIN program
MAIN    MOVLW   0x40        ; Initialize STKPTR with arbitrary value of 0x40
        MOVWF   STKPTR
        CLRF    TRISD       ; PORTD is output
        CLRF    TRISB       ; PORTB is output
        SETF    TRISC       ; PORTC is input
        CLRF    PORTB       ; rs=0 rw=0 en=0
        MOVLW   D'10'       ; 20 msec delay
        CALL    DELAY
        MOVLW   0x0C        ; Display on, Cursor off
        CALL    CMD
        MOVLW   D'10'       ; 20 msec delay
        CALL    DELAY
        MOVLW   0x01
        CALL    CMD         ; Clear Display
        MOVLW   D'10'       ; 20 msec delay
        CALL    DELAY
        MOVLW   0x06        ; Shift Cursor to the right
        MOVLW   D'10'       ; 20 msec delay
        CALL    DELAY
        MOVLW   0x80        ; Move cursor to the start of the first line
        CALL    CMD
        MOVLW   D'10'       ; 20 msec delay
        CALL    DELAY
        MOVLW   A'S'        ; Send ASCII S
        CALL    LCDDATA
        MOVLW   A'w'        ; Send ASCII w
        CALL    LCDDATA
        MOVLW   A'i'        ; Send ASCII i
        CALL    LCDDATA
        MOVLW   A't'        ; Send ASCII t
        CALL    LCDDATA
        MOVLW   A'c'        ; Send ASCII c
        CALL    LCDDATA
        MOVLW   A'h'        ; Send ASCII h
        CALL    LCDDATA
        MOVLW   A' '        ; Send ASCII space
        CALL    LCDDATA
        MOVLW   A'V'        ; Send ASCII V
        CALL    LCDDATA
        MOVLW   A'a'        ; Send ASCII a
        CALL    LCDDATA
        MOVLW   A'l'        ; Send ASCII l
        CALL    LCDDATA
        MOVLW   A'u'        ; Send ASCII u
```

FIGURE 8.24 Assembly language program for the PIC18F4321-LCD interface

```
         CALL     LCDDATA
         MOVLW    A'e'      ; Send ASCII e
         CALL     LCDDATA
         MOVLW    A':'      ; Send ASCII :
         CALL     LCDDATA
AGAIN    MOVF     PORTC, W  ; Move switch value to WREG
         ANDLW    0x0F      ; Mask lower 4 bits
         IORLW    0x30      ; Convert to ASCII data by Oring with 0x30
         CALL     LCDDATA   ; Display switch value on screen
         MOVLW    0x10
         CALL     CMD
         BRA      AGAIN
CMD      MOVWF    PORTD     ; Command is sent to PORTD
         MOVLW    0x04
         MOVWF    PORTB     ; rs=0 rw=0 en=1
         MOVLW    D'10'     ; 20 msec delay
         CALL     DELAY
         CLRF     PORTB     ; rs=0 rw=0 en=0
         RETURN
LCDDATA  MOVWF    PORTD     ; Data sent to PORTD
         MOVLW    0x05      ; rs=1 rw=0 en=1
         MOVWF    PORTB
         MOVLW    D'10'     ; 20 msec delay
         CALL               DELAY
         MOVLW    0x01
         MOVWF    PORTB     ; rs=1 rw=0 en=0
         RETURN
DELAY    MOVWF    0x20
LOOP1    MOVLW    D'255'    ; LOOP2 provides 2 msec delay with a count of 255
         MOVWF    0x21
LOOP2    DECFSZ   0X21
         GOTO     LOOP2
         DECFSZ   0x20
         GOTO     LOOP1
         RETURN
         END
```

FIGURE 8.24 Assembly language program for the PIC18F4321-LCD interface (continued)

8.5.1 Basics of Keyboard and Display Interface to a Microcontroller

A common method of entering programs into a microcontroller is via a keyboard. An inexpensive way of displaying microcontroller results is by using seven-segment displays. The main functions to be performed for interfacing a keyboard are

- Sense a key actuation.
- Debounce the key.
- Decode the key.

Let us now elaborate on keyboard interfacing concepts. A keyboard is arranged

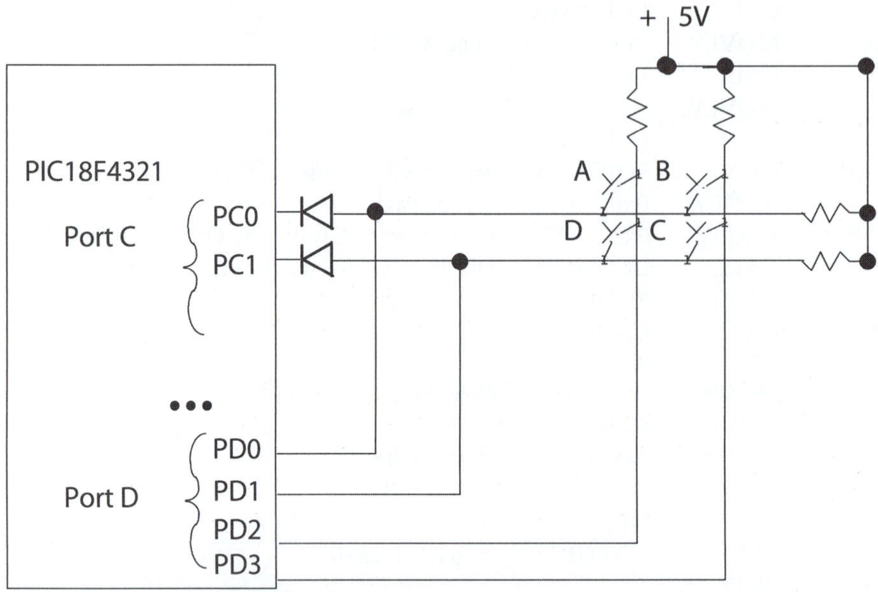

Figure 8.25 PIC18F4321 interface to a 2 x 2 keyboard

in rows and columns. Figure 8.25 shows a 2 × 2 keyboard interfaced to a typical microcontroller such as the PIC18F4321. In Figure 8.25 the columns are normally at a HIGH level. A key actuation is sensed by sending a LOW (closing the diode switch) to each row one at a time via PC0 and PC1 of Port C. The two columns can then be input via PD2 and PD3 of Port D to see whether any of the normally HIGH columns are pulled LOW by a key actuation. If so, the rows can be checked individually to determine the row in which the key is down. The row and column code for the key pressed can thus be found.

The next step is to debounce the key. *Key bounce* occurs when a key is pressed or released, it bounces for a short time before making the contact. When bounce occurs, it may appear to the microcontroller that the same key has been actuated several times instead of just once. This problem can be eliminated by reading the keyboard after about 20 ms and then verifying to see if it is still down. If it is, the key actuation is valid. The next step is to translate the row and column code into a more popular code, such as hexadecimal or ASCII. This can easily be accomplished by a program. Certain characteristics associated with keyboard actuation must be considered while interfacing to a microcontroller. Typically, these are two-key lockout and *N*-key rollover. The two-key lockout ensures that only one key is pressed. An additional key depressed and released does not generate any codes. The system is simple to implement and most often used. However, it might slow down the typing because each key must be released fully before the next one is pressed down. On the other hand, the *N*-key rollover will ignore all keys pressed until only one remains down.

Now let us elaborate on the interfacing characteristics of typical displays. The following functions are typically performed for displays:

- Output the appropriate display code.
- Output the code via right entry or left entry into the displays if there are more than one display.

These functions can easily be realized by a microcontroller program. If there is more than one display, the displays are typically arranged in rows. A row of four displays

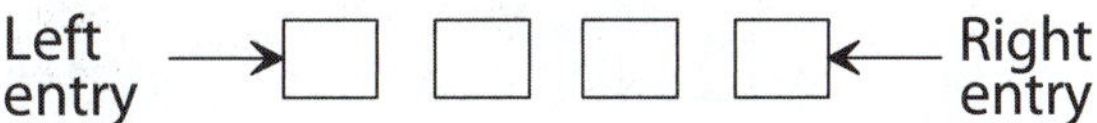

Figure 8.26 Row of four displays

is shown in Figure 8.26. In the figure, one has the option of outputting the display code via right entry or left entry. If the code is entered via right entry, the code for the least significant digit of the four-digit display should be output first, then the next-digit code, and so on. The program outputs to the displays are so fast that visually all four digits will appear on the display simultaneously. If the displays are entered via left entry, the most significant digit must be output first and the rest of the sequence is similar to that of right entry.

Two techniques are typically used to interface a hexadecimal display to the microcontroller: nonmultiplexed and multiplexed. In nonmultiplexed methods, each hexadecimal display digit is interfaced to the microcontroller via an I/O port. Figure 8.27 illustrates this method. BCD-to-seven-segment conversion is done in software. The microcontroller can be programmed to output to the two display digits in sequence. However, the microcontroller executes the display instruction sequence so fast that the displays appear to the human eye at the same time. Figure 8.28 illustrates the multiplexing method of interfacing the two hexadecimal displays to the microcontroller. In the multiplexing scheme, appropriate seven-segment code is sent to the desired displays on seven lines common to all displays. However, the display to be illuminated is grounded. Some displays, such as Texas Instrument's TIL 311, have an on-chip decoder. In this case, the microcontroller is required to output 4 bits to a display. Note that the TIL311 displays a hex digit (A - F) based on a 4-bit input.

8.5.2 PIC18F4321 Interface to a Hexadecimal Keyboard and a Seven-segment Display

In this section, the basic concepts associated with interfacing a hexadecimal keyboard along with a seven-segment display to the PIC18F4321 is provided in a simplified manner. The PIC18F4321 microcontroller is designed to display a hex digit (0 - F) entered

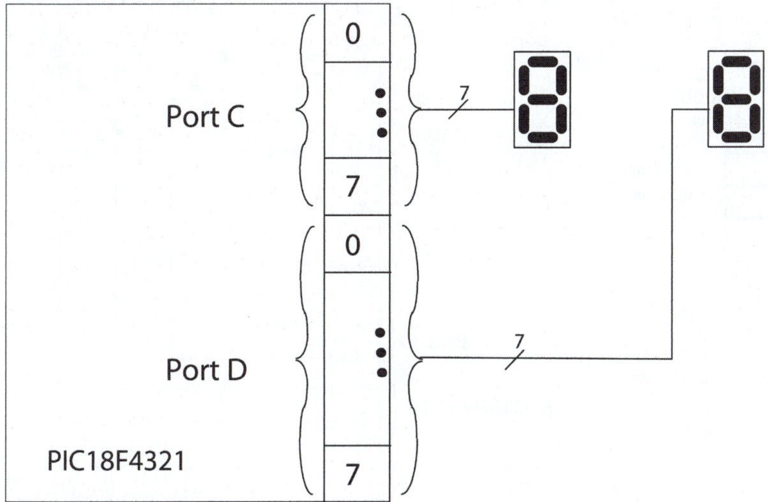

Figure 8.27 Nonmultiplexed hexadecimal displays

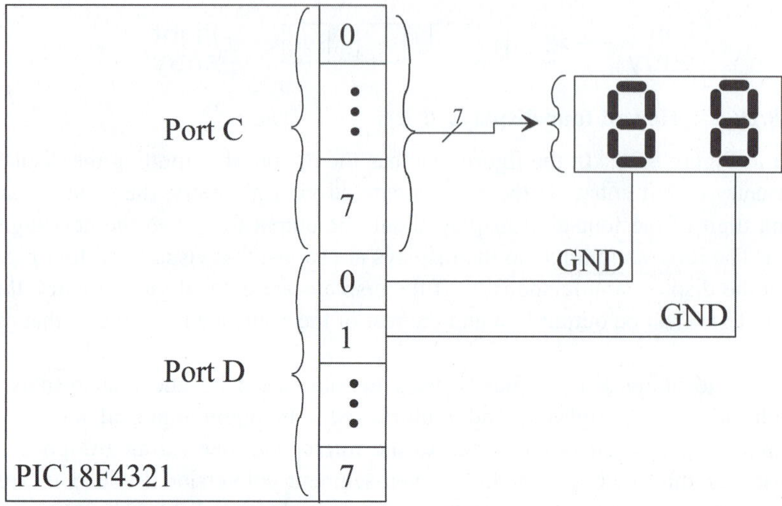

FIGURE 8.28 Multiplexed hexadecimal displays

via a hexadecimal keypad (16 keys). The user will push one of the hex digits from 0 to F using the keys on the hexadecimal keyboard. The PIC18F4321 will input these data via PORTD, and output to a seven-segment display with an on-chip decoder such as the TIL311. Figure 8.29 shows the hardware schematic.

Three 8-bit I/O ports (Port B, Port C, Port D) of the PIC18F4321 are used in the design. Ports B, C, and D are configured as follows:

- Port B is configured as an output port to display the key(s) pressed.
- Port C is configured as an output port to output zeros to the rows to detect a key actuation.

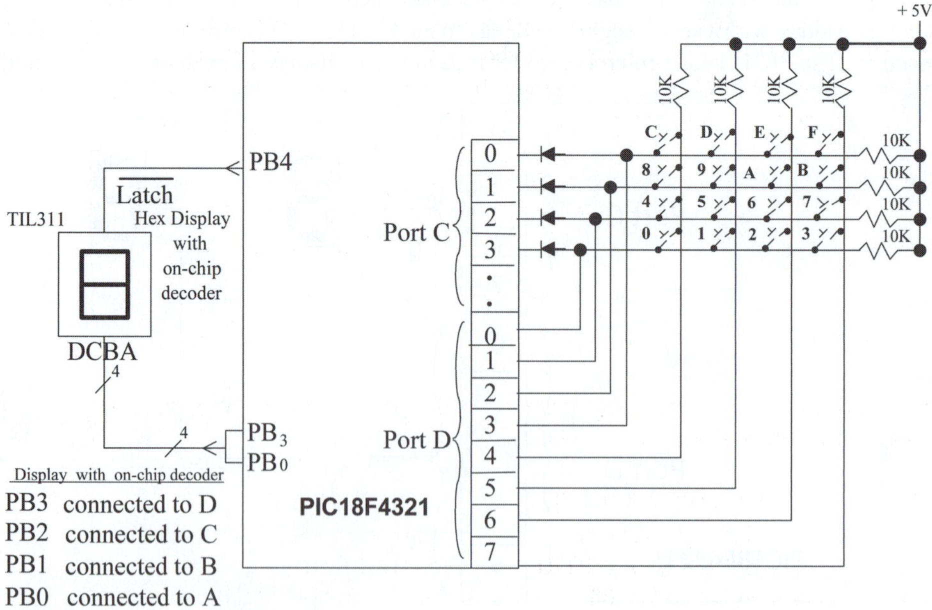

FIGURE 8.29 PIC18F4321 interface to keyboard and display

- Port D is configured as an input port to receive the row–column code.

The PIC18F4321 default crystal frequency of 1 MHz is assumed. Debouncing is provided to avoid unwanted oscillation caused by the opening and closing of key contacts. To ensure stability for the input signal, a delay of 20 ms is used for debouncing the input. Using the 2 msec delay routine from the previous section as the inner loop, the following subroutine can be used for a delay routine of 20 ms:

```
DELAY       MOVLW       D'10'
            MOVWF       0x20
LOOP1       MOVLW       D'255'   ;LOOP2 provides 2 msec delay with a count of 255
            MOVWF        0x21
LOOP2       DECFSZ      0X21
            GOTO        LOOP2
            DECFSZ      0x20
            GOTO        LOOP1
            RETURN
```

A PIC18F assembly program is written for the keyboard/display interface. The program scans all 16 keys for key actuation. As soon as a key actuation is detected, the program will debounce the key using the DELAY routine, and then determine the key pressed using a decode table stored in memory. The decode table contains seven-segment code for the hex digits from 0 to F.

Texas Instrument's TIL311 hex-to-seven-segment decoder is used for the display. The TIL311 includes an on-chip decoder. It has four inputs (D, C, B, A, with D as the most significant bit and A as the least significant bit), and seven outputs (a-g). In order to display a hex digit such as F in Figure 8.29, the PIC18F4321 can be programmed to output 1111 via bits 3-0 of PORTB on TIL311's DCBA pins, and a LOW via bit 4 of PORTB on the TIL311's $\overline{\text{LATCH}}$ pin. The TIL311 will then display F on the seven-segment display.

The PIC18F assembly language program written at address 0x100 for interfacing the PIC18F4321 to a hexadecimal keyboard and a seven-segment display is provided below. Note that to explain the program, line numbers are included using # symbol with the comments in the following:

```
            INCLUDE <P18F4321.INC>
            ORG       0x100          ; #1Starting address of the program
OPEN        EQU       0xF0           ; #2Row/column codes if all
                                     ; keys
                                     ; are open
COUNTER     EQU       0x80

; Transfer  keyboard decode table at the end of this program starting from program memory
; address 0x200  to data memory address 0x50
            MOVLW   0x00             ; #3 Move upper 5 bits (00H) of address
            MOVWF   TBLPTRU          ; to TBLPTRU
            MOVLW   0x02             ; Move bits 15-8 (02H) of address
            MOVWF   TBLPTRH          ; to  TBLPTRH
            MOVLW   0x00             ; Move bits 7-0 (00H) of address
            MOVWF   TBLPTRL          ; #4 to TBLPTRL
```

```
            LFSR      1, 0x50       ; Initialize FSR1 to 0x50 to be used as
                                    ; destination pointer in data memory
            MOVLW     D'16'         ; Initialize COUNTER with 16
            MOVWF     COUNTER       ; Move [WREG] into COUNTER
LOOP        TBLRD*+                 ; Read data from program memory into
                                    ; TABLAT, increment TBLPTR by 1
            MOVF      TABLAT, W     ; Move [TABLAT] into WREG
            MOVWF     POSTINC1      ; Move [WREG] into data memory pointed to
                                    ; by FSR1, and then increment FSR1 by 1
            DECF      COUNTER, F    ; Decrement COUNTER BY 1
            BNZ       LOOP          ; #5 Branch  if  Z = 0, else Stop
; Perform initializations
            CLRF      TRISB         ; #6 Configure Port B as an output port
            CLRF      TRISC         ; #7 Configure Port C as an  output port
            SETF      TRISD         ; #8Configure Port D as an input port
            MOVLW     0x25          ; STKPTR is initialized with arbitrary value
            MOVWF     STKPTR        ; #9since subroutine DELAY is used later
; Detect a key actuation, debounce it, decode, and display
            MOVLW     0             ; #10 Send 0 to enable display and then
            MOVWF     PORTB         ; #11 Initialize display with 0
SCAN_KEY    MOVWF     PORTC         ; #12 Output 0s to rows of the keyboard
            MOVLW     OPEN          ; #13 Move 0xF0 to 0x30
            MOVWF     0x30
KEY_OPEN    MOVF      PORTD,W       ; #14 Read PORTD into WREG
            SUBWF     0x30, W       ; #15 Are all keys opened?
            BNZ       KEY_OPEN      ; #16 Repeat if closed
            CALL      DELAY         ; #17 Debounce for 20 ms
KEY_CLOSE   MOVF      PORTD, W      ; #18 Read PORTD into WREG
            SUBWF     0x30, W       ; #19 Are all keys closed?
            BZ        KEY_CLOSE     ; #20 Repeat if opened
            CALL      DELAY         ; #21 Debounce again for 20 ms
            SETF      0x35          ; #22 Set 0x35 contents to all 1's
            BCF       STATUS, C     ; #23 Clear Carry Flag
NEXT_ROW    RLCF      0x35, F       ; #24 Set up row mask
            MOVFF     0x35, 0x36    ; #25 Save row mask  in 0x36
            MOVFF     0x35, PORTC   ; #26 Output 0 to a row
            MOVF      PORTD, W      ; #27 Read PORTD into WREG
            MOVWF     0x31          ; #28 Save row/column codes in 0x31
            MOVLW     0xF0          ; Move data for masking
            ANDWF     0x31, W       ; #29 Mask row code
            CPFSEQ    0x30          ; #30 Is column code affected?
            BRA       DECODE        ; #31 If affected, row found
                                    ; 0x31 has row and column code
            MOVFF     0x36, 0x35    ; #32 Restore row mask in 0x35
            BSF       STATUS, C     ; #33 Clear Carry flag to 0
            GOTO      NEXT_ROW      ; #34 Check next row
DECODE      MOVLW     D'16'         ; #35 Initialize 0x32 with 16 decimal since there
            MOVWF     0x32          ; #36 are 16 hex digits
```

```
            MOVWF   0x33            ; Move 16 to 0x33
            DECF    0x33, F         ; Decrement 0x33 by 1 to contain hex digits
                                    ; F to 0
            LFSR    0, 0x50         ; #37 Initialize FSR0 with 0x50
            MOVF    0x31, W         ; #38 Move row code to WREG
SEARCH      CPFSEQ  POSTINC0        ; #39 Compare and skip if equal
            BRA     SEARCH1         ; #40 Loop if not found
            MOVFF   0x33, PORTB     ; #41 Get character along with LOW enable
            BRA     NEXT1           ; #42 Branch to NEXT1
SEARCH1     MOVF    0x31, W
            DECF    0x33, F         ; #43 Decrement 0x32
            DECF    0x32, F         ; Decrement 0x33

            BNZ     SEARCH          ; #44 Branch to SEARCH if not 0
NEXT1       GOTO    SCAN_KEY        ; #45 Return to scan another key
DELAY       MOVLW   D'10'           ; #46 20 msec delay routine
            MOVWF   0x20
 LOOP1      MOVLW   D'255'          ; #47 LOOP2 provides 2 msec delay
            MOVWF   0x21
LOOP2       DECFSZ  0x21
            GOTO    LOOP2
            DECFSZ  0x20
            GOTO    LOOP1
            RETURN
            ORG     0x200           ; #48 Keyboard decode table
TABLE       DB      0x77            ; Code for F
            DB      0xB7            ; Code for E
            DB      0xD7            ; Code for D
            DB      0xE7            ; Code for C
            DB      0x7B            ; Code for B
            DB      0xBB            ; Code for A
            DB      0xDB            ; Code for 9
            DB      0xEB            ; Code for 8
            DB      0x7D            ; Code for 7
            DB      0xBD            ; Code for 6
            DB      0xDD            ; Code for 5
            DB      0xED            ; Code for 4
            DB      0x7E            ; Code for 3
            DB      0xBE            ; Code for 2
            DB      0xDE            ; Code for 1
            DB      0xEE            ; Code for 0
            END
```

In the program, a decode table for keys 0 through F is stored at address 0x200 (chosen arbitrarily). The codes for the hexadecimal numbers 0 through F are obtained by inspecting Figure 8.29. For example, consider key F. When key F is pressed and if a LOW is output by the program to bit 0 of Port C, the second row and second column of the keyboard will be LOW. This will make the content of Port D:

Bit Number:	Bit 7	Bit 6	Bit 5	Bit 4	Bit 3	Bit 2	Bit 1	Bit 0	
Data:	0	1	1	1	0	1	1	1	$= 77_{16}$

Thus, a code of 77_{16} is obtained at Port D when key F is pressed. Diodes are connected at the 4 bits (bits 0-3) of Port C. This is done to make sure that when a 0 is output by the program to one of these bits (row of the keyboard), the diode switch will close and will generate a LOW on that row.

Now, if a key is pressed on a particular row that is LOW, the column connected to this key will also be LOW. This will enable the programmer to obtain the appropriate key code for each key.

Next, the assembly language program will be explained using some of the line numbers included in the comment field.

Line #1 is the starting address of the program at 0x100. This address is chosen arbitrarily. Line #2 equates label OPEN to data 0xF0. This is because when all keys are up (no keys are pushed) and 0's are output to the rows via PORTC in Figure 8.29, data input at PORTD will be 11110000 (0xF0). Note that bits 0 - 3 of PORTD are connected to rows and bits 4-7 of PORTD are connected to columns of the keyboard.

Line #'s 3 through 5 transfer the keyboard decode table located at line# 48 from program memory address 0x200 to data memory address 0x50. Line #'s 6 through 9 configure ports and initialize STKPTR . Line #'s 10 and 11 initialize the seven-segment display by outputting 0.

Line #'s 12 through 16 check to see if any key is pushed. This is done by outputting 0's to all rows via PORTC, and then inputting PORTD. If all keys are open, data at PORTD will be 0xF0. Hence, 0xF0 stored in data memory address 0x30 is subtracted from data at PORTD in WREG. If Z = 0, the program waits in a loop with label KEY_OPEN until a key is pushed. When a key is closed, Z = 1, and the program comes out of the loop. Line #17 CALLs the DELAY routine to debounce the key by providing a delay of 20 ms.

Line #'s 18 through 20 detect a key closure. The program inputs Port D into WREG, and subtracts 0xF0 stored in 0x30 from [WREG]. If Z = 1, the program waits in a loop with label KEY_CLOSE until a key is closed. If Z = 0, the program leaves the loop. Line #21 CALLs the DELAY routine to debounce as soon as a key closure is detected. It is necessary to determine exactly which key is pressed. This is accomplished by outputting a '0' to a row while outputting 1's to the other three rows. Hence, a sequence of row-control codes (0xFE, 0xFD, 0xFB, and 0xF7, where the upper 4 bits 'F' are don't cares in this case) are output via PORTC . Line #'s 22 through 25 initialize 0x35 to all 1's, clear the C-bit to 0, and rotate [0x35] through C once to the left to contain the appropriate row control code.

For example, after the first RLCF in line #24, 0x35 will contain 11111110 (0xFE). Note that the low 4 bits are the row-control code (the upper 4 bits are don't cares) for the first pass in the loop, labeled NEXT_ROW. Line #26 outputs these data to PORTC to make the top row of the keyboard zero. The row–column code is input via PORTD to determine if the column code changes corresponding to each different row code. Line #'s 27 and 28 input PORTD into 0x31 via WREG. The top row of the keyboard will be 0 if C or D or E or F is pushed.

Lines 29 through 31 make the lower 4 bits 0's and retain the upper 4 bits. The columns are checked for equality by comparing with 0xF0 (contents of register 0x30) using the instruction "CPFSEQ 0x30" at line 30. If not equal, the row is found, and register 0x31 will contain the row and column code. If the column code is not 0xF0 (changed), the input key is identified. The program then goes through a lookup table to match the row–column code saved in 0x31. If the code is found, the corresponding index value, which equals the input key's value (a single hex digit), is displayed. However, if no key in the top row is pushed, a 0 is output to the second row, and the process continues. The program is written such that it will scan continuously for an input key and update the display for each new input.

Suppose that key F is pushed when the program branches to DECODE at line #35. Line #'s 35 through 37 initialize data memory addresses 0x32 and 0x33 with 16 (total number of hex digits), decrement [0x33] by 1 which will initially hold F, and will contain the hex digit to be displayed. Line #37 will load the starting address 0x50 of the decode table into FSR0 to be used as an indirect pointer. Line #38 moves the row code 0x77 (for F) saved in data memory address 0x31 (Line #28) into WREG. The instruction "CPFSEQ POSTINC0" at line 39 will compare the code for F at address 0x50 (starting address of the table) with [WREG] since it is assumed that 'F' is pushed. Hence, there will be a match; the instruction at line #40 will be skipped. The instruction "MOVFF 0x33, PORTB" at line #41 will be executed which will output 1111 to the DCBA inputs along with a LOW on the $\overline{\text{LATCH}}$ enable line of the TIL311 via PORTB, and the digit F will be displayed on the seven-segment display. The instruction "BRA NEXT1" at line #42 will branch to label NEXT1 (Line #45) which, in turn, will go back and repeat the process.

Questions and Problems

8.1 Identify the power and ground pins on the PIC18F4321. What is the purpose of using multiple power and ground pins?

8.2 What is the default clock frequency of the PIC18F4321?

8.3 How does the PIC18F obtain the address of the first instruction to be executed? What is this address?

8.4 What is the difference between power-on reset and manual reset?

8.5 List the PIC18F4321 I/O ports along with their sizes.

8.6 Write a single PIC18F instruction to configure:

(a) all bits of Port C as inputs (b) all bits of Port D as outputs

(c) bits 0 through 4 of Port B as inputs (d) all bits of Port A as outputs

8.7 Assume PIC18F4321. Suppose that three switches are connected to bits 0–2 of Port C and an LED to bit 6 of Port D. If the number of HIGH switches is even, turn the LED ON; otherwise, turn the LED OFF. Write a PIC18F assembly language program to accomplish this.

8.8 The PIC18F4321 microcontroller is required to drive the LEDs connected to bit 0 of Ports A and B based on the input conditions set by switches connected to bit 1 of Ports A and B. The I/O conditions are as follows:

- If the input at bit 1 of Port A is HIGH and the input at bit 1 of Port B is LOW, the LED at Port A will be ON and the LED at Port B will be OFF.
- If the input at bit 1 of Port A is LOW and the input at bit 1 of Port B is HIGH, the LED at Port A will be OFF and the LED at Port B will be ON.
- If the inputs of both Ports A and B are the same (either both HIGH or both LOW), both LEDs of Ports A and B will be ON.

Write a PIC18F assembly language program to accomplish this.

8.9 The PIC18F4321 microcontroller is required to test a NAND gate. Figure P8.9 shows the I/O hardware needed to test the NAND gate. The microcontroller is to be programmed to generate the various logic conditions for the NAND inputs, input the NAND output, and turn the LED ON connected to bit 3 of Port D if the NAND gate chip is found to be faulty. Otherwise, turn the LED ON connected to bit 4 of Port D. Write a PIC18F assembly language program at address 0x100 to accomplish this.

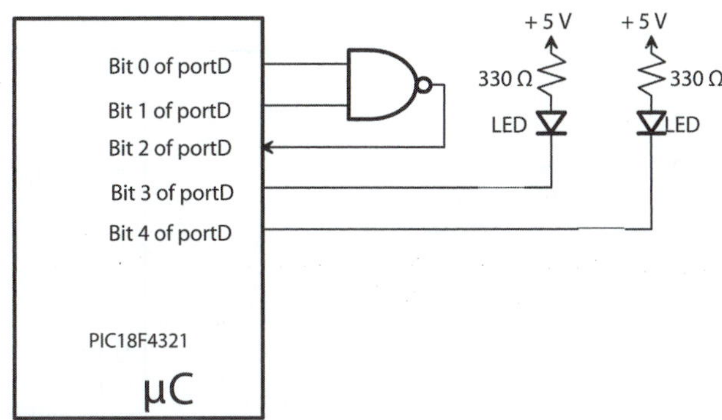

FIGURE P 8.9 (Assume that both LEDs are OFF initially.)

8.10 The PIC18F4321 microcontroller is required to add two 3-bit numbers stored in the lowest 3 bits of data registers 0x20 and 0x21 and output the sum (not to exceed 9) to a common-cathode seven-segment display connected to Port C as shown in Figure P8.10. Write a PIC18F assembly language program at address 0x200 to accomplish this by using a look-up table.

8.11 The PIC18F4321 microcontroller is required to input a number from 0 to 9 from an ASCII keyboard interfaced to it and output to an EBCDIC printer. Assume that the keyboard is connected to Port C and the printer is connected to Port D. Store the EBCDIC codes for 0 to 9 starting at an address 0x30, and use this lookup table to write a PIC18F assembly language program at address 0x100 to accomplish this.

8.12 In Figure P8.12, the PIC18F4321 is required to turn on an LED connected to bit 1 of Port C if the comparator voltage Vx > Vy; otherwise, the LED will be turned off. Write a PIC18F assembly language program at address 0x200 to accomplish this using conditional or polled I/O.

8.13 Repeat Problem 8.12 using Interrupt I/O by connecting the comparator output to INT1. Note that RB1 is also multiplexed with INT1. Write main program at 0x80 and interrupt service routine at 0x150 in PIC18F assembly language. The main program will configure the I/O ports, enable interrupt INT1, initialize STKPTR to 0x30, turn the LED OFF, and then wait for interrupt. The interrupt service routine will turn the LED ON and return to the main program at the appropriate location so that the LED is turned ON continuously until the next interrupt.

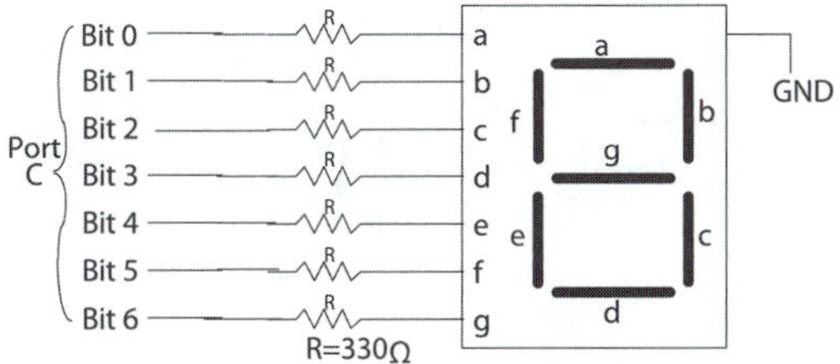

FIGURE P8.10

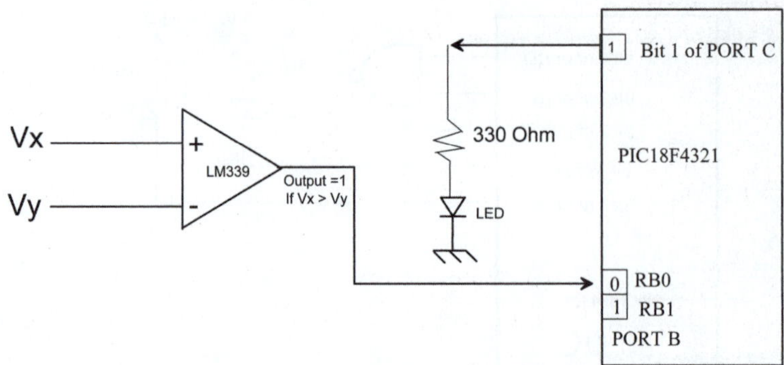

FIGURE P8.12

8.14 In Figure P8.14, if $V_x > V_w$, turn an LED ON connected at bit 3 of Port C. If $V_y > V_z$, turn the LED OFF. Assume that $V_x > V_w$ and $V_y > V_z$ will not occur at the same time. Using ports, registers, and memory locations as needed and INT0 interrupt:

(a) Draw a neat block diagram showing the PIC18F4321 microcontroller and the connections to ports in the diagram in Figure P8.14.

(b) Write the main program at 0x150 and the service routine at 0x200 in PIC18F assembly language. The main program will initialize STKPTR to 0x20, initialize the ports and wait for an interrupt. The service routine will accomplish the task and stop.

8.15 What is the interrupt address vector upon power-on reset?

8.16 Identify the PIC18F4321 external interrupts as maskable or nonmaskable.

8.17 What are the interrupt address vectors for high-priority and low-priority interrupts?

8.18 What are the priority levels for INT0 through INT2 external interrupts of the PIC18F4321 upon power-on reset?

8.19 What is the difference between PIC18F "RETFIE" and "RETFIE 1" instructions?

8.20 Write PIC18F instruction sequence tin PIC18F assembly language to set interrupt priority of INT1 as the high level, and interrupt priority for INT2 level as low level.

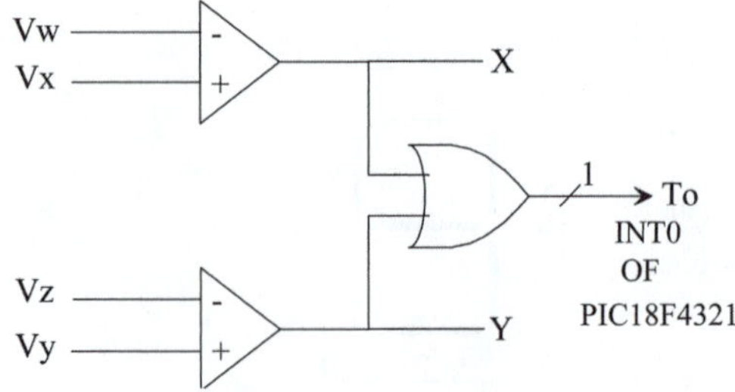

FIGURE P8.14

8.21 What is the interrupt priority level of INT0?

8.22 What are the triggering levels (rising or falling edge) of INT0 - INT2 upon power-on reset? Write a PIC18F assembly language program to activate the triggering level of INT0 by rising edge, and the INT1 and INT2 interrupts by falling edge.

8.23 How many Interrupt-on-Change pins are provided on the PIC18F4321? Are they activated by rising or falling edge?

8.24 What is the purpose of RS, $R/\overline{W}$, and EN pins on the Optrex DMC16249 LCD. Describe briefly how these signals are used to display data on the LCD via D0-D7 pins.

8.25 Assume the PIC18F4321- DMC 16249 interface of Figure 8.23. Write a program in PIC18F assembly language program at address 0x200 to display the phrase "PIC18F" on the LCD as soon as the four input switches connected to Port C are all HIGH.

8.26 What are the factors to be considered for interfacing a hex keyboard to a microcontroller?

8.27 What is meant by two-key lockout and N-key rollover?

9

PIC18F HARDWARE AND INTERFACING: PART 2

In this chapter we describe the second part of hardware aspects of the PIC18F4321. Topics include PIC18F4321's on-chip timers, analog interfaces (ADC and DAC), serial I/O, and CCP (Capture/Compare/Pulse Width Modulation).

9.1 PIC18F Timers

The PIC18F microcontroller family contains four to five on-chip hardware timers. The PIC18F4321 microcontroller includes four timers, namely, Timer0, Timer1, Timer2, and Timer3. These timers can be used to generate time delays using on-chip hardware. Note that the basic hardware inside each of these timers is a register that can be incremented or decremented at the rising or falling edge of a clock. The register can be loaded with a count for a specific time delay. Time delay is computed by subtracting the initial starting count from the final count in the register, and then multiplying the subtraction result by the clock frequency.

These timers can also be used as event counters. Note that an event counter is basically a register with the clock replaced by an event such as a switch. The counter is incremented or decremented whenever the switch is activated. Thus the number of times the switch is activated (occurrence of the event) can be determined. The basic features associated with these timers will now be explained.

Finally, the PIC18F CCP module utilizes these timers to perform capture, compare, or PWM (pulse width modulation) functions. These topics will be discussed later in this chapter.

7	6	5	4	3	2	1	0	
TMR0ON	T08BIT	T0CS	T0SE	PSA	T0PS2	T0PS1	T0PS0	T0CON

bit 7 **TMR0ON:** Timer0 On/Off Control bit
1 = Enables Timer0
0 = Stops Timer0

bit 6 **T08BIT**: Timer0 8-Bit/16-Bit Control bit
1 = Timer0 is configured as an 8-bit timer/counter
0 = Timer0 is configured as a 16-bit timer/counter

bit 5 **T0CS**: Timer0 Clock Source Select bit
1 = External clock connected to RA4/T0CKI pin (pin 6; Timer0 external clock input)
0 = Internal clock from crystal oscillator (divide by 4; crystal frequency can vary from 4 to 25 MHz)

bit 4 **T0SE**: Timer0 Source Edge Select bit
1 = Increment on high-to-low transition on T0CKI pin
0 = Increment on low-to-high transition on T0CKI pin

bit 3 **PSA**: Timer0 Prescaler Assignment bit
1 = TImer0 prescaler is NOT assigned. Timer0 clock input bypasses prescaler.
0 = Timer0 prescaler is assigned. Timer0 clock input comes from prescaler output.

bit 2-0 **T0PS2:T0PS0**: Timer0 Prescaler Select bits
111 = 1:256 prescale value
110 = 1:128 prescale value
101 = 1:64 prescale value
100 = 1:32 prescale value
011 = 1:16 prescale value
010 = 1:8 prescale value
001 = 1:4 prescale value

FIGURE 9.1 T0CON (TIMER0 Control) Register

9.1.1 Timer0

The Timer0 can operate as a timer or as a counter in 8-bit or 16-bit mode. The Timer0 uses the internal clock when used as a timer, and external clock (T0CK1) when used as a counter.

Timer0 as a timer The Timer0 can be used as a timer by setting the TMR0ON (bit 7 of T0CON register of Figure 9.1) to 1 using the PIC18F instruction "BSF T0CON, TM0ON." After the Timer0 is started, it counts up by incrementing the contents of the register (TMR0L for 8-bit timer mode or TMR0H:TMR0L for 16-bit timer mode) by 1 at each instruction cycle. The TMR0L counts up until the TMR0L reaches 0xFF in the 8-bit mode. The TMR0IF (interrupt on overflow) flag bit in the INTCON register is set to 1 when the TMR0L rolls over from 0xFF to 0x00. In the 8-bit mode, only the TMR0L register is used; the TMR0H register is not used, and contains a value of 0.

Similarly, in the 16-bit, after the Timer0 is started, the TMR0H:TMR0L register

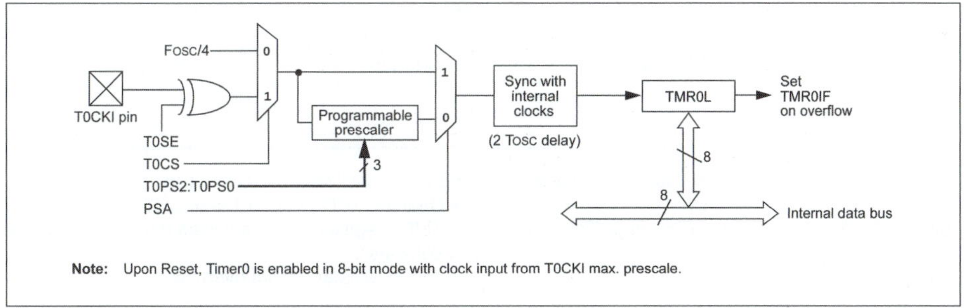

FIGURE 9.2 TIMER0 block diagram (8-bit mode)

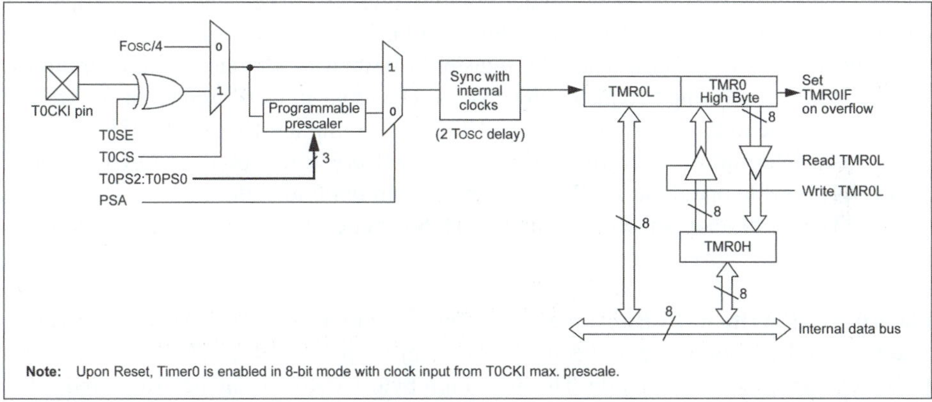

FIGURE 9.3 TIMER0 block diagram (16-bit mode)

pair counts up until the TMR0H:TMR0L reaches 0xFFFF in the 16-bit timer mode. The TMR0IF (interrupt on overflow) flag bit in the INTCON register is set to 1 when the TMR0H:TMR0L rolls over from 0xFFFF to 0x0000.

The timer can be stopped in either 8-bit or 16-bit mode by clearing the TMR0ON (bit 7 of T0CON register of Figure 9.1) to 0 using the PIC18F instruction "BCF T0CON, TM0ON." One of the PIC18F Special Function Registers (SFRs) called the T0CON register shown in Figure 9.1 controls all aspects of the Timer0 operation, including the prescale selection. Timer0 is both readable and writable.

Timer0 as a counter The Timer0 can be configured as a counter by setting the T0CS bit (bit 5 in the T0CON register of Figure 9.1) to 1. This will enable the PIC18F4321 to use an external clock connected to the T0CK1 pin. The T0SE bit (bit 4 of the T0CON register can then be cleared to 0 to increment the Timer0 register on the rising edge of the clock or set to one to increment the Timer0 register on the falling edge of the clock.

Timer0 block diagrams Figure 9.2 and Figure 9. 3 show the simplified block diagrams of the Timer0 module in 8-bit and 16-bit modes, respectively. In Figure 9.2, T0SE (bit 4 of T0CON register of Figure 9.1) is Exclusive-ORed with the T0CK1 clock (pin 6 of the PIC18F4321, Figure 8.1). The output of the Exclusive-OR gate is selected when T0CS (bit5 of T0CON of Figure 9.1) is one. When T0CS is 0, the internal oscillator (Fosc/4) is used and the Timer0 operates as a timer; otherwise, the external clock is selected and the Timer0 operates as a counter. TheTimer0 bypasses the prescaler if the PSA bit (bit 3 of

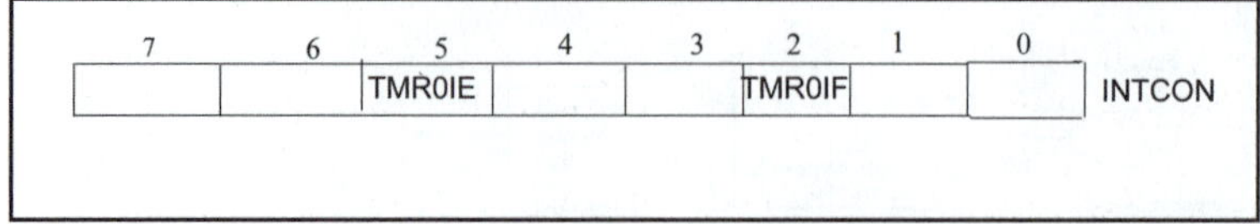

Bit 5 **TMR0IE** (TMR0 Overflow Interrupt Enable bit): 1 = Enables the TMR0 overflow interrupt
0 = Disables the TMR0 overflow interrupt.

Bit 2 **TMR0IF** (TMR0 Overflow Interrupt Flag bit): 1 = TMR0 register has overflowed (must be cleared in software),
0 = TMR0 register did not overflow

FIGURE 9.4 INTCON register with the TMR0IE and TMR0IF bits

T0CON register of Figure 9.1) is 1; otherwise, the prescaler is selected (PSA = 0), and is specified by T0PS2:T0PS0 (T0CON of Figure 9.1). The next block provides a two-cycle (Tosc) delay. This is because when the register is written to, the increment operation is inhibited for the following two instruction cycles. The user can work around this by writing an adjusted value to the timer register called TMR0. Example 9.1 illustrates this.

An interrupt on overflow indicated by TMR0IF (bit 2 of INTCON register in Figure 9.4) is set to one if the TMR0L rolls over from 0xFF to 0x00.

The block diagram for the Timer0 16-bit mode of Figure 9.3 can similarly be explained.

Timer0 Read/Write in 16-Bit Mode Two 8-bit registers (TMR0H and TMRH) are used to hold the 16-bit value in this mode. The register pair TMR0H:TMR0L is used as a 16-bit register in the 16-bit mode. The 8-bit high byte TMR0H is latched (buffered). It is not the actual high byte of Timer0 in 16-bit mode, and is not directly readable or writable (refer to Figure 9.3). Since TMR0H is not the actual high byte register of Timer0, one should initialize TMR0H before TMR0L to avoid any errors. Note that the upper 8-bit value of the timer is stored in the latched register, and loaded into actual TMR0H when TMR0L is loaded.

Similarly, a write to the high byte of Timer0 must also take place through the TMR0H Buffer register. The high byte is updated with the contents of TMR0H when a write occurs to TMR0L. This allows all 16 bits of Timer0 to be updated at once.

Prescaler An 8-bit counter is available as a prescaler for the Timer0. The value of the prescaler is set by the PSA (bit 3 of T0CON) and T0PS2:T0PS0 bits (bits 0 through 3 of T0CON) which determine the prescaler assignment and prescale ratio. Clearing the PSA bit assigns the prescaler to the Timer0. When it is assigned, prescale values from 1:2 through 1:256 in power-of-2 increments are selectable. This is shown in Figure 9.1.

When assigned to the Timer0 module, all instructions writing to the TMR0 register (e.g., CLRF TMR0, MOVWF TMR0, BSF TMR0) clear the prescaler count. The prescaler assignment is fully under software control and can be changed "on-the-fly" during program execution.

Timer0 Interrupt and Timer0 Overflow Flag bits The Timer0 counts up in increments of one from 0x00 to 0xFF for 8-bit mode and from 0x0000 to 0xFFFF for 16-bit mode. Hence, any value between 0x00 and 0xFF can be loaded into TMR0L register for 8-bit mode. For 16-bit mode, any value between 0x0000 and 0xFFFF can be loaded into the 16-bit register TMR0H:TMR0L.

Figure 9.4 (redrawn from Figure 8.17) shows the INTCON register with the TMR0IE and TMR0IF bits. The TMR0 interrupt is generated (if enabled by setting TMR0IE to one using BSF INTCON, TMR0IE) when the TMR0 register overflows from 0xFF to 0x00 in 8-bit mode, or from 0xFFFF to 0x0000 in 16-bit mode. This overflow sets the TMR0IF flag bit shown in Figure 9.4 (bit 2 of INTCON). The interrupt can be masked by clearing the TMR0IE bit. Before reenabling the interrupt, the TMR0IF bit must be cleared in software by the Interrupt Service Routine. Note that the PIC18F instruction "BCF INTCON, TMR0IF" will clear timer interrupt flag bit.

In the 8-bit mode, the TMR0IF bit is set to one when the timer value in 8-bit register TMR0L is incremented from 0xFF to 0x00 (overflow). In the 16-bit mode, the TMR0IF bit is set to one when the timer value in 16-bit register TMR0H:TMR0L is incremented from 0xFFFF to 0x0000 (overflow). An extra clock is required when Timer0 rolls over from 0xFF to 0x00 in 8-bit mode or from 0xFFFF to 0x0000.

Example 9.1 Assuming a 4 MHz crystal oscillator, calculate the time delay for the following PIC18F instruction sequence:

```
MOVLW    0xD4
MOVWF    T0CON    ; Initialize T0CON with 0xD4
MOVLW    0x80     ; Load 8-bit timer with count 0x80
MOVWF    TMR0L
```

Solution

The above PIC18F instruction sequence loads 0xD4 into the T0CON register. Note that 0xD4 = 11010100_2. Hence, from Figure 9.1, the T0CON register can be drawn with the binary data, as shown in Figure 9. 5. Comparing data of Figure 9. 5 with data of Figure 9.1, the following results are obtained:
TMR0ON = 1 meaning TIMER0 is ON, T08BIT = 1 meaning 8-bit timer, T0CS = 0 meaning internal instruction clock, PSA = 0 meaning prescaler enabled, and T0PS2 T0PS1 T0PS0 = 100 meaning 1:32 prescale value.

Clock period = 1/(4 MHz) = 0.25 μ sec, Instruction cycle clock period = 4 x 0.25 μ sec = 1 μ sec. Since the prescaler multiplies the Instruction cycle clock period by the prescaler value,
Time Delay = (Instruction cycle clock perod) x (Prescaler value) x (Counter value)
= (1 μ sec) x (32) x (128) = 4096 μ sec = 4.096 msec

Note that, in the above, Counter value = 0x80 = 128 in decimal. This value determines the desired time delay. Also, the last two instructions, MOVLW and MOVWF, account for the two instruction cycles, during which the increment operation is inhibited before writing to the TMR0L register.

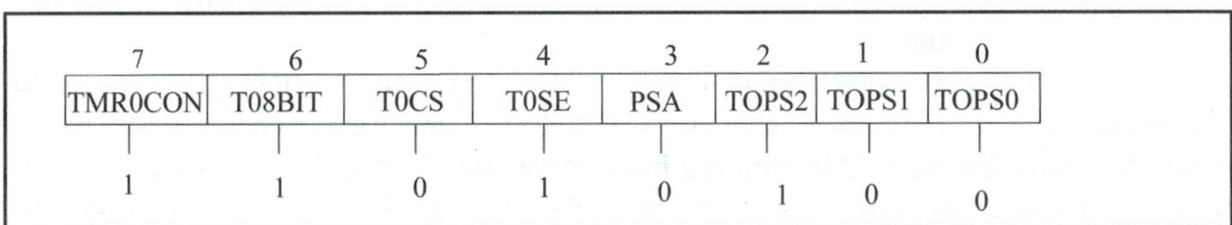

FIGURE 9.5 T0CON register with binary data 11010100_2

Example 9.2 Using Timer0 in 16-bit mode, write a PIC18F assembly language program to obtain a time delay of 1 ms. Assume 8 MHz crystal, and a prescale value of 1:128.

Solution

Since the timer works with divide by 4, crystal frequency = (8MHz)/4 = 2 MHz. Instruction cycle clock period = (1/2 MHz) = 0.5 μ sec.

The bits in register T0CON of Figure 9.1 are as follows:
TMR0ON(bit 7) = 0, T08BIT (bit 6) = 0, T0CS (bit 5) = 0, PSA (bit 3) = 0, and
T0PS2 T0PS1 T0PS0 = 110 for a prescale value of 1:128. Hence, the T0CON register will be initialized with 0x06.

Time delay = Instruction cycle x Prescale value x Count
Hence, Count = (1 ms) / (0. 5 μ sec x 128) = 15.625 which can be approximated to an integer value of 16 (0x0010). The timer counts up from an initialized value to 0xFFFF, and then rolls over (increments) to 0000H. The number of counts for rollover is (0xFFFF - 0x0010) = 0xFFEF.

Note that an extra cycle is needed for the rollover from 0xFFFF to 0x0000, and the TMR0IF flag is then set to 1. Because of this extra cycle, the total number of counts for rollover = 0xFFEF + 1 = 0xFFF0.

The following PIC18F assembly language program will provide a time delay of 1 ms:

```
            INCLUDE <P18F4321.INC>
            MOVLW   0x06              ; Initialize T0CON
            MOVWF   T0CON
            MOVLW   0xFF              ; Initialize TMR0H first with 0xFF
            MOVWF   TMR0H
            MOVLW   0xF0              ; Initialize TMR0L next
            MOVWF   TMR0L
            BCF     INTCON, TMR0IF    ; Clear Timer0 flag bit
            BSF     T0CON, TMR0ON     ; Start Timer0
BACK        BTFSS   INTCON, TMR0IF    ; Check Timer0 flag bit for 1
            GOTO    BACK              ; Wait in loop
            BCF     T0CON, TMR0ON     ; Stop Timer0
FINISH      BRA     FINISH            ; Halt
            END
```

9.1.2 Timer1

The Timer1 can be used as a 16-bit timer or a counter. It consists of two 8-bit registers, namely, TMR1H and TMR1L. An interrupt on overflow occurs when the Timer1 overflows from 0xFFFF to 0x0000. An extra cycle is required when the Timer1 rolls over from 0xFFFF to 0x0000.

Timer1 is controlled through the T1CON Control register (Figure 9.6). It also contains the Timer1 Oscillator Enable bit (T1OSCEN). Timer1 can be enabled or disabled by setting or clearing the TMR1ON (bit 0 of T1CON) control bit.

7	6	5	4	3	2	1	0	
RD16	T1RUN	T1CKPS1	T1CKPS0	T1OSCEN	$\overline{\text{T1SYNC}}$	TMR1CS	TMR1ON	T1CON

bit 7 **RD16:** 16-Bit Read/Write Mode Enable bit
1 = Enables register read/write of TImer1 in one 16-bit operation
0 = Enables register read/write of Timer1 in two 8-bit operations

bit 6 **T1RUN:** Timer1 System Clock Status bit
1 = Device clock is derived from Timer1 oscillator
0 = Device clock is derived from another source

bit 5-4 **T1CKPS1:T1CKPS0:** Timer1 Input Clock Prescale Select bits
11 = 1:8 prescale value
10 = 1:4 prescale value
01 = 1:2 prescale value
00 = 1:1 prescale value

bit 3 **T1OSCEN:** Timer1 Oscillator Enable bit
1 = Timer1 oscillator is enabled
0 = Timer1 oscillator is shut off

bit 2 **T1SYNC:** Timer1 External Clock Input Synchronization Select bit
When TMR1CS = 1:
1 = Do not synchronize external clock input
0 = Synchronize external clock input
When TMR1CS = 0:
This bit is ignored. Timer1 uses the internal clock when TMR1CS = 0.

bit 1 **TMR1CS:** Timer1 Clock Source Select bit
1 = External clock from pin RC0/T1OSO/T13CKI (on the rising edge)
0 = Internal clock (FOSC/4)

bit 0 **TMR1ON:** Timer1 On bit
1 = Enables Timer1
0 = Stops Timer1

FIGURE 9.6 T1CON (Timer1 Control) Register

Timer1 Operation Timer1 can operate as a timer, a synchronous counter, or an asynchronous counter. The operating mode is determined by the clock select bit, TMR1CS (bit 1 of T1CON). When TMR1CS is cleared to 0, Timer1 operates as a timer using the internal clock, and increments on every internal instruction cycle (Fosc/4). When the TMR1CS bit is set to 1, Timer1 increments on every rising edge of the Timer1 external clock input or the Timer1 oscillator, if enabled. Note that the on-chip crystal oscillator circuit can be enabled by setting the Timer1 Oscillator Enable bit, T1OSCEN (bit 3 of T1CON). The oscillator is a low-power circuit rated for 32 kHz crystals. Finally, the timer1 is enabled by setting the TMR1ON (bit 0 of T1CON register) to 1.

Timer1 interrupts The TMR1 register pair (TMR1H:TMR1L) increments from 0x0000 to 0xFFFF and rolls over to 0x0000. The Timer1 interrupt, if enabled, is generated on overflow which is latched in interrupt flag bit, TMR1IF (bit 0 of PIR1), shown in Figure 9.7. The TMR1 overflow interrupt bit can be enabled or disabled by setting or clearing the

7	6	5	4	3	2	1	0	
PSPI F	ADIF	RCIF	TXIF	SSPI F	CCP1I F	TMR2IF	TMR1IF	PIR1

bit 7 PSPIF: Parallel Slave Port Read/Write Interrupt Flag bit
1 = A read or a write operation has taken place (must be cleared in software)
0 = No read or write has occurred

bit 6 ADIF: A/D Converter Interrupt Flag bit
1 = An A/D conversion completed (must be cleared in software), 0 = The A/D conversion is not complete

bit 5 RCIF: EUSART Receive Interrupt Flag bit
1 = The EUSART receive buffer, RCREG, is full (cleared when RCREG is read)
0 = The EUSART receive buffer is empty

bit 4 TXIF: EUSART Transmit Interrupt Flag bit
1 = The EUSART transmit buffer, TXREG, is empty (cleared when TXREG is written)
0 = The EUSART transmit buffer is full

bit 3 SSPIF: Master Synchronous Serial Port Interrupt Flag bit
1 = The transmission/reception is complete (must be cleared in software)
0 = Waiting to transmit/receive

bit 2 CCP1IF: CCP1 Interrupt Flag bit
Capture mode:
1 = A TMR1 register capture occurred (must be cleared in software), 0 = No TMR1 register capture occurred
Compare mode:
1 = A TMR1 register compare match occurred (must be cleared in software)
0 = No TMR1 register compare match occurred
PWM mode:
Unused in this mode.

bit 1 TMR2IF: TMR2-to-PR2 Match Interrupt Flag bit
1 = TMR2-to-PR2 match occurred (must be cleared in software), 0 = No TMR2-to-PR2 match occurred

bit 0 TMR1IF: TMR1 Overflow Interrupt Flag bit,
1 = TMR1 register overflowed (must be cleared in software)
0 = TMR1 register did not overflow

FIGURE 9.7 PIR1 (Peripheral Interrupt Request) Register1

Timer1 Interrupt Enable bit, TMR1IE (bit 0 of PIE1), shown in Figure 9.8. The other bits in the PIR1 and PIE1 registers contain the individual flag and enable bits for the peripheral interrupts. These bits will be discussed as the related topics are covered.

Example 9.3 Write a PIC18F assembly language program to provide a delay of 1 msec using Timer1 with an internal clock of 4 MHz. Use 16-bit mode of Timer1 and the prescaler value of 1:4.

Solution

For 4 MHz clock, each instruction cycle = 4 x (1/4 MHz) = 1 μ sec
Total instruction cycles for 1 msec delay = $(1 \times 10^{-3}/10^{-6}) = 1000$

With the prescaler value of 1:4, instruction cycles = 1000 / 4 = 250
Number of counts for rollover = 65535_{10} - 250_{10} = 65285_{10} = 0xFF05
An extra cycle is required for rollover from 0xFFFF to 0x0000 which sets the TMR1IF to 1.
Hence, total number of counts = 0xFF05 + 1 = 0xFF06
Therefore, TMR1H must be loaded with 0xFF, and TMR1L with 0x06.
The PIC18F assembly language program for one msec delay is provided below:

```
        INCLUDE   <P18F4321.INC>
        MOVLW     0xC1              ; 16-bit mode, 1:4 prescaler, enable Timer1
        MOVWF     T1CON             ; Load into T1CON register
        MOVLW     0xFF              ; Initialize TMR1H with 0xFF
        MOVWF     TMR1H
```

7	6	5	4	3	2	1	0	
PSPI E	ADIE	RCIE	TXIE	SSPI E	CCPI1 E	TMR2IE	TMR1IE	PIE1

bit 7 **PSPIE:** Parallel Slave Port Read/Write Interrupt Enable bit
1 = Enables the PSP read/write interrupt
0 = Disables the PSP read/write interrupt

bit 6 **ADIE:** A/D Converter Interrupt Enable bit
1 = Enables the A/D interrupt
0 = Disables the A/D interrupt

bit 5 **RCIE:** EUSART Receive Interrupt Enable bit
1 = Enables the EUSART receive interrupt
0 = Disables the EUSART receive interrupt

bit 4 **TXIE:** EUSART Transmit Interrupt Enable bit
1 = Enables the EUSART transmit interrupt
0 = Disables the EUSART transmit interrupt

bit 3 **SSPIE:** Master Synchronous Serial Port Interrupt Enable bit
1 = Enables the MSSP interrupt
0 = Disables the MSSP interrupt

bit 2 **CCP1IE:** CCP1 Interrupt Enable bit
1 = Enables the CCP1 interrupt
0 = Disables the CCP1 interrupt

bit 1 **TMR2IE:** TMR2-to-PR2 Match Interrupt Enable bit
1 = Enables the TMR2-to-PR2 match interrupt
0 = Disables the TMR2-to-PR2 match interrupt

bit 0 **TMR1IE:** TMR1 Overflow Interrupt Enable bit
1 = Enables the TMR1 overflow interrupt
0 = Disables the TMR1 overflow interrupt

FIGURE 9.8 PIE1 (Peripheral Interrupt Enable) Register 1

```
        MOVLW   0x06            ; Initialize TMR1L with 0x06
        MOVWF   TMR1L
        BCF     PIR1, TMR1IF    ; Clear Timer1 overflow flag in PIR1
BACK    BTFSS   PIR1, TMR1IF    ; If TMR1IF=1, skip next instruction to halt
        GOTO    BACK
HERE    BRA     HERE            ; Halt
        END
```

Note that external loop can be used with the above 1 msec delay routine as the inner loop to obtain higher time delays.

9.1.3 Timer2

Timer2 contains an 8-bit timer register and 8-bit period register (TMR2 and PR2). The Timer2 can be programmed with prescale values of 1:1, 1:4, and 1:16, and postscale values of 1:1 through 1:16.

The module is controlled through the T2CON register shown in Figure 9.9. The T2CON register enables or disables the timer and configures the prescaler and postscaler. Timer2 can be shut off by clearing control bit, TMR2ON (bit 2 of T2CON), to minimize power consumption.

Timer2 Operation In normal operation, the PR2 register is initialized to a specific value, and the 8-bit timer register (TMR2) is incremented from 0x00 on each internal clock (FOSC/4). A 4-bit counter/prescaler on the clock input gives direct input, divide-by-4 and divide-by-16 prescale options. These are selected by the prescaler control bits, T2CKPS1:T2CKPS0 (bits 1, 0 of T2CON). The value of TMR2 is compared to that of the

7	6	5	4	3	2	1	0	
--------------	T2OUTPS3	T2OUTPS2	T2OUTPS1	T2OUTPS0	TMR2ON	T2CKPS1	T2CKPS0	T2CON

bit 7 **Unimplemented:** Read as '0'

bit 6-3 **T2OUTPS3:T2OUTPS0:** Timer2 Output Postscale Select bits
0000 = 1:1 postscale
0001 = 1:2 postscale
•
•
•
1111 = 1:16 postscale

bit 2 **TMR2ON:** Timer2 On bit
1 = Timer2 is on
0 = Timer2 is off

bit 1-0 **T2CKPS1:T2CKPS0:** Timer2 Clock Prescale Select bits
00 = Prescaler is 1
01 = Prescaler is 4
1x = Prescaler is 16

FIGURE 9.9 T2CON (Timer2 Control) Register

8-bit period register, PR2, on each clock cycle. When the two values match, the Timer2 outputs a HIGH on the TMR2IF flag in the PIR1 register, and also resets the value of TMR2 to 0x00 on the next cycle.The output frequency is divided by a counter/postscale value (1:1 to 1:16) as specified in the T2CON register. Note that the interrupt is generated and the TMR2IF flag bit in the PIR1 register (Figure 9.7) is set to 1, indicating the match between TMR2 and PR2 registers. The TMR2IF bit must be cleared to 0 using software.

Timer2 Interrupt Timer2 can also generate an optional device interrupt. The Timer2 output signal (TMR2-to-PR2 match) provides the input for the 4-bit output counter/postscaler. This counter generates the TMR2 match interrupt flag which is latched in TMR2IF (bit 1 of PIR1, Figure 9.7). The interrupt is enabled by setting the TMR2 Match Interrupt Enable bit, TMR2IE (bit 1 of PIE1, Figure 9.8). A range of 16 postscale options (from 1:1 through 1:16 inclusive) can be selected with the postscaler control bits, T2OUTPS3:T2OUTPS0 (bits 6-3 of T2CON, Figure 9.9).

Example 9.4 Write a PIC18F assembly language program using Timer2 to turn on an LED connected at bit 0 of Port D after 10 sec. Assume an internal clock of 4 MHz, a prescaler value of 1:16, and a postscaler value of 1:16.

Solution

For 4 MHz clock, each instruction cycle = 4 x 1/(4MHz) = 1 μ sec. TMR2 is incremented every 1 μ sec. When the TMR2 value matches with the value in PR2, the value in TMR2 is cleared to 0 in one instruction cycle. Since the PR2 is 8-bit wide, we can have a maximum counter value of 255. Let us calculate the delay with this PR2 value.

Delay = (Instruction cycle) x (Prescale value) x (Postscale value) x (Counter value + 1)
= (1 μ sec) x (16) x (16) (255 + 1)
= 65.536 msec

Note that, in the above, one is added to the Counter value since an additional clock is needed when it rolls over from 0xFF to 0x00, and sets the TMR2IF to 1.

External counter value for 10 sec delay using 65.536 msec as the inner loop = (10 sec)/(65.536 msec), which is approximately 153 in decimal.

The PIC18F assembly language is provided below:

```
            INCLUDE    <P18F4321.INC>
EXT_CNT     EQU        0x50
            BCF        TRISD, 0        ; Configure bit 0 of PORT D as an output
            BCF        PORTD, 0        ; Turn LED OFF
            MOVLW      0x7A            ; 1:16 prescaler, 1:16 postscaler Timer1 off
            MOVWF      T2CON           ; Load into T2CON register
            MOVLW      0x00            ; Initialize TMR2 with 0x00
            MOVWF      TMR2
            MOVLW      D'153'          ; Initialize EXT_CNT with 153
            MOVWF      EXT_CNT
LOOP        MOVLW      D'255'          ; Load PR2 with 255
            MOVWF      PR2
            BCF        PIR1, TMR2IF    ; Clear Timer2 interrupt flag in PIR1
```

7	6	5	4	3	2	1	0	
RD16	T3CCP2	T3CKPS1	T3CKPS0	T3CCP1	$\overline{\text{T3SYNC}}$	TMR3CS	TMR3ON	T3CON

bit 7 **RD16:** 16-Bit Read/Write Mode Enable bit
1 = Enables register read/write of Timer3 in one 16-bit operation
0 = Enables register read/write of Timer3 in two 8-bit operations

bit 6,3 **T3CCP2:T3CCP1:** Timer3 and Timer1 to CCPx Enable bits
1x = Timer3 is the capture/compare clock source for the CCP modules
01 = Timer3 is the capture/compare clock source for CCP2;
Timer1 is the capture/compare clock source for CCP1
00 = Timer1 is the capture/compare clock source for the CCP modules

bit 5-4 **T3CKPS1:T3CKPS0**: Timer3 Input Clock Prescale Select bits
11 = 1:8 prescale value
10 = 1:4 prescale value
01 = 1:2 prescale value
00 = 1:1 prescale value

bit 2 **$\overline{\text{T3SYNC}}$:** Timer3 External Clock Input Synchronization Control bit
(not usable if the device clock comes from Timer1/Timer3)
When TMR3CS = 1:
1 = Do not synchronize external clock input
0 = Synchronize external clock input
When TMR3CS = 0:
This bit is ignored. Timer3 uses the internal clock when TMR3CS = 0.

bit 1 **TMR3CS:** Timer3 Clock Source Select bit
1 = External clock input from Timer1 oscillator or T13CKI (on the rising edge after the first falling edge)
0 = Internal clock (FOSC/4)

bit 0 **TMR3ON:** Timer3 On bit
1 = Enables Timer3
0 = Stops Timer3

FIGURE 9.10 T3CON (Timer3 Control) Register

```
        BSF     T2CON, TMR2ON ; Set TMR2ON bit in T2CON to start timer
BACK    BTFSS   PIR1, TMR2IF  ; If TMR2IF=1, skip next instruction
        GOTO    BACK
        DECF    EXT_CNT
        BNZ     LOOP
        BSF     PORTD, 0      ; Turn LED ON
        BCF     T2CON, TMR2ON ; Turn off Timer2
FINISH  GOTO    FINISH        ; Halt
        END
```

In the above program, the execution times associated with some of the instructions such as MOVLW D'153', MOVWF EXT_CNT, DECF EXT_CNT, and BNZ LOOP are discarded. These execution times are very small compared to 10 sec delay.

9.1.4 Timer3

Timer3 is a 16-bit register, and can be used as a 16-bit timer or a 16-bit counter. Although Timer3 consists of two 8-bit registers, namely, TMR3H (high byte) and TMR3L (low byte), it can only be programmed in 16-bit mode. The Timer3 module is controlled through the T3CON register (Figure 9.10). Some of the bits of the T3CON register are associated with the CCP module. This topic will be discussed later in this chapter.

Timer3 Operation Timer3 can operate in one of three modes, namely, timer, synchronous counter, and asynchronous counter. The operating mode is determined by the clock select bit, TMR3CS (bit 1 of T3CON, Figure 9.10). When TMR3CS is cleared to 0, Timer3 increments on every internal instruction cycle (FOSC/4). When the bit is set to 1, Timer3 increments on every rising edge of the Timer1 external clock input or the Timer1 oscillator, if enabled.

Timer3 Interrupt The TMR3 register pair (TMR3H:TMR3L) increments from 0x0000 to 0xFFFF and overflows to 0x0000. The Timer3 interrupt, if enabled, is generated on overflow and is latched in interrupt flag bit TMR3IF (bit 1 of PIR2, Figure 9.11). This interrupt can be enabled or disabled by setting or clearing the Timer3 Interrupt Enable bit, TMR3IE (bit 1 of PIE2, Figure 9.12).

9.2 Analog Interface

A/D (Analog-to-Digital) and D/A (Digital-to-Analog) converters are widely used these days for performing data acquisition and control. Separate A/D and D/A converter chips are commercially available. These chips are interfaced externally in microprocessor-based applications. The PIC18F4321 includes an on-chip A/D. However, an external D/A chip needs to be interfaced if digital-to-analog conversion is desired. Figure 9.13 (redrawn from Figure 1.1 for convenience) shows a typical control application where A/D and D/A converters are used.

Suppose that it is necessary to maintain the temperature of a furnace to a desired level to maintain the quality of a product. Assume that the designer has decided to control this temperature by adjusting the fuel. This can be accomplished using a typical microcontroller such as the PIC18F4321 along with the interfacing components as follows. Temperature is an analog (continuous) signal. It can be measured by a temperature-sensing (measuring) device such as a thermocouple. The thermocouple provides the measurement in millivolts (mV) equivalent to the temperature.

Since microcontrollers only understand binary numbers (0's and 1's), each analog mV signal must be converted to a binary number using the microcontroller's on-chip A/D converter. Note that the PIC18F contains an on-chip A/D converter. The PIC18F does not include an on-chip D/A converter. However, the D/A converter chip can be interfaced to the PIC18F externally.

First, the millivolt signal is amplified by a mV/V amplifier to make the signal compatible for A/D conversion. A microcontroller such as the PIC18F4321 can be programmed to solve an equation with the furnace temperature as an input. This equation compares the temperature measured with the temperature desired which can be entered into the microcontroller using the keyboard. The output of this equation will provide the appropriate opening and closing of the fuel valve to maintain the appropriate temperature. Since this output is computed by the microcontroller, it is a binary number. This binary output must be converted into an analog current or voltage signal.

The D/A converter chip inputs this binary number and converts it into an analog current (*I*). This signal is then input into the current/pneumatic (*I/P*) transducer for opening or closing the fuel input valve by air pressure to adjust the fuel to the furnace. The furnace temperature desired can thus be achieved. Note that a transducer converts one form of

7	6	5	4	3	2	1	0	
OSCFIF	CMIF	-----------	EEIF	BCLIF	HLVDIF	TMR3IF	CCP2IF	PIR2

bit 7 **OSCFIF:** Oscillator Fail Interrupt Flag bit
1 = Device oscillator failed, clock input has changed to INTOSC (must be cleared in software)
0 = Device clock operating

bit 6 **CMIF:** Comparator Interrupt Flag bit
1 = Comparator input has changed (must be cleared in software)
0 = Comparator input has not changed
bit 5 **Unimplemented:** Read as ‘0’

bit 4 **EEIF:** Data EEPROM/Flash Write Operation Interrupt Flag bit
1 = The write operation is complete (must be cleared in software)
0 = The write operation is not complete or has not been started

bit 3 **BCLIF:** Bus Collision Interrupt Flag bit
1 = A bus collision occurred (must be cleared in software)
0 = No bus collision occurred

bit 2 **HLVDIF:** High/Low-Voltage Detect Interrupt Flag bit
1 = A high/low-voltage condition occurred; direction determined by VDIRMAG bit (HLVDCON<7>)
0 = A high/low-voltage condition has not occurred

bit 1 **TMR3IF:** TMR3 Overflow Interrupt Flag bit
1 = TMR3 register overflowed (must be cleared in software)
0 = TMR3 register did not overflow

bit 0 **CCP2IF:** CCP2 Interrupt Flag bit
Capture mode:
1 = A TMR1 register capture occurred (must be cleared in software)
0 = No TMR1 register capture occurred
Compare mode:
1 = A TMR1 register compare match occurred (must be cleared in software)
0 = No TMR1 register compare match occurred
PWM mode:
Unused in this mode.

FIGURE 9.11 PIR2 (Peripheral Interrupt Request) Register 2

7	6	5	4	3	2	1	0	
OSCFIE	CMIE	-----------	EEIE	BCLIE	HLVDIE	TMR3IE	CCP2IE	PIE2

bit 7 **OSCFIE:** Oscillator Fail Interrupt Enable bit
1 = Enabled
0 = Disabled

bit 6 **CMIE:** Comparator Interrupt Enable bit
1 = Enabled
0 = Disabled

bit 5 **Unimplemented:** Read as '0'

bit 4 **EEIE:** Data EEPROM/Flash Write Operation Interrupt Enable bit
1 = Enabled
0 = Disabled

bit 3 **BCLIE:** Bus Collision Interrupt Enable bit
1 = Enabled
0 = Disabled

bit 2 **HLVDIE:** High/Low-Voltage Detect Interrupt Enable bit
1 = Enabled
0 = Disabled

bit 1 **TMR3IE:** TMR3 Overflow Interrupt Enable bit
1 = Enabled
0 = Disabled

bit 0 **CCP2IE:** CCP2 Interrupt Enable bit
1 = Enabled
0 = Disabled

FIGURE 9.12 PIE2 (Peripheral Interrupt Enable) Register 2

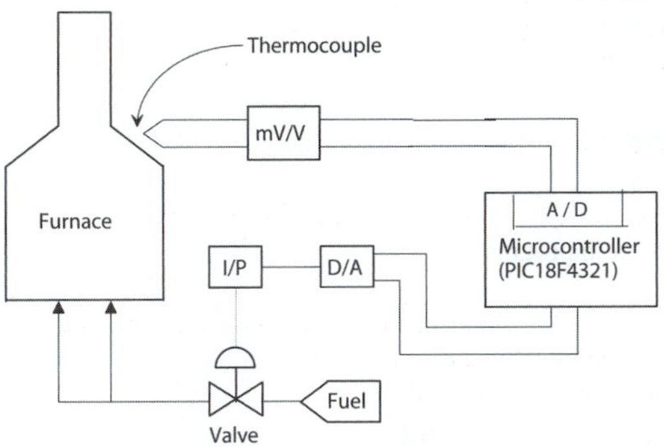

FIGURE 9.13 Furnace temperature control

energy (electrical current, in this case) to another form (air pressure, in this example).

9.2.1 On-chip A/D Converter

The PIC 18F4321 contains an on-chip A/D converter (or sometimes called ADC) module with 13 channels (AN0-AN12). An analog input can be selected as an input on one of these 13 channels, and can be converted to a corresponding 10-bit digital number. Three control registers, namely, ADCON0 through ADCON2, are used to perform the conversion.

The ADCON0 register, shown in Figure 9.14, controls the operation of the A/D module. The ADCON0 register can be programmed to select one of 13 channels using bits CHS3 through CHS0 (bits 5 through 2). The conversion can be started by setting the GO/$\overline{\text{DONE}}$ (bit 1) to 1. Once the conversion is completed, this bit is automatically cleared to 0 by the PIC18F4321.

The ADCON1 register, shown in Figure 9.15, configures the functions of the port pins as Analog (A) input or Digital (D) I/O. The table shown in Figure 9.15 shows how the port bits are defined as analog or digital signals by programming the PCFG3 through PCFG0 bits (bits 3 through 0) of the ADCON1 register. This register can also be

7	6	5	4	3	2	1	0	
--------------	-----------	CHS3	CHS2	CHS1	CHS0	GO/$\overline{\text{DONE}}$	ADON	ADCON0

bit 7-6 **Unimplemented:** Read as '0'
bit 5-2 **CHS3:CHS0:** Analog Channel Select bits
0000 = Channel 0 (AN0)
0001 = Channel 1 (AN1)
0010 = Channel 2 (AN2)
0011 = Channel 3 (AN3)
0100 = Channel 4 (AN4)
0101 = Channel 5 (AN5)
0110 = Channel 6 (AN6)
0111 = Channel 7 (AN7)
1000 = Channel 8 (AN8)
1001 = Channel 9 (AN9)
1010 = Channel 10 (AN10)
1011 = Channel 11 (AN11)
1100 = Channel 12 (AN12
1101 = Unimplemented
1110 = Unimplemented
1111 = Unimplemented

bit 1 **GO/ $\overline{\text{DONE}}$:** A/D Conversion Status bit

When ADON = 1:
1 = A/D conversion in progress
0 = A/D idle

bit 0 **ADON:** A/D On bit
1 = A/D converter module is enabled
0 = A/D converter module is disabled

FIGURE 9.14 ADCON0 (A/D Control Register0)

programmed to select the reference voltages for the A/D.

The ADCON2 register, shown in Figure 9.16, configures the A/D clock source, programmed acquisition time and justification. The A/D conversion time per bit is defined as TAD. The A/D conversion requires 11 TAD per 10-bit conversion. The source of the A/D conversion clock is software selectable. For correct A/D conversions, the A/D conversion clock (TAD) must be as short as possible, but greater than the minimum requirement of 0.7 μ sec for a Tosc-based clock with Vref $\geq$3V After conversion, the 10-bit binary output of the A/D is placed in a 16-bit register (two 8-bit register pair) ADRESH:ADRESL. Since six bits of the 16-bit register will not be used, the ADFM bit (bit 7) of the ADCON2 can be set to 1 or cleared to 0 to provide the conversion reading, respectively, as right or left justified with unused bits as 0's.

The PIC18F4321 contains a 10-bit on-chip A/D with 13 channels. Figure 9.17 shows a block diagram of the PIC18F4321 A/D. The ADRESH and ADRESL registers contain the result of the A/D conversion. When the A/D conversion is complete, the result is loaded into the ADRESH:ADRESL register pair, the GO/$\overline{\text{DONE}}$ bit (ADCON0 register) is cleared, and the A/D Interrupt Flag bit, ADIF, is set.

7	6	5	4	3	2	1	0	
-------------	-------------	VCFG1	VCFG0	PCFG3	PCFG2	PCFG1	PCFG0	ADCON1

bit 7-6 Unimplemented: Read as '0'
bit 5 VCFG1: Voltage Reference Configuration bit (VREF - source)
1 = V REF - (AN2)
0 = V SS
bit 4 VCFG0: Voltage Reference Configuration bit (VREF + source)
1 = V REF + (AN3)
0 = V DD
bit 3-0 PCFG3:PCFG0: A/D Port Configuration Control bits
bit 7-6 Unimplemented: Read as '0'
bit 5 VCFG1: Voltage Regerence Configuration bit (VREF-source)
1 = VREF- (AN2)
0 = VSS
bit 4 VCFG0: Voltage Reference Configuration bit (VREF + source)
1 = VREF + (AN3)
0 = VDD
bit 3-0 (decimal) in Column 1: PCFG3:PCFG0: A/D Port Configuration Control bits

	AN12	AN11	AN10	AN9	AN8	AN7	AN6	AN5	AN4	AN3	AN2	AN1	AN0
0	A	A	A	A	A	A	A	A	A	A	A	A	A
1	A	A	A	A	A	A	A	A	A	A	A	A	A
2	A	A	A	A	A	A	A	A	A	A	A	A	A
3	D	A	A	A	A	A	A	A	A	A	A	A	A
4	D	D	A	A	A	A	A	A	A	A	A	A	A
5	D	D	D	A	A	A	A	A	A	A	A	A	A
6	D	D	D	D	A	A	A	A	A	A	A	A	A
7	D	D	D	D	D	A	A	A	A	A	A	A	A
8	D	D	D	D	D	D	A	A	A	A	A	A	A
9	D	D	D	D	D	D	D	A	A	A	A	A	A
10	D	D	D	D	D	D	D	D	A	A	A	A	A
11	D	D	D	D	D	D	D	D	D	A	A	A	A
12	D	D	D	D	D	D	D	D	D	D	A	A	A
13	D	D	D	D	D	D	D	D	D	D	D	A	A
14	D	D	D	D	D	D	D	D	D	D	D	D	A
15	D	D	D	D	D	D	D	D	D	D	D	D	D

A = Analog input D = Digital I/O

FIGURE 9.15 ADCON1 (A/D Control Register 1)

7	6	5	4	3	2	1	0	
ADFM	------------	ACQT2	ACQT1	ACQT0	ADCS2	ADCS!	ADCS0	ADCON2

bit 7 **ADFM:** A/D Result Format Select bit
1 = Right justified; 10-bits in lower 2 bits of ADRESH with upper 6 bits as 0's and in 8 bits of ADRESL
0 = Left justified; 10-bits in 8-bits of ADRESH and in upper 2 bits of ADRESL with lower 6 bits as 0's
bit 6 **Unimplemented:** Read as '0'
bit 5-3 **ACQT2:ACQT0:** A/D Acquisition Time Select bits
111 = 20 TAD
110 = 16 TAD
101 = 12 TAD
100 = 8 TAD
011 = 6 TAD
010 = 4 TAD
001 = 2 TAD
000 = 0 TAD(1)
bit 2-0 **ADCS2:ADCS0:** A/D Conversion Clock Select bits
111 = FRC (clock derived from A/D RC oscillator)(1)
110 = FOSC/64
101 = FOSC/16
100 = FOSC/4
011 = FRC (clock derived from A/D RC oscillator)(1)
010 = FOSC/32
001 = FOSC/8
000 = FOSC/2
Note 1: If the A/D FRC clock source is selected, a delay of one TCY (instruction cycle) is
added before the A/D clock starts. This allows the SLEEP instruction to be executed before starting a conversion.

FIGURE 9.16 ADCON2 (A/D Control Register 2)

The following steps should be followed to perform an A/D conversion:

1. Configure the A/D module:
- Configure analog pins, voltage reference, and digital I/O (ADCON1)
- Select A/D input channel (ADCON0)
- Select A/D acquisition time (ADCON2)
- Select A/D conversion clock (ADCON2)
- Turn on A/D module (ADCON0)

2. Configure A/D interrupt (if desired):
- Clear ADIF bit (bit 6 of PIR1, Figure 9.7)
- Set ADIE bit (bit 6 of PIE1, Figure 9.8)
- Set GIE bit (bit 7) and PEIE (bit 6) of INTCON register, Figure 8.17(a)
- All interrupts including A/D Converter interrupt, branch to address 0x000008 (default) upon power-on reset. However, the A/D Converter interrupt can be configured as low priority by setting the ADIP bit (bit 6) of the IPRI register (See Microchip manual) to branch to address 0x000018. The instruction, BSF IPR1, ADIP can be used for this purpose.

3. Wait for the required acquisition time (if required).
4. Start conversion:

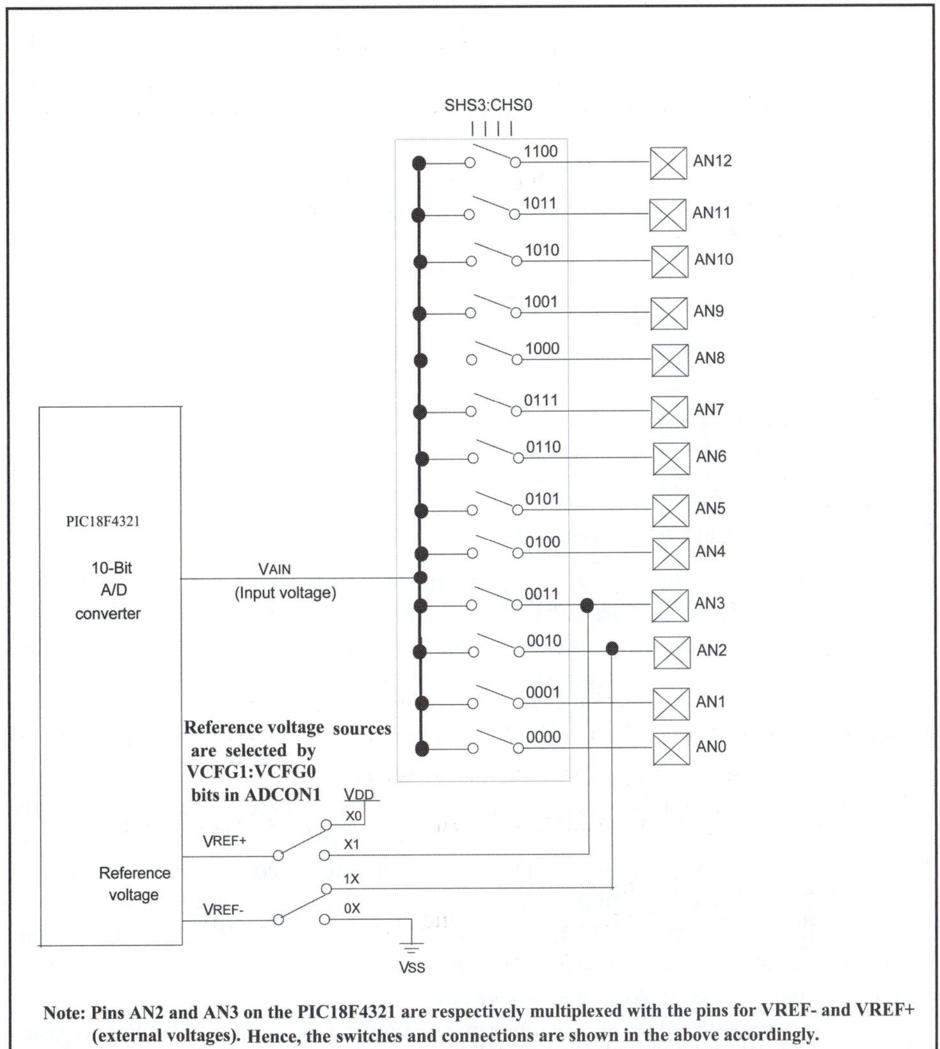

FIGURE 9.17 Block diagram of the PIC18F4321 A/D

- Set GO/$\overline{\text{DONE}}$ bit (ADCON0 register) to 1

5. Wait for A/D conversion to complete, by either

- Polling for the GO/$\overline{\text{DONE}}$ bit to be cleared to 0 (conversion completed) or waiting for the A/D interrupt

6. Read A/D Result registers (ADRESH:ADRESL); clear bit ADIF, if required.

7. For next conversion, go to step 1 or step 2, as required. The A/D conversion time per bit is defined as TAD. A minimum wait of 2 TAD is required before the next acquisition starts.

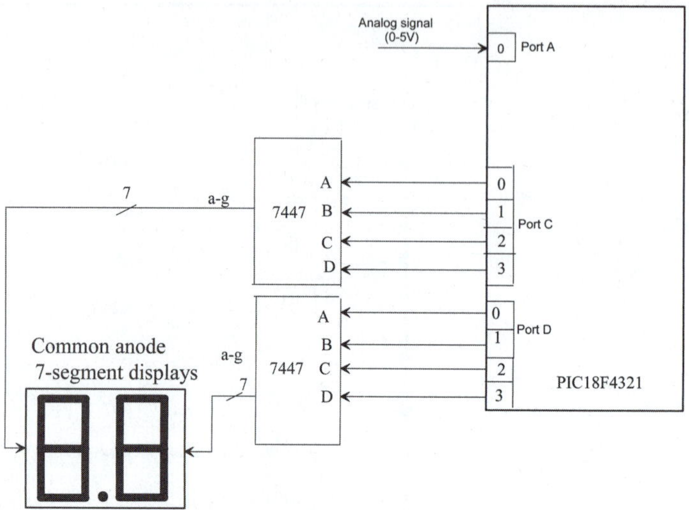

FIGURE 9.18 Figure for Example 9.5

Example 9.5 A PIC18F4321 microcontroller shown in Figure 9.18 is used to implement a voltmeter to measure voltage in the range 0 to 5 V and display the result in two decimal digits: one integer part and one fractional part. Using polled I/O, write a PIC18F assemble language program to accomplish this.

Solution

In order to design the voltmeter, the PIC18F4321 on-chip A/D converter will be used. Three registers, ADCON0-ADCON2, need to be configured. In ADCON0, bit 0 of PORT A (RA0/AN0) is designated as the analog signal to be converted. Hence, CHS3-CHS0 bits (bits 5-2) are programmed as 0000 to select channel 0 (AN0). The ADCON0 register is also used to enable the A/D, start the A/D, and then check the "end of conversion" bit. In the PIC18F assembly language program provided below, the ADCON0 is loaded with 0x01 which will select AN0, and enable A/D.

The reference voltages are chosen by programming the ADCON1 register. In this example, V_{DD} (by clearing bit 4 of of ADCON1 to 0), and V_{SS} (by clearing bit 5 of ADCON1 to 0) will be used. Note that V_{DD} and V_{SS} are already connected to the PIC18F4321. The ADCON1 register is also used to configure AN0 (bit 0 of Port A) as an analog input by writing 1101 (13 decimal in Figure 9.15) at PCFG3-PCFG0 (bits 3-0 of ADCON1). Note that there are several choices to configure AN0 as an analog input. In the program, the ADCON1 is loaded with 0x0D which will select VSS and VDD as reference voltage sources, and AN0 as analog input.

In the program, the ADCON2 is loaded with 0xA9 which will provide the 8-bit result right justified, select 12 TAD (requires at least 11 TAD for 10-bit conversion), and select Fosc/8.

The ADCON2 is used to set up the acquisition time, conversion clock, and, also, if the result is to be left or right justified. In this example, 8-bit result is assumed. The A/D result is configured as right justified, and, therefore, the 8-bit register ADRESL will contain the result. The contents of ADRESH are ignored.

Note that the maximum decimal value that can be accommodated in 8 bits of

ADRESH is 255_{10} (FF_{16}). Hence, the maximum voltage of 5 V will be equivalent to 255_{10}. This means that 1 volt = 51 (decimal). The display (D) in decimal is given by

$$D = 5 \times \text{ (input/255)}$$
$$= \text{input/51}$$
$$= \underbrace{\text{quotient}}_{\text{Integer part}} + \text{remainder}$$

This gives the integer part. The fractional part in decimal is

$$F = (\text{ remainder/51}) \times 10$$
$$\simeq (\text{remainder})/5$$

For example, suppose that the decimal equivalent of the 8-bit output of A/D is 200.

$$D = 200/51 \Rightarrow \text{quotient} = 3, \text{ remainder} = 47$$
$$\text{integer part} = 3$$
$$\text{fractional part, F} = 47/5 = 9$$

Therefore, the display will show 3.9 V.

From these equations, the final result will be in BCD. Both integer and fractional parts of the result will be output to two 7447s (BCD to seven-segment decoder) in order to display them on two seven-segment displays arranged in a row, as shown in Figure 9.18. The PIC18F assembly language program for the voltmeter is provided below:

```
             INCLUDE<P18F4321.INC>
D0           EQU     0x30          ;Contains data for right (fractional) 7-seg
D1           EQU     0x31          ;Contains data for left (integer) 7-seg
ADCONRESULT  EQU     0x34          ;Contains 8-bit A/D result
             ORG     0x100         ;Starting address of the program
             MOVLW   0x32          ;Initialize STKPTR to 0x32 (arbitrary value)
             MOVWF   STKPTR        ;Since subroutines are used
             CLRF    TRISC         ;Configure PortC as output
             CLRF    TRISD         ;Configure PortD as output
             SETF    TRISA         ;Configure PortA as input
             MOVLW   0x01
             MOVWF   ADCON0        ;Select AN0 for input and enable ADC
             MOVLW   0x0D
             MOVWF   ADCON1        ;Select VDD and VSS as reference
                                   ;voltages and AN0 as analog input.
             MOVLW   0xA9
             MOVWF   ADCON2        ;Select right justified 12TAD and Fosc/8
START        BSF     ADCON0, GO    ;Start A/D conversion
INCONV       BTFSC   ADCON0, DONE  ;Wait until A/D conversion is done
             BRA     INCONV
```

```
            MOVFF   ADRESL,ADCONRESULT ;Move ADRESL of result into
                                       ;ADCONRESULT register
            CALL    DIVIDE             ;Call the divide subroutine
            CALL    DISPLAY            ;Call display subroutine
            BRA     START
DIVIDE      CLRF    D0                 ;Clears  D0
            CLRF    D1                 ;Clears D1
            MOVLW   D'51'              ;#1 Load 51 into WREG
EVEN        CPFSEQ  ADCONRESULT        ;#2
            BRA     QUOTIENT           ;#3
            INCF    D1, F              ;#4
            SUBWF   ADCONRESULT, F     ;#5
QUOTIENT    CPFSGT  ADCONRESULT        ;#6 Checks if ADCONRESULT
                                       ;still greater than 51
            BRA     DECIMAL            ;#7
            INCF    D1, F              ;#8 Increment D1 for each time
                                       ;ADCONRESULT is greater
                                       ;than 51
            SUBWF   ADCONRESULT, F     ;#9 Subtract 51 from
                                       ;ADCONRESULT
            BRA     EVEN               ;#10
DECIMAL     MOVLW   0x05               ;#11
REMAINDER   CPFSGT  ADCONRESULT        ;#12 Checks if ADCONRESULT
                                       ;greater than 5
            BRA     DIVDONE            ;#13
            INCF    D0, F              ;#14 Increment D0
            SUBWF   ADCONRESULT, F     ;#15 Subtract 5
                                       ;from ADCONRESULT
            BRA     REMAINDER
DIVDONE     RETURN                     ;#16
DISPLAY     MOVFF   D1, PORTC          ;#17 Output D1  on integer 7-seg
            MOVFF   D0, PORTD          ;#18 Output D0  on fractional 7-seg
            RETURN
            END
```

In the above, since the PIC18F does not have any unsigned division instruction, a subroutine called DIVIDE is written to perform unsigned division using repeated subtraction. In the DIVIDE subroutine, the output of the A/D contained in the ADCONRESULT register is subtracted by 51. Each time the subtraction result is greater than 51, the contents of register D1 (address 0x31) is incremented by one, this will yield the integer part of the answer. Once the contents of the ADCONRESULT reaches a value below 51, the remainder part of the answer is determined. This is done by subtracting the number in ADCONRESULT subtracted by 5. Each time the subtraction result is greater than 5, register D0 (address 0x30) is incremented by one. Finally, the integer value is placed in D1 and the remainder part is placed in D0. Now the only task left is to display the result on the seven-segment display.

The # symbol along with a number in the comment field is used in some of the lines in the above program in order to explain the program logic. Line#1 moves 51 (decimal) into WREG. The CPFSEQ at Line#2 compares the A/D's 8-bit result in ADCONRESULT with 51 for equality. Suppose that the analog input voltage at AN0 is one volt, which is 51 in decimal. Since [WREG] = [ADCONRESULT] = 51, the program branches to line #4, and increments [D1] by 1, storing 1 in D1. The SUBWF ADCONRESULT, F at Line #5 subtracts [WREG] from [ADCONRESULT] and stores the result in ADCONRESULT. Since the subtraction result is 0 in this case, '0' is stored in ADCONRESULT. The CPFSGT ADCONRESULT instruction at Line #6 compares [WREG] with [ADCONRESULT] to check whether [ADCONRESULT] > [WREG]. Since [WREG] = 51 and [ADCONRESULT] = 0, the program executes BRA DECIMAL at Line #7, and branches to label DECIMAL at Line #11 where 5 is moved into WREG.

The CPFSGT ADCONRESULT at Line #12 is then executed to check whether [ADCONRESULT] > [WREG]. Since [WREG] = 5 and [ADCONRESULT] = 0, the program executes BRA DIVDONE at Line #13, and branches to label DIVDONE at Line #16 where the RETURN instruction is executed. The program returns to the "CALL DISPLAY"— one instruction after "CALL DIVIDE." The program pushes the address of the next instruction "BRA START" onto the hardware stack, and executes the subroutine called DISPLAY (Line #17). The instruction "MOVFF D1, PORTC" at Line #17 outputs [D1] = 0x01 to PORTC. Hence, '1' is displayed on the integer display.

Note that the BCD number '1' of the integer part is contained in the low four bits of D1 which are output to the DCBA inputs of the integer 7447 to display a '1' on the left (integer) seven-segment display. The instruction "MOVFF D0, PORTD" outputs [D0] = 0x00 to PORTD, and a '0' is displayed on the right (fractional) display. Outputting to integer and fractional displays using instructions in sequence are executed so fast by the PIC18F that the displays appear to human eyes at the same time. Finally, "1.0" indicating 1.0 volt is displayed on the two seven-segment displays.

Next, suppose that the decimal value contained in the ADCONRESULT (A/D converter's output) is 200 (decimal) which is equivalent to 3.9 volts. Line#1 moves 51 (decimal) into WREG. The CPFSEQ at Line#2 compares the A/D's 8-bit result in ADCONRESULT with 51 for equality. Since [WREG] = 51, and [ADCONRESULT] = 200, the program executes the instruction "BRA QUOTIENT" at line #3, and branches to Line #6. The CPFSGT ADCONRESULT instruction at Line #6 compares [WREG] with [ADCONRESULT] to check whether [ADCONRESULT] > [WREG]. Since [WREG] = 51 and [ADCONRESULT] = 200, the program branches to Line #8, and increments [D1] by 1, and stores the result in D1. The instruction "SUBWF ADCONRESULT, F" at Line #9 is then executed, [WREG] is subtracted from [ADCONRESULT], and the result is stored in ADCONRESULT. Since [WREG] = 51 and [ADCONRESULT] = 200, the subtraction result 149 will be stored in ADCONRESULT. The instruction "BRA EVEN" at Line #10 is executed next. The program branches to label EVEN at line #2 to execute the instruction "CPFSEQ ADCONRESULT", and the loop is repeated until the result of subtraction in ADCONRESULT is less than 51. This will happen in this case when [D1] = 3, and [ADCONRESULT] = 47 = subtraction result of "[ADCONESULT]- [WREG]" after going through the loop three times. As soon as [ADCONRESULT] < [WREG], the instruction "BRA DECIMAL' at Line #7 is executed where the fractional part '9' is determined in the same manner as the last example. The rest of the logic is very similar to that in the last example. Finally, "3.9" will be displayed on the two seven-segment displays.

9.2.2 Interfacing an External D/A (Digital-to-Analog) Converter to the PIC18F4321

Most microcontrollers such as the PIC18F4321 do not have any on-chip D/A converter (or sometimes called DAC). Hence, external D/A converter chip is interfaced to the PIC18F4321 to accomplish this function. Some microcontrollers such as the Intel/ Analog Devices 8051 include an on-chip D/A converter. In order to illustrate the basic concepts associated with interfacing a typical D/A converter such as the Maxim MAX5102 to the PIC18F4321, consider Figure 9.19.

The MAX5102 is a 16-pin chip. In this example, the PIC18F4321 microcontroller is interfaced with the MAX5102 chip to convert an 8-bit binary input to an analog voltage from 0 to 5 V. Eight switches connected to PORTD of the PIC18F4321 will provide a value between 0 and 255 that will be converted by the MAX5102 into a DC voltage between 0 V and 5 V. This analog voltage will then appear on the OUTA or OUTB pin of the MAX5102 D/A converter.

The MAX5102 contains two independent D/A converters, namely, DAC A and DAC B. These D/A converters are selected by the A0 input pin on the MAX5102. The two converters share the same 8-bit input pins, D0-D7. The $\overline{\text{WR}}$ input pin on the MAX5102 when HIGH latches the 8-bit input data for DAC A and DAC B for conversion to analog voltage. For example, A0 = 0 and $\overline{\text{WR}}$ = 0, the 8-bit data on DAC A input is transparent while A0 = 1 and $\overline{\text{WR}}$ = 0, the 8-bit data on DAC B input is transparent. The analog voltage output for either DAC A or DAC B will be available when $\overline{\text{WR}}$ = 1. One must make sure that 8-bit input data are valid before $\overline{\text{WR}}$ goes to 0 to get rid of any glitches.

The manufacturer recommends that for proper operation of the MAX5102, the VREF should be connected to ground via a 0.1 μF capacitor. Note that the programmer must ensure that the timing requirements for $\overline{\text{WR}}$, A0, and D0-D7 are met according to the manufacturer's specification. Hence, each time the PIC18F4321 outputs 8-bit new data on the data pins of the MAX5102, a delay of a few milliseconds (for example, 2 msec) may be required so that the data will be valid before outputting a LOW on the $\overline{\text{WR}}$ pin.

Example 9.6 Assume the block diagram of Figure 9.19. Write a PIC18F assembly language program that will input eight switches via PORTD of the PIC18F4321, and output the byte to D0-D7 input pins of the MAX5102 D/A converter. The microcontroller will send appropriate signals to the $\overline{\text{WR}}$ and A0 pins so that the D/A converter will convert the input byte to an analog voltage between 0 and 5 V, and output the converted voltage on its OUTA pin.

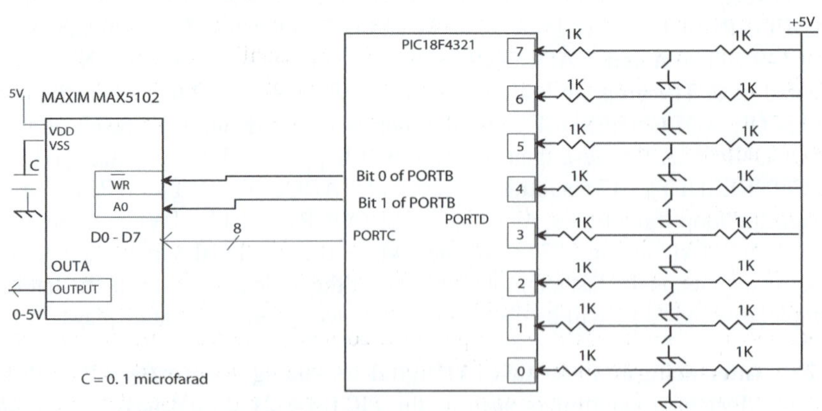

FIGURE 9.19 Figure for Example 9.6

Solution

The steps for writing a PIC18F assembly language program for the D/A converter interface of Figure 9.19 are provided in the following:

1. Configure PORTB and PORTC as outputs, and PORTD as input.
2. Output a LOW to A0 Pin of the D/A via bit 1 of PORTB to select OUTA.
3. Output a LOW to $\overline{WR}$ pin of the D/A via bit 0 of PORTB.
4. Input the switches via PORTD, and output to PORTC.
5. Output a HIGH to $\overline{WR}$ pin of the A/D via bit 0 of PORTB to latch 8-bit input data for converting to analog voltage. No delay is needed since the program will be written to input one byte of data from the switches.

The PIC18F assembly language program is provided below:

```
       INCLUDE  <P18F4321.INC>
       ORG      0x100
       CLRF     TRISB          ; Configure PORTB as output
       CLRF     TRISC          ; Configure PORTC as output
       SETF     TRISD          ; Configure PORTD as input
       BCF      PORTB, 1       ; Clear A0 to 0 to select OUTA
       BCF      PORTB, 0       ; Output LOW on bit 0 of PORTB
       MOVFF    PORTD, PORTC   ; Input switches, output to PORTD
       BSF      PORTB, 0       ; Latch data for conversion
FINISH BRA      FINISH         ; Halt
       END
```

9.3 Serial Interface

In various instances, it is desirable to transmit binary data from one microcontroller to another. In such situations, data can be transmitted using either parallel or serial transmission techniques. In parallel transmission, each bit of the binary data is transmitted over a separate wire or line.

In serial transmission, only one line is used to transmit the complete binary data bit by bit. Hence, the transmitting device such as a microcontroller must convert parallel data into a string of serial bits. The receiving device such as another microcontroller must convert data from serial to parallel. Data are usually sent starting with the least significant bit. In order to differentiate among various bits, a clock signal is used. Serial data transmission can be divided into two types: synchronous and asynchronous. We now briefly describe them.

9.3.1 Synchronous Serial Data Transmission

The basic feature of synchronous serial data transmission is that data are transmitted or received based on a clock signal. After deciding on a specific rate of data transmission, commonly known as "baud rate" (bits per second), the transmitting device sends a data bit at each clock pulse. In order to interpret data correctly, the receiving device must know the start and end of each data unit. Therefore, in synchronous serial data transmission, the receiver must know the number of data units to be transferred. Also, the receiver must be synchronized with data boundaries. Usually, one or two SYNC characters (a string of bits) are used to indicate the start of each synchronous data stream.

The data unit normally contains error bits such as parity. In some transmissions, the least significant bit is used as a parity bit. The synchronous receiver usually waits in a "hunt" mode while looking for data. As soon as it matches one or more SYNC characters based on the number of SYNC characters used, the receiver starts interpreting the data. In synchronous serial transmission, the transmitting device needs to send data continuously to the receiving device. However, if data are not ready to be transmitted, the transmitter will pad with SYNC characters until data are available.

As mentioned before, in synchronous serial transfer, the receiver must know the number of SYNC characters used, and the number of data units to be transferred. Once the receiver matches the SYNC characters, it receives the specified number of data units, and then goes into a "hunt" mode for matching the SYNC pattern for next data.

9.3.2 Asynchronous Serial Data Transmission

In this type of data transfer, the transmitting device does not need to be synchronized to the receiving device. The transmitting device can send one or more data units when it has data ready to be sent. Each data unit must be formatted. In other words, each data unit must contain "start" and "stop" bits, indicating the beginning and the end of each data unit. The interface circuits between the transmitting device and the receiving device must perform the following functions:

1. Converts an 8-bit parallel data unit from the transmitting device into serial data for transmitting them to the receiving device.
2. Converts serial data from the receiving device into parallel data for sending them back to the transmitting device assuming two-way (full duplex) transmission.

Each data unit can be divided into equal time intervals, called "bit intervals." A data bit can be either HIGH or LOW during each bit interval. For example, 8-bit data will have eight bit intervals. Each data bit will correspond to one of the eight bit intervals.

The format for asynchronous serial data typically contains the following information:

1. A LOW START bit.
2. 5-8 bits, denoting the actual data being transferred.
3. An optional parity bit for either odd or even parity.
4. 1, 1½, or 2 STOP bits having HIGH levels. Note that 1½ STOP bits mean a HIGH level with a duration of 1.5 times the bit interval.

9.3.3 PIC18F Serial I/O

Serial I/O is typically fabricated as an on-chip module with microcontrollers. This will facilitate interfacing microcontrollers with other microcontrollers or peripheral devices. Several protocol (rules) standards for serial data transmission have been introduced over the years. Two such standards implemented include SPI (Serial Peripheral Interface) developed by Motorola and I^2C (Inter-Integrated Circuit) developed by Philips. Both protocols are based on synchronous serial data transmission.

The SPI is a protocol established for data transfer between a master and a slave device. The master device can be a microcontroller while the slave device can be devices such as another microcontroller, EEPROMs, and A/D converters. The I^2C protocol, on the other hand, is widely used for transferring data among the ICs (Integrated Circuits) on PCBs (Printed Circuit Boards).

The PIC18F4321 contains an on-chip Master Synchronous Serial Port (MSSP) module which is a serial interface, useful for communicating with other peripheral or microcontroller devices. The MSSP module can operate in either SPI or I^{2}C mode. We will cover the PIC18F SPI in this section.

PIC18F4321 pins and signals for the SPI mode The PIC18F SPI primarily uses three pins of the PIC18F4321. They are SCK (Serial Clock, pin 18), SDI (SPI Data In, pin 23), and SDO (SPI Data Out, pin 24). A fourth pin, namely, $\overline{SS}$ (SPI Slave select input, pin 7), is provided for applications requiring multiple slave devices.

PIC18F4321 registers in SPI mode The MSSP module uses four registers for SPI mode operation. These are:

- MSSP Control Register 1 (SSPCON1)
- MSSP Status Register (SSPSTAT)
- Serial Receive/Transmit Buffer Register (SSPBUF)
- MSSP Shift Register (SSPSR) – Not directly accessible

Figures 9.20 and 9.21 show the SSPCON1 and SSPSTAT registers, respectively. The SSPCON1 and SSPSAT are the control and status registers in SPI mode operation. The SSPCON1 can be used to enable and configure the serial port pins by setting the SSPEN bit (bit 5) of the SSPCON1 register. The SSPCON1 can also be used to select the master or slave mode using the SSPM3-SSPM0 bits (bits 3-0), and the clock polarity using the CKP bit (bit 4).

Both SPI and I^{2}C modes use the SSPSTAT register. This register can be used to select the SPI mode (master or slave using bit 7), SPI clock (bit 6), and the buffer full status of the buffer register of the receiver (BF bit, bit 0). Bits 1-5 of the SSPSTAT are used in the I^{2}C mode.

Operation When initializing the SPI, several options need to be specified. This is done by programming the appropriate control bits (bits 0-5 of SSPCON1 and bits 7-6 of SSPSTAT). These control bits allow the following to be specified:

- Master mode (SCK is the clock output)
- Slave mode (SCK is the clock input)
- Clock Polarity (idle state of SCK)
- Data Input Sample Phase (middle or end of data output time)
- Clock Edge (output data on rising/falling edge of SCK)
- Clock Rate (master mode only)
- Slave Select mode (slave mode only)

The MSSP consists of a transmit/receive shift register (SSPSR) and a buffer register (SSPBUF). The SSPSR shifts the data in and out of the device, the most significant bit (MSB) first. The SSPBUF holds the data that were written to the SSPSR until the received data are ready. Once the 8 bits of data have been received, that byte is moved to the SSPBUF register. Then, the Buffer Full detect bit, BF (bit 0 of SSPSTAT), and the interrupt flag bit, SSPIF, are set to 1. This double-buffering of the received data (SSPBUF) allows the next byte to start reception before reading the data that were just received. Any write to the SSPBUF register during transmission/reception of data will be ignored and the write collision detect bit, WCOL (bit 7 of SSPCON1), will be set. User software must clear the WCOL bit so that it can be determined if the following write(s) to the SSPBUF register completed successfully. When the application software is expecting to receive valid data, the SSPBUF should be read before the next byte of data to transfer is written to the SSPBUF. The BF indicates when SSPBUF has been loaded with the received data (transmission is complete). When the SSPBUF is read, the BF bit is cleared. These data may be irrelevant if the SPI is only a transmitter. Generally, the MSSP interrupt is used to determine when the transmission/reception has completed. The SSPBUF must be read and/or written. If the interrupt method is not going to be used, then software polling can be done to ensure that a write collision does not occur.

Enabling SPI I/O To enable the serial port, MSSP Enable bit, SSPEN (bit 5 of SSPCON1) must be set. To reset or reconfigure SPI mode, clear the SSPEN bit, reinitialize

7	6	5	4	3	2	1	0	
WCOL	SSPOV	SSPEN	CKP	SSPM3	SSPM2	SSPM1	SSPM0	SSPCON1

bit 7 **WCOL:** Write Collision Detect bit (Transmit mode only)
1 = The SSPBUF register is written while it is still transmitting the previous word (must be cleared in software)
0 = No collision

bit 6 **SSPOV:** Receive Overflow Indicator bit
SPI Slave mode:
1 = A new byte is received while the SSPBUF register is still holding the previous data. In case of overflow, the data in SSPSR is lost. Overflow can occur only in Slave mode. The user must read the SSPBUF, even if only transmitting data, to avoid setting overflow (must be cleared in software).
0 = No overflow
Note: In Master mode, the overflow bit is not set since each new reception (and transmission) is initiated by writing to the SSPBUF register.

bit 5 **SSPEN:** Synchronous Serial Port Enable bit
1 = Enables serial port and configures SCK, SDO, SDI, and SS as serial port pins
0 = Disables serial port and configures these pins as I/O port pins
Note: When enabled, these pins must be properly configured as input or output.

bit 4 **CKP:** Clock Polarity Select bit
1 = Idle state for clock is a high level
0 = Idle state for clock is a low level

bit 3-0 **SSPM3:SSPM0:** Synchronous Serial Port Mode Select bits
0101 = SPI Slave mode, clock = SCK pin, SS pin control disabled, SS can be used as I/O pin
0100 = SPI Slave mode, clock = SCK pin, SS pin control enabled
0011 = SPI Master mode, clock = TMR2 output/2
0010 = SPI Master mode, clock = FOSC/64
0001 = SPI Master mode, clock = FOSC/16
0000 = SPI Master mode, clock = FOSC/4
Note: Bit combinations not specifically listed here are either reserved or implemented in I2C™ mode only.

FIGURE 9.20 SSPCON1 (MSSP CONTROL) Register 1 in SPI mode

the SSPCON registers, and then set the SSPEN bit. This configures the SDI, SDO, SCK, and $\overline{\text{SS}}$ pins as serial port pins. For the pins to behave as the serial port function, the data direction bits (in the TRIS register) must be appropriately programmed as follows:

- SDI is automatically controlled by the SPI module.
- SDO must have TRISC (bit 5) bit cleared to 0.
- SCK (Master mode) must have TRISC (bit 3) bit cleared to 0.
- SCK (Slave mode) must have TRISC (bit 3) bit set to 1.
- $\overline{\text{SS}}$ must have TRISA (bit 5) bit set to 1 for multiple slaves. Note that RA5 is multiplexed with $\overline{\text{SS}}$.

Any serial port function that is not desired may be overridden by programming the corresponding data direction (TRIS) register to the opposite value.

Figure 9.22 shows a simplified block diagram of SPI Master/Slave connection between two PIC18F4321s.

In Figure 9.22, the master PIC18F4321 initiates the data transfer by sending the SCK signal. The master can initiate the data transfer at any time because it controls the

7	6	5	4	3	2	1	0	
SMP	CKE	D/$\overline{\text{A}}$	P	S	R/$\overline{\text{W}}$	UA	BF	SSPSTAT

bit 7 SMP: Sample bit
SPI Master mode:
1 = Input data sampled at end of data output time
0 = Input data sampled at middle of data output time
SPI Slave mode:
SMP must be cleared when SPI is used in Slave mode.

bit 6 CKE: SPI Clock Select bit
1 = Transmit occurs on transition from active to Idle clock state
0 = Transmit occurs on transition from Idle to active clock state
Note: Polarity of clock state is set by the CKP bit (SSPCON1<4>).

bit 5 D/A: Data/Address bit
Used in I2C™ mode only.

bit 4 P: Stop bit
Used in I2C mode only. This bit is cleared when the MSSP module is disabled, SSPEN is cleared.

bit 3 S: Start bit
Used in I2C mode only.

bit 2 R/W: Read/Write Information bit
Used in I2C mode only.

bit 1 UA: Update Address bit
Used in I2C mode only.

bit 0 BF: Buffer Full Status bit (Receive mode only)
1 = Receive complete, SSPBUF is full
0 = Receive not complete, SSPBUF is empty

FIGURE 9.21 SSPSTAT (MSSP Status Register) in SPI mode

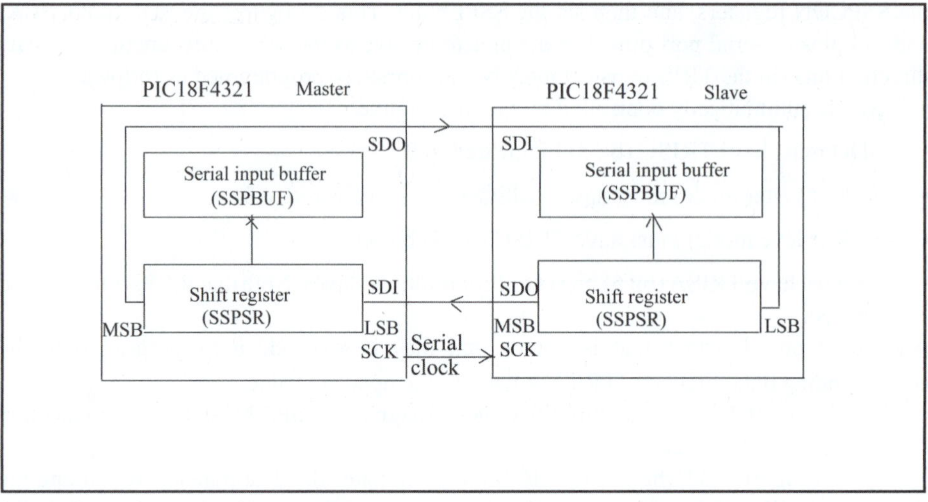

FIGURE 9.22 SPI Master/Slave interface between two PIC18F4321s

SCK. The master determines when the slave is to broadcast data by the software protocol. Data are shifted out of both shift registers on their programmed clock edge and latched on the opposite edge of the clock. Both processors should be programmed to the same Clock Polarity (CKP); then both controllers would send and receive data at the same time.

Since read and write operations must be performed on each data byte, some data may not be useful. Note that whether the data are meaningful (or dummy data) depends on the application software. This leads to three types of data transmission:

- Master sends data, and Slave sends dummy data
- Master sends data, and Slave sends data
- Master sends dummy data, and Slave sends data

Example 9.7 Figure 9.23 shows a block diagram for interfacing two PIC18F4321 in SPI mode. One of the microcontrollers is the master while the other is the slave. The master PIC18F4321 will input four switches via bits 0-3 of PORTB, and then transmit the 4-bit data using its SDO pin to the slave's SDI pin. The slave PIC18F4321 will output these data to four LEDs, and turn them ON or OFF based on the switch inputs.

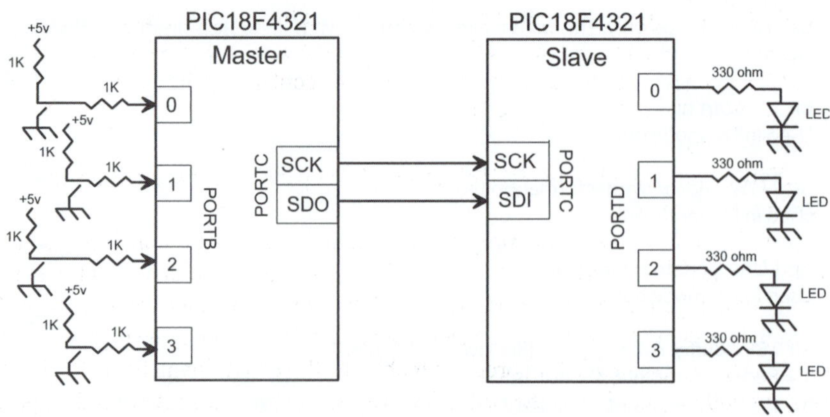

FIGURE 9.23 Figure for Example 9.7

Write a PIC18F assembly language at 0x70 for the master PIC18F4321 that will configure PORTB and PORTC, initialize STKPTR to 0x30, initialize SSPSTAT and SSPCON1, input switches, and call a subroutine called SERIAL_WRITE to place this data into its SSPBUF register.

Also, write a PIC18F assembly language program at 0x100 for the slave PIC18F4321 that will configure PORTC and PORTD, initialize SSPSTAT and SSPCON1 registers, input data from its SDI pin, places the data in the slave's SSPBUF, and then output to the LEDs.

Solution

The PIC18F assembly language programs for the master PIC18F4321 and the slave PIC18F4321 in Figure 9.24 are written using the following steps as the guidelines:

Master PIC18F4321

1. Configure PORTB as input and SD0 and SCK as outputs.
2. Select CKE (SPI clock select bit) using the master's SSPSTAT register.
3. Enable serial functions, select master mode with clock such as fosc/4 using the SSPCON1 register.
4. Input switches into WREG, and then CALL a subroutine called SERIAL_WRITE to move switch data into the master's Serial Buffer register (SSPBUF).
5. Wait in a loop, and check whether BF bit in the master's SSPSTAT register is 1, indicating completion of transmission.
6. As soon as BF = 1, the program returns from the subroutine, and branches to Step 4.

Slave PIC18F4321

1. Initialize SDI and SCK pins as inputs, and PORTD as output. Note that the SCK is controlled by the master, and, therefore, it is configured as an input by the slave.
2. Select CKE same as the master CKE (high to low clock in this example) using the slave's SSPSTAT register.
3. Enable serial functions, disable the $\overline{SS}$ pin, and select slave mode using the slave's SSPCON1 register. Note that the $\overline{SS}$ pin is used by multiple slaves.
4. Wait in a loop, and check whether BF = 1 in the slave's SSPSTAT register.
5. If BF = 0; wait. However, if BF = 1, output the contents of the slave's Serial Buffer register (SSPBUF) to slave's PORTD.
6. Go to Step 5.

Figure 9.24 provides the PIC18F assembly language programs for the master and the slave microcontrollers.

Let us now explain the program of Figure 9.24. First, consider the master PIC18F4321. The CKE bit (bit 6) in the SSPSTAT is set to one so that data transmission will occur from an active to an idle (HIGH to LOW) clock. Next, the register SSPCON1 is configured in order to set up the parameters for serial transmission. The bit SSPEN (bit 5) in the SSPCON1 is set to HIGH in order to enable the three pins, namely, SCK, SDO, and SDI. Writing 0000 to bits 3-0 of the SSPCON1 register define the master mode operation with a clock of Fosc/4.

```
;Program for  the master PIC18F4321
                INCLUDE <P18F4321.INC>
                ORG     0x00            ;Reset
                GOTO    MAIN
                ORG     0x70
MAIN            BCF     TRISC, RC5      ;Configure RC5/SD0  as output
                BCF     TRISC, RC3      ;Configure RC3/SCK as output
                MOVLW   0x0F
                MOVWF   ADCON1          ;Make PORTB digital input
                MOVLW   0x30            ;Initialize STKPTR to 0x30 since subroutine
                MOVWF   STKPTR          ;called SERIAL_WRITE is used in the
                                        ;program
                MOVLW   0x40
                MOVWF   SSPSTAT         ;Set data transmission on high to low clock
                MOVLW   0x20
                MOVWF   SSPCON1         ;Enable serial functions and set to
                                        ; master device, and Fosc/4
GET_DATA        MOVF    PORTB,W         ;Move switch value to WREG
                CALL    SERIAL_WRITE ;Call SERIAL_WRITE function
                BRA     GET_DATA
SERIAL_WRITE MOVWF      SSPBUF          ;Move switch value to serial buffer
WAIT            BTFSS   SSPSTAT, BF     ;Wait until transmission is complete
                BRA     WAIT
                RETURN
                END
; Program for the slave PIC18F4321
                INCLUDE <P18F4321.INC>
                ORG     0x00            ;Reset
                GOTO    MAIN
                ORG     0x100
MAIN            BSF     TRISC, RC4      ;Configure RC4/SDI  as input
                BSF     TRISC, RC3      ;Configure RC3/SCK as input
                CLRF    TRISD           ;Configure PORTD as output
                MOVLW   0x40
                MOVWF   SSPSTAT         ;Set data transmission on high to low clock
                MOVLW   0x25
                MOVWF   SSPCON1         ;Enable serial functions and set to the slave
WAIT            BTFSS   SSPSTAT, BF     ;Wait until transmission is complete (BF=1)
                BRA     WAIT            ;If BF=0, wait
                MOVFF   SSPBUF, PORTD ;Output serial buffer data to PORTD LEDs
                BRA     WAIT
                END
```

FIGURE 9.24 PIC18F assembly language program for Example 9.7

The following PIC18F instructions accomplish this:

```
MOVLW    0x20
MOVWF    SSPCON1   ;Enable serial functions and set to master and Fosc/4
```

Next, consider the PIC18F assembly language program for the slave; the four bits (bits 3-0) of the slave's SSPCON1 are initialized with 0101. This will place the microcontroller in the slave mode, and, also, the $\overline{SS}$ pin will be disabled. since there is only one serial device in this example. Note that the $\overline{SS}$ pin is required if multiple slave devices are used. Also, the SCK pin will be used as the clock.

Let us now briefly explain the program logic. The assembly language program for the master, the PIC18F4321, will first perform all initializations, input the switches, and place the master in the WREG. The program will then call a subroutine called SERIAL_WRITE. The subroutine moves the switch inputs into the SSPBUF register. As soon as the serial parameters for the master such as the SCK clock pin is set up, data are automatically transmitted to the slave device. Once all the data have been written, the BF bit (bit 0) in the SSPSTAT register of the master microcontroller will go to HIGH, indicating completion of transmission.

The program for the slave microcontroller waits in a loop until the BF flag in its SSPSTAT register goes to HIGH, indicating that the transmission is completed. The switch values from the slave's SSPBUF register are output to the LEDs connected at PORTD using the MOVFF instruction as follows:

```
MOVFF     SSPBUF, PORTD  ;Move serial buffer value to PORTD
```

After programming the master and the slave with the programs of Figure 9.24, upon hardware reset, the master PIC18F4321 jumps to address 0x70 (arbitrarily chosen address) while the slave PIC18F4321 jumps to 0x100 (arbitrarily chosen address). Note that both processors do not need to be reset at the same time. Also, both the master and the slave start executing the respective programs. The master microcontroller performs initalizations, moves switch data input continuously, and waits in the GET_DATA loop. The slave microcontroller also performs initializations, and then waits in the WAIT loop until BF = 1. As soon as the serial communication is established between the master and the slave, the master transmits the contents of SSPBUF via its SDO pin to the slave's SDI pin using the SCK clock. The switch data are transferred to the slave's SSPBUF register. After completion of the transfer, the slave's BF bit in the SSPSTAT register becomes 1. The slave then outputs these data to the LEDs via PORTD.

This example has been successfully implemented in the laboratory. This example can also be implemented using the $\overline{SS}$ pin. In that case, the slave's $\overline{SS}$/RA5 pin should be connected to the master's one of the I/O port bits. The I/O port bit must be configured as an output by the master via programming. Also, the $\overline{SS}$/RA5 pin must be configured as an input pin. The SSPCON1 should be loaded with 0x24, which will initialize the slave PIC18F4321 in slave mode, and enable its $\overline{SS}$ pin.

9.4 PIC18F4321 Capture/Compare/PWM (CCP) Modules

The CCP module is implemented in the PIC18F4321 as an on-chip feature to provide measurement and control of time-based pulse signals.

Capture mode causes the contents of an internal 16-bit timer to be written in special function registers upon detecting an nth rising or falling edge of a pulse. Compare mode generates an interrupt or change on output pin, when Timer1 matches a preset comparison value. PWM mode creates a re-configurable square wave duty cycle output at a user set frequency. The application software can change the duty cycle or period by modifying the value written to specific special function registers.

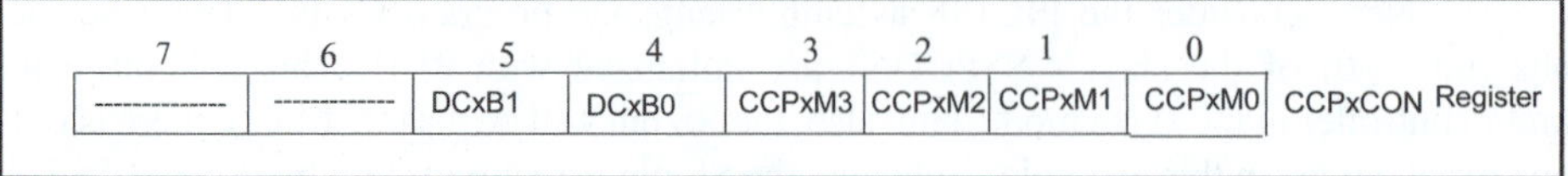

bit 7-6 **Unimplemented:** Read as '0'

bit 5-4 **DCxB1:DCxB0**: PWM Duty Cycle bit 1 and bit 0 for CCP Module x
Capture mode:
Unused.
Compare mode:
Unused.
PWM mode:
These bits are the lower two bits (bit 1 and bit 0) of the 10-bit PWM duty cycle. The higher eight bits (DCx9:DCx2) of the duty cycle are found in CCPRxL.

bit 3-0 **CCPxM3:CCPxM0**: CCPx Module Mode Select bits
0000 = Capture/Compare/PWM disabled (resets CCP module)
0001 = Reserved
0010 = Compare mode, toggle output on match (CCPxIF bit is set)
0011 = Reserved
0100 = Capture mode, every falling edge
0101 = Capture mode, every rising edge
0110 = Capture mode, every 4th rising edge
0111 = Capture mode, every 16th rising edge
1000 = Compare mode: initialize CCP pin low; on compare match, force CCP pin high (CCPxIF bit is set)
1001 = Compare mode: initialize CCP pin high; on compare match, force CCP pin low (CCPxIF bit is set)
1010 = Compare mode: generate software interrupt on compare match (CCPxIF bit is set, CCP pin reflects I/O state)
1011 = Compare mode: trigger special event, reset timer, start A/D conversion on CCPx match (CCPxIF bit is set)
11xx = PWM mode

FIGURE 9.25 CCPxCON register

The PIC18F4321 contains two CCP modules, namely, CCP1 and CCP2. The CCP1 module of the PIC18F4321 is implemented as a standard CCP with enhanced PWM capabilities for better DC motor control. Hence, the CCP1 module in the PIC18F4321 is also called ECCP (Enhanced CCP). Note that the CCP2 module is provided with standard capture, compare, and PWM features. The CCP1 and CCP2 modules will be referred to as CCPx in the following discussion.

9.4.1 CCP Registers

Each CCP module is associated with an 8-bit control register (CCPxCON) shown in Figure 9.25. The CCPxCON can be used to select one of the three modes, namely, Compare, Capture, or PWM.

Each CCP module also contains a 16-bit data register (CCPRx). The 16-bit data register, in turn, is comprised of two 8-bit registers: CCPRxL (low byte) and CCPRxH (high byte). This 16-bit data register can operate as a 16-bit Capture register, a 16-bit Compare register, or an 8-bit PWM register holding the 8-bit decimal part of the duty cycle.

9.4.2 CCP Modules and Associated Timers

The CCP modules utilize Timers 1, 2, or 3, depending on the mode selected. Timer1 and Timer3 are available to modules in Capture or Compare modes, while Timer2 is available for modules in PWM mode. The assignment of a particular timer to a module

TABLE 9.1 Assignment of timers for the PIC18F4321 CCP mode

CCP mode selected	Timer
Capture mode	Timer1 or Timer3
Compare mode	Timer1 or Timer3
PWM mode	Timer2

is determined by the Timer to CCP enable bits in the T3CON register (Figure 9.10). Both modules may be active at any given time and may share the same timer resource if they are configured to operate in the same mode (Capture, Compare, or PWM) at the same time. The assignment of the timers is summarized in Table 9.1.

9.4.3 PIC18F4321 Capture Mode

In Capture mode, the CCPRxH:CCPRxL register pair captures the 16-bit value of the TMR1 or TMR3 registers when an event (such as every rising or falling edge) occurs on the corresponding CCPx pin. The event is selected by the mode select bits, CCPxM3:CCPxM0 (bits 3-0 of CCPxCON, Figure 9.25). When a capture is made, the interrupt request flag bit, CCPxIF, is set; it must be cleared in software. If another capture occurs before the value in register CCPRx is read, the old captured value is overwritten by the new captured value.

In Capture mode, the appropriate CCPx pin (RC2/CCP1/P1A, pin 17 or RC1/T1OSI/CCP2, pin 16) of the PIC18F4321 should be configured as an input by setting the corresponding TRIS direction bit. Also, the timers that are to be used with the capture feature (Timer1 and/or Timer3) must be running in Timer mode or Synchronized Counter mode. In Asynchronous Counter mode, the capture operation will not work. The timer to be used with each CCP module is selected in the T3CON register (Figure 9.10).

When the Capture mode is changed, a false capture interrupt may be generated. The user should keep the CCPxIE interrupt enable bit clear to avoid false interrupts. The interrupt flag bit, CCPxIF, should also be cleared following any such change in operating mode.

In summary, the following steps can be used to program the PIC18F4321 in capture mode to determine the period of a waveform (assume CCP1; similar procedure for CCP2):

1. Load the CCP1CON register (Figure 9.25) with appropriate data for capture mode.
2. Configure RC2/CCP1/P1A as an input pin using the TRISC register.
3. Select Timer1 and/or Timer3 by loading appropriate data respectively into T1CON register (Figure 9.6) and/or T3CON register (Figure 9.10).
4. Clear the interrupt request flag, CCP1IF for CCP1 (Register PIR1 of Figure 9.7) or CCP2IF for CCP2 (Register PIR2 of Figure 9.11), after a capture so that the next capture can be made.
5. Clear the interrupt enable bit, CCP1IE for CCP1 (Register PIE1 of Figure 9.8) or CCP2IE for CCP2 (Register PIE2 of Figure 9.12), to avoid false interrupts.
6. Clear CCPR1H and CCPR1L to 0.
7. Check CCP1IF flag in PIR1 and wait in a loop until CCP1IF is 1 for the first rising edge. As soon as the first rising edge is detected, start Timer1 (or Timer3).
8. Save CCPR1H and CCPR1L in data memory such as REGX and REGY.
9. Clear CCP1IF to 0.
10. Check CCP1IF flag in PIR1 and wait in a loop until CCP1IF is 1 for the second rising edge. As soon as the second rising edge is detected, stop Timer1 (or Timer3).

11. Disable capture by clearing CCP1CON register.
12. Perform 16-bit subtraction: [CCPR1H:CCPR1L] - [REGX:REGY].
13. 16-bit result in register pair [REGX:REGY] will contain the period of the incoming waveform in terms of the number of clock cycles.

Typical applications of the capture mode include:

- measurement of the pulse width of an unknown periodic signal by capturing the subsequent leading (rising) and trailing (falling) edges of a pulse.
- measurement of the period of a signal by capturing two subsequent leading or trailing edges.
- measurement duty cycle . Note that the duty cycle is defined as (t1/T) x 100 where t1 is the fraction of the time the signal is HIGH in a period T.

Example 9.8 Assume PIC18F4321. Write a PIC18F assembly language program at address 0x200 to measure the period (in terms of the number of clock cycles) of an incoming periodic waveform connected at RC2/CCP1/P1A pin. Store result in registers 0x21 (high byte) and 0x20 (low byte). Use Timer3, and capture mode of CCP1.

Solution

The PIC18F assembly language program is provided below:

```
        INCLUDE<P18F4321.INC>
        ORG     0x200
        MOVLW   B'00000101'       ;Select capture mode rising edge
        MOVWF   CCP1CON
        BSF     TRISC, CCP1       ;Configure RC2/CCP1/P1A pin as input
        MOVLW   B'01000000'       ;Select TIMER3 as clock source for capture
        MOVWF   T3CON             ;Select TIMER3 internal clock, 1:1 prescale
                                  ;TIMER3 OFF
        BCF     PIE1, CCP1IE      ;Disable CCP1IE to avoid false interrupt
        MOVLW   0X00
        MOVWF   CCPR1H            ;Clear CCPR1H to 0
        MOVWF   CCPR1L            ;Clear CCPR1L to 0
        BCF     PIR1, CCP1IF      ;Clear CCP1IF
WAIT    BTFSS   PIR1, CCP1IF      ;Wait for the first rising edge
        GOTO    WAIT
        BSF     T3CON, TMR3ON     ;Turn Timer3 ON
        MOVFF   CCPR1L, 0x20      ;Save CCPR1L in 0x20 at 1st rising edge
        MOVFF   CCPR1H, 0x21      ;Save CCPR1H in 0x21 at 1st rising edge
        BCF     PIR1, CCP1IF      ;Clear CCP1IF
WAIT1   BTFSS   PIR1, CCP1IF      ;Wait for next rising edge
        GOTO    WAIT1
        BCF     T3CON, TMR3ON     ;Turn OFF Timer3
        CLRF    CCP1CON           ;Disable capture
        MOVF    0x20, W           ;Move 1st low byte to WREG
        SUBWF   CCPR1L, F         ;Subtract WREG from 2nd low byte
                                  ;Result in 0x20
        MOVF    0x21, W           ;Move 1st High byte to WREG
        SUBWFB  CCPR1H, F         ;Subtract WREG with borrow
```

```
                                    ;from 2nd high byte, result in 0x21
HERE    BRA      HERE               ;Halt
        END
```

9.4.4 PIC18F4321 Compare Mode

In Compare mode, the 16-bit CCPRx (CCPR1H:CCPR1L for CCP1 or CCPR2H: CCPR2L) register value is constantly compared against the value in either the TMR1 or the TMR3 register. When a match occurs, the CCPx pin (RC2/CCP1/P1A pin or RC1/ T1OSI/CCP2 pin of the PIC18F4321 PORTC) can be:

- driven high
- driven low
- toggled (high-to-low or low-to-high)
- remain unchanged (that is, reflects the state of the I/O latch)

The action on the pin is based on the value of the mode select bits (CCPxM3:CCPxM0) in CCPxCON register (Figure 9.25). As soon as a match occurs, the interrupt flag bit, CCPxIF, is set to one. The user must configure the CCPx pin as an output by clearing the appropriate TRIS bit. Timer1 and/or Timer3 must be running in Timer mode or Synchronized Counter mode if the CCP module is using the compare feature. In Asynchronous Counter mode, the compare operation may not work.

When the Generate Software Interrupt mode is chosen (CCPxM3:CCPxM0 = 1010), the corresponding CCPx pin is not affected. Only a CCP interrupt is generated, if enabled and the CCPxIE bit is set. Both CCP modules are equipped with a Special Event Trigger. This is an internal hardware signal generated in Compare mode to trigger actions by other modules. The Special Event Trigger is enabled by selecting the Compare Special Event Trigger mode (CCPxM3:CCPxM0 = 1011). For either CCP module, the Special Event Trigger resets the Timer register pair for whichever timer resource is currently assigned as the module's time base. This allows the CCPRx registers to serve as a programmable period register for either timer. The Special Event Trigger for CCP2 can also start an A/D conversion. In order to do this, the A/D converter must already be enabled.

Typical applications of the compare mode include generation of a certain time delay, a pulse train, or a waveform with a specific duty cycle.

The following steps can be used to program the PIC18F4321 in capture mode to provide time delay or determine the period of a waveform:

1. Load the CCP1CON (or CCP2CON) register (Figure 9.25) with appropriate data for compare mode.
2. Configure the RC2/CCP1/P1A pin (or RC1/T1OSI/CCP2 pin) of PORTC as an output.
3. Load the CCPR1H:CCPRIL (or CCPR2H:CCPR2L) register pair with appropriate values.
4. Load Timer1 (or Timer3) in the timer mode or synchronized counter mode by loading appropriate data into T1CON (or T3CON) register.
5. Initialize Timer1H:Timer1L (or Timer3H:Timer3L) to 0.
6. Clear CCP1IF in PIR1 (or CCP2IF in PIR2).
7. Start Timer1 (or Timer3).
8. Wait in a loop until the CCP1IF (or CCP2IF) is HIGH.

9. As soon as match occurs (CCP1IF or CCP2IF HIGH), stop Timer1 (or Timer3).

Example 9. 9 Assume PIC18F4321 with an internal crystal clock of 20 MHz. Write a PIC18F assembly language program at address 0x100 that will toggle the RC2/CCP1/P1A pin after a time delay of 10 msec. Use Timer3, and compare mode of CCP1.

Solution

With 20 MHz internal crystal, Fosc = 20 MHz. Since Timer3 uses Fosc/4, Timer clock frequency = Fosc/4 = 5 MHz. Hence, clock period of Timer3 = 0.2 μ sec. Counter value = (10 msec)/(0.2 μ sec) = $500_{10} = 01F4_{16}$. Hence, CCPR1H :CCPR1L should be loaded with 0x01F4 for the PIC18F4321 compare mode.

The PIC18F assembly language program is provided below:

```
       INCLUDE <P18F4321.INC>
       ORG     0x100
       MOVLW   0x02              ;Select compare mode, toggle CCP1 pin
       MOVWF   CCP1CON           ;on match
       BCF     TRISC, CCP1       ;Configure CCP1 pin as output
       MOVLW   0x40              ;Select TIMER3 as clock source for
                                 ;compare
       MOVWF   T3CON             ;Select TIMER3 internal clock, 1:1 prescale
                                 ;TIMER3 OFF
       MOVLW   0x01              ;Load CCPR1H with 0x01
       MOVWF   CCPR1H
       MOVLW   0xF4              ;Load CCPR1L  with 0xF4
       MOVWF   CCPR1L
       CLRF    TMR3H             ;Initialize TMR3H to 0
       CLRF    TMR3L             ;Initialize TMR3L to 0
       BCF     PIR1, CCP1IF      ;Clear CCP1IF
       BSF     T3CON, TMR3ON     ;Start Timer3
WAIT   BTFSS   PIR1, CCP1IF      ;Wait in a loop until CCP1IF is 1. CCP1 pin
       BRA     WAIT              ;toggles when match occurs
       BCF     T3CON, TMR3ON     ;Stop Timer3
HERE   BRA     HERE              ; Halt
       END
```

9.4.5 PIC18F4321 PWM (Pulse Width Modulation) Mode

In PWM mode, the CCPx pin can be configured as an output to generate a periodic waveform with a specified frequency, and a 10-bit duty cycle. The PWM duty cycle is specified by writing to the upper eight bits of the CCPRxL register; the lower two bits are written to bits 5 and 4 of the CCPxCON register. Timer2 is used for the PWM mode. The PWM period is specified by writing to the 8-bit PR2 register in the CCP module.

When TMR2 is equal to PR2, the following three events occur on the next increment cycle:

- TMR2 is cleared.
- The CCPx pin is set (exception: if PWM duty cycle = 0%, the CCPx pin will not be set).

- The PWM duty cycle is latched from CCPRxL into CCPRxH.

The PWM period is specified by writing to the PR2 register. From the data sheet, the PWM period can be calculated using the following formula:
PWM Period = [(PR2) + 1] x 4 x Tosc x (TMR2 Prescale Value)
where Tosc = (1/Fosc), Fosc is the crystal frequency, and TMR2 Prescale Value can be initialized as 1, 4, or 16 using the T2CON register.
Hence, PR2 = [(Fosc)/(4 x Fpwm x TMR2 Prescale Value)] - 1
Note that PWM frequency (Fpwm) is defined as 1/[PWM period].

As mentioned before, the PWM duty cycle is specified by writing to the CCPRxL register and to the CCPxCON<5:4> bits. Up to 10-bit resolution is available. The CCPRxL contains the eight most significant bits, and the CCPxCON (bits 5 and 4) contains the two least significant bits. This 10-bit value is represented by CCPRxL:CCPxCON (bits 5 and 4). The following equation is used to calculate the PWM duty cycle in time:

PWM Duty Cycle = (CCPRXL:CCPXCON<5:4>) x Tosc x (TMR2 Prescale Value).

As mentioned before, the duty cycle is defined as the percentage of the time the pulse is high in a clock period. Note that the upper eight bits in the CCPRxL are the decimal part of the duty cycle while bits 5 and 4 of the CCPxCON register contain the fractional part of the duty cycle. For example, consider 25% duty cycle. Since duty cycle is a fraction of the PR2 register value, decimal value for the duty cycle with a PR2 value of 30 is 7.5 (0.25 x 30). Hence, the 8-bit binary number 00000111_2 must be loaded into CCPRxL, and $10_2(0.5_{10})$ must be loaded for DCxB1 and DCxB0 bits in the CCPxCON register (Figure 9.25).

CCPRxL and CCPxCON (bits 5, 4) can be written to at any time, but the duty cycle value is not latched into CCPRxH until after a match between PR2 and TMR2 occurs (i.e., the period is complete). In PWM mode, CCPRxH is a read-only register.

The following procedure should be followed when configuring the CCP module for PWM operation:

1. The PR2 register should be initialized with the PWM period.
2. Load the PWM duty cycle by writing to the CCPRxL register for higher eight bits, and bits 5, 4 of CCPxCON (Figure 9.25) for lower two bits.
3. Make the CCPx pin an output by clearing the appropriate TRIS bit.
4. Set the TMR2 prescale value, then enable Timer2 by writing to T2CON.
5. Initialize TMR2 register to 0.
6. Set up the CCPx module for PWM operation, and turn Timer2 ON.

Example 9.10 Write a PIC18F assembly language program at 0x100 to generate a 4 KHz PWM with a 50% duty cycle on the RC2/CCP1/P1A pin of the PIC18F4321. Assume 4 MHz crystal.

Solution

PR2 = [(Fosc)/(4 x Fpwm x TMR2 Prescale Value)] - 1
PR2 = [(4 MHz)/(4 x 4 KHz x 1)] - 1 assuming Prescale value of 1
PR2 = 249. With 50% duty cycle, decimal value of the duty cycle = 0.5 x 249 = 124.5. Hence, the CCPR1L register will be loaded with 124, and bits DC1B1:DC0B0 (CCP1CON

register) with 10 (binary).

The PIC18F assembly language program is provided below:

```
        INCLUDE   <P18F4321.INC>
        ORG       0x100
        MOVLW     D'249'            ;Initialize PR2 register
        MOVWF     PR2
        MOVLW     D'124'            ;Initialize CCPR1L
        MOVWF     CCPR1L
        MOVLW     0x20              ;CCP1 OFF,
        MOVWF     CCP1CON           ;DC1B1:DC0B0=10
        BCF       TRISC, CCP1       ;Configure CCP1 pin as output
        CLRF      T2CON             ;1:1 prescale, Timer2 OFF
        MOVLW     0x2C              ;PWM mode
        MOVWF     CCP1CON
        CLRF      TMR2              ;Clear Timer2 to 0
BACK    BCF       PIR1, TMR2IF      ;Clear TMR2IF to 0
        BSF       T2CON, TMR2ON     ;Turn Timer2 ON
WAIT    BTFSS     PIR1, TMR2IF      ;Wait until TMR2IF is HIGH
        GOTO      WAIT
        BRA       BACK
        END
```

In the above program, the value of TMR2 is compared to that of the period register PR2 on each clock cycle. When the two values match, the comparator generates a match signal as the timer output. This signal also resets the value of TMR2 to 0x00 on the next cycle. In the above program, the last instruction BRA BACK branches to the label where the TMR2IF flag in the PIR1 is cleared to 0. The program does not have to go back to clear TMR2 to 0 since the TMR2 is automatically cleared after each match.

9.5 DC Motor Control

Typical applications of the PWM mode include DC motor control. The speed of a DC motor is directly proportional to the driving voltage. The speed of a motor increases as the voltage is increased. In earlier days, voltage regulator circuits were used to control the speed of a DC motor. But voltage regulators dissipate lots of power. Hence, the PIC18F in the PWM mode is used to control the speed of a DC motor. In this scheme, power dissipation is significantly reduced by turning the driving voltage to the motor ON and OFF. The speed of the motor is a direct function of the ON time divided by the OFF time.

Sometimes, it is desirable to change direction of rotation of the DC motor. This can be accomplished by reversing the direction of the motor via software by interfacing a device called an H-Bridge to an I/O port of the PIC18F. Note that the speed of the motor, on the other hand, can be controlled using the PWM mode, and by connecting the DC motor to a PWM pin such as the PIC18F CCP1. The basic concepts associated with the DC motor control using the PIC18F4321's PWM mode will be illustrated in Example 9.11.

Microcontrollers such as the PIC18F4321 are not capable of outputting the required large current and voltage to control a typical DC motor. Hence, a driver such as the CNY17F Optocoupler is needed to amplify the current and voltage provided by the PIC18F's output, and provide appropriate levels for the DC motor. One of the many useful applications for employing a PWM signal is its ability to control a mechanical device, such as a motor.

Note that the motor will run faster or slower based on the duty cycle of the PWM signal. The motor runs faster as the duty cycle of the PWM signal at the CCPx pin is increased. To illustrate this concept, two different duty cycles will be used in the following example (Example 9.11).

Example 9.11 Figure 9.26 shows a simplified diagram interfacing the PIC18F4321 to a DC motor via the CNY17F Optocoupler. The purpose of this example is to control the speed of a DC motor by inputting two switches connected at bit 0 and bit 1 of PORTD. The motor will run faster or slower based on the switch values (00 or 11), but will not provide any measure of the exact RPM of the motor.

When both switches are closed (00), a PWM signal at the CCP1 pin of the PIC18F4321 with 50% duty cycle will be generated. When both switches are open (11), a PWM signal at the CCP1 pin of the PIC18F4321 with 75% duty cycle will be generated. Otherwise, the motor will stop, and the program will wait in a loop.

If switches are closed (00), the motor will run using the 4 KHz PWM pulse of Example 9.10 with 50% duty cycle. If both switches are open (11), the motor will run using the same PWM pulse at a faster speed with a duty cycle of 75%. The program will first perform initializations, and wait in a loop until the switches are 00 or 11.
Write a PIC18F assembly language program to accomplish this.

Solution

The schematic of Figure 9.26 uses a CNY17F Optocoupler which serves two purposes. The first purpose is to protect the PIC18F4321 microcontroller by isolating the motor from the microcontroller. The second purpose the optocoupler serves is allowing the user to take a 0-5 V PWM signal and boost it to a 0-12 V source, where any voltage could be used that is safe for the optocoupler.

From Example 9.10, PR2 = 249. With 50% duty cycle, Count = 0.5 x 249 = 124.5. Hence, the CCPR1L register will be loaded with 124, and bits DC1B1:DC0B0 (CCP1CON register) with 10_2.

With 75% duty cycle, Count = 0.75 x 249 = 186.75. Hence, the CCPR1L register will be loaded with 186, and bits DC1B1:DC0B0 (CCP1CON register) with 11_2.

The PIC18F assembly language program is provided below:

```
        INCLUDE <P18F4321.INC>
        ORG     0x100
        MOVLW   D'249'          ;Initialize PR2 register
        MOVWF   PR2
        BCF     TRISC, CCP1     ;Configure CCP1 pin as output
        BSF     TRISD, RD0      ;Configure RD0 as an input bit
        BSF     TRISD, RD1      ;Configure RD1 as an input bit
        CLRF    T2CON           ;1:1 prescale, Timer2 OFF
        MOVLW   0x3C            ;PWM mode,DC1B1:DC0B0=11
        MOVWF   CCP1CON
```

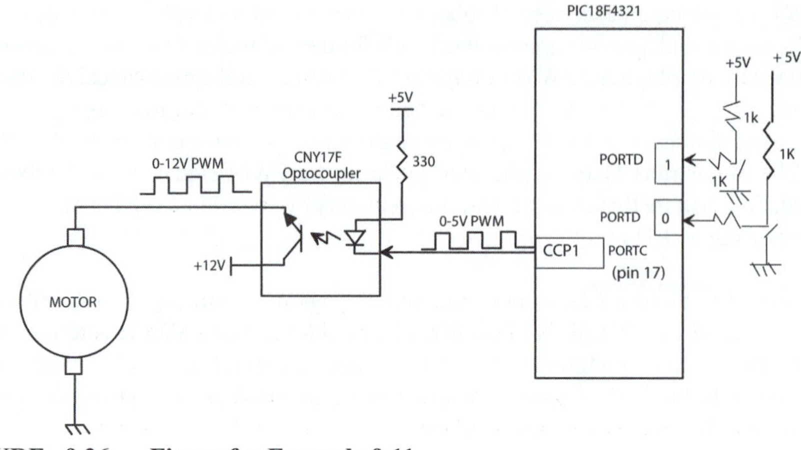

FIGURE 9.26 Figure for Example 9.11

```
SWITCH  BTFSC  PORTD, RD0      ;If switch0 is  LOW, check switch1
                               ;for LOW
        BRA    SWITCH1         ;If switch0 is HIGH,  branch to
                               ;check switch1 for HIGH
        BTFSC  PORTD, RD1      ;If both switches are LOW, branch
                               ;to DUTY50 and generate
                               ;PWM with 50%  duty cycle;
        BRA    SWITCH          ;else, go back and wait
                               ;Both switches are HIGH, go to
                               ;75% duty cycle
        BRA    DUTY50          ;Both switches LOW, go to 50%
                               ;duty cycle
SWITCH1 BTFSS  PORTD, RD1      ;If both switches are HIGH, branch
                               ;to DUTY75, and generate PWM
                               ;with 75% duty cycle;
        BRA    SWITCH          ;else, go back and wait
DUTY75  MOVLW  D'186'          ;For 75% duty cycle
        MOVWF  CCPR1L
        MOVLW  0x3C            ;PWM mode,DC1B1:DC0B0=11
        MOVWF  CCP1CON
        BRA    TIMER
DUTY50  MOVLW  D'124'          ;For 50% duty cycle
        MOVWF  CCPR1L
        MOVLW  0x2C            ;PWM mode,DC1B1:DC0B0=10
        MOVWF  CCP1CON         ;Initialize CCP1CON
TIMER   CLRF   TMR2            ;Clear Timer2 to 0
BACK    BCF    PIR1, TMR2IF    ;Clear TMR2IF to 0
        BSF    T2CON, TMR2ON   ;Turn Timer2 ON
WAIT    BTFSS  PIR1, TMR2IF    ;Wait until TMR2IF is HIGH
        BRA    WAIT            ;(end of period)
        BRA    SWITCH          ;Repeat to initialize and read switch
                               ;inputs

        END
```

Questions and Problems

9.1 Find the contents of T0CON register to program Timer0 in 8-bit mode with 1:16 prescaler using the external clock, and incrementing on negative edge.

9.2 Write a PIC18F assembly language instruction sequence to initialize Timer0 as an 8-bit timer to provide a time delay with a count of 100. Assume 4 MHz internal clock with a prescaler value of 1:16.

9.3 Write a PIC18F assembly language program to generate a square wave with a period of 4 ms on bit 0 of PORTC using a 4 MHz crystal. Use Timer0.

9.4 Write a PIC18F assembly language program to generate a square wave with a period of 4 ms on bit 7 of PORTD using a 4 MHz crystal. Use Timer1.

9.5 Write a PIC18F assembly language program to turn an LED ON connected at bit 0 of PORTC when the TMR2 register reaches a value of 200. Assume a 4 MHz crystal. Use prescaler and postscaler values of 1:16.

9.6 Write a PIC18F assembly language program to generate a square wave on pin 3 of PORTC with a 4 ms period using Timer3 in 16-bit mode with a prescaler value of 1:8. Use a 4 MHz crystal.

9.7 Repeat Example 9.5 using A/D converter's interrupt bit indicating completion of conversion. Use addresses, and other parameters of your choice.

9.8 Design and develop hardware and software for a PIC18F4321-based system (Figure P9.8) that would measure, compute, and display the Root-Mean-Square (RMS) value of a sinusoidal voltage. The system is required to:

1. Sample a 5 V (zero-to-peak voltage), 60 Hz sinusoidal voltage 128 times.

2. Digitize the sampled value using the on-chip ADC of the PIC18F4321 along with its interrupt upon completion of conversion signal.

3, Compute the RMS value of the waveform using the formula,
RMS Value = SQRT $\left[\sum_{n=1}^{N}(X_n^2)/N\right]$, where X_n's are the samples, and N

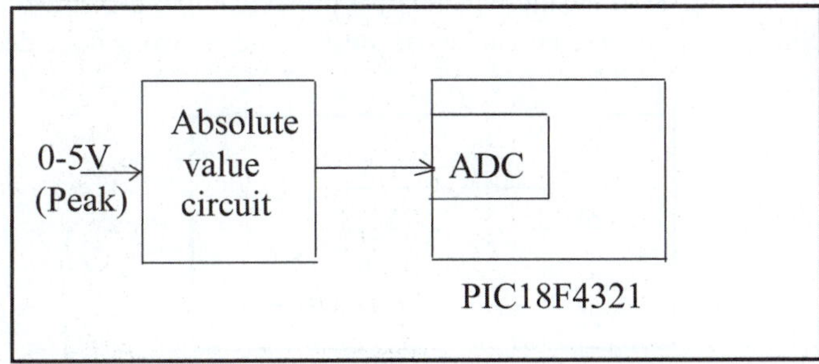

FIGURE P9.8

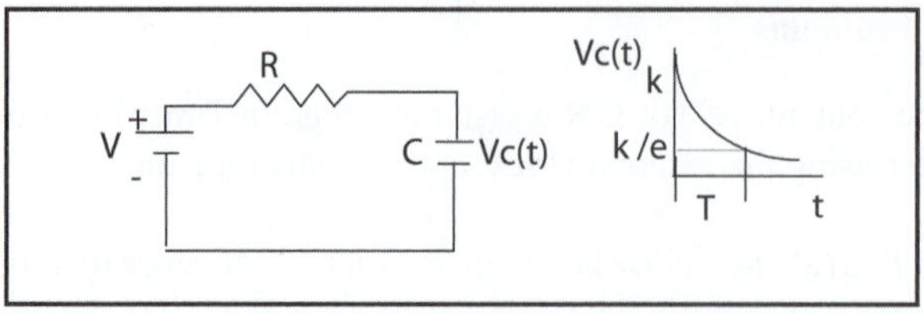

FIGURE P9.9

is the total number of samples. Display the RMS value on seven-segment displays.

(a) Flowchart the problem.

(b) Convert the flowchart to a PIC18F assembly language program.

9.9 *Capacitance meter.* Consider the RC circuit of Figure P9.9. The voltage across the capacitor is Vc (t) = k $e^{-t/RC}$. In one-time constant RC, this voltage is discharged to the value k/e. For a specific value of R, value of the capacitor C = T/R, where T is the time constant that can be counted by the PIC18F4321. Design the hardware and software for the PIC18F4321 to charge a capacitor by using a pulse to a voltage of your choice. The PIC18F4321 will then stop charging the capacitor, measure the discharge time for one time constant, and compute the capacitor value.

(a) Draw a hardware schematic.

(b) Write a PIC18F assembly language program to accomplish the above.

9.10 Design a PIC18F4321-based digital clock. The clock will display time in hours, minutes, and seconds. Write a PIC18F assembly language program to accomplish this.

9.11 Design a PIC18F4321-based system to measure the power absorbed by a 2K resistor (Figure P9.11). The system will input the voltage (V) across the 2K resistor, convert it to an 8-bit input using the PIC18F4321's on-chip A/D converter, and then compute the power using V^2/R.

9.12 Design a PIC18F4321-based system (Figure P9.12) as follows: The system will drive two seven-segment digits, and monitor two key switches. The system will start displaying 00. If the increment key is pressed, it will increment the display by one. Similarly, if the decrement key is pressed, the display will be decremented by

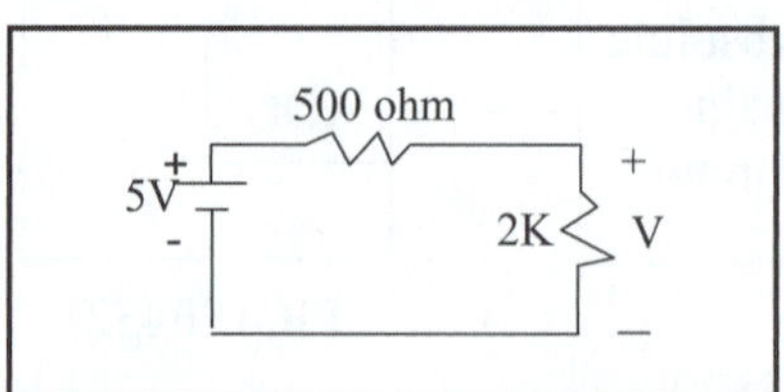

FIGURE P9.11

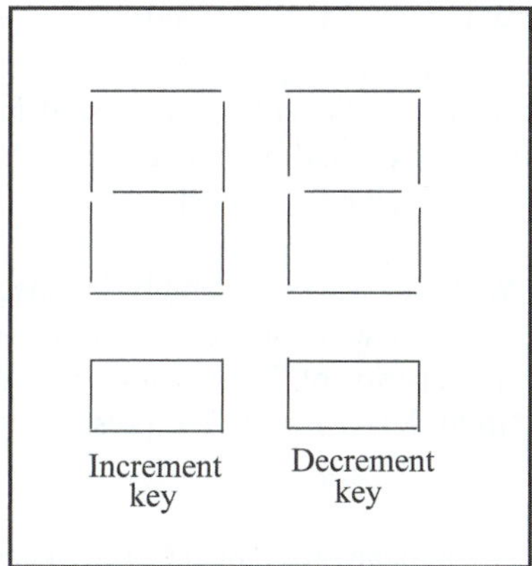

FIGURE P9.12

one. The display will go from 00 to 09, and vice versa.
Write a PIC18F assembly language program to accomplish the above. Use ports and data memory addresses of your choice. Draw a block diagram of your implementation.

9.13 It is desired to implement a PIC18F4321-based system as shown in Figure P9.13. The system will scan a hex keyboard with 16 keys, and drive three seven-segment displays. The PIC18F4321 will input each key pressed, scroll them in from the right side of the displays, and keep scrolling as each key is pressed. The leftmost digit is just discarded. The system continues indefinitely. Write a PIC18F assembly language program at address 0x100 to accomplish the above. Use ports and data memory addresses of your choice.

9.14 Assume that two PIC18F4321s are interfaced in the SPI mode. A switch is connected to bit 0 of PORTD of the master PIC18F4321 and, an LED is connected to bit 5 of PORTB of the slave PIC18F4321. Write PIC18F assembly language programs to input the switch via the master, and output it to the LED of the slave PIC18F4321. If the switch is open, the LED will be turned ON while the LED will be turned OFF if the switch is closed.

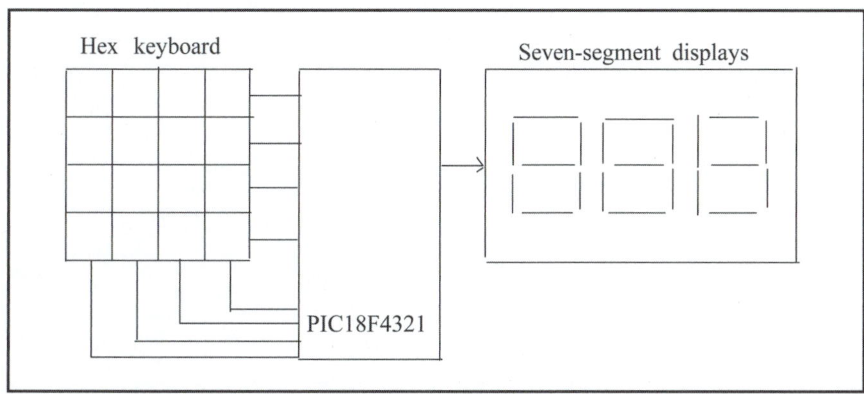

FIGURE P9.13

9.15 Assume PIC18F4321. Write a PIC18F assembly language program at address 0x200 that will measure the period of a periodic pulse train on the CCP1 pin using the capture mode. The 16-bit result will be performed in terms of the number of internal (Fosc/4) clock cycles, and will be available in the TMR1H:TMR1L register pair. Use 1:1 prescale value for Timer1.

9.16 Assume PIC18F4321. Write a PIC18F assembly language program at address 0x200 that will generate a square wave on the CCP1 pin using the Compare mode. The square wave will have a period of 20 ms with a 50% duty cycle. Use Timer1 internal clock (Fosc/4 from XTAL) with 1:2 prescale value. Assume 4-MHz crystal.

9.17 Write a PIC18F assembly language program at 0x100 to generate a 16 KHz PWM with a 75% duty cycle on the RC2/CCP1/P1A pin of the PIC18F4321. Assume 10 MHz crystal.

9.18 It is desired to change the speed of a DC motor by dynamically changing its pulse width using a potentiometer connected at bit 0 of PORTB (Figure P9.18). Note that the PWM duty cycle is controlled by the potentiometer. Write a PIC18F assembly language program that will input the potentiometer voltage via the PIC18F4321's on-chip A/D converter using interrupts, generate the PWM waveform on the CCP1 pin, and then change the speed of the motor as the potentiometer voltage is varied.

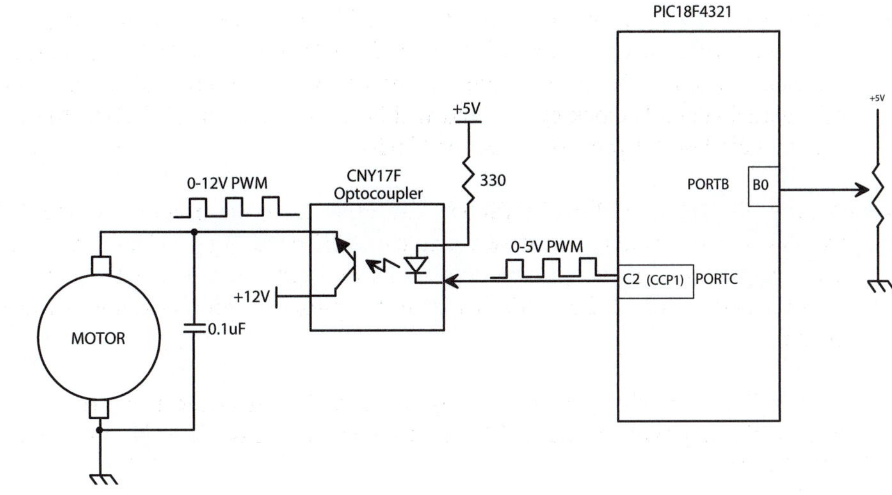
PIC18F4321
+5V
330
0-12V PWM
CNY17F
Optocoupler
+5V
PORTB
B0
0-5V PWM
+12V
C2 (CCP1)
PORTC
MOTOR
0.1uF

FIGURE P9.18

10

BASICS OF PROGRAMMING THE PIC18F USING C

In this chapter we describe basics of writing C language program for the PIC18F microcontroller family. Topics include C-basics, loops in C, functions, bit-wise operations, structures, unions, and bit fields. Finally, several worked-out I/O examples written in PIC18F assembly language in Chapter 9 are converted into C in this chapter in order to illustrate how to program the PIC18F using C.

10.1 Introduction to C Language

As mentioned in Chapter 3, a programmer's efficiency increases significantly with assembly language compared to machine language. However, the programmer needs to be well acquainted with the microcontroller's architecture and its instruction set. Furthermore, the programmer has to provide an opcode for each operation that the microcontroller has to carry out in order to execute a program. As an example, for adding two numbers, the programmer would instruct the microcontroller to load the first number into a register, add the second number to the register, and then store the result in memory. However, the programmer might find it tedious to write all of the steps required for a large program. Also, to become a reasonably good assembly language programmer, one needs to have a lot of experience. Also, it takes a long time to debug assembly code.

High-level language programs composed of English-language-type statements rectify all of these deficiencies of machine and assembly language programming. The programmer does not need to be familiar with the internal microcontroller structure or its instruction set. Also, each statement in a high-level language corresponds to a number of assembly or machine language instructions. For example, consider the statement c = a + b; written in a high-level language such as C or JAVA. This single statement adds the contents of a with b and stores the result in c. This is equivalent to a number of steps in machine or assembly language, as mentioned before. It should be pointed out that the letters a, b, and c do not refer to particular registers within the microcontroller. Rather, they are memory locations.

The C language is widely used at present. Typical microcontrollers such as the PIC18F family can be programmed using this high-level language. A high-level language is a problem-oriented language. The programmer does not have to know the details of the architecture of the microcontroller and its instruction set. Basically, the programmer follows the rules of the particular language being used to solve the problem at hand. A second advantage is that a program written in a particular high-level language can be executed by two different microcontrollers, provided that they both understand that language. For example, a program written in C for a PIC18F microcontroller will run on the HC12 microcontroller because both microcontrollers have a compiler to translate the C language into their particular machine language; minor modifications are required for I/O programs. C is a high-level language that includes I/O instructions.

Compilers normally provide inefficient machine codes because of the general guidelines that must be followed for designing them. However, compiled codes generate many more lines of machine code than does an equivalent assembly language program. Therefore, the assembled program will take up less memory space and will execute much faster than the compiled C. Although C language includes I/O instructions, applications involving I/O are normally written in assembly language. One of the main uses of assembly language is in writing programs for real-time applications. *Real time* indicates that the task required by the application must be completed before any other input to the program can occur that would change its operation. Typical programs involving non-real-time applications and extensive mathematical computations may be written in C.

The C Programming language was developed by Dennis Ritchie of Bell Labs in 1972. C has become a very popular language for many engineers and scientists, primarily because it is portable except for I/O and, however, can be used to write programs requiring I/O operations with minor modifications. This means that a program written in C for the PIC18F4321 will run on the Texas Instruments MSP430 with some modifications related to I/O as long as C compilers for both microcontrollers are available.

C is a general-purpose programming language and is found in numerous applications as follows:

- **Systems Programming.** Many operating systems (such as UNIX and its variant LINUX), compilers, and assemblers are written in C. Note that an operating system typically is included with the personal computer when it is purchased. The operating system provides an interface between the user and the hardware by including a set of commands to select and execute the software on the system.
- **Computer-Aided Design (CAD) Applications.** CAD programs are written in C. Typical tasks to be accomplished by a CAD program are logic synthesis and simulation.
- **Numerical Computation.** Software written in C is used to solve mathematical problems such as solving linear system of equations and matrix inversion. Industry standard MATLAB software is written in C.
- **Other Applications.** These include programs for printers and floppy disk controllers, and digital control algorithms such as PI (Proportional Integral) and PID (Proportional Integral Derivative) algorithms using microcontrollers.

A C program may be viewed as a collection of functions. Execution of a C program will always begin by a call to the function called "main." This means that all C programs should have its main program named as **main**. However, one can give any name to other functions.

A simple C program that prints "I wrote a C-program" is

```
/* First C-program */
# include <stdio.h>
void main ( )
{
        printf ("I  wrote a C-program") ;
}
```

Here, `main ()` is a function of no arguments, indicated by ( ). The parenthesis must be present even if there are no arguments. The braces { } enclose the statements that make up the function. The line **`printf("I wrote a C-program");`** is a function call that calls a function named **`printf`**, with the argument **`I wrote a C-program.`** **`printf`** is a library function that prints output on the terminal. Note that `/* */` is used to enclose comments. These are not translated by the compiler. C++ compilers are used to compile C programs these days. Hence, // followed by comment can be used instead of /* */. Note that the comments in C++ are written after // and it spans until the end of the line. A variation of the C program just described is

```
// Another C program
# include <stdio.h>
void main ( )
{
        printf ("I wrote");
        printf ("a C-");
        printf ("program);
        printf ("\n");
}
```

Here, `# include` is a preprocessor directive for the C language compiler. These directives give instructions to the compiler that are performed before the program is compiled. The directive `#include <stdio.h>` inserts additional statements in the program. These statements are contained in the file stdio.h. The file `stdio.h` is included with the standard C library. The `stdio.h` file contains information related to the input/output statement.

The `\n` in the last line of the program is C notation for the newline character. Upon printing, the cursor moves forward to the left margin on the next line. `printf` never supplies a newline automatically. Therefore, multiple `printf`'s may be used to output "I wrote a C-program" on a single line in a few steps. The escape sequence `\n` can be used to print three statements on three different lines. An illustration is given in the following:

```
# include <stdio.h>
void main (void )
{
        printf ("I wrote  a C-program  \n");
        printf ("This will be printed on a new line \n");
        printf ("So also is this line \ n" );
}
```

All variables in C must be declared before their use. The compiler provides an error message if one forgets a declaration. A declaration includes a type and a list of variables that have that type. For example, the declaration `int a,b` implies that the variables `a` and `b` are integers.

Next, write a program to add two integers a and b where a = 100 and b = 200. The C program is

```
# include <stdio.h>
void main ( )
{         int a = 0x64, b = 0xc8;  // a and  b  are specified in hex

          printf ("The sum is  %x\n", a+b);

}
```

This program shows how to declare two integers and initialize them with hexadecimal numbers. The format specifier %x allows the sum to be printed as a hexadecimal number. This program will print the sum as 0x12C which is 300 in decimal. The scanf allows the programmer to enter data from the keyboard. A typical expression for `scanf` is `scanf("%d%d", &a, &b);`

This expression indicates that the two values to be entered via the keyboard are in decimal. These two decimal numbers are to be stored in addresses `a` and `b`. Note that the symbol `&` is an address operator.

The C program for adding and subtracting two integers `a` and `b` using `scanf` is

```
// C program that performs basic I/O
# include <stdio.h>
void main ( )
{
        int a, b;
        printf ("Input two integers: ");
        scanf ("%d%d" , &a, &b) ;
        printf ("Their sum is: %d\n", a+b);
        printf ("The difference is: %d\n" , a - b);
}
```

In summary, writing a working C program involves four steps as follows:

Step 1: Using a text editor, prepare a file containing the C code. This file is called the "source file."

Step 2: Preprocess the code. The preprocessor makes the code ready for compiling. The preprocessor looks through the source file for lines that start with a #. In the previous programming examples, `#include <stdio.h>` is a preprocessor directive. This directive copies the contents of the standard header file `stdio.h` into the source code. This header file `stdio.h` describes typical input/output functions such as scanf( ) and `printf( )` functions.

Step 3: The compiler translates the preprocessed code into machine code. The output from the compiler is called object code.

Step 4: The linker combines the object file with code from the C libraries. For instance, in the examples shown here, the actual code for the library function `printf( )` is inserted from the standard library to the object code by the linker. The linker generates an executable file. Thus, the linker makes a complete program.

Before writing C programs, the programmer must make sure that the computer runs either the UNIX or MS-DOS operating system. Two essential programming tools are required. These are a text editor and a C compiler. The text editor is a program provided with a computer system to create and modify compiler files. The C compiler is also a program that translates C code into machine code.

In summary, the C language offers the following features:

- provides support to structured programming
- is portable and small size
- includes many operators for low and high level operations
- provides data structures such as arrays, strings, structures, and unions

10.2 Data Types

The data types in C language include `char,` `int,` `float`, and `double`. A variable declared as a `char` (character) usually holds eight bits of data. A variable of `int` (integer) type, on the other hand, can hold 16 or 32 bits of data. The type `float` specifies a 32-bit single precision floating point number. The type `double` can be used to declare a data as a 64-bit double precision floating point number. We will use only `char` and `int` data types in this book. Note that `float` and `double` data types are not needed in most of the microcontroller-based applications. In addition, floating-point computations are too costly in terms of space and time.

The qualifiers `unsigned` and `signed` can be used with `char` and `int` data types. The `unsigned char` is always positive, and covers a range of values from 0 to 255. Typical examples of `unsigned char` include age and memory address which are always positive.

The `signed char` covers a range of values from -128 to +127 (0 being positive). The C compilers use `signed char` as default. Hence, using `char` instead of `signed char` will specify the data as a `signed character`. Typical examples of `signed char` include voltage and temperature which can be positive and negative.

The `unsigned integer` covers a range of values from 0 to 65535 while the `signed integer` covers a range of values from - 32,768 to + 32767 (0 being positive). The `signed int` or simply `int` (default for C compilers) can be used to specify values from -32,768 to +32,767 (0 being positive).

The lengths of integers can be modified using the qualifiers `shortlong` and `long`. For typical C compilers, shortlong is 24-bit while long is 32-bit.

Examples of declaring `char` and `int` data types are provided below:

```
unsigned char i; /* specifies i as an unsigned 8-bit number*/
char x; /*declares x as a signed 8-bit number */
unsigned int b; /*declares b as unsigned 16-bit integer*/
int a; /* defines variable as 16-bit signed integer*/
```

10.3 Bit Manipulation Operators

C provides six bit manipulation operators as shown in Table 10.1. The applications is of these bit manipulation operators are discussed in Chapters 4 and 6.

Typical examples for each of these operators are provided below:

0x24 & 0x0F = 0x04
0x70 | 0x02 = 0x72
0xE1 ^ 0xFF = 0x1E
~ 0x25 = 0xDA
0x27 >>2 = 0x09

TABLE 10.1 Bit manipulation operations in C

Logic operators	Operation performed
&	AND
|	OR
^	XOR
! or ~	NOT
>>	Right Shift
<<	Left Shift

0xA1 << 3 = 0x08

The bit manipulation operators are very common in I/O operations. Hence, some examples showing their applications in bit manipulation for PIC18F I/O ports are provided in the following.

The AND operator is typically used for clearing one or more bits to 0. For example, the C statement `PORTC = PORTC & 0x7F;` will clear bit 7 of PORTC to 0 without changing the other bits of PORTC. The OR operator is typically used to set one or more bits to 1.

For example, the C statement `PORTD = PORTD | 0x05;` will insert 1's at bits 0 and 2 of PORTD without changing the other bits of PORTD. The XOR is typically used to find one's complement of (toggle) one or more bits. For example, the C statement `PORTC = PORTC ^ 0xFF;` will toggle all bits of PORTC.

The above three statements can also be specified in a compact form as shown below:

```
PORTC  &= 0x7F;
PORTD  |= 0x05;
PORTC  ^= 0xFF;
```

Next, let us discuss some of the applications of the logic operators.

The left shift operation is very useful for multiplying an unsigned number by 2^n by shifting it *n* times to left provided a '1' is not shifted out of the most significant bit. For example, consider y = 10*x. Note that 10_{10} in binary is 1010_2.
Also, 10*x = 8*x + 2*x. This means that 8*x = x<<3, and 2*x = x<<1.
Hence, y = (x<<3) + (x<<1).

The right shift operation, on the other hand, is very convenient for dividing an unsigned number by 2^n by shifting it *n* times to the right, provided a '1' is not shifted out of the least significant bit. As an example, consider y = (a + b)/2.

Note that y = (a + b)/2 = (a+b) >>1

Similarly, y = (a + b +c +d)/4 = (a + b + c + d) >>2

In order illustrate the Exclusive-OR (XOR) operator, consider X == Y. This expression is the same as !(X^Y). When X equals Y then X^Y = 0x00, and !(0x00) evaluates to true. The XOR can also be used to swap two variables without the need for a temporary variable. The following example illustrates this:

<u>SWAP with temporary variable</u>

```
temp = x;
    x = y;
    y = temp;
```

<u>SWAP without temporary variable</u>

```
y = x ^ y;
x = x ^ y;
y = x ^ y;
```

Let us verify the above using a numerical example. With x = 1001 and y = 0111, y = x ^ y = (1001)^(0111) = 1110, x = x ^ y = (1001) ^ (1110) = 0111, y = x ^ y = (0111)^(1110) = 1001. Hence, y = x. The above swapping of x with y works because of the following identity,

$p \oplus (p \oplus q) = q$

Also, the above statements can be represented in compact form as:

```
y ^= x;
x ^= y;
y ^= x;
```

Example 10.1 Write a C program to convert a 16-bit number, each byte containing an ASCII digit into packed BCD. The 16-bit number is stored in two consecutive locations (from LOW to HIGH) in data memory with the low byte pointed to by address 0x40, and the high byte pointed to by address 0x41. Store the packed BCD result in 0x70.

Solution

```
# include   <p18f4321.h>
void        main  ()
   {
            unsigned char  a, b, c;
            unsigned char  addr1 = 0x40;
            unsigned char  addr2 = 0x41;
            unsigned char  addr3 = 0x70;
            addr1 = addr1 &  0x0f;           // Mask  off upper four bits of  the low byte
            addr2 =  addr2 & 0x0f;           // Mask  off upper four bits of the high byte
            addr2 =  addr2 << 4;             // Shift high byte  4 times to left
            addr3 = addr2 | addr1;           // Packed BCD byte in addr3
   }
```

10.4 Control Structures

Control structures allow programmers to modify control flow which is a sequential flow by default. Structures allow one to make decisions and create loops which make the hardware to replicate execution of statements. Typical structured control structures in C include `if-else`, `switch`, `while`, `for` and, `do-while`.

10.4.1 The `if-else` Construct

The syntax for the `if-else` construct is as follows:

```
if (cond)
   statement1;
else
   statement2;
```

Figure 10.1 shows the flowchart for the `if-else` construct.

This is a one-entry-one-exit structure in that if the condition is true, the statement1 is executed; else (if the condition is false), statement1 is skipped, and the statement2 is executed. An example of the `if-else` structure (flowchart in Figure 10.2) is provided in the following:

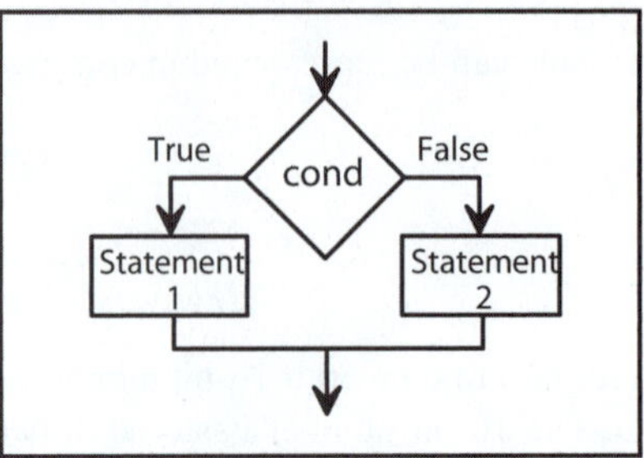

FIGURE 10.1 The `if-else` construct

```
unsigned char x, y, z;
if  (x < y)
    z = x + y;
else
z = x - y;
```

In the above, if x is less than y, the unsigned 8-bit numbers x and y are added, and the 8-bit result is stored in z. On the other hand, if x is not less than y, then the statement z = x - y is executed. As another example, consider the following. This code finds the larger of the two 8-bit unsigned numbers, a and b, and saves the result in big.

```
unsigned   char a, b, big;
  if (a>b)
    big = a;
else
    big = b;
```

The flowchart in Figure 10.3 illustrates the above example.

Braces are required if multiple statements need to be executed if the condition is true or false. Consider x and y as two unsigned numbers. If x>y, then find x+y, and x/y and store them in u and v respectively. Otherwise, find x-y and increment w by 4, and store them in z and w respectively. Finally, compute x*y, and store in t.

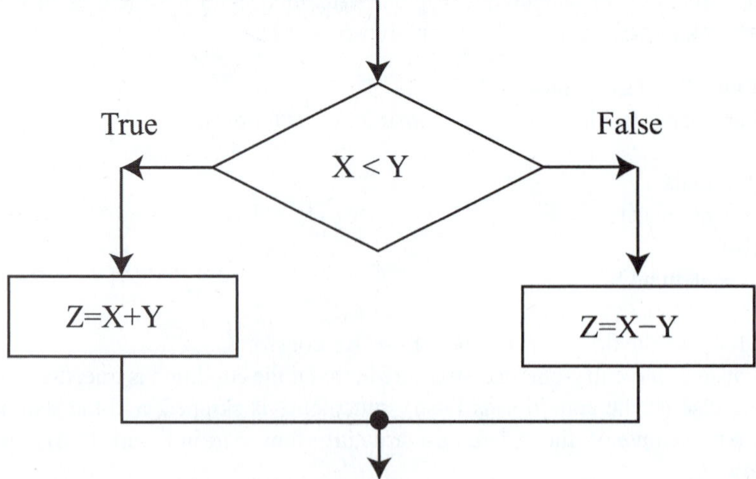

FIGURE 10.2 An example of if-else structure

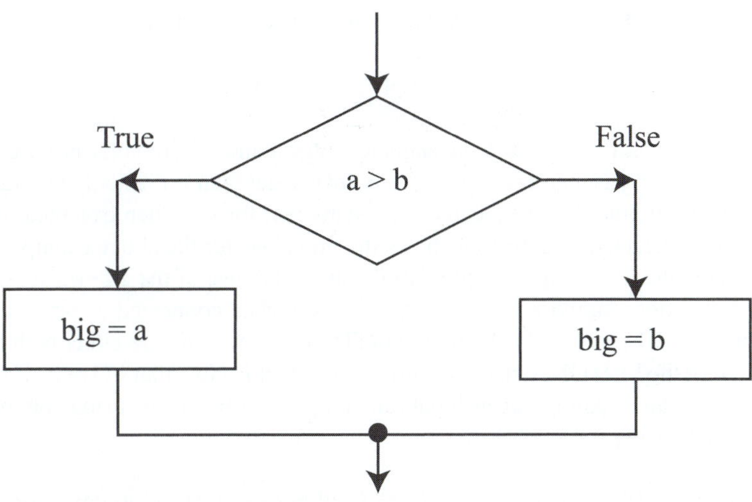

FIGURE 10.3 The if-else structure for finding the larger of two unsigned numbers

```
unsigned char t, u, v, w, x, y, z;
if   (x < y) {
        u = x + y;              // Add x with y, and store in u
        v = x / y;              // Divide x by y and store in v
}
else  {
        z = x - y;              // Subtract y from x and store result in z
```

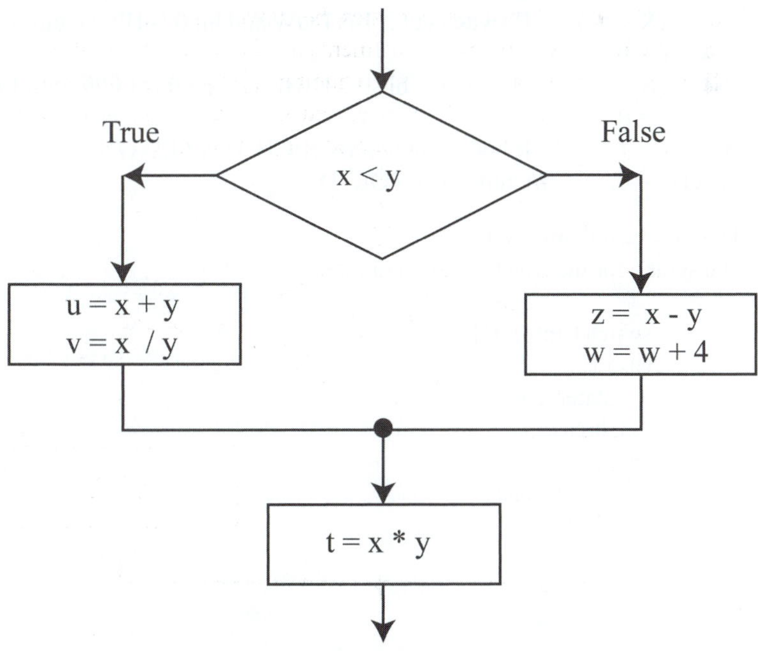

FIGURE 10.4 The if-else example for multiple statements

```
        w+= 4;                // Value of w is incremented by 4
    }
        t = x * y             // Multiply x by y and store in t
```

In the above, if x < y, then the statements for u and v in the braces are executed. The statement t = x * y is then executed. If x is not greater than y, then only the statements for z and w+ in the braces are executed. The statement for t is then executed. In either case, t = x * y is executed. Figure 10.4 shows the flowchart for the above example.

Finally, the following example will illustrate the use of the `if-else` construct in an I/O application. Suppose that there are two switches connected at bits 0 and 1 of PORTC, and an LED connected at bit 0 of PORTD. If the switches are either both LOW or both HIGH, turn the LED ON; otherwise, turn the LED OFF. Note that I/O ports C and D of the PIC18F4321 can be configured as inputs and outputs, respectively, using the following C language statements:

```
TRISC = 0xFF;          // Configure PORTC as an input port
TRISD = 0x00;          // Configure PORTD as an output port
```

Next, using the else-if construct, the following C program will accomplish this:

```
unsigned char    X, Y;
TRISC = 0xFF;    //Configure PORTC as an input port
TRISD = 0x00;    //Configure PORTD as an output port
X = PORTC;       //Input switches via PORTC
X& = 0x03;       //Mask all bits except bits 0, 1 and retain switch values
if (X == 0)      //If both switches are LOW, turn LED ON
      Y = 1;
else if (X == 1) //If switch at bit 1 is LOW and bit 0 is HIGH, turn LED OFF
      Y = 0;
else if (X == 2) //If switch at bit 1 is HIGH and bit 0 is LOW, turn LED OFF
      Y = 0;
else   Y = 1;    //If both switches are HIGH, turn LED ON
PORTD = Y;       //Output Y to PORTD
```

10.4.2 The `switch` Construct

The syntax for the `switch` expression is

```
switch ( integer){
case 1:
    statements;
    break;
case 2:
    statements;
    break:
        ---
        ---
        ---
```

```
    case n:
        statement;
    }
```

In the above, the integer included with the switch statement is compared with the each of the integers included with the case statements. If the values match, the statements associated with that case statement are executed. For example, consider switch (2). If integer = 2, then the statements with case 2 are executed. The break exits from the switch construct. Example 10.3 on seven-segment display illustrates the use of the switch construct. The if-else statements of the switch/LED example of the last section can be replaced using the switch construct as follows:

```
switch (X) {
 Case 0: Y = 1; break;
 Case 1: Y = 0; break;
 Case 2: Y = 0; break;
 Case 3: Y = 1;
}
```

10.4.3 The `while` Construct

The `while` construct allows a programmer to describe loops. The syntax for the `while` construct is provided below:

```
while (condition) {
      statements
}
next statement;
```

In the above, if the condition is true, the statements are executed, and then control is returned to the top of the loop. The condition is tested again. If the condition is true again, statements are executed, and control is returned to the top of the loop. The process is repeated as long as the condition is true. However, if the condition is false at the start or during repeating the process when the condition is checked, the statements in the braces are not executed. The next statement following the second brace is executed. Note that braces are not required for a single statement. Figure 10.5 depicts the flowchart for the while construct.

As an example, consider the following with a single statement in the loop:

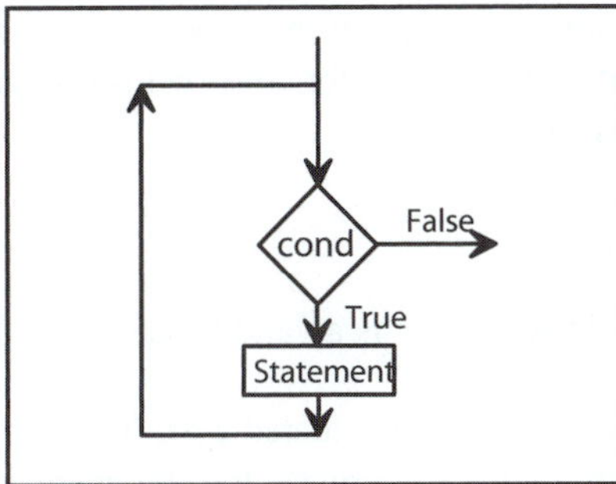

FIGURE 10.5 The `while` construct

```
int  n = 4;
while  ( n <= 16)
        n += 4;
z = n;
```

In the above, n is 4 before entering the while loop. The condition n <= 16 is true the first time. The expression n = n + 4 continuously changes the value of n in increments of 4 to 8, 12, 16, 20. As soon as n = 16, the condition n <= 16 is false, and the next statement z = n is executed, which assigns z with the value of 20.

Finally, note that infinite loop occurs when the condition in the `while` construct is always true. Note that

```
while  (1)
   ;
```

describes an infinite loop in C , and is equivalent to
`here bra here` in PIC18F assembly language.

The `while` loops can also be used to write software delay routines. A simple delay loop using the `while` construct is provided below:

```
unsigned  int  k = 1000; // initialize  k  to 1000
   while (k>10)
         k -- ;
```

A nested delay loop using the while construct is provided below:

```
unsugned  int i, j, k ;
               i  = 0; k = 1000;
               while (i < k) {
                  j = 1;
                  while (j < 100)
                  j++;
                  i++;
                  }
```

10.4.4 The `for` Construct

The `for` construct is another loop structure supported by the C language. It is more flexible than the while construct. Hence, the for construct is often a preferred choice. The syntax of a `for` construct is as follows:

```
for  (e1; e2; e3)
      s;
```

where e1, e2, and e3 are C expressions and 's' is a valid C statement. Figure 10.6 shows the flowchart for the `for` loop.

In Figure 10.6, the expression e1 is evaluated first. The expression e2 is then evaluated; if it is false, the loop terminates. Otherwise, the statement 's' is executed. The

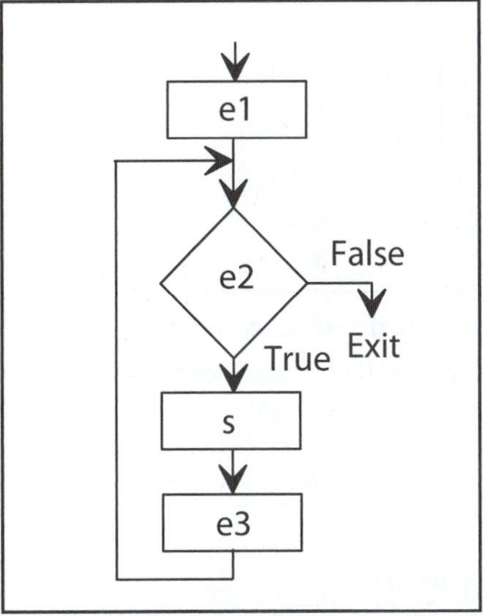

FIGURE 10. 6 Flowchart for the `for` loop

expressions e3 and e2 are then evaluated, and the process continues. The expression e1 normally contains code to initialize the loop. The expression e2 describes the exit condition. The purpose of the expression e3 is to modify the exit condition so that the loop terminates at some point. The following 'while' loop

```
int  n = 4;
while  ( n <= 16)
     n = n + 4;
     z = n;
```

described earlier can be replaced with the 'for' loop provided below:

```
int  n;
for  ( n = 4; n<= 16; n+ = 4)
     ;
     z = n;
```

Note that for (; ;) is an infinite loop in C, and is equivalent to

`here   bra   here` in PIC18F assembly language.

The single loop and nested loop delay routines can be written using the `for` construct provided in the following:

Simple delay routine using `for` with a single loop:

```
          unsigned int k;
for ( k = 1000 ;  k>0 ; k--)
          ;
```

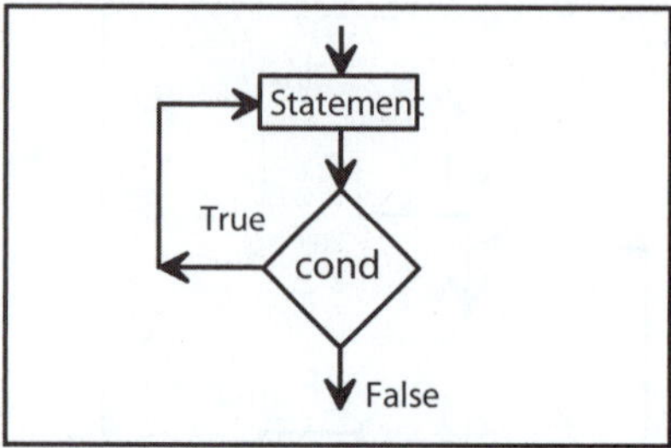

FIGURE 10.7 Figure for the `do-while` construct

Delay routine using 'for' with nested loop:

```
unsigned int i, j, k = 1000;
          for (i = 0 ; i < k ; i++)
              for (j = 0 ; j<100 ; j++)
                        ;
```

10.4.5 The `do-while` Construct

The C language provides another loop structure called the `do-while`. The syntax for the `do-while` construct is provided below:

```
do {
        statements
} while (condition);
```

Figure 10.7 shows the flowchart for the `do-while` construct.

The `do-while` loop is a post-checked loop in the sense that the exit condition is evaluated after executing all statements. If the condition evaluates to False, the loop terminates. Otherwise, the process continues until the condition becomes false. Whatever is accomplished by a `do-while` construct can be achieved by a `for` loop construct. Hence, this loop structure is not as popular as `while` or `for` loop structures.

10.5 Structures and Unions

In addition to built-in data types such as `char` and `int`, C supports user-defined nonhomogeneous collections. A structure permits the programmer to access a group of different data types using a common user-defined name. The structure can be declared using the keyword `struct` followed by a user-defined name. An example of the structure declaration is provided below:

```
struct    struct_name {
          int a;
          char b ;
} my_struct ;
```

Note that in the above, the `name` is optional. However, if it is present, it defines the tag or user-defined name of the structure, and the tag can be used later. The variable name `my_struct` is of type `struct_name`. Also, Size of the structure = Sum of the sizes of its componenets. Hence, Size of the above structure = Size(int) + Size (char) = 2 bytes + 1 byte = 3 bytes.

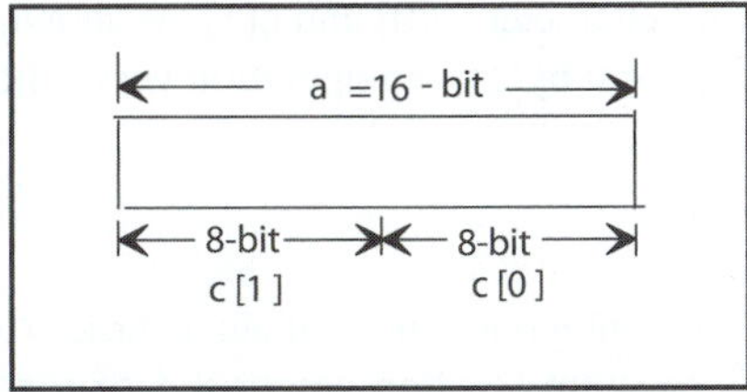

FIGURE 10.8 Example of space saving with union

Union is a space-saving structure. The memory referenced by the keyword `union` can store different types of data with the restriction that at any one time, the memory holds a single type of data. Note that different data types of the union share the same memory space. The data in a union must be referenced by a member of the proper data type. As an example, a union can be declared as follows:

```
union   num    {
        char   x ;
        int    a ;
} ;
```

Note that Size of a union = Size of the largest data type held by the union. In the above, Size of the union "num" = Max {size of (int), size of (char)} = Max {2, 1} = 2 bytes.

In the above code, any variable of type `union` num can hold a char or an int. The names 'x' and 'a' specify which type of data is referenced. To specify a variable `y` of type `union num`, the following statement can be used:

```
union num y;
```

Now, a character B can be assigned to x using the following:

```
y . x = 'B' ;
```

Next, an integer 2756 can be assigned to y using the following:

```
y . a = 2756;
```

Since at any one time y holds data of a single type, the assignment of 2756 to y cancels the assignment of 'B' to y.

As mentioned before, Union offers a space saving structure. For union variable, memory space is allocated to hold the largest number. This memory can then be used to hold the smaller numbers. As an example. consider the following:

```
union   {
        int a ;
        char c[2] ;
} myu ;
```

In the above union, 16 bits of memory space are allocated to integer 'a'; the same storage space is shared by 8-bit characters c[0] and c[1], as shown in Figure 10.8. This union allows a prgrammer to perform byte swap operation very efficiently.

10.6 Functions in C

A function in C allows a programmer to encapsulate a task. Hence, a function is a task-specialized module. A C program is often comprised of a collection of functions. Functions are written and tested separately before they are added to the library. The end user can use the functions in a library as many times as needed. Since a function is tested thoroughly prior to placing in the library, its use not only reduces program development time, but also increases the software reliability.

As an example, consider the add function shown below:

```
int add (int p , int q){
int  t ;
        t = p + q;
        return  t;
}
```

This function inputs two integers (p and q), and returns the sum of p and q via the local variable "t" of the function. Note that the function add (10, 12) will return 22 as the answer.

As another example, consider the following function which returns the number of ones in a given byte x:

```
unsigned char count (unsigned char x) {
unsigned char c;
for (c=0 ; x! = 0 ; x>>1)
        if (x &0x01)
                c++;
return c;
}
```

In many situations, a function does not have to return a value. A typical example is the software delay loop provided below:

```
void delay (unsigned int p) {
     unsigned int i ;
  for ( i = 0; i<p; i++)
        ;
}
```

The reserve word void indicates that this function does not return any result to the caller.

10.7 Macros

Macros can make programming in C easier by reducing the amount of code the programmer has to actually write. This is accomplished by letting the compiler produce those redundant pieces of program that are used routinely in various places in the program. Basically a macro is a set of codes used repeatedly throughout the program. When the macro is written, it is assigned a name. Rather than writing the same sequence of codes each time they are needed, the name of the macro is inserted in their place. During compiling, whenever the name of the macro is encountered, the compiler will insert the sequence of codes that this name represents.

The macro is in some way similar to a function, in that the usage is repeated. As an example, consider the following macro:

```
#define   MPY10 (X)  (X << 3) + (X<< 1)
```

This macro multiplies X by 10 using two left shift operations discussed earlier. A typical macro call is Z = MPY10 (Y);

Note that there is no transfer of control in macro. In contrast, a function call is performed via transfer of control in the program. The C compiler expands the macro in the program. For example,

```
Z = MPY10 (A)  + MPY10 (B)       is expanded as
Z = (A<<3) + (A<< 1)  + (B<<3)  + (B<<1);
```

Macros are faster than functions or subroutines since there is no overhead in macro associated with transfer of control.

10.8 Configuring PIC18F4321 I/O Ports Using C

In Chapter 8, we discussed how to configure the PIC18F4321 I/O ports as inputs and outputs in assembly language. For example, the SETF or CLRF instructions can be used to make all bits of Port C and Port D as inputs and outputs as follows:

```
SETF      TRISC  ; Set all bits in TRISC to 1's and configure
                 ; Port C as an input port.
CLRF      TRISD  ; Clear all bits in TRISD to 0's and configure
                 ; Port D as an output port
```

Using C language, the PIC18F assembly language instruction sequence can be replaced by the following statements:

```
TRISC = 0xFF;        // Configure PORT C as an input port
TRISD = 0 ;          // Configure PORT D as an output port
```

As mentioned in Chapter 8, configuring Port A, Port B and Port E is different than configuring Port C and Port D. This is because certain bits of Port A, Port B, and Port E are multiplexed with analog inputs. For example, bits 0 through 3 and bit 5 of Port A are multiplexed with analog inputs AN0 through AN4, bits 0 through 4 of Port

B are multiplexed with analog inputs AN8 through AN12, and bits 0 through 2 of Port E are multiplexed with analog inputs AN5 through AN7 (Figure 8.1). When a port bit is multiplexed with an analog input, then bits 0-3 of a special function register (SFR) called ADCON1 (A/D Control Register 1 with mapped data memory address 0xFC1) must be used to configure the port bit as input. The other bits in ADCON1 are associated with the A/D converter. Figure 8.9 shows the ADCON1 register along with the associated bits for digital I/O. When bits 0 through 3 of the ADCON register are loaded with 1111, the analog inputs (AN0- AN12) multiplexed with the associated bits of Port A, Port B, and Port E are configured as digital I/O. This will also make these port bits as inputs automatically; the corresponding TRISx registers are not required to configure the ports. However, for configuring these ports as outputs, the corresponding TRISx bits must be loaded with 0's; the ADCON1 register is not required for configuring these port bits as outputs. The following examples will illustrate this.

For example, the following C statement will configure all 13 port bits multiplexed with AN0 - AN12 as inputs:

```
ADCON1 = 0x0F ; // Configure 13 bits of Ports A, B, and E as inputs
```

Note that the TRISx registers associated with Ports A, B, and E can be used to configure these ports as outputs.

Next, in order to configure PORTA and bit 4 of PORTB as outputs in PIC18F assembly language, the following instruction sequence can be used:

```
BCF          TRISA, 1     ; Configure bit 1 of PORTA as output
```

The MPLAB C18 compiler provides built-in unions for configuring a port bit. This allows the programmer to address a single bit in a port without changing the other bits in the port. For example, bit 2 of Port C can be configured as an output by writing a '1' at bit 2 of TRISC as follows:

```
# define portbit  PORTCbits.RC2   // Declare a bit (bit 2) of Port C
TRISCbits.TRISC2 = 0 ;            // Configure bit 2 of Port C as an output
```

Now, a '1' can be output to bit 2 of Port C using the following statement:

```
portbit = 1;
```

Similarly, the statement, `portbit = 0;` will output a '0' to bit 2 of Port C.

Next, bit 3 of Port D can be configured as an input by writing a '0' at bit 3 of TRISD as follows:

```
# define portbit  PORTDbits.RD3   // Declare a bit (bit 3) of Port D
TRISDbits. TRISD3 = 1 ;           // Configure bit 3 of Port D as an input
```

Example 10. 2 Assume PIC18F4321. Suppose that three switches are connected to bits 0-2 of Port C and an LED to bit 6 of Port D. If the number of HIGH switches is even, turn the LED on; otherwise, turn the LED off. Write a C language program to accomplish this.

Solution

The C language program is shown below:

```
#include <p18f4321.h>
#define   portc0  PORTCbits.RC0
#define   portc1  PORTCbits.RC1
#define   portc2  PORTCbits.RC2
void main (void)
{
        unsigned char mask = 0x07;              // Data for masking off upper 5 bits
                                                // of Port C
        unsigned char masked_in;
        unsigned char xor_bit;
        TRISC = 0xFF;                           // Configure Port C as an input port
        TRISD = 0;                              // Configure Port D as an output port
        while(1)
        {
            masked_in  =  PORTC ^ mask;         // Mask input bits
            if (masked_in == 0)
            PORTD = 0x40;                       // For all low switches (even), turn led on
            else
            xor_bit = portc0 ^ portc1 ^ portc2; // Xor input bits
            if (xor_bit == 0)                   // For even # of high switches,
            PORTD = 0x40;                       // turn led on

        else
            PORTD = 0;                          // For odd # of high switches, turn led off
        }
}
```

Example 10.3 Assume PIC18F4321. Suppose that it is desired to input a switch connected to bit 4 of Port C, and then output it to an LED connected to bit 2 of Port D.

Solution

The following C code will accomplish this:

```
# include  <P18F4321.h>
# define portc_bitin  PORTCbits.RC4     // Declare a bit (bit 4) of Port C
# define portd_bitout  PORTDbits.RD2  // Declare a bit (bit 2) of Port D
        void main  (void )
        {
        TRISCbits.TRISC4 = 1;           // Configure bit 4 of Port C as an input bit
        TRISDbits.TRISD2 = 0;           // Configure bit 2 of Port D as an output bit
        while  (1)                      //  Halt
                                        {
        portd_bitout = portc_bitin;     // Output switch to LED
```

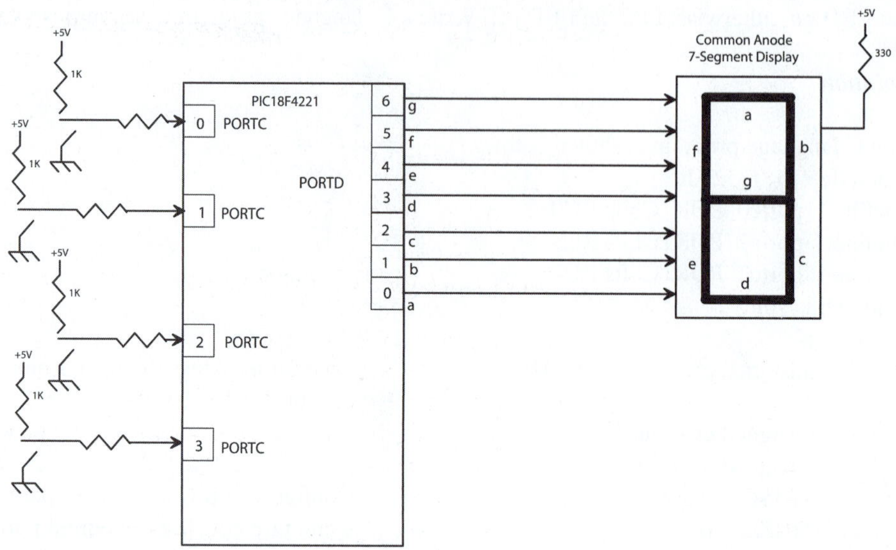

FIGURE 10.9 Figure for Example 10.4

}

}

Example 10. 4 The PIC18F4321 microcontroller shown in Figure 10.9 is required to output a BCD digit (0 to 9) to a common-anode seven-segment display connected to bits 0 through 6 of Port D. The PIC18F4321 inputs the BCD number via four switches connected to bits 0 through 3 of Port C. Write a C language program that will display a BCD digit (0 to 9) on the seven-segment display based on the switch inputs.

Solution

The C code is provided below:

```
#include <p18f4321.h>
void main ()
unsigned char input;
unsigened char code[10] = {0x40, 0x79, 0x24, 0x30, 0x19, 0x12, 0x03, 0x78, 0x00, 0x18};
unsigned char oput;
TRISD = 0;              //Configure PortD as Output
TRISC = 0xFF;           //Configure PortC as Input
while (1) {
        input = PORTC & 0x0F;
        oput = Code [input];
        PORTD = oput;
        }
```

In the above, the last two lines can be combined as PORTD = Code [input];

In the above program, first the PORTB is set as an output port and PORTC is set as an input port. A variable 'input' is then declared. The program moves to an infinite 'while' loop where it will first take the input from the four switches via PORTC, and mask the

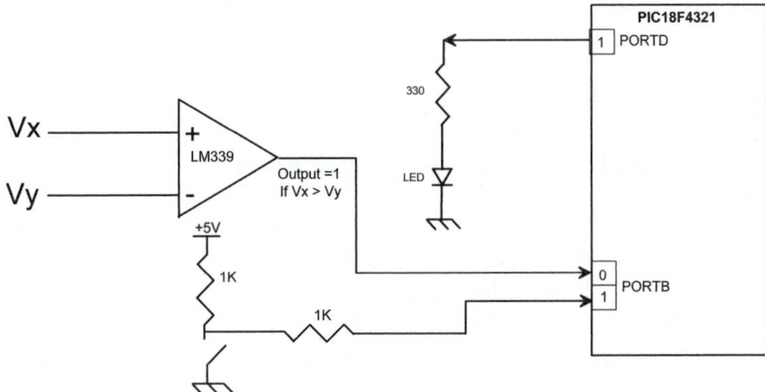

FIGURE 10.10 Figure for Example 10.5

first four bits. An unsigned char array code is set up in order to contain the seven-segment code for each decimal digit from 0 through 9. The input is used as the index to the array, and the corresponding LED code is sent to PORTD. The code then repeats this process and displays the proper digit on the seven-segment display based on switch inputs.

Example 10.5 Assume that the PIC18F4321 micrococontroller shown in Figure 10.10 is required to perform the following:

If Vx > Vy , turn the LED ON if the switch is open; otherwise, turn the LED OFF. Write a C program to accomplish the above by inputting the comparator output via bit 0 of port B.

Solution

The C program is provided below:

```
#include <p18f4321.h>

void main (void)
{
TRISD = 0; //PORTD is output
ADCON1 = 0x0F; //Configure for PORTB to be digital input
PORTD = 0; //Turn LED OFF
while(1)
{
        PORTD = 0; //Turn LED OFF
        while(PORTBbits.RB0 ==1) //While Vx > Vy
        {
                if(PORTBbits.RB1==1)
                        PORTD = 1; //Turn LED ON
                else if (PORTBbits.RB1== 0)
                        PORTD = 0; //Turn LED OFF
        }
}
}
```

In the above code, the register ADCON1 is used to configure Port B. Within the infinite `while` loop, the code checks to see when the comparator output is one indicating Vx > Vy. The LED is then turned ON or OFF based on the state of the switch.

10.9 Programming PIC18F4321 Interrupts Using C

The PIC18F4321 interrupts are covered in detail in Section 8.3 of chapter 8. The PIC18F4321 interrupts can be classified into two groups: high-priority interrupt levels and low-priority interrupt levels. The high-priority interrupt vector is at address 0x000008 and the low-priority interrupt vector is at address 0x000018 in the program memory. High-priority interrupt events will interrupt any low-priority interrupts that may be in progress.

As mentioned before, upon power-on reset, the interrupt address vector is 0x000008 (default), and no interrupt priorities are available. The IPEN bit (bit 7 of the RCON register) of the RCON register in Figure 8.5 can be programmed to assign interrupt priorities. Upon power-on reset, IPEN is automatically cleared to 0, and the PIC18F operates as a high-priority interrupt (single interrupt) system. Hence, the interrupt vector address is 0x000008. During normal operation, the IPEN bit can be set to one by executing the RCONbits.IPEN=1; to assign priorities in the system.

When interrupt priority is enabled (IPEN = 1), there are two bits which enable interrupts globally. Setting the GIEH bit (bit 7 of INTCON register of Figure 8.16) enables all interrupts that have the priority bit set (high priority). Setting the GIEL bit (bit 6 of INTCON register of Figure 8.16) enables all interrupts that have the low priority. When the interrupt flag, enable bit, and appropriate global interrupt enable bit are set, the interrupt will vector immediately to address 0x000008 or 0x000018, depending on the priority bit setting. Individual interrupts can be disabled through their corresponding enable bits.

Note that the C18 compiler does not allow the program to automatically jump to the interrupt service routine from the interrupt address vector. Hence, the PIC18F assembly language instructions `GOTO` or `BRA` must be used to jump to the interrupt service routine.

If interrupt priority levels are used, high-priority interrupt sources can interrupt a low priority interrupt. Low-priority interrupts are not processed while high-priority interrupts are in progress. The return address is pushed onto the stack and the PC is loaded with the interrupt vector address (0x000008 or 0x000018). Once in the Interrupt Service Routine, the source(s) of the interrupt must be determined for the priority interrupt system by polling the interrupt flag bits. The interrupt flag bits must be cleared in software before re-enabling interrupts to avoid recursive interrupts. In order to jump to the interrupt service routine from the interrupt address vector such as 0x000008 or 0x000018, the programmer should first check the interrupt flag bits to find the source of interrupt in a priority interrupt system, and then use the GOTO or BRA instruction of the assembly language to jump to the interrupt service routine.

Based upon discussion on interrupts in Section 8.3 of Chapter 8, INT0 can be initialized to recognize interrupts using the following C code:

```
ADCON1=0x0F;            //Configure PORTB to be digital input
                        //since PORTB contains interrupt pins
INTCONbits.INT0IE=1;    //Enable external interrupt
INTCONbits.INT0IF=0;    //Clear the external interrupt flag
INTCONbits.GIE=1;       //Enable global interrupts
```

Based upon detailed coverage of interrupts in Section 8.3 of Chapter 8, INT0 (High Priority) and INT1 (Low Priority) can be initialized to recognize interrupts using the following C code:

```
ADCON1=0x0F;                    //Configure PORTB to be digital input
                                //Since PORTB contains interrupt pins
INTCONbits.INT0IE=1;            //Enable external interrupt INT0
INTCON3bits.INT1IE=1;           //Enable external interrupt INT1
INTCONbits.INT0IF=0;            //Clear INT0 external interrupt flag
INTCON3bits.INT1IF=0;           //Clear INT1 external interrupt flag
INTCON3bits.INT1IP=0;           //Set INT1 to low priority interrupt
RCONbits.IPEN=1;                //Enable priority interrupts
INTCONbits.GIEH=1;              //Enable global high priority interrupts
INTCONbits.GIEL=1;              //Enable global low priority interrupts
```

10.9.1 Specifying Interrupt Address Vector using the C18 Compiler

As mentioned before, using the MPLAB assembler, the programmer uses the ORG directive to specify the starting address of a program or data. Using the C18 C compiler, the programmer can use the directive `#pragma code begin` to specify an address to a program or to data at address `begin`. For example, the C statement `#pragma code int_vect = 0x000008` will assign the address 0x000008 to label `int_vect`. Note that `pragma` and `code` are keywords of the C18 compiler.

10.9.2 Assigning Interrupt Priorities Using the C18 Compiler

The C18 compiler uses the keywords `interrupt` and `interruptlow` to specify high- and low-priority interrupt levels. Note that the PIC18F interrupt address vectors for the high- and low-priority levels are 0x000008 and 0x000018 respectively. The programmers can use these keywords which allow a program to branch automatically from the respective interrupt address vector to a different program to find the source of the interrupt (for multiple interrupts with priorities), and then to the appropriate service routine.

10.9.3 A Typical Structure for Interrupt Programs Using C

The default interrupt INT0 with vector address 0x000008 is used in the following to illustrate the interrupt programs using C. Typical structures for the main program and the service routine are provided below:

```
#include <P18F4321.h>
void     ISR (void);
#pragma code Int=0x08  //At interrupt code jumps here
void Int(void)
{
_asm  //Using assembly language
GOTO ISR
_endasm
}
#pragma code //  Main program
void main( )
```

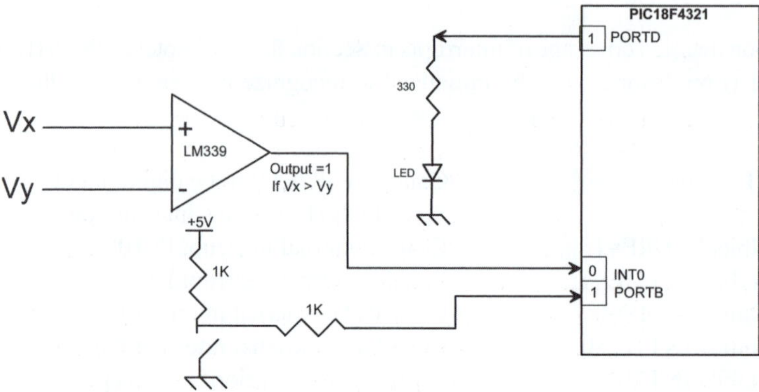

FIGURE 10.11 Figure for Example 10.6

```
{
// Typically configure ports, enable INT0IE, clear INT0IF

while(1){          // Wait in infinite loop for the interrupt to occur

}

#pragma   interrupt  ISR
void ISR(void)   // Start of Interrupt service routine
        {
                  // clear INT0IF, and then write the service
        }
}
```

In the above, the `#pragma` directive will place the code fragments at specific locations in memory. The main program typically configures ports, enables interrupt, clears interrupt flag bit and waits in an infinite loop for the interrupt to occur. When the interrupt occurs and is recognized by the PIC18F4321, the program will automatically jump to memory location 0x000008. Note that the C18 compiler does not allow the program to automatically jump to the interrupt service routine from the interrupt address vector 0x000008. Hence, the PIC18F assembly language instructions `GOTO` or `BRA` must be used to jump to the interrupt service routine. The `GOTO ISR` is used for this purpose. The program must include the statement `#pragma interrupt` ISR. The keyword `interrupt` will jump to the ISR service routine. The keyword `interrupt` will also insert RETFIE at the end of the service routine which will return control to the main program.

Example 10.6 Assume that the PIC18F4321 micrococontroller shown in Figure 10.11 is required to perform the following:

If Vx > Vy , turn the LED ON if the switch is open; otherwise, turn the LED OFF. Write a C program to accomplish the above by interrupting the PIC18F4321 by the comparator output via INT0. Also, write the main program in C which will initialize Port B and Port D, and then wait for interrupt in an infinite loop.

Solution

In this example, an LM339 comparator is interfaced with the PIC18F4321 using C and external interrupts. An external interrupt allows the microcontroller to trigger an interrupt from a source outside the PIC18F4321 such as a comparator. As with the previous interrupt example, the code starts with the #pragma command which will place code fragments at specific locations in memory, and when the interrupt is triggered, the microcontroller will automatically jump to memory location 0x000008, and then to COMP_ISR.

In the main program, PORTD is configured as an output, and PORTB is configured as a digital input. The external interrupt flag is cleared to 0, and the global interrupt is enabled. The main program then waits in an infinite 'while' loop that turns the LED OFF until the comparator output is HIGH, interrupting the microcontroller. After recognizing the interrupt, the code will automatically jump to address 0x000008, and then jump to the service routine at COMP_ISR via the code at COMP_int. Within the service routine, the code will continue to take the switch data from PORTB and output the data to the LED via PORTD. It will continue to do this as long as the comparator output stays HIGH. When Vx is lower than Vy, the comparator will output 0V and the code will return to the infinite 'while' loop and turn the LED OFF.
The PIC18F program using C is provided below:

```
#include <P18F4321.h>
void COMP_ISR (void);
#pragma code COMP_Int=0x08  //At interrupt code jumps here
void COMP_Int(void)
{
_asm  //Using assembly language
GOTO COMP_ISR
_endasm
}
#pragma code
void main( )                    //Start of the main program
{
TRISD=0x00;                     //PORTD is output
ADCON1=0x0F;                    //Configure PORTB to be digital input
INTCONbits.INT0IE=1;            //Enable external interrupt
INTCONbits.INT0IF=0;            //Clear the external interrupt flag
INTCONbits.GIE=1;               //Enable global interrupts
PORTD=0;                        //Turn off LED;

while(1){                       //Wait in an infinite loop for the interrupt to occur
        PORTD=0;                //LED is off
}

}
#  pragma  interrupt COMP_ISR
void COMP_ISR(void)     //Start of the Comparator interrupt service routine
        {
```

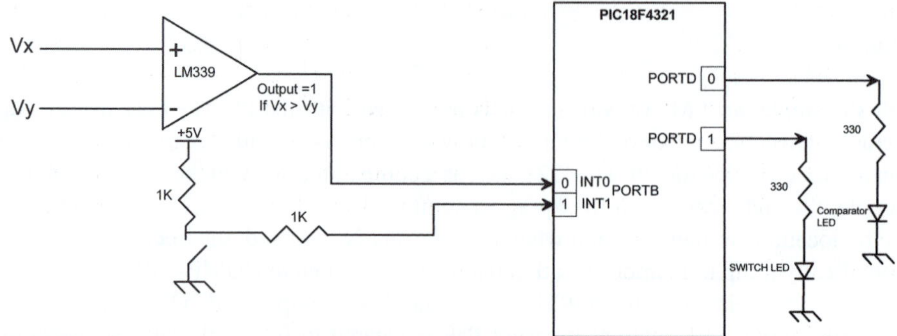

FIGURE 10.12 **Figure for Example 10.7**

```
        INTCONbits.INT0IF=0; //Clear external interrupt flag
                while(PORTBbits.RB0==1){         //Check if comparator is high
                        PORTD = PORTB;           //Move PORTB into PORTD
        }
}
```

Example 10.7 In Figure 10.12, if Vx > Vy, the PIC18F4321 is interrupted via INT0. On the other hand, opening the switch will interrupt the microcontroller via INT1. Note that in the PIC18F4321, INT0 has the higher priority than INT1. Write the main program in C that will perform the following:

- Configure PORTB as interrupt inputs.
- Clear interrupt flag bits of INT0 and INT1.
- Set INT1 as low priority interrupt.
- Enable IEN in INTCON3
- Enable global HIGH and LOW interrupts.
- Turn both LEDs at PORTD OFF.
- Wait in an infinite loop for one or both interrupts to occur.

Also, write a service routine for the high priority interrupt (INT0) in C that will perform the following:

- Check to see if the comparator output is still 1. If it is, turn LED at bit 0 of PORTD ON. If the comparator output is 0, return.

Finally, write a service routine for the low priority interrupt (INT1) in C that will perform the following:

- Clear interrupt flag for INT1
- Check to see if the switch is still 1. If it is, turn LED at bit 1 of PORTD ON. If the switch input is 0, return.

Solution

This example will demonstrate the interrupt priority system of the PIC18F microcontroller. Using interrupt priority, the user has the option to have various interrupts assigned as either low-priority or high-priority interrupts. If a low-priority interrupt and a high-priority interrupt occur at the same time, the PIC18F will always service the high priority interrupt first.

In the above example, the high priority is assigned to the comparator while the switch is assigned with the low priority. Hence, if both interrupts were triggered at the same time, the LED associated with the comparator would be turned ON first, and then the LED associated with the switch will be turned ON. Note that the external interrupt INT0 can only be a high-priority interrupt. Hence, INT0 is connected to the comparator output while the switch is connected to INT1 since it has the low priority. At the end of the code provided below, it can be seen that there are two interrupt service routines, HP_COMP_ISR and LP_SWITCH_ISR, which are the high-priority and low-priority service routines.

The following code implements priority interrupts on the PIC18F using C:

```
#include <P18F4321.h>

void HP_COMP_ISR (void);
void LP_SWITCH_ISR(void);

#pragma code High_Priority_COMP_Int=0x08  //High interrupt code jumps here
void COMP_Int(void)
{
_asm //Using assembly language
GOTO  HP_COMP_ISR
_endasm
}

#pragma code Low_Priority_SWITCH_Int=0x018  //Low interrupt code jumps here
void Switch_Int(void)
{
_asm  //Using assembly language
GOTO  LP_SWITCH_ISR
_endasm
}

#pragma code

void main( )
{
TRISBbits.TRISB0=1;            //Set pin 0 of PORTB as input
TRISBbits.TRISB1=1;            //Set pin 1 of PORTB as input
TRISD=0x00;                    //PORTD is output
ADCON1=0x0F;                   //Configure PORTB to be digital input
INTCONbits.INT0IE=1;           //Enable external interrupt INT0
INTCON3bits.INT1IE=1;          //Enable external interrupt INT1
INTCONbits.INT0IF=0;           //Clear INT0 external interrupt flag
INTCON3bits.INT1IF=0;          //Clear INT1 external interrupt flag
INTCON3bits.INT1IP=0;          //Set INT1 to low priority interrupt
RCONbits.IPEN=1;               //Enable priority interrupts
INTCONbits.GIEH=1;             //Enable global high priority interrupts
INTCONbits.GIEL=1;             //Enable global low priority interrupts
PORTD=0;                       //Turn off LED;

while(1){
```

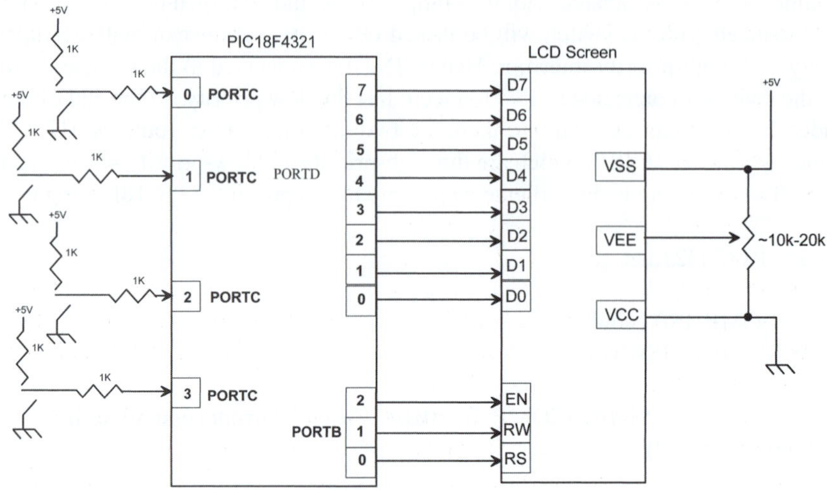

FIGURE 10.13 PIC18F4321 interface to LCD and switches

```
        PORTD=0;                //LED is off
}

}
#pragma interrupt HP_COMP_ISR
void HP_COMP_ISR(void){                 //High-priority interrupt service
        INTCONbits.INT0IF=0;            //Clear external interrupt flag
        while(PORTBbits.RB0==1)         //Check if comparator is high
{
                PORTD=0x01;             //Turn on LED
        }
}

#pragma interrupt low LP_SWITCH_ISR
void LP_SWITCH_ISR(void){               //Low-priority interrupt service
        INTCON3bits.INT1IF=0;           //Clear external interrupt flag
        while(PORTBbits.RB1==1)         //Check if switch is still on
         PORTD=0x02;                    //Turn on LED

}
```

10.10 Programming the PIC18F4321 Interface to LCD Using C

The PIC18F4321 is interfaced to the Optrex DMC 16249 LCD in Section 8.4 of Chapter 8, and the programs are written using PIC18F assembly language. In this section, the same program will be written using C. For convenience, some of the concepts described in Chapter 8 will be repeated in this section.

Note that the PIC18F4321 is also interfaced to the seven-segment LED display in Chapter 8. The seven-segment LEDs are easy to use, and can display only numbers and limited characters. An LCD is very useful for displaying numbers and several ASCII characters along with graphics. Furthermore, the LCD consumes low power. Because of inexpensive price of the LCDs these days, they have been becoming popular. The LCDs are widely used in notebook computers.

Figure 10.13 (same as Figure 8.23, redrawn for convenience) shows the PIC18F4321's interface to a typical LCD display such as the Optrex DMC16249 LCD with a 2-line x 16-character display screen. As with the PIC18F assembly language program, the C program is written to display the phrase "Switch Value:" along with the numeric BCD value (0 through 9) of the four switch inputs.

The Optrex DMC16249 LCD shown in Figure 10.13 contains 14 pins. The VCC pin is connected to +5 V and the VSS pin is connected to ground the VEE pin is the contrast control for brightness of the display. VEE is connected to a potentiometer with a value between 10k and 20k. The seven data pins (D0-D7) are used to input data and commands to display desired the message on the screen.

The three control pins, EN, $R/\overline{W}$, and RS, allow the user to let the display know what kind of data is sent. The EN pin latches the data from the D0-D7 pins into the LCD display. Data on D0-D7 pins will be latched on the trailing edge (high-to-low) of the EN pulse. The EN pulse must be at least 450 ns wide. The $R/\overline{W}$ (read/write) pin, allows the user to either write to the LCD or read data from the LCD. In this example, the $R/\overline{W}$ pin will always be zero since only a string of ASCII data is written to the LCD. The $R/\overline{W}$ pin is set to one for reading data from the LCD.

The command or data can be output to the LCD in two ways. One way is to provide time delays of a few milliseconds before outputting the next command or data. The second approach utilizes a busy flag to determine whether the LCD is free for the next data or command. For example, in order to display ASCII characters one at a time, the LCD must be read by outputting a HIGH on the $R/\overline{W}$ pin. The busy flag can be checked to ensure whether the LCD is busy or not before outputting another string of data. Note that the busy flag can thus be used instead of time delays.

Finally, the RS (Register Select) pin is used to determine whether the user is sending command or data. The LCD contains two 8-bit internal registers. They are command register and data register. When RS = 0, the command register is accessed, and typical LCD commands such as clear cursor left (hex code 0x04) can be used. Table 8.4 shows a list of some of the commands. Note that the busy flag is bit 7 of the LCD's command register. The busy bit can be read by outputting 0 to RS pin, 1 to $R/\overline{W}$ pin, and a leading edge (LOW to HIGH) pulse to the EN pin.

When attempting to send data or commands to the LCD, the user must make sure that the values of EN, $R/\overline{W}$, and RS are correct, along with appropriate timing. A PIC18F assembly language program can be written to output appropriate values to these pins via I/O ports. For example, in order to send the 8-bit command code to the LCD, write a PIC18F assembly language program to perform the following steps:

- output the command value to the PIC18F4321 I/O port that is connected to the LCD's D0-D7 pins.
- send 0 to RS pin and 0 to $R/\overline{W}$ pin.
- Send a '1' and then a '0' to the EN pin to latch the LCD's D0-D7 code.

As mentioned earlier, the example in Figure 10.10 will display the phrase "Switch Value:" along the BCD value of the four switch inputs. Four switches are connected to bits 0 through 3 of PORTC. The D0-D7 pins of the LCD are connected to bits 0 through 7 of PORTD. The RS, $R/\overline{W}$, and EN pins of the LCD are connected to bits 0, 1, and 2 of

PORTB of PIC18F4321.

The complete LCD program in C is shown in the following. Note that time delay rather than the busy bit is used before outputting the next character to the LCD. Two functions are used: one for outputting command code, and the other for the delay. PORTB and PORTD are configured as input ports, and PORTC is set up as an input port. Also, assume 1-MHz default crystal frequency for the PIC18F4321.

As an example, let us consider the code for outputting a command code such as the command "move cursor to the beginning of the first line" (Start at line 1 position 0) to the LCD. From Table 9.1, the command code for this is 0x80. From the LCD program shown below, the statement `cmd(0x80);` will execute the following C code:

```
void cmd(unsigned char value)
{
        PORTD=value;            //Command is sent to PORTD
        PORTB=0x04;             //rs=0 rw=0 en=1
        delay(10);              //20msec delay
        PORTB=0x00;             //rs=0 rw=0 en=0
}
```

The above CMD function first outputs the value=0x80 to PORTD. Since PORTD is connected to LCD's D0-D7 pins, these data will be available to be latched by the LCD. The following few lines of the above code of the CMD function are for outputting 0's to RS and $R/\overline{W}$ pins, and a trailing edge (1 to 0) pulse to EN pin along with a delay of 20 msec. Hence, the LCD will latch 0x80, and the cursor will move to the start of the first line.

The following C loop will provide 2 msec delay:

```
void delay(unsigned int itime)          //2 msec delay
{
        unsigned int i,j;
        for(i=0; i<itime; i++)
                for(j=0; j<255;j++);
}
```

The above C code along with the statement `delay(10);` will provide 20 msec delay.

Similarly, the program logic (shown below) for outputting other ASCII characters and switch input data can be explained.

The complete LCD program using C is provided below:

```
#include <P18F4321.h>
void cmd(unsigned char);
void data(unsigned char);
void delay(unsigned int);
void main(void)
{
unsigned char input,output;

        TRISD=0;                //PORTD is output
```

```
        TRISB=0;                //PORTB is output
        TRISC=0xFF;             //PORTC is input
        PORTB=0x00;             //rs=0 rw=0 en=0
        delay(10);              //20msec delay
        cmd(0x0C);              //Display On, Cursor Off
        delay(10);              //20msec delay
        cmd(0x01);              //Clear Display
        delay(10);              //20msec delay
        cmd(0x06);              //Shift cursor to the right
        delay(10);              //20msec delay
        cmd(0x80);              //Start at line 1 position 0
        delay(10);              //20msec delay
unsigned char tstr [14] = {'s', 'w', 'i', 't', 'c', 'h', ' ', 'i', 'n', 'p', 'u', 't', ':'}
unsigned char i;
for (i = 0; i<14; i++)
        data (tstr [i]);
while(1)
{
        input= PORTC&0x0F;      //Mask  switch value
        output=0x30 | input;    //Logically OR switch inputs with 0x30 to obtain
                                //the ASCII code
        data(output);           //Display switch value on screen
        delay(10);              //20msec delay
        cmd(0x10);              //Shift cursor left one
}
}
void cmd(unsigned char value)
{
        PORTD=value;            //Command is sent to PORTD
        PORTB=0x04;             //rs=0 rw=0 en=1
        delay(10);              //20msec delay
        PORTB=0x00;             //rs=0 rw=0 en=0
}

void data(unsigned char value)
{
        PORTD=value;            //Data sent to PORTD
        PORTB=0x05;             //rs=1 rw=0 en=1
        delay(10);              //20msec delay
        PORTB=0x01;             //rs=1 rw=0 en=0
        }
void delay(unsigned int itime)  //2 msec delay
{
        unsigned int i,j;
        for(i=0; i<itime; i++)
                for(j=0; j<255;j++);
}
```

10.11 PIC18F on-chip Timers

As mentioned in Section 9.1 of Chapter 9, the PIC18F microcontroller family contains four to five on-chip hardware timers. The PIC18F4321 microcontroller includes four timers, namely, Timer0, Timer1, Timer2, and Timer3. These timers can be used to generate time delays using on-chip hardware. Note that the basic hardware inside each of these timers is a register that can be incremented or decremented at the rising or falling edge of a clock. The register can be loaded with a count for a specific time delay. Time delay is computed by subtracting the initial starting count from the final count in the register, and then multiplying the subtraction result by the clock frequency.

These timers can also be used as event counters. Note that an event counter is basically a register with the clock replaced by an event such as a switch. The counter is incremented or decremented whenever the switch is activated. Thus the number of times the switch is activated (occurrence of the event) can be determined. The basic features associated with these timers along with PIC18F assembly language programming examples were provided in Chapter 9. In this section, same examples will be included using C.

Finally, the PIC18F CCP (Compare/Capture/PWM) module utilizes these timers to perform capture, compare or PWM (Pulse Width Modulation) functions. These topics will be discussed later in this chapter along with programming examples in C.

Example 10.8 Using Timer0 in 16-bit mode, write a C language program to obtain a time delay of 1 ms. Assume 8-MHz crystal, leading edge clock, and a prescale value of 1:128.

Solution

Since the timer works with a divide-by-4 crystal, timer frequency = (8MHz)/4 = 2 MHz.
Instruction cycle clock period = (1/2 MHz) = 0.5 μ sec.

The bits in register T0CON of Figure 9.1 are as follows:
TMR0ON(bit 7) = 0, T08BIT (bit 6) = 0, T0CS (bit 5) = 0, T0SE (bit 4) = 0, and PSA (bit 3) = 0.
T0PS2 T0PS1 T0PS0 = 100 for a prescale value of 1: 128. Hence, the T0CON register will be initialized with 0x06.

Time delay = Instruction cycle x Prescale value x Count
Hence, Count = (1 ms) / (0. 5 μ sec x 128) = 15.625 which can be approximated to an integer value of 16 (0x0010). The timer counts up from an initialized value to 0xFFFF, and then rolls over (increments) to 0000H. The number of counts for rollover is (0xFFFF - 0x0010) = 0xFFEF.

Note that an extra cycle is needed for the rollover from 0xFFFF to 0x0000, and the TMR0IF flag is then set to 1. Because of this extra cycle, number of counts for the rollover is 17 cycles which means that the total number of counts for rollover will be 0xFFF0.

The following C language program will provide a time delay of 1 ms:

```
#include<p18f4321.h>
void main(void)
    {
```

```
        T0CON=0x06;                      // Initialize T0CON
        TMR0H=0xFF;                      // Initialize TMR0H first with 0xFF
        TMR0L=0xF0;                      // Initialize TMR0L next
        INTCONbits.TMR0IF=0;             // Clear Timer0 flag bit
        T0CONbits.TMR0ON=1;              // Start Timer0
        while(INTCONbits.TMR0IF==0);     // Wait for Timer0 flag bit to be 1
        T0CONbits.TMR0ON=0;              // Stop Timer0
        while(1);                        // Halt
}
```

Example 10.9 Write a PIC18F assembly language program to provide a delay of 1 msec using Timer1 with an internal clock of 4 MHz. Use 16-bit mode of Timer1 and the prescale value of 1:4.

Solution

For 4 MHz clock, each instruction cycle = 4 x (1/4 MHz) = 1 μsec
Total instruction cycles for 1 msec delay = $(1 \times 10^{-3} / 10^{-6})$ = 1000
With the prescale value of 1:4, instruction cycles = 1000 / 4 = 250
Counter value = $65536_{10} - 250_{10} = 65286_{10}$ = 0xFF06
Hence, TMR1H must be loaded with 0xFF, and TMR1L with 0x06

The C language program for one msec delay is provided below:

```
#include<p18f4321.h>
void main(void)
{
        T1CON=0xC1;                    //16-bit mode, 1:4 prescaler, enable Timer1
        TMR1H=0xFF;                    //Initialize TMR1H with 0xFF
        TMR1L=0x06;                    //Initialize TMR1L with 0x06
        PIR1bits.TMR1IF=0;             //Clear Timer1 overflow flag in PIR1
        T1CONbits.TMR1ON=1;            //Start Timer1
        while(PIR1bits.TMR1IF==0);     //Wait for Timer1 interrupt to be 1
        T1CONbits.TMR1ON=0;            //Stop Timer1
        while(1);                      //Halt
}
```

Example 10.10 Write a PIC18F assembly language program using Timer2 to turn an LED connected at bit 0 of PORT D after 10 sec. Assume an internal clock of 4 MHz. a prescale value of 1:16, and a postscale value of 1:16.

Solution

For 4 MHz clock, each instruction cycle = 4 x 1/(4MHz) = 1 μ sec. TMR2 is incremented every 1 μ sec. When the TMR2 value matches with the value in PR2, the value in TMR2 is cleared to 0 in one instruction cycle. Since, the PR2 is 8-bit wide, we can have a maximum counter value of 255. Let us calculate the delay with this PR2 value.

Delay = (Instruction cycle) x (Prescale value) x (Postscale value) x (Counter value + 1)
= (1 μ sec) x (16) x (16) (255 + 1)
= 65.536 msec

Note that, in the above, one is added to the Counter value since an additional clock is needed when it rolls over from 0xFF to 0x00, and sets the TMR2IF to 1. External counter value for 10 sec delay using 65.536 msec as the inner loop = (10 sec)/(65.536 msec), which is approximately 153 in decimal.

The C language is provided below:

```
#include<p18f4321.h>
void main(void)
{
    unsigned char i;
    TRISDbits.TRISD0=0;           // Configure bit 0 of Port D as an output
    PORTDbits.RD0=0;              // Turn LED OFF
    T2CON=0x7A;                   // 1:16 prescaler, 1:16 postscaler Timer1 off
    TMR2=0x00;                    // Initialize TMR2H with 0x00
    for(i=0;i<153;i++)
    {
    PR2=255;                      // Load PR2 with 255
    PIR1bits.TMR2IF=0;            // Clear Timer2 interrupt flag in PIR1
    T2CONbits.TMR2ON=1;           // Set TMR2ON bit in T2CON to start timer
    while(PIR1bits.TMR2IF==0);    // Wait for Timer2 interrupt to be 1
    }
    PORTDbits.RD0=1;              // Turn LED ON
    T2CONbits.TMR2ON=0;           // Turn off Timer2
    while(1);                     // Halt
}
```

10.12 Programming the PIC18F4321 on-chip A/D Converter Using C

The PIC18F4321 on-chip A/D converter is covered in detail in Section 9.2.1 of Chapter 9. Also, the concepts associated with the PIC18F A/D converter interface were illustrated in Chapter 9 using examples in PIC18F assembly language.

In summary, the PIC18F4321 contains an on-chip A/D converter (or sometimes called ADC) module with 13 channels (AN0-AN12). An analog input can be selected as an input on one of these 13 channels, and can be converted to a corresponding 10-bit digital number. Three control registers, namely, ADCON0 through ADCON2, are used to perform the conversion. Example 10.11 illustrates these concepts by designing a voltmeter and writing the programs in C.

Example 10.11 A PIC18F4321 microcontroller shown in Figure 10.14 is used to implement a voltmeter to measure voltage in the range 0 to 5 V and display the result in two decimal digits: one integer part and one fractional part. Using polled I/O, write a C language program to accomplish this.

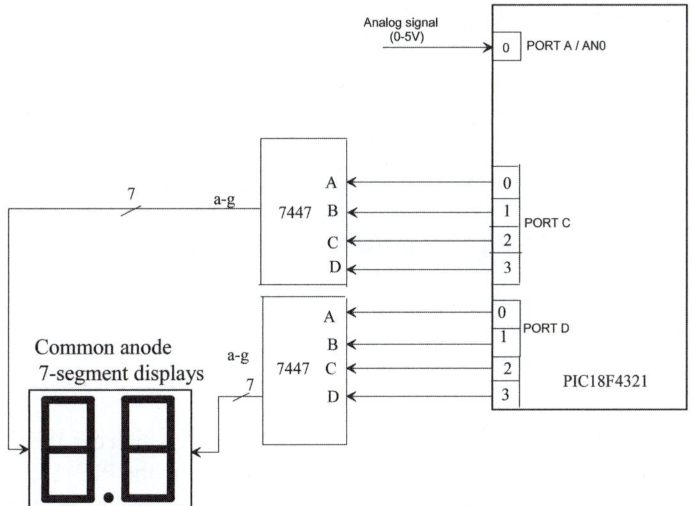

FIGURE 10.14 Figure for Example 10.11

Solution

In order to design the voltmeter, the PIC18F4321 on-chip A/D converter available will be used. Three registers, ADCON0-ADCON2, need to be configured. In ADCON0, bit 0 of Port A (RA0/AN0) is designated as the analog signal to be converted. Hence, CHS3-CHS0 bits (bits 5-2) are programmed as 0000 to select channel 0 (AN0). The ADCON0 register is also used to enable the A/D, start the A/D, and then check the "End of conversion" bit.

The reference voltages are chosen by programming the ADCON1 register. In this example, V_{DD} (by clearing bit 4 of of ADCON1 to 0) and V_{SS} (by clearing bit 5 of ADCON1 to 0) will be used. Note that V_{DD} and V_{SS} are already connected to the PIC18F4321. The ADCON1 register is also used to configure AN0 (bit 0 of Port A) as an analog input by writing 1101 at PCFG3-PCFG0 bits (bits 3-0 of ADCON1). Note that there are several choices to configure AN0 as an analog input.

The ADCON2 is used to set up the acquisition time, conversion clock, and, also, if the result is to be left or right justified. In this example, 8-bit result is assumed. The A/D result is configured as right justified, and, therefore, the 8-bit register ADRESL will contain the result. The contents of ADRESH are ignored. Note that the following concepts are repeated from Chapter 9 for convenience.

Because the maximum decimal value that can be accommodated in 8 bits of ADRESH is 255_{10} (FF_{16}), the maximum voltage of 5 V will be equivalent to 255_{10}. This means the display in decimal is given by

$$
\begin{aligned}
D &= 5 \times (\text{input}/255) \\
&= \text{input}/51 \\
&= \underbrace{\text{quotient}}_{\text{Integer part}} + \text{remainder}
\end{aligned}
$$

This gives the integer part. The fractional part in decimal is

$$F = (\text{ remainder}/51\ )\times 10$$

$$\simeq (\text{remainder })/5$$

For example, suppose that the decimal equivalent of the 8-bit output of A/D is 200.

$$D = 200/51 \ \Rightarrow \text{quotient} = 3, \text{remainder} = 47$$

$$\text{integer part} = 3$$

$$\text{fractional part, } F = 47/5 = 9$$

Therefore, the display will show 3.9 V.

From these equations, the final result will be in BCD, which can then be sent to the 7447 decoder. Since there is no unsigned division instruction in the PIC18F, a subroutine called DIVIDE is written to perform unsigned division using repeated subtraction.

In the DIVIDE subroutine, the output of the A/D contained in the ADCONRESULT register is subtracted by 51. Each time the subtraction result is greater than 51, the contents of register D1 (address 0x31) are incremented by one; this will yield the integer part of the answer. Once the contents of the ADCONRESULT reaches a value below 51, the remainder part of the answer is determined. This is done by subtracting the number in ADCONRESULT by 5. Each time the subtraction result is greater than 5, register D0 (address 0x30) is incremented by one. Finally, the integer value is placed in D1 and the remainder part is placed in D0. Now the only task left is to display the result on the seven-segment display.

The C language program for the voltmeter is provided below:

```
#include <P18F4321.h>
unsigned int FINAL,ADCONRESULT; //Initialize variables
unsigned char D1,D0;
void CONVERT(void);

void main ( )
{
unsigned int i;
TRISD = 0;              // Port D is Output
TRISC = 0;              // Port C is Output
ADCON0 = 0x01;          // Configure the ADC registers
ADCON1 = 0x0D;
ADCON2 = 0xA9;
D0=0;                   // Data to display '0' on integer 7-seg display
D1=0;
while(1)                // Data to display '0' on fractional 7-seg display
{
ADCON0bits.GO = 1;      // Start the ADC

while(ADCON0bits.DONE == 1); //Display until conversion complete
```

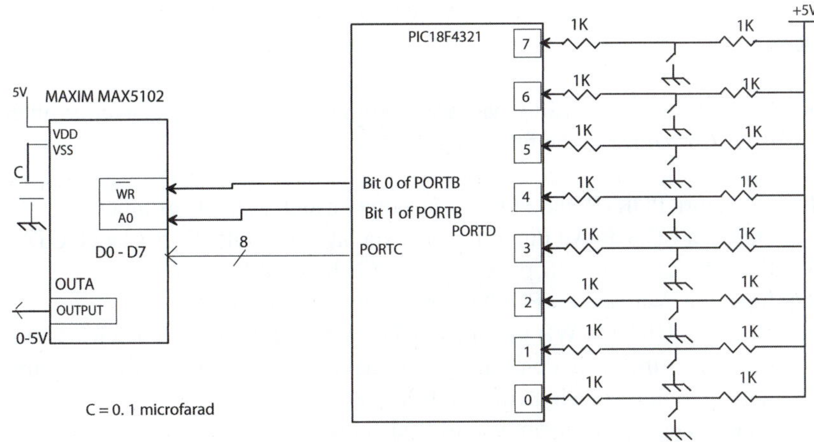

FIGURE 10.15 Figure for Example 10.12

```
        PORTC = D1;     // Output D1 to integer 7-segment display
        PORTD = D0;     // Output D0 to fractional 7-segment

ADCONRESULT = ADRESL; // Move the ADC result into ADCONRESULT
FINAL = (ADCONRESULT*10)/51; // Conversion factor
CONVERT( );
}

}

void CONVERT( )
{
D1 = FINAL/10;
D0 = FINAL% 10;             // D0 is remainder of FINAL divided by 10
}
```

10.13 Interfacing an External D/A (Digital-to-Analog) Converter Using C

As discussed in Chapter 9, most microcontrollers such as the PIC18F4321 do not have any on-chip D/A converter (or sometimes called DAC). Hence, an external D/A converter chip is interfaced to the PIC18F4321 to accomplish this function. Some microcontrollers such as the Intel/Analog Devices 8051 include an on-chip D/A converter. In order to illustrate the basic concepts associated with interfacing a typical D/A converter such as the Maxim MAX5102 was interfaced to the PIC18F4321 as discussed in Section 9.2.2 of Chapter 9.

Example 10.12 Assume the block diagram of Figure 10.15 . Write a PIC18F assembly language program that will input eight switches via PORTD of the PIC18F4321, and output the byte to D0-D7 input pins of the MAX5102 D/A converter. The microcontroller will send appropriate signals to the $\overline{WR}$ and A0 pins so that the D/A converter will convert the input byte to an analog voltage between 0 and 5 V, and output the converted voltage on its OUTA pin.

Solution

The steps for writing a PIC18F C language program for the D/A converter interface of Figure 10.15 is provided in the following:

1. Configure PORTB and PORTC as outputs, and PORTD as input.
2. Output a LOW to A0 Pin of the D/A via bit 1 of PORTB to select OUTA.
3. Output a LOW to $\overline{WR}$ pin of the D/A via bit 0 of PORTB.
4. Input the switches via PORTD, and output to PORTC.
5. Output a HIGH to $\overline{WR}$ pin of the A/D via bit 0 of PORTB to latch 8-bit input data for converting to analog voltage. No delay is needed since the program will be written to input one byte of data from the switches.

   The C language program is provided below:

```
#include <p18f4321.h>
void main (void)
{
TRISB=0x00;                 // Configure PORTB as output
TRISC=0x00;                 // Configure PORTC as output
TRISD=0xFF;                 // Configure PORTD as input
PORTBbits.RB1=0;            // Clear A0 to 0 to select OUTA
PORTBbits.RB0=0;            // Output LOW on bit 0 of PORTB
PORTC=PORTD;                // Input switches, output to PORTD
PORTBbits.RB0=1;            // Latch data for conversion
        while(1);           // Halt
}
```

10.14 PIC18F SPI Mode for Serial I/O Using C

As mentioned in Chapter 9, serial I/O is typically fabricated as an on-chip module with microcontrollers. This will facilitate interfacing microcontrollers with other microcontrollers or peripheral devices. Several protocol (rules) standards for serial data transmission have been introduced over the years. Two such standards implemented include SPI (Serial Peripheral Interface) developed by Motorola and I^2C (Inter-Integrated Circuit) developed by Philips. Both protocols are based on synchronous serial data transmission.

The SPI is a protocol established for data transfer between a master and a slave device. The master device can be a microcontroller while the slave device can be devices such as another microcontroller, EEPROMs, and A/D converters. The I^2C protocol, on the other hand, is widely used for transferring data among the ICs (Integrated Circuits) on PCBs (Printed Circuit Boards).

The PIC18F4321 contains an on-chip Master Synchronous Serial Port (MSSP) module which is a serial interface, useful for communicating with other peripheral or microcontroller devices. The MSSP module can operate in either SPI or I^2C mode.

The PIC18F SPI mode using assembly language was described in detail in Section 9.3.3 of Chapter 9. We will provide the same example of Chapter 9 (Example 9.7)

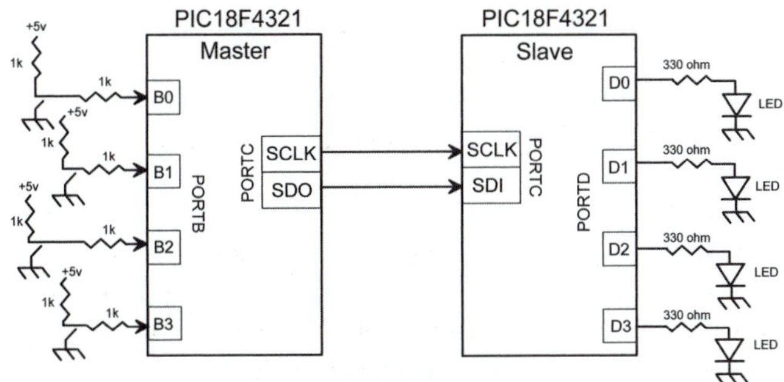

FIGURE 10.16 Figure for Example 10.13

for programming the PIC18F SPI using C in this section.

Example 10.13 Figure 10.16 shows a block diagram for interfacing two PIC18F4321s in SPI mode. One of the microcontrollers is the master while the other is the slave. The master PIC18F4321 will input four switches via bits 0-3 of PORTB, and then transmit the 4-bit data using its SDO pin to the slave's SDI pin. The slave PIC18F4321 will output these data to four LEDs, and turn them ON or OFF based on the switch inputs.
Write a C language for the master PIC18F4321 that will configure PORTB and PORTC, initialize SSPSTAT and SSPCON1, input switches, and place these data into its SSPBUF register.

Also, write a C language program for the slave PIC18F4321 that will configure PORTC and PORTD, initialize SSPSTAT and SSPCON1 registers, input data from its SDI pin, place the data in the slave's SSPBUF, and then output to the LEDs.

Solution

The following code is used to program the master PIC18F4321 device:

```
#include <p18f4321.h>
void SPI_out(unsigned char);

void main (void)
{
unsigned char output;
TRISCbits.TRISC5 = 0;           // RC5 is output
TRISCbits.TRISC3 =0;            // RC3 is output
ADCON1=0x0F;                    // Configure PORTB to be digital input
SSPSTAT= 0x40;                  // Transmission occurs on high to low clock
SSPCON1 = 0x20;                 // Enable serial functions and set as master device
        while(1){
        output = PORTB;         // Move switch value to output
        SPI_out(output);        // Send variable 'output' to SPI_out
        }
}
void SPI_out(unsigned char SPI_data)
{
        SSPBUF = SPI_data;      // Place switch value into the serial buffer
```

```
        while (SSPSTATbits.BF == 0);        // Wait for transmission to finish
}
```

The following code is used on the slave PIC18F4321 device:

```
#include <p18f4321.h>

void main (void)
{
TRISCbits.TRISC4 = 1;          // RC4 is input
TRISCbits.TRISC3 =1;           // RC3 is input
TRISD=0x00;                    // PORTD is output
SSPSTAT= 0x40;                 // Transmission occurs on high to low clock
SSPCON1 = 0x25;                // Enable serial functions and disable the  slave device
        while(1){
        while (SSPSTATbits.BF == 0);        // Wait for transmission to finish
        PORTD=SSPBUF;                       // Move serial buffer to PORTD
        }
}
```

Note that the above program is explained thoroughly in Example 9.7 of Chapter 9.

10.15 Programming the PIC18F4321 CCP Modules Using C

As mentioned in Chapter 9, the CCP module is implemented in the PIC18F4321 as an on-chip feature to provide measurement and control of time-based pulse signals. The basic concepts associated with the PIC18F CCP are explained in Chapter 9. Some of them will be repeated here for convenience.

Capture mode causes the contents of an internal 16-bit timer to be written in special function registers upon detecting an nth rising or falling edge of a pulse. Compare mode generates an interrupt or change on output pin when Timer1 matches a preset comparison value. PWM mode creates a re-configurable square wave duty cycle output at a user set frequency. The application software can change the duty cycle or period by modifying the value written to specific special function registers.

The PIC18F4321 contains two CCP modules, namely, CCP1 and CCP2. The CCP1 module of the PIC18F4321 is implemented as a standard CCP with enhanced PWM capabilities for better DC motor control. Hence, the CCP1 module in the PIC18F4321 is also called ECCP (Enhanced CCP). Note that the CCP2 module is provided with standard capture, compare, and PWM features.

The CCP module is describe in detail in Section 9.4 of Chapter 9. In this section, PIC18F assembly language programs (Examples 9.8 through 9.10 of Chapter 9) will be converted to C programs to illustrate how to program the PIC18F4321 CCP module using C.

Example 10.14 Assume PIC18F4321. Write a C language program to measure the period (in terms of the number of clock cycles) of an incoming periodic waveform connected at RC2/CCP1/P1A pin. Store result in registers 0x21 (high byte) and 0x20 (low byte). Use Timer3, and capture mode of CCP1.

Solution

```
#include<p18f4321.h>

void main(void)
{
     unsigned char FIRST_CCPR1L, FIRST_CCPR1H, HIGH_BYTE, LOW_BYTE;
     CCP1CON=0x05;              // Select capture mode rising  edge
     TRISCbits.TRISC2=1;        // Configure RC2/CCP1/P1A pin as input
     T3CON=0x40;                // Select TIMER3 as clock source for capture
     PIE1bits.CCP1IE=0;         // Disable CCP1IE  to avoid false interrupt
     CCPR1H=0x00;               // Clear CCPR1H to 0
     CCPR1L=0x00;               // Clear CCPR1L to 0
     PIR1bits.CCP1IF=0;         // Clear CCP1IF

     while(PIR1bits.CCP1IF==0);  // Wait for the first rising edge

     T3CONbits.TMR3ON=1;        // Turn Timer3 ON
     FIRST_CCPR1L=CCPR1L;       // Save CCPR1L in FIRST_CCPR1L at 1st rising edge
     FIRST_CCPR1H=CCPR1H;       // Save CCPR1H in FIRST_ at 1st rising edge
     PIR1bits.CCP1IF=0;         // Clear CCP1IF

     while(PIR1bits.CCP1IF==0);  // Wait for next rising edge

     T3CONbits.TMR3ON=0;        // Turn OFF  Timer3
     CCP1CON=0x00;              // Disable capture
     LOW_BYTE=CCPR1L-FIRST_CCPR1L;  // Low byte of result
     HIGH_BYTE=CCPR1H-FIRST_CCPR1L;  // High byte of result

     while(1);                  // Halt
}
```

Example 10.15 Assume PIC18F4321 with an internal crystal clock of 20 MHz. Write a C language program at address 0x100 that will toggle the RC2/CCP1/P1A pin after a time delay of 10 msec. Use Timer3, and compare mode of CCP1.

Solution

With 20 MHz internal crystal, Fosc = 20 MHz. Since Timer3 uses Fosc/4. Timer clock frequency = Fosc/4 = 5 MHz. Hence, clock period of Timer3 = 0.2 μ sec Counter value = (10 msec)/(0.2 μ sec) = 500_{10}= 0x01F4. Hence, CCPR1H :CCPR1L should be loaded with 0x01F4 for the PIC18F4321 compare mode.
The C language program is provided below:

```
#include<p18f4321.h>
void main(void)
{
     CCP1CON=0x02;              // Select compare mode, toggle CCP1 pin on match
```

```
        TRISCbits.TRISC2=0;           // Configure CCP1 pin as output
        T3CON=0x40;                   // Select TIMER3 as clock for compare, 1:1 prescale
        CCPR1H=0x01;                  // Load CCPR1H with 0x01
        CCPR1L=0xF4;                  // Load CCPR1L  with 0xF4
        TMR3H=0;                      // Initialize TMR3H to 0
        TMR3L=0;                      // Initialize TMR3L to 0
        PIR1bits.CCP1IF=0;            // Clear CCP1IF
        T3CONbits.TMR3ON=1;           // Start Timer3

        while(PIR1bits.CCP1IF==0); // Wait in a loop until CCP1IF is 1

        T3CONbits.TMR3ON=0;           // Stop Timer3

        while(1);                     // Halt
    }
```

Example 10.16 Write a C language program at 0x100 to generate a 4 KHz PWM with a 50% duty cycle on the RC2/CCP1/P1A pin of the PIC18F4321. Assume 4 MHz crystal.

Solution

PR2 = [(Fosc)/(4 x Fpwm x TMR2 Prescale Value)] - 1
PR2 = [(4 MHz)/(4 x 4 KHz x 1)] - 1 assuming Prescale value of 1
PR2 = 249. With 50% duty cycle, decimal value of the duty cycle = 0.5 x 249 = 124.5.Hence, the CCPR1L register will be loaded with 124, and bits DC1B1:DC0B0 (CCP1CON register) with 10 (binary).

The C language program is provided below:

```
#include<p18f4321.h>

    void main(void)
    {
        PR2=249;                      // Initialize PR2 register
        CCPR1L=124;                   // Initialize CCPR1L
        CCP1CON=0x20;                 // CCP1 OFF, DC1B1:DC0B0=10
        TRISCbits.TRISC2=0;           // Configure CCP1 pin as output
        CCP1CON=0x2C;                 // PWM mode
        TMR2=0;                       // Clear Timer2 to 0

        while(1)
        {
        PIR1bits.TMR2IF=0;            // Clear TMR2IF to 0
        T2CONbits.TMR2ON=1;           // Turn Timer2 ON
        while(PIR1bits.TMR2IF==0);    // Wait until TMR2IF is HIGH
        }
    }
```

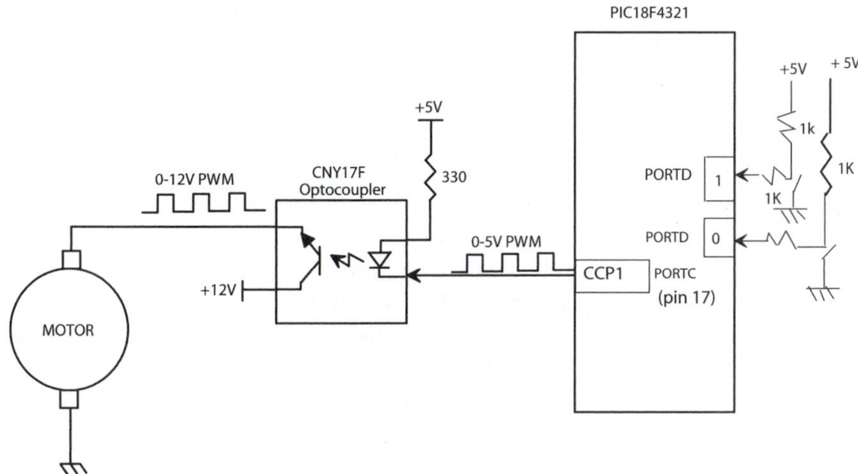

FIGURE 10.17 Figure for Example 10.17

10.16 DC Motor Control Using PWM Mode and C

As mentioned in Chapter 9, typical applications of the PWM mode include DC motor control. The speed of a DC motor is directly proportional to the driving voltage. The speed of a motor increases as the voltage is increased. In earlier days, voltage regulator circuits were used to control the speed of a DC motor. But voltage regulators dissipate lots of power. Hence, the PIC18F in the PWM mode is used to control the speed of a DC motor. In this scheme, power dissipation is significantly reduced by turning the driving voltage to the motor ON and OFF. The speed of the motor is a direct function of the ON time divided by the OFF time.

Microcontrollers such as the PIC18F4321 are not capable of outputting the required large current and voltage to control a typical DC motor. Hence, a driver such as the CNY17F Optocoupler is needed to amplify the current and voltage provided by the PIC18F's output, and provide appropriate levels for the DC motor. One of the many useful applications for using a PWM signal is its ability to control a mechanical device, such as a motor.

Note that the motor will run faster or slower based on the duty cycle of the PWM signal. The motor runs faster as the duty cycle of the PWM signal at the CCPx pin is increased. To illustrate this concept, two different duty cycles will be used in Example 10.17 converts the PIC18F assembly language program of Chapter 9 (Example 9.11) into C.

Example 10.17 Figure 10.17 shows a simplified diagram interfacing the PIC18F4321 to a DC motor via the CNY17F Optocoupler. The purpose of this example is to control the speed of a DC motor by inputting two switches connected at bit 0 and bit 1 of PORTD. The motor will run faster or slower based on the switch values (00 or 11), but will not provide any measure of the exact RPM of the motor.

When both switches are closed (00), a PWM signal at the CCP1 pin of the PIC18F4321 with 50% duty cycle will be generated. When both switches are open (11), a PWM signal at the CCP1 pin of the PIC18F4321 with 75% duty cycle will be generated. Otherwise, the motor will stop, and the program will wait in a loop.

If switches are closed (00), the motor will run using the 4 KHz PWM pulse of Example 10.16 with 50% duty cycle. If both switches are open (11), the motor will run at a faster speed with a duty cycle of 75%. The program will first perform initializations, and wait in a loop until the switches are 00 or 11.
Write a C language program to accomplish this.

Solution

From Example 10.16, PR2 = 249. With 50% duty cycle, Count = 0.5 x 249 = 124.5. Hence, the CCPR1L register will be loaded with 124, and bits DC1B1:DC0B0 (CCP1CON register) with 10_2 when the switch values are 00 (both switches are closed).

With 75% duty cycle, Count = 0.75 x 249 = 186.75. Hence, the CCPR1L register will be loaded with 186, and bits DC1B1:DC0B0 (CCP1CON register) with 11 when the switch values are 11 (both switches are open).

The C program is provided below:

```
#include<p18f4321.h>
void main(void)
{
     PR2=249;                    // Initialize PR2 register
     TRISCbits.TRISC2=0;         // Configure CCP1 pin as output
     TRISDbits.TRISD0=1;         // Configure RD0 as an input bit
     TRISDbits.TRISD1=1;         // Configure RD1 as an input bit
     T2CON=0x00;                 // 1:1 prescale, Timer2 OFF

     CCP1CON=0x3C; //PWM mode,DC1B1:DC0B0=11
          while(1)
          {
               if(PORTDbits.RD0==1 && PORTDbits.RD1==1)
                              // If switches are  HIGH, 75% duty cycle
                    {
                    CCPR1L=186;                   // For 75% duty cycle
                    CCP1CON=0x3C;      // PWM mode,DC1B1:DC0B0=11
                    TMR2=0;            // Clear Timer2 to 0
                    PIR1bits.TMR2IF=0; // Clear TMR2IF to 0
                    T2CONbits.TMR2ON=1;        // Turn Timer2 ON
          while(PIR1bits.TMR2IF==0);       // Wait until TMR2IF is HIGH
                    }

               if(PORTDbits.RD0==0 && PORTDbits.RD1==0)
                              // If switches are LOW, 50% duty cycle
                    {
                    CCPR1L=124;                   // For 50% duty cycle
                    CCP1CON=0x2C;
                    TMR2=0;            // Clear Timer2 to 0
                    PIR1bits.TMR2IF=0;            // Clear TMR2IF to 0
                    T2CONbits.TMR2ON=1;          // Turn Timer2 ON
```

```
                    while(PIR1bits.TMR2IF==0); // Wait until TMR2IF is HIGH
                    }

               if(PORTDbits.RD0==0 && PORTDbits.RD1==1)
                    {
                    T2CONbits.TMR2ON=0;
                    }
               if(PORTDbits.RD0==1 && PORTDbits.RD1==0)
                    {
                    T2CONbits.TMR2ON=0;
                    }
          }

     }
```

Questions and Problems

10.1 Write a C language statement to configure

(a all bits of Port C as inputs

(b) all bits of Port D as outputs

(c) bits 0 through 4 of Port B as inputs

(d) all bits of Port A as outputs

10.2 The PIC18F4321 microcontroller is required to drive the LEDs connected to bit 0 of Ports A and B based on the input conditions set by switches connected to bit 1 of Ports A and B. The I/O conditions are as follows:

- If the input at bit 1 of Port A is HIGH and the input at bit 1 of Port B is LOW, the LED at Port A will be ON and the LED at Port B will be OFF.
- If the input at bit 1 of Port A is LOW and the input at bit 1 of Port B is HIGH, the LED at Port A will be OFF and the LED at Port B will be ON.
- If the inputs of both Ports A and B are the same (either both HIGH or both LOW), both LEDs of Ports A and B will be ON.

Write a C language program to accomplish this.

10.3 The PIC18F4321 microcontroller is required to test a NAND gate. Figure P10.3 shows the I/O hardware needed to test the NAND gate. The microcomputer is to be programmed to generate the various logic conditions for the NAND inputs, input the NAND output, and turn the LED ON connected to bit 3 of Port D if the NAND gate chip is found to be faulty. Otherwise, turn the LED ON connected to bit 4 of Port D. Write a C language program to accomplish this.

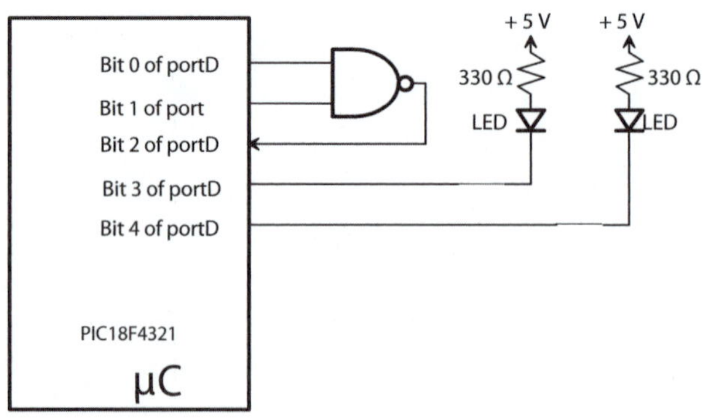

FIGURE P10.3

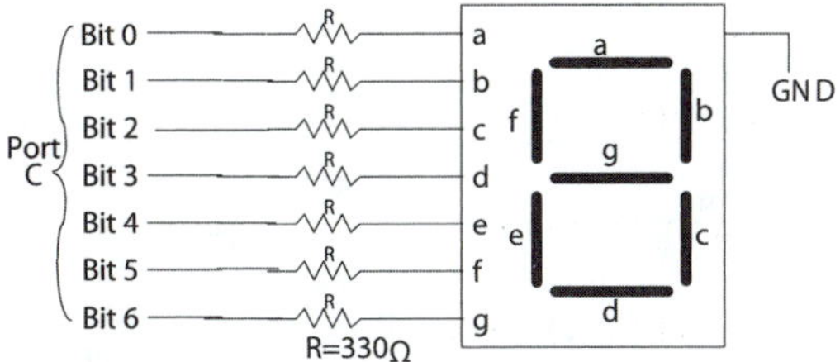

FIGURE P10.4

10.4 The PIC18F4321 microcontroller is required to add two 3-bit numbers stored in the lowest 3 bits of data registers 0x20 and 0x21 and output the sum (not to exceed 9) to a common-cathode seven-segment display connected to Port C as shown in Figure P10.4. Write a C language program to accomplish this by using a lookup table. Assume that the lookup table is already stored in data memory.

10.5 The PIC18F4321 microcontroller is required to input a number from 0 to 9 from an ASCII keyboard interfaced to it and output to an EBCDIC printer. Assume that the keyboard is connected to Port C and the printer is connected to Port D. Store the EBCDIC codes for 0 to 9 starting at an address 0x30, and use this lookup table to write a C language program to accomplish this. Note that decimal numbers 0 through 9 are represented by F0H through F9H in EBCDIC code, and by 30H through 39H in ASCII code as mentioned in Chapter 1. Assume that the lookup table for EBCDIC codes is already stored in data memory.

10.6 In Figure P10.6, the PIC18F4321 is required to turn on an LED connected to bit 1 of Port D if the comparator voltage Vx > Vy; otherwise, the LED will be turned off. Write a C language program to accomplish this using conditional or polled I/O.

10.7 Repeat Problem 10.6 using Interrupt I/O by connecting the comparator output to INT1. Note that RB1 is also multiplexed with INT1. Write main program and interrupt service routine in C language. The main program will configure the I/O ports, enable interrupt INT1, turn the LED OFF, and then wait for interrupt. The interrupt service routine will turn the LED ON and return to the main program at the appropriate location so that the LED is turned ON continuously until the next interrupt.

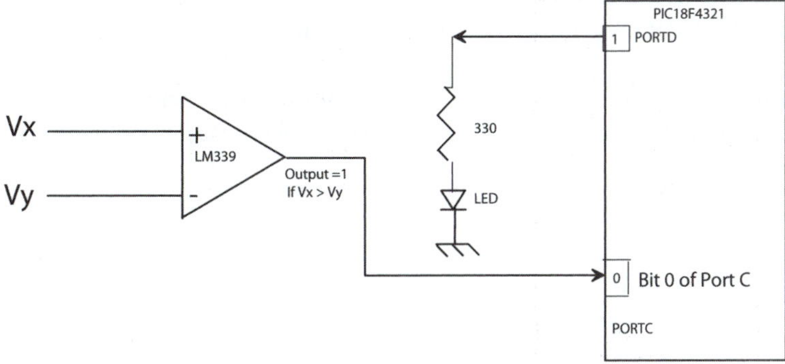

FIGURE P10.6

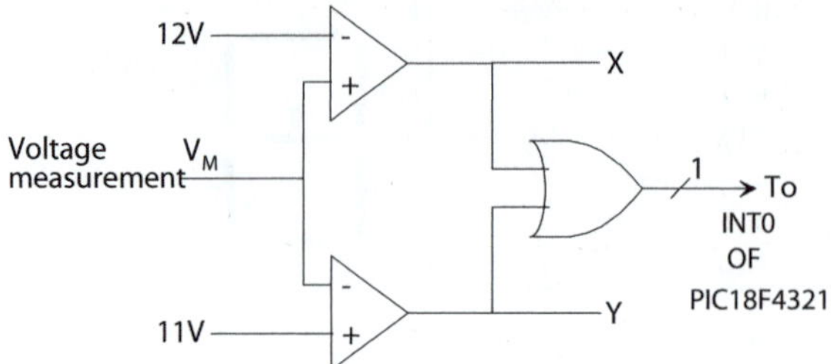

FIGURE P10. 8

10.8 In Figure P10.8, if $V_M > 12$ V, turn an LED ON connected to bit 3 of port A. If $V_M < 11$ V, turn the LED OFF. Using ports, registers, and memory locations as needed and INT0 interrupt:

(a) Draw a neat block diagram showing the PIC18F4321 microcontroller and the connections to ports in the diagram in Figure P10.8.

(b) Write the main program and the service routine in C language. The main program will initialize the ports and wait for an interrupt. The service routine will accomplish the task and stop.

10. 9 Write C language program to set interrupt priority of INT1 as the high level, and interrupt priority for INT2 level as low level.

10.10 Assume the PIC18F4321- DMC 16249 interface of Figure 10.10. Write a C program to display the phrase "PIC18F" on the LCD as soon as the four input switches connected to Port C are all HIGH.

10.11 Write a C program to initialize Timer0 as an 8-bit timer to provide a time delay with a count of 100. Assume 4 MHz internal clock with a prescaler value of 1:16.

10.12 Write a C language program to generate a square wave with a period of 4 ms on bit 0 of PORTC using a 4 MHz crystal. Use Timer0.

10.13 Write a C language program to generate a square wave with a period of 4 ms on bit 7 of PORTD using a 4 MHz crystal. Use Timer1.

10.14 Write a C language program to turn an LED ON connected at bit 0 of PORTC when the TMR2 reaches a value of 200. Assume a 4 MHz crystal.

10.15 Write a C language program to generate a square wave on pin 3 of PORTC with a 4 ms period using Timer3 in 16-bit mode with a prescaler value of 1:8. Use a 4 MHz crystal.

10.16 Assume PIC18F4321 with Fosc = 1 MHz. Consider the following C program:

```
#include <p18f4321.h>
void    delay(void);
void    main(void){
        TRISBbits.TRISB=0;
        while (1){
                PORTBbits.RB4^=1;
                delay( );
        }
}
void    delay( ){
        T0CON=0x01;
        TMR0H=0xC2;
        TMR0L=0xF7;
        T0CONbits.TMR0ON=1;
        while(INTCONbits.TM0IF = = 0)
                ;
        T0CONBits.TMR0ON = 0;
        INTCONbits.TMR0IF = 0;
}
```

(a) What type of signal is generated on the RB4 pin? What is its frequency?

(b) What is the frequency of the signal that appears on the RB4 pin?

(c) Repeat part (b) if T0CON is initialized with 0x42, and TMR0H = 0xC2 is deleted from the program

10.17 Repeat Example 10.11 and write a C program using A/D converter's interrupt bit indicating completion of conversion. Use addresses, and other parameters of your choice.

10.18 Design and develop hardware and software for a PIC18F4321-based system (Figure P10.18) that would measure, compute, and display the Root-Mean-Square (RMS) value of a sinusoidal voltage. The system is required to

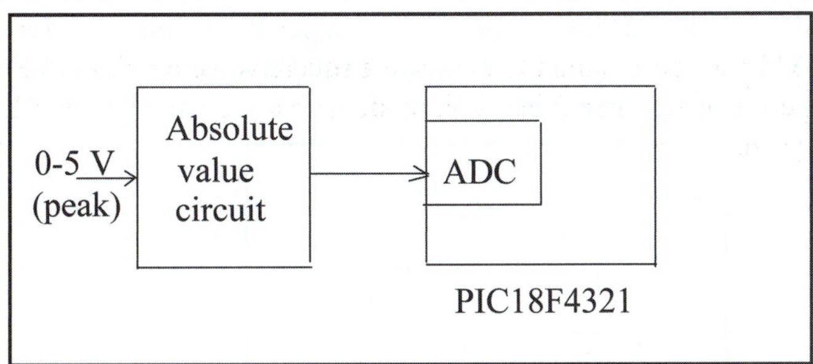

FIGURE P10.18

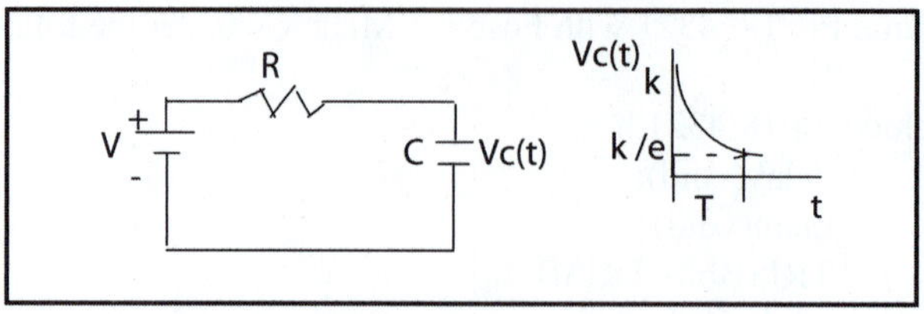

FIGURE P10.19

1. Sample a 5 V (zero-to-peak voltage), 60 Hz sinusoidal voltage 128 times.
2. Digitize the sampled value using the on-chip ADC of the PIC18F4321 along with its interrupt upon completion of conversion signal.
3. Compute the RMS value of the waveform using the formula,

   RMS Value = SQRT $[(\sum_{n=1}^{N}(X_n^2)/N]$, where X_n's are the samples and N is the total number of samples. Display the RMS value on two seven-segment displays (one for integer part, and the other for fractional part).

(a) Draw a hardware block diagram.

(b) Write C language program to accomplish the above.

10.19 *Capacitance meter.* Consider the RC circuit of Figure P10. 19. The voltage across the capacitor is Vc (t) = k $e^{-t/RC}$. In one-time constant RC, this voltage is discharged to the value k/e. For a specific value of R, value of the capacitor C = T/R, where T is the time constant that can be counted by the PIC18F4321. Design the hardware and software for the PIC18F4321 to charge a capacitor by using a pulse to a voltage of your choice via an amplifier. The PIC18F4321 will then stop charging the capacitor, measure the discharge time for one time constant, and compute the capacitor value.

(a) Draw a hardware schematic.

(b) Write a C program to accomplish the above.

10.20 Design a PIC18F4321-based digital clock. The clock will display time in hours, minutes, and seconds. Write a C program to accomplish this.

10.21 Design a PIC18F4321-based system to measure the power absorbed by a 2K resistor (Figure P10.21). The system will input the voltage (V) across the 2K resistor, convert it to an 8-bit input using the PIC18F4321's on-chip A/D converter, and then compute the power using V^2/R.

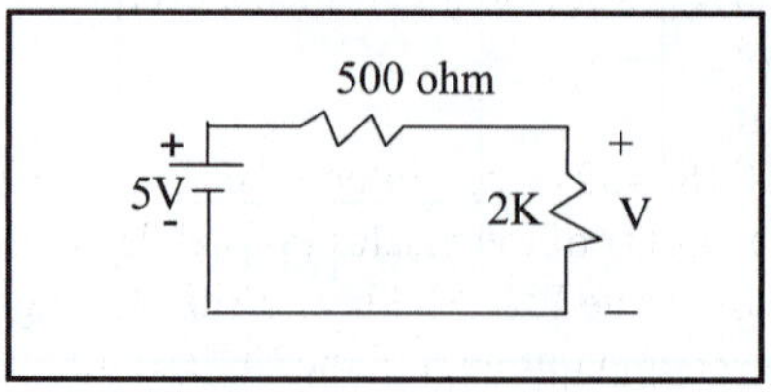

FIGURE P10.21

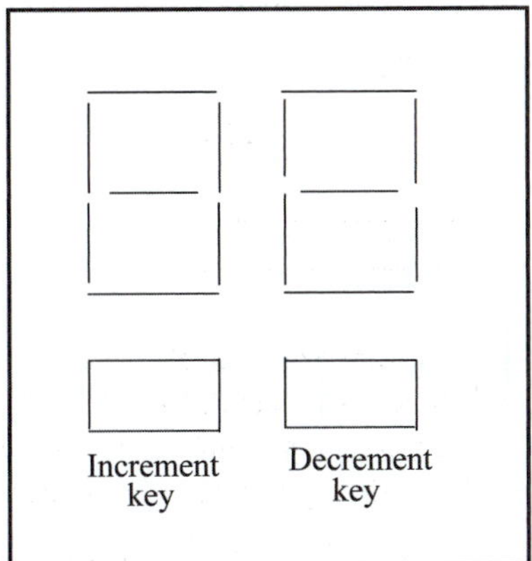

FIGURE P10.22

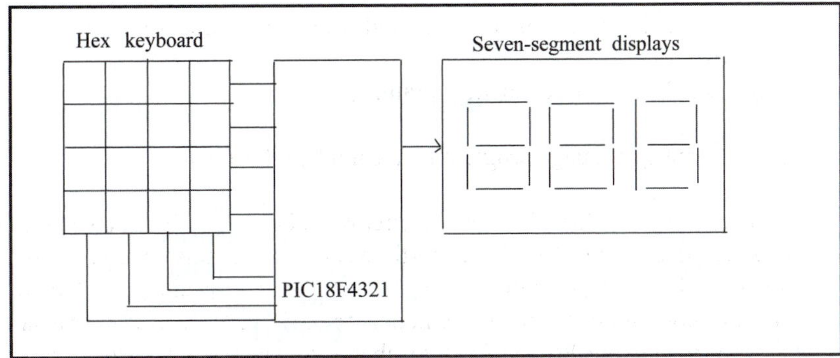

FIGURE P10.23

10.22 Design a PIC18F4321-based system (Figure P10.22) as follows: The system will drive two seven-segment digits, and monitor two key switches. The system will start displaying 00. If the increment key is pressed, it will increment the display by one. Similarly, if the decrement key is pressed, the display will be decremented by one. The display will go from 00 to 09, and vice versa.
Write a C language program to accomplish the above. Use ports and data memory addresses of your choice. Draw a block diagram of your implementation.

10.23 It is desired to implement a PIC18F4321-based system as shown in Figure P10.23. The system will scan a hex keyboard with 16 keys, and drive three seven-segment displays. The PIC18F4321 will input each key pressed, scroll them in from the right side of the displays, and keep scrolling as each key is pressed. The leftmost digit is just discarded. The system continues indefinitely. Write a C language program at address 0x100 to accomplish the above. Use ports and data memory addresses of your choice.

10.24 Assume that two PIC18F4321s are interfaced in the SPI mode. A switch is connected to bit 0 of PORTD of the master PIC18F4321 and an LED is connected to bit 5 of PORTB of the slave PIC18F4321. Write C language programs to input the switch via the master, and output it to the LED of the slave PIC18F4321. If the

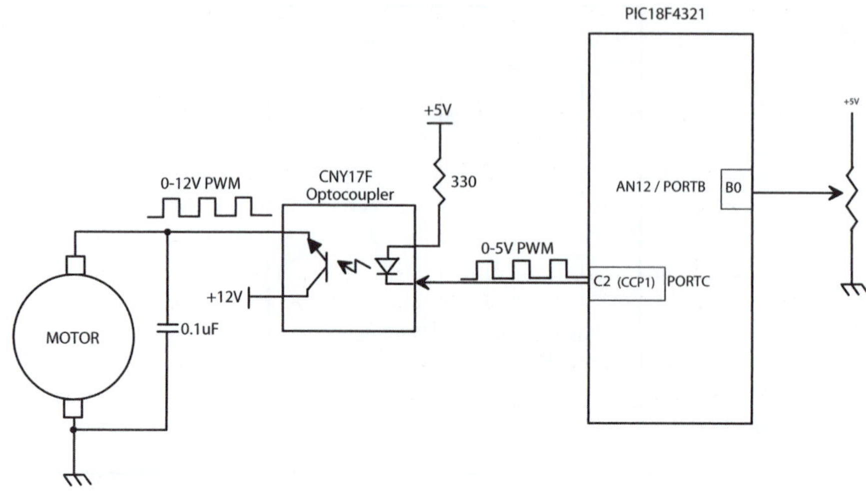

FIGURE P10.28

switch is open, the LED will be turned ON while the LED will be turned OFF if the switch is closed.

10.25 Assume PIC18F4321. Write a C language program that will measure the period of a periodic pulse train on the CCP1 pin using the capture mode. The 16-bit result will be performed in terms of the number of internal (Fosc/4) clock cycles, and will be available in the TMR1H:TMR1L register pair. Use 1:1 prescale value for Timer1. Store the 16-bit result in CCPR1H:CCPR1L.

10.26 Assume PIC18F4321. Write a C language program that will generate a square wave on the CCP1 pin using the Compare mode. The square wave will have a period of 20 ms with a 50% duty cycle. Use Timer1 internal clock (Fosc/4 from XTAL) with 1:2 prescale value.

10.27 Write a C language program to generate a 16 KHz PWM with a 75% duty cycle on the RC2/CCP1/P1A pin of the PIC18F4321. Assume 10 MHz crystal.

10.28 It is desired to change the speed of a DC motor by dynamically changing its pulse width using a potentiometer connected at bit 0 of PORTB (Figure P10.28). Note that the PWM duty cycle is controlled by the potentiometer. Write a C language program that will input the potentiometer voltage via the PIC18F4321's on-chip A/D converter using interrupts, generate the PWM waveform on the CCP1 pin, and then change the speed of the motor as the potentiometer voltage is varied.

10.29 It is desired to implement a traffic light controller using the PIC18F4321 as follows:

Step 1: Make North-South light Green and East-West light Red for 10 seconds. Check to see if any waiting car is trying to go from east to west and vice versa. If there is a waiting car, go to step 2; otherwise, repeat this step.

Step 2: Make North-South light Yellow and East-West light Red for 2 seconds, and go to Step 3.

Step 3: Make North-South light Red and East-West light Green for 5 seconds, and then go to Step 4.

Step 4: Make North-South light Red, and East-West light Yellow for 2 seconds, and then go to Step 1.

Also, include provision for an emergency input. When this input is asserted, a flashing RED light in both directions will be activated.

Write a C program for the above state machine using Timer0 in 16-bit mode. Draw an ASM chart showing all inputs and outputs.

APPENDIX

ANSWERS TO SELECTED PROBLEMS

Chapter 2

2.1 A single chip microcomputer contains CPU, memory, and I/O on a single chip. A typical microcontroller contains the CPU, memory, I/O, timers, A/D converter—all on a single chip.

2.4 (a) sign = 0, carry = 0, zero = 0, overflow = 0.
(d) sign = 1, carry = 0, zero = 0, overflow = 1.

2.6 (a) 20BE
(b) (20BE) = 05, (20BF) = 02.

2.7 To load the program counter with the address of the first instruction to be executed.

Chapter 3

3.2 16 MB

3.5 (a) 16,384
(b) 128 chips
(c) 4 bits

3.7 (a) 20
(b) 6 x 64 decoder

3.8 14 unused address pins available.
Maximum directly addressable memory = 16 megabytes

3.10 Memory chip #1 EC00H - EDFFH

Memory chip #2 F200H - F3FFH

3.11 (a) ROM Map: 0000H - 07FFH
RAM Map: 2000H - 20FFH

3.16 Using standard I/O, the microcontroller uses an output pin such as an $M/\overline{IO}$ pin to distinguish between memory and I/O. Also, the microcontroller uses IN and OUT instructions to perform I/O operation in standard I/O.
Using memory-mapped I/O, the microcontroller uses an unused address pin to distinguish between memory and I/O. The ports are mapped as memory locations. Memory-oriented instructions are used for performing I/O operation.
The PIC18F uses only memory-mapped I/O.

3.17 Memory-mapping provides the physical addresses for the microcontroller's main memory while memory-mapped I/O maps port addresses into memory locations.

3.20 Interrupt address vector is the starting address of the service routine.

Chapter 4

4.2 Yes.

4.3 No.

4.7 Use the following identities:
$a \oplus a = 0$ and $a \oplus 0 = a$ and $(a \oplus b) \oplus a = b$

4.8 Product = $0000\ 0000\ 0000\ 0100_2$

4.9 Quotient = -8, remainder = -1. The sign of the remainder is the same as the sign of the dividend unless remainder is zero.

Chapter 5

5.2 Flash memory.

5.4 SRAM.

5.9 (a) PC contains addresses of instructions in program memory whereas the FSRs point indirectly to data memory.

5.10 4002H.

5.17 (a) Implied mode.

5.17 (b) Literal mode.

Chapter 6

6.1
```
MOVF    0x30, W
ADDWF   0x40, W
MOVWF   0x50
```
6.3 (a) [0x20] = FFH

6.6
```
CLRF   0x20
SETF   0x22
```

6.10 (a) MOVLW 0

6.13 Assume N1 and N2 are already loaded into registers 0x20 and 0x21, respectively.
```
INCLUDE   <P18F4321.INC>
ORG       0x100
SWAPF     0x21, F    ; Swap nibbles of N2 in 0x21
MOVF      0x20, W    ; Move [0x20] into WREG
ADDWF     0x21, W    ; Add [WREG] with [0x21], store result in WREG
MOVWF     0x30       ; Store result in 0x30
SLEEP
END
```

6.21 Assume that the unsigned 16-bit number is 0x0124 (arbitrarily chosen). Since the remainder can be discarded, unsigned division can be accomplished by logically shifting the 16-bit unsigned number 0x0124 once to the right.

```
INCLUDE   <P18F4321.INC>
ORG       0x100
MOVLW     01H          ; Load high byte into 0x20
MOVWF     0x20
MOVLW     0x24         ; Load low byte into 0x21
MOVWF     0x21
BCF       STATUS, C    ; Clear carry flag to 0
RRCF      0x20, F      ; Right shift [0x20][0x21] once
RRCF      0x21, F
SLEEP
END
```

Chapter 7

7.2 Assume data are already loaded into 0x30.

```
INCLUDE   <P18F4321.INC>
ORG       0x200
MOVFF     0x30, 0x40   ; Copy data in 0x40
SWAPF     0x30, F      ; Move data into low 4 bits
MOVLW     0x0F         ; Move mask data into WREG
```

```
      ANDWF    0x30, F        ; One unsigned 8-bit data set in 0x30
      ANDWF    0x40, W        ; Another unsigned data set in WREG
      MULWF    0x30           ; Unsigned  multiply data
                              ; Since result will be 8-bit maximum,
                              ; PRODL will contain result
      MOVWF    PRODL, 0x31    ; Result in 0x31
      SLEEP                   ; HALT
      END
```

7.7 Assume arrays x[i] and y[i] are already loaded into 0x20 and 0x30, respectively. Use MULWF for unsigned multiplication.

```
      INCLUDE  <P18F4321.INC>
      ORG      0x100
      CLRF     0x50           ; Clear  sum  to 0
      LFSR     0, 0x0020      ; Load 0x0020 into FSR0
      LFSR     1, 0x0030      ; Load 0x0030  into FSR1
      MOVLW    D'10'          ; Move 10 (decimal) into counter 0x75
      MOVWF    0x75
LOOP  MOVF     POSTINC0, W    ; Move  x[i] into WREG, increment pointer
      MULWF    POSTINC1       ; Unsigned multiply in x[i] * y[i]
      MOVF     PRODL, W       ; Move 8-bit product to WREG
      ADDWF    0x50, F        ; Sum in 0x50
      DECF     0x75, F        ; Decrement counter by 1
      BNZ      LOOP           ; Repeat if Z = 0
      SLEEP                   ; 0x70 contains the result and  halt
      END
```

7.18 Q = 120

Chapter 8

8.2 1 MHz

8.6 (a) SETF TRISC

8.6 (b) CLRF TRISD

8.7

```
      INCLUDE  <P18F4321.INC>
      ORG      0x100
      SETF     PORTC          ; Configure  PORTC as an input port
      BCF      TRISD, 6       ; Configure bit 6 of PORTD as output
      BCF      PORTD, 6       ; Turn LED  OFF
      MOVF     PORTC, F       ; Input PORTC
      MOVLW    0x07
```

```
          ANDWF    PORTC, F      ; Retain low three bits
          MOVLW    0x00          ; Check for no high switches, 0 is an even number
          SUBWF    PORTC, W
          BZ       LED           ; If no high switches, turn LED ON
          MOVLW    0x03          ; Check for two high switches
          SUBWF    PORTC, W
          BZ       LED           ; If  two high switches, turn LED ON
          MOVLW    0x05          ; Check for two high switches
          SUBWF    PORTC, W
          BZ       LED           ; If  two high switches, turn LED ON
          MOVLW    0x06          ; Check for two high switches
          SUBWF    PORTC, W
          BZ       LED           ; If  two high switches, turn LED ON
FINISH    SLEEP    ; Halt
LED       BSF      PORTD, 6      ; Turn LED  ON
          BRA      FINISH
          END
```

8.12

```
          INCLUDE   <P18F4321.INC>
          ORG       0x200
          MOVLW     0x0F          ; Configure PORTB as an input port
          MOVWF     ADCON1
          BCF       TRISC, 1      ; Configure bit 1 of PORTC as an output
BACK      MOVF      PORTB, F      ; Input PORTB
          RLNCF     PORTB, W      ; Rotate left once to align output  data
          MOVWF     PORTC         ; Output  to  LED
          BRA       BACK
          END
```

Chapter 9

9.1

Bit 7: Set to 0 so that TMR0 is off
Bit 6: Set to 1 in order to enable the 8-bit mode of TMR0
Bit 5: Set to 1 so that an external crystal oscillator can be used
Bit 4: Set to 1 so the timer will increment when the clock is going from high to low (negative edge).
Bit 3: Set to 0 in order to enable the prescaler function
Bit 2-0: Set to 011 to enable a 1:16 prescaler
Hence, T0CON=0x73

9.4

```
          INCLUDE   <P18F4321.INC>
          ORG       0x70
          BCF       TRISD, RD7         ;Bit 7 of PORTD is output
```

```
        MOVLW    0xC8
        MOVWF    T1CON           ; Set up Timer1 as 16-bit no prescaler
        MOVLW    0xA0
        MOVWF     TMR1L          ; Value placed in TMR1L
        MOVLW    0x0F
        MOVWF    TMR1H           ; Value place in TMR1H
        BCF      PIR1, TMR1IF    ; Clear Timer1 interrupt flag
        BSF      PIE1, TMR1IE    ; Enable Timer1 interrupt
LOOP    BSF      T1CON, TMR1ON   ; Turn on Timer1
WAIT    BTFSS    PIR1, TMR1IF    ; Wait until Timer1 interrupt occurs
        BRA      WAIT
        BCF      T1CON, TMR1ON   ; Turn Timer1 off
        COMF     PORTD, RD7      ; one's complement of bit 7 of PORTD
        MOVLW     0xA0
        MOVWF    TMR1L           ; Value placed in TMR1L
        MOVLW    0x0F
        MOVWF    TMR1H           ; Value placed in TMR1H
        BCF      PIR1, TMR1IF    ; Clear Timer1 interrupt flag
        BSF      T1CON, TMR1ON   ; Turn on Timer1
        BRA      LOOP
        END
```

9.10

```
        INCLUDE  <P18F4321.INC>
        SEC      EQU    0x20
        MIN      EQU    0x21
        HOUR     EQU    0x22
        ORG      0X70
        MOVLW    0x30              ; Initialize STKPTR since subroutines
        MOVWF    STKPTR ; are used
        MOVLW    0x03
        MOVWF    T0CON             ; 1:16 prescale internal oscillator
        MOVLW    0x0B
        MOVWF    TMR0L             ; Value placed in TMR0L
        MOVLW    0xDC
        MOVWF    TMR0H             ; Value placed in TMR0H
        BCF      INTCON, TMR0IF    ; Clear Timer0 interrupt flag
        BSF      INTCON, TMR0IE    ; Enable Timer1 interrupt
        CLRF     SEC               ; Clear SEC register
        CLRF     MIN               ; Clear MIN register
        CLRF     HOUR              ; Clear hour register
        BSF      T0CON, TMR0ON     ; Turn on Timer0
LOOP    BTFSS    INTCON, TMR0IF
        GOTO     LOOP
        INCF     SEC
        MOVLW    D'60'             ; Move 60 into WREG
        CPFSLT   SEC               ; Compare value in SEC to 60 and skip
                                   ; if less
```

```
            CALL     INC_MIN          ; than; otherwise, CALL INC_MIN
                                      ; subroutine
            BSF      T0CON,TMR0ON
            BRA      LOOP
            ORG      0x100
INC_MIN     CLRF     SEC              ; Clear SEC register
            INCF     MIN              ; Increment the MIN register
            MOVLW    D'60'            ; Move 60 into WREG
            CPFSLT   MIN              ; Compare value in MIN to 60 and skip
                                      ; if less
CALL        INC_HOUR                  ; Otherwise, CALL  INC_HOUR
                                      ; subroutine
            RETURN
            ORG      0x200
INC_HOUR    CLRF     MIN              ; Clear MIN register
            INCF     HOUR             ; Increment the HOUR register
            MOVLW    D'25'            ; Move 25 into WREG
            CPFSLT   HOUR             ; Compare value in HOUR to 25 skip
                                      ; if less than
            CLRF     HOUR             ; Clear HOUR register
            RETURN
            END
```

Chapter 10

10.1 (a) TRISC = 0xFF;

10.1 (b) TRISA = 0;

10.12

```
#include <p18f4321.h>

void main(void)
{
        TRISC=0x00;                    // PortC is output
        T0CON=0x08;                    // Timer0 is 16-bit no prescaler
        TMR0L=0xA0;                    // Value placed in lower 8 bits of TMR0
        TMR0H=0x0F;                    // Value placed in upper 8-bits of TMR0
        INTCONbits.TMR0IF=0;           // Clear TMR0 interrupt flag
        INTCONbits.TMR0IE=1;           // Enable TMR0 interrupt
        T0CONbits.TMR0ON=1;            // Turn on TMR0
        while(1)
        {
        while(INTCONbits.TMR0IF==0);
        T0CONbits.TMR0ON=0;            // Turn off TMR0
```

```
        PORTCbits.RC0=~PORTCbits.RC0;// Change output of square wave
        TMR0L=0xA0;                  // Value placed in lower 8 bits of TMR0
        TMR0H=0x0F;                  // Value placed in upper 8-bits of TMR0
        INTCONbits.TMR0IF=0;         // Clear TMR0 interrupt flag
        T0CONbits.TMR0ON=1;          // Turn on TMR0
        }

}
```

10.16 (a)

A square wave with 50% duty cycle or a symmetrical square wave will be generated on the pin RB4.

10.25

```
#include <P18F4321.h>

void main()
{
    TRISC=1;                       // PORTC is input
    CCP1CON=0x05;                  // Capture mode, event on rising edge
    T1CON=0xC8;                    // Internal clock, no prescale
    TMR1L=0;                       // Clear TMR1L register
    TMR1H=0;                       // Clear TMR1H register
    PIR1bits.CCP1IF=0;             // Clear CCP1 interrupt flag

    while(PIR1bits.CCP1IF==0);    // Wait for first rising edge
    T1CONbits.TMR1ON=1;            // Turn on TMR1
    PIR1bits.CCP1IF=0;             // Clear CCP1 interrupt flag
    while(PIR1bits.CCP1IF==0);    // Wait for second rising edge
    T1CONbits.TMR1ON=0;            // Turn off TMR1
    while(1);                      // Period is found in registers CCPR1L and CCPR1H

}
```

APPENDIX

B

GLOSSARY

Absolute Addressing: This addressing mode specifies the address of data with the instruction.

Accumulator: Register used for storing the result after most ALU operations; available with 8-bit microcontrollers.

Address: A unique identification number (or locator) for source or destination of data. An address specifies the register or memory location of an operand involved in the instruction.

Addressing Mode: The manner in which a microcontroller determines the effective address of source and destination operands in an instruction.

Address Register: A register used to store the address (memory location) of data.

Address Space: The number of storage location in a microcontroller's memory that can be directly addressed by the CPU. The addressing range is determined by the number of address lines on the CPU.

American Standard Code for Information Interchange (ASCII): An 8-bit code commonly used with microcontrollers for representing alphanumeric codes. Decimal numbers 0 through 9 are represented by 30 (Hex) through 39 (Hex) in EBCDIC.

Analog-to-Digital (A/D) Converter: Transforms an analog voltage into its digital equivalent. The PIC18F microcontroller contains an on-chip A/D converter.

Arithmetic and Logic Unit (ALU): A digital circuit that performs arithmetic and logic operations on two *n*-bit numbers.

Assembler: A program that translates an assembly language program into a machine language program.

Assembly Language: A type of microcontroller programming language that uses a semi-English-language statement.

Asynchronous Operation: The execution of a sequence of steps such that each step is initiated upon completion of the previous step.

Base Address: An address that is used to convert all relative addresses in a program to absolute (machine) addresses.

Baud Rate: Rate of data transmission in bits per second.

Big Endian: This convention is used to store a 16-bit number such as 16-bit data in two bytes of memory locations as follows: the low memory address stores the high byte while the high memory address stores the low byte. The Motorola/Freescale HC11 8-bit microcontroller follows the big Endian format.

Binary-Coded Decimal (BCD): The representation of 10 decimal digits, 0 through 9, by their corresponding 4-bit binary number.

Bit: An abbreviation for a binary digit. A unit of information equal to one of two possible states (one or zero, on or off, true or false).

Branch: The branch instruction allows the computer to skip or jump out of program sequence to a designated instruction either unconditionally or conditionally (based on conditions such as carry or sign).

Breakpoint: Allows the user to execute the section of a program until one of the breakpoint conditions is met. It is then halted. The designer may then single step or examine memory and registers. Typically breakpoint conditions are program counter address or data references. Breakpoints are used in debugging assembly language programs.

Buffer: A temporary memory storage device designed to compensate for the different data rates between a transmitting device and a receiving device (for example, between a CPU and a peripheral). Current amplifiers are also referred to as buffers.

Bus: A collection of wires that interconnects microcontroller modules.

Bus Arbitration: Bus operation protocols (rules) that guarantee conflict-free access to a bus. Arbitration is the process of selecting one respondent from a collection of several candidates that concurrently request service.

Bus Cycle: The period of time in which a microcontroller carries out read or write operations.

Central Processing Unit (CPU): The brain of a microcontroller containing the ALU, register section, and control unit.

Chip: An Integrated Circuit (IC) package containing digital circuits.

CISC: Complex Instruction Set Computer. The Control unit is designed using microprogramming. Contains a large instruction set. Difficult to pipeline compared to RISC.

Clock: Timing signals providing synchronization among the various components in a microcontroller. Analogous to heart beats of a human being.

CMOS: Complementary MOS. Dissipates low power, offers high density and speed compared to TTL.

Compiler: A program that translates the source code written in a high-level programming language into machine language that is understandable to the microcontroller.

Computer: The basic blocks of a computer are the central processing unit (CPU), the memory, and the input/output (I/O).

Condition Code Register: Contains information such as carry, sign, zero, and overflow based on ALU operations.

Control Unit: Part of the CPU; its purpose is to translate or decode instructions read (fetched) from the main memory into the Instruction Register.

Data: Basic elements of information represented in binary form (that is, digits consisting of bits) that can be processed or produced by a microcontroller. Data represents any group of operands made up of numbers, letters, or symbols denoting any condition, value, or state. Typical microcontroller operand sizes include: a byte (8 bits), or a word which typically contains 2 bytes (16-bits).

Debugger: A program that executes and debugs the object program generated by the assembler or compiler. The debugger provides a single stepping, breakpoints, and program tracing.

Decoder: A chip, when enabled, that selects one of 2^n output lines based on *n* inputs.

Digital-to-Analog (D/A) Converter: Converts binary number to analog signal.

Diode: Two terminal electronic switch.

Directly Addressable Memory: The memory address space in which the microcontroller can directly execute programs. The maximum directly addressable memory is determined by the number of the microcontroller's address pins.

DRAM: See Dynamic RAM.

Duty Cycle: The duty cycle of a periodic waveform is defined as the percentage of the time the pulse is high in a clock period.

Dynamic RAM: Stores data as charges in capacitors and, therefore, must be refreshed since capacitors can hold charges for a few milliseconds. Hence, requires refresh circuitry.

EAROM (Electrically Alterable Read-Only Memory): Same as EEPROM or E^2 PROM. Can be programmed one line at a time without removing the memory from its sockets. This memory is also called read-mostly memory since it has much slower write times than read times.

Editor: A program that produces an error-free source program, written in assembly or high-level languages.

EEPROM or **E^2PROM:** Same as EAROM (see EAROM).

Effective Address: The final address used to carry out an instruction. Determined by the addressing mode.

EPROM (Erasable Programmable Read-Only Memory): Can be programmed. All programs in an EPROM chip can be erased using ultraviolet light. The chip must be removed from the circuit board for programming.

Exclusive-OR: The output is 0, if inputs are same; otherwise, the output is 1.

Extended Binary-Coded Decimal Interchange Code (EBCDIC): An 8-bit code sometimes used with computers for representing alphanumeric codes. Normally used by IBM. Decimal numbers 0 through 9 are represented by F0 (Hex) through F9 (Hex) in EBCDIC.

Firmware: Microprogram is sometimes referred to as firmware to distinguish it from hardwired control (purely hardware method).

Flag(s): An indicator, often a single bit, to indicate some conditions such as trace, carry, zero, and overflow.

Flash Memory: Utilizes a combination of EPROM and EEPROM technologies. Used in cellular phones and digital cameras. Also, used to hold program memory on the PIC18F microcontroller.

Flip-Flop: One-bit memory.

Gate: Digital circuits that perform logic operations.

Handshaking: Data transfer via exchange of control signals between the microprocessor and an external device.

Hardware: The physical electronic circuits (chips) that make up the microcontroller.

Hardwired Control: Used for designing the control unit using all hardware.

Harvard CPU Architecture: The CPU uses separate instruction and data memory units along with separate buses for instructions and data.

HCMOS: High-speed CMOS. Provides high density and consumes low power.

Hexadecimal Number System: Base-16 number system.

High-Level Language: A type of programming language that uses a more understandable human-oriented language such as C.

HMOS: High-density MOS reduces the channel length of the NMOS transistor and provides increased density and speed in VLSI circuits.

Immediate Address: An address that is used as an operand by the instruction itself.

Implied Address: An address is not specified, but is contained implicitly in the instruction.

Index: A number (typically 8-bit signed or 16-bit unsigned) is used to identify a particular element in an array (string). The index value typically contained in a register is utilized by the indexed addressing mode.

Indexed Addressing: The effective address of the instruction is determined by the sum of the address and the contents of the index register. Used to access arrays.

Index Register: A register used to hold a value when indexing data, such as when a value is used in indexed addressing to increment a base address contained within an instruction.

Indirect Address: A register holding a memory address to be accessed.

Instruction: Causes the microcontroller to carry out an operation on data. A program contains instructions and data.

Instruction Cycle: The sequence of operations that a microcontroller has to carry out while executing an instruction.

Instruction Register (IR): A register storing instructions.

Instruction Set: Lists all of the instructions that the microcontroller can execute.

Internal Interrupt: Activated internally by exceptional conditions such as completion of A/D conversion.

Interpreter: A program that executes a set of machine language instructions in response to each high-level statement in order to carry out the function.

Interrupt I/O: An external device can force the microcontroller to stop executing the current program temporarily so that it can execute another program known as the interrupt service routine.

Interrupts: A temporary break in a sequence of a program, initiated externally or internally, causing control to jump to a routine, which performs some action while the program is stopped.

I/O (Input/Output): Describes that portion of a microcontroller that exchanges data between the microcontroller system and an external device.

I/O Port: A register that contains control logic and data storage used to connect a microcontroller to external peripherals.

Inverting Buffer: Performs NOT operation. Current amplifier.

Keyboard: Has a number of push button-type switches configured in a matrix form (rows x columns).

Keybounce: When a mechanical switch opens or closes, it bounces (vibrates) for a small period of time (about 10-20 ms) before settling down.

Large-Scale Integration (LSI): An LSI chip contains 100 to 1000 gates.

LCD: Liquid Crystal Display. Displays numbers and several ASCII characters along with graphics. Furthermore, the LCD consumes low power. Because of inexpensive price of the LCD these days, they have been becoming popular. The LCDs are widely used in notebook computers.

LED: Light Emitting Diode. Typically, a current of 10 ma to 20 ma flows at 1.7 to 2.4 V drop across it.

Little Endian: This convention is used to store a 16-bit number such as 16-bit data in two bytes of memory locations as follows: the low memory address stores the low byte while the high memory address stores the high byte. The PIC18F microcontroller follows the little endian format.

Loops: A programming control structure where a sequence of microcontroller instructions are executed repeatedly (looped) until a terminating condition (result) is satisfied.

Machine Code: A binary code (composed of 1's and 0's) that a microcontroller understands.

Machine Language: A type of microntroller programming language that uses binary or hexadecimal numbers.

Macroinstruction: Commonly known as an instruction; initiates execution of a complete microprogram. Example includes assembly language instructions.

Macroprogram: The assembly language program.

Mask: A pattern of bits used to specify (or mask) which bit parts of another bit pattern are to be operated on and which bits are to be ignored or "masked" out. Uses logical AND operation.

Mask ROM: Programmed by a masking operation performed on the chip during the manufacturing process; its contents cannot be changed by user.

Maskable Interrupt: Can be enabled or disabled by executing typically the interrupt instructions.

Memory: Any storage device that can accept, retain, and read back data.

Memory Access Time: Average time taken to read a unit of information from the memory.

Memory Address Register (MAR): Stores the address of the data.

Memory Cycle Time: Average time lapse between two successive read operations.

Memory Map: A representation of the physical locations within a microcontroller's addressable main memory.

Memory-Mapped I/O: I/O ports are mapped as memory locations, with every connected device treated as if it were a memory location with a specific address. Manipulation of I/O data occurs in "interface registers" (as opposed to memory locations); hence there are no input (read) or output (write) instructions used in memory-mapped I/O.

Microcode: A set of instructions called "microinstructions" usually stored in a ROM in the control unit of a microcontroller's CPU to translate instructions of a higher-level programming language such as assembly language programming.

Microcomputer: Consists of a microprocessor, a memory unit, and an input/output unit.

Microcontroller: Typically includes a CPU, memory, I/O, timer, A/D (analog-to-digital) converter in the same chip.

Microinstruction: Some microcontrollers have an internal memory called control memory. This memory is used to store a number of codes called microinstructions. These microinstructions are combined to design the instruction set of the microcontroller.

Microprocessor: CPU on a single chip. The Central Processing Unit (CPU) of a microcomputer.

Microprogramming: Some microcontrollers use microprogramming to design the instruction set. Each instruction in the instruction register initiates execution of a microprogram stored typically in ROM inside the control unit to perform the required operation.

Multiplexer: A hardware device that selects one of *n* input lines and produces it on the output.

Nested Subroutine: A commonly used programming technique in which one subroutine calls another subroutine.

Nibble: A 4-bit word.

Non-inverting Buffer: Input is same as output. Current amplifier.

Nonmaskable Interrupt: Occurrence of this type of interrupt cannot be ignored by microcontroller, and even though interrupt capability of the microcontroller is disabled, its effect cannot be disabled by instruction.

Non-Multiplexed: A non-multiplexed microcontroller pin that assigns a unique function as opposed to a multiplexed microcontroller pin defining two functions on a time-shared basis.

Object Code: The binary (machine) code into which a source program is translated by a compiler, assembler, or interpreter.

One's Complement: Obtained by changing 1's to ' 0's, and 0's to 1's of a binary number.

One-Pass Assembler: This assembler goes through the assembly language program once and translates it into a machine language program. This assembler has the problem of defining forward references. See Two-Pass Assembler.

Opcode (Operation Code): Part of an instruction defining the operation to be performed.

Operand: A datum or information item involved in an operation from which the result is obtained as a consequence of defined addressing modes. Various operand types contain information, such as source address, destination address, or immediate data.

Operating System: Consists of a number of program modules to provide resource management. Typical resources include CPU, disks, and printers.

Page: Some microcontrollers divide the memory locations into equal blocks. Each of these blocks is called a page and contains several addresses.

Parallel Operation: Any operation carried out simultaneously with a related operation.

Parallel Transmission: Each bit of binary data is transmitted over a separate wire.

Parity: The number of 1's in a word is odd for odd parity and even for even parity.

Peripheral: An I/O device capable of being operated under the control of a CPU through communication channels. Examples include disk drives, keyboards, CRTs, printers, and modems.

Personal Computer: Low-cost, affordable microcomputer normally used by an individual for word processing and Internet applications.

Physical Address Space: Address space is defined by the address pins of the microcontroller.

Pipeline: A technique that allows a microcontroller processing operation to be broken down into several steps (dictated by the number of pipeline levels or stages) so that the individual step outputs can be handled by the microcontroller in parallel. Often used to fetch the processor's next instruction while executing the current instruction, which considerably speeds up the overall operation of the microcontroller. Overlaps instruction fetch with execution.

Pointer: A storage location (usually a register within a microcontroller) that contains the address of (or points to) a required item of data or subroutine.

Polled Interrupt: A software approach for determining the source of interrupt in a multiple interrupt system.

POP Operation: Reading from the top or bottom of a stack.

Port: A register through which the microcontrollers communicate with peripheral devices.

Primary or Main Memory: Storage that is considered internal to the microcontroller. The microcontroller can directly execute all instructions in the main memory. The maximum size of the main memory is defined by the number of address pins in the CPU.

Processor Memory: A set of CPU registers for holding temporary results when a computation is in progress.

Program: A self-contained sequence of computer software instructions (source code) that, when converted into machine code, directs the computer to perform specific operations for the purpose of accomplishing some processing task. Contains instructions and data.

Program Counter (PC): A register that normally contains the address of the next instruction to be executed in a program.

Programmed I/O: The microcontroller executes a program to perform all data transfers between the microcontroller system and external devices.

PROM (Programmable Read-Only Memory): Can be programmed by the user by using proper equipment. Once it is programmed, its contents cannot be altered.

Protocol: A list of data transmission rules or procedures that encompass the timing, control, formatting, and data representations by which two devices are to communicate. Also known as hardware "handshaking," which is used to permit asynchronous communication.

PUSH Operation: Writing to the top or bottom of a stack.

Random Access Memory (RAM): A read/write memory. RAMs (static or dynamic) are volatile in nature (in other words, information is lost when power is removed).

Read-Only-Memory (ROM): A memory in which any addressable operand can be read from, but not written to, after initial programming. ROM storage is nonvolatile (information is not lost after removal of power).

Reduced Instruction Set Computer (RISC): A simple instruction set is included. The RISC architecture maximizes speed by reducing clock cycles per instruction. The control unit is designed using hardwired control. Easier to implement pipelining.

Register: A high-speed memory usually constructed from flip-flops that are directly accessible to the CPU. It can contain either data or a specific location in memory that stores word(s) used during arithmetic, logic, and transfer operations.

Register Indirect: Uses a register that contains the address of data.

Relative Address: An address used to designate the position of a memory location in a routine or program.

RISC: See Reduced Instruction Set Computer.

Routine: A group of instructions for carrying out a specific processing operation. Usually refers to part of a larger program. A routine and subroutine have essentially the same meaning, but a subroutine could be interpreted as a self-contained routine nested within a routine or program.

SDRAM: Synchronous DRAM. This chip contains several DRAMs internally. The control signals and address inputs are sampled by the SDRAM by a common clock.

Secondary Memory Storage: An auxiliary data storage device that supplements the main (primary) memory of a computer. It is used to hold programs and data that would otherwise exceed the capacity of the main memory. Although it has a much slower access time, secondary storage is less expensive. Examples include floppy and hard disks.

Sequential Circuit: Combinational circuit with memory.

Serial Transmission: Only one line is used to transmit the complete binary data bit by bit.

Seven-Segment LED: Contains an LED in each of the seven segments. Can display numbers.

Signed Number A signed binary number includes both positive and negative numbers. It is represented in the microcontroller in two's complement form. For example, the decimal number +15 is represented in 8-bit two's complement form as 00001111 (binary) or 0F (hexadecimal). The decimal number -15 can be represented in 8-bit two's complement form as 11110001 (binary) or F1 (hexadecimal). Also, the most significant bit (MSB) of a signed number represents the sign of the number. For example, bit 7 of an 8-bit number, bit 15 of a 16-bit number, and bit 31 of a 32-bit number represent the signs of the respective numbers. A "0" at the MSB represents a positive number; a "1" at the MSB represents a negative number.

Single-Chip Microcomputer: Microcomputer (CPU, memory, and input/output) on a chip.

Single-chip Microprocessor: Microcomputer CPU (microprocessor) on a chip.

Single Step: Allows the user to execute a program one instruction at a time and examine contents of memory locations and registers.

Software: consists of a collection of programs that contain instructions and data for performing a specific task in a microcontroller.

Source Code: The assembly language program written by a programmer using assembly language instructions. This code must be translated to the object (machine) code by the assembler before it can be executed by the microcontroller.

SRAM: See Static RAM.

Stack: An area of read/write memory typically used by a microcontroller during subroutine calls or occurrence of an interrupt.The microcontroller saves in the stack the contents of the program counter before executing the subroutine or program counter contents, and other status information before executing the interrupt service routine. Thus, the microcontroller can return to the main program after execution of the subroutine or the interrupt service routine. The stack is a last in/first out (LIFO) read/write memory (RAM) that can also be manipulated by the programmer using PUSH and POP instructions.

Stack Pointer: A register used to address the stack.

Standard I/O: Utilizes a control pin on the CPU typically called the $M/\overline{IO}$ pin in order to distinguish between input/output and memory; IN and OUT instructions are used for input/output operations.

Static RAM: Also known as **SRAM**. Stores data in flip-flops; does not need to be refreshed. Information is lost upon power failure unless backed up by battery.

Status Register: A register that contains information concerning the flags in a microcontroller.

Subroutine: A program carrying out a particular function and which can be called by another program known as the main program. A subroutine needs to be placed in memory only once and can be called by the main program as many times as the programmer wants.

Synchronous Operation: Operations that occur at intervals directly related to a clock period.

Tracing: Allows single stepping. A dynamic diagnostic technique permits analysis (debugging) of the program's execution.

Tristate Buffer: Has three output states: logic 0, 1, and a high-impedance state. This chip is typically enabled by a control signal to provide logic 0 or 1 outputs. This type of buffer can also be disabled by the control signal to place it in a high-impedance state.

Two's Complement: The two's complement of a binary number is obtained by replacing each 0 with a 1 and each 1 with a 0 and adding one to the resulting number.

Two-Pass Assembler: This assembler goes through the assembly language program twice. In the first pass, the assembler assigns binary addresses to labels. In the second pass, the assembly program is translated into the machine language. No problem with forward branching.

Unsigned Number: An unsigned binary number has no arithmetic sign and, therefore, is always positive. Typical examples are your age or a memory address, which are always positive numbers. An 8-bit unsigned binary integer represents all numbers from 00 through FF (0 through 255 in decimal).

Very-Large-Scale Integration (VLSI): A VLSI chip contains more than 1000 gates. More commonly, a VLSI chip is identified by the number of transistors rather than the gate count.

von Neumann (Princeton) CPU Architecture: Uses a single memory unit and the same bus for accessing both instructions and data.

Word: The bit size of a microcontroller refers to the number of bits that can be processed simultaneously by the basic arithmetic and logic circuits of the CPU. A number of bits taken as a group in this manner is called a word.

APPENDIX C: PIC18F INSTRUCTION SET (Alphabetical Order)

Instruction	*Example*	*Operation*
ADDLW data8	ADDLW 0x07	[WREG] + 0x07 → [WREG]
ADDWF F, d, a	ADDWF 0x20, W	[WREG] + [0x20] → [WREG]
	ADDWF 0x20, F	[WREG] + [0x20] → [0x20]
ADDWFC F, d, a	ADDWFC 0x40, W	[WREG] + [0x40] + Carry → [WREG]
	ADDWFC 0x40, F	[WREG] + [0x40] + Carry → [0x40]
ANDLW data8	ANDLW 0x02	[WREG] AND 0x02 → [WREG]
ANDWF F, d, a	ANDWF 0x30, W	[WREG] AND [0x30] → [WREG]
	ANDWF 0x30, F	[WREG] AND [0x30] → [0x30]
BC d8	BC START	Branch to START if C = 1 where START is an 8-bit signed #
BCF F, b, a	BCF 0x30, 2	Clear bit number 2 to 0 in data register 0x30, store result in 0x30
	BCF STATUS, C	Clear the Carry Flag to 0 in the status register.
BN d8	BN START	Branch to START if N = 1 where START is an 8-bit signed #
BNC d8	BNC START	Branch to START if C ≠ 1 where START is an 8-bit signed #
BNN d8	BNN START	Branch to START if N = 0 where START is an 8-bit signed #
BNOV d8	BBNOV START	Branch to START if OV = 0 where START is an 8-bit signed #
BNZ d8	BNZ START	Branch to START if Z = 0 where START is an 8-bit signed #
BOV d8	BOV START	Branch to START if OV = 1 where START is an 8-bit signed #
BRA d8	BRA START	Branch always to START where START is an 8-bit signed #
BSF F, b, a	BSF 0x20, 7	Set bit number 7 to 1 in data register 0x20, store result in 0x20.
	BSF STATUS, C	Set the Carry Flag to 1 in the status register.
BTFSC F, b, a	BTFSC 0x50, 3	If bit number 3 in data register 0x50 is 0, skip the next instruction; otherwise, the next instruction is executed.

Instruction	*Example*	*Operation*
BTFSS F, b, a	BTFSC 0x40, 0	If bit number 0 in data register 0x40 is 1, skip the next instruction; otherwise, the next instruction is executed.
BTG F, b, a	BTG 0x20, 2	Invert (ones complement) bit number 2 in data register 0x20.
BZ d8	BZ START	Branch to START if Z = 1 where START is an 8-bit signed #
CALL k, s	CALL BEGIN	This is a two-word instruction. The simplest way to CALL a subroutine is when s = 0 (default); pushes current program counter (PC+4) which is also the return address , and loads PC with BEGIN which is the starting address of the subroutine.
CLRF F, a	CLRF 0x40	Clear the contents of data register to 0.
CLRWDT	CLRWDT	Reset the watchdog timer.
COMF F, d, a	COMF 0x30, F	One's complement each bit of [0x30], and store the result in 0x30.
	COMF 0x30, W	One's complement each bit of [0x30], and store the result in WREG.
CPFSEQ F, a	CPFSEQ 0x30	Unsigned comparison. If [0x30] =[WREG], skip the next instruction; else, the next instruction is executed.
CPFSGT F, a	CPFSGT 0x50	Unsigned comparison. If [0x50] >[WREG], skip the next instruction; else, the next instruction is executed.
CPFSLT F, a	CPFSLT 0x60	Unsigned comparison. If [0x60] <[WREG], skip the next instruction; else, the next instruction is executed.
DAW	DAW	Decimal Adjust [WREG] resulting from earlier addition of two packed BCD digits providing correct packed BCD result.
DECF F, d, a	DECF 0x20, W	Decrement [0x20] by 1, and store result in WREG.
	DECF 0x20, F	Decrement [0x20] by 1, and store result in 0x20.
DECFSZ F, d, a	DECFSZ 0x30, W	Decrement [0x30] by 1, and store result in WREG. If [WREG] = 0, skip the next instruction; else, execute the next instruction.
	DECFSZ 0x30, F	Decrement [0x30] by 1, and store the result in 0x30. If [0x30] = 0, skip the next instruction; else, execute the next instruction.

Instruction	*Example*	*Operation*
DECFSNZ F, d, a	DECFSNZ 0x50,W	Decrement [0x50] by 1, and store result in WREG. If [WREG] $\neq$ 0, skip the next instruction; else, execute the next instruction.
	DECFSNZ 0x50, F	Decrement [0x50] by 1, and store the result in 0x50. If [0x50] $\neq$ 0, skip the next instruction; else, execute the next instruction.
GOTO k	GOTO START	Unconditional branch to address START
INCF F, d, a	INCF 0x20, W	Increment [0x20] by 1, and store result in WREG.
	INCF 0x20, F	Increment [0x20] by 1, and store result in 0x20.
INCFSZ F, d, a	INCFSZ 0x30, W	Increment [0x30] by 1, and store result in WREG. If [WREG] = 0, skip the next instruction; else, execute the next instruction.
	INCFSZ 0x30, F	Increment [0x30] by 1, and store the result in 0x30. If [0x30] = 0, skip the next instruction; else, execute the next instruction.
INCFSNZ F, d, a	INCFSNZ 0x50, W	Increment [0x50] by 1, and store result in WREG. If [WREG] $\neq$ 0, skip the next instruction; else, execute the next instruction.
	INCFSNZ 0x50, F	Increment [0x50] by 1, and store the result in 0x50. If [0x50] $\neq$ 0, skip the next instruction; else, execute the next instruction.
IORLW k	IORLW 0x54	The contents of WREG are logically ORed with 0x54, and the result is stored in WREG.
IORWF F, d, a	IORWF 0x50, W	The contents of WREG are logically ORed with the contents of 0x50, and the result is stored in WREG.
	IORWF 0x50, F	The contents of WREG are logically ORed with the contents of 0x50, and the result is stored in 0x50.
LFSR F, k	LFSR 0, 0x0080	Load 00H into FSR0H, and 80H into FSR0L.
MOVF F, d, a	MOVF 0x30, W	The contents of 0x30 are loaded into WREG
	MOVF 0x30, F	The contents of 0x30 are copied into 0x30
MOVFF Fs, Fd	MOVFF 0x50,0x60	Move [0x50] into [0x60]. The contents of 0x50 are unchanged.

Instruction	*Example*	*Operation*
MOVLB k	MOVLF 0x04	Load BSR with 04H.
MOVLW k	MOVLW 0x21	Load WREG with 21H.
MOVWF F, a	MOVWF 0x50	Move the contents of WREG into 0x50.
MULLW k	MULLW 0xF1	[WREG] x F1H → [PRODH] [PRODL]; unsigned multiplication.
MULWF F, a	MULWF 0x30	[WREG] x [0x30] → [PRODH] [PRODL] ; unsigned multiplication.
NEGF F, a	NEGF 0x20	Negate the contents of 0x20 using two's complement.
NOP	NOP	No Operation
POP	POP	Discard top of stack pointed by SP, and decrement PC by 1.
PUSH	PUSH	Push or write PC onto the stack, and increment SP by 1.
RCALL n	RCALL START	Relative subroutine CALL. One-word instruction. Pushes PC+2 onto the hardware stack. START is an 11-bit signed number. Jumps to a subroutine located at an address (PC +2) + 2 x START.
RESET	RESET	Reset all registers and flags that are affected by a $\overline{\text{MCLR}}$ reset.
RETFIE	RETFIE	Return from Interrupt.
RETLW k	RETLW k	WREG is loaded with the 8-bit literal k, and PC is loaded with the return address from the top of stack.
RETURN	RETURN	Return from subroutine.
RLCF F, d, a	RLCF 0x20, W	Rotate [0x20] once to the left through Carry. Store result in WREG.
	RLCF 0x20, F	Rotate [0x20] once to the left through Carry. Store result in register 0x20.
RLNCF F, d, a	RLNCF 0x30, W	Rotate [0x30] once to the left without Carry. Store result in WREG.
	RLNCF 0x30, F	Rotate [0x30] once to the left without Carry. Store result in register 0x30.
RRCF F, d, a	RRCF 0x50, W	Rotate [0x50] once to the right through Carry. Store result in WREG.
	RRCF 0x50, F	Rotate [0x50] once to the right through Carry. Store result in register 0x50.
RRNCF F, d, a	RRNCF 0x60, W	Rotate [0x60] once to the right without Carry. Store result in WREG.
	RRNCF 0x60, F	Rotate [0x60] once to the right without Carry. Store result in register 0x60.
SETF F, a	SETF 0x30	The contents of 0x30 are set to 1's.
SLEEP	SLEEP	Enter Sleep mode.
SUBFWB F, d, a	SUBFWB 0x20, W	[WREG] –[0x20] – Carry → [WREG]

Instruction	*Example*	*Operation*
	SUBFWB 0x20, F	[WREG] –[0x20] – Carry → [0x20]
SUBLW k	SUBLW 0x05	[0x05] – [WREG] → [WREG]
SUBWF F, d, a	SUBWF 0x50, W	[0X50] – [WREG] → [WREG]
	SUBWF 0x50, F	[0X50] – [WREG] → [0x50]
SUBWFB F, d, a	SUBWFB 0x32, W	[0x32] – [WREG] – Carry → [WREG]
	SUBWFB 0x32, F	[0x32] – [WREG] – Carry → [0x32]
SWAPF F, d, a	SWAPF 0x30, W	The upper and lower 4 bits of register 0x30 are exchanged. The result is stored in WREG.
	SWAPF 0x30, F	The upper and lower 4 bits of register 0x30 are exchanged. The result is stored in register 0x30.
TBLRD	TBLRD	Table Read
TBLWT	TBLWT	Table Write
TSTFSZ F, a	TSTFSZ 0x50	If [0x50] = 0, skip the next instruction; else, execute the next instruction.
XORLW k	XORLW 0xF2	[WREG] XOR F2H → [WREG]
XORWF F, d, a	XORWF 0x30, W	[WREG] XOR [0x30] → [WREG]
	XORWF 0x30, F	[WREG] XOR [0x30] → [0x30]

APPENDIX D: PIC18F INTRUCTION SET — DETAILS

24.0 INSTRUCTION SET SUMMARY

PIC18F4321 family devices incorporate the standard set of 75 PIC18 core instructions, as well as an extended set of 8 new instructions, for the optimization of code that is recursive or that utilizes a software stack. The extended set is discussed later in this section.

24.1 Standard Instruction Set

The st andard PIC 18 ins truction s et ad ds many enhancements to the prev ious PIC® i nstruction sets, while maintaining an easy m igration from thes e PIC instruction sets. Most instructions are a single program memory word (16 bits), but there are four instructions that require two program memory locations.

Each si ngle-word in struction is a 16-bit word d ivided into an opcode, which specifies the instruction type and one or m ore op erands, which furth er sp ecify the operation of the instruction.

The instruction set is highly orthogonal and is grouped into four basic categories:

- **Byte-oriented** operations
- **Bit-oriented** operations
- **Literal** operations
- **Control** operations

The PIC18 instruction set summary in Table 24-2 lists **byte-oriented**, **bit-oriented**, **literal,** a nd **control** operations. T able 24-1 sh ows the o pcode fiel d descriptions.

Most **byte-oriented** instructions have three operands:

1. The file register (specified by 'f')
2. The destination of the result (specified by 'd')
3. The accessed memory (specified by 'a')

The f ile reg ister d esignator 'f' s pecifies w hich fil e register is to be used by the instruction. The destination designator 'd' specifies where the result of the operation is to be placed. If 'd' is zero, the result is placed in the WREG register. If 'd' is one, the result is placed in the file register specified in the instruction.

All **bit-oriented** instructions have three operands:

1. The file register (specified by 'f')
2. The bit in the file register (specified by 'b')
3. The accessed memory (specified by 'a')

The bit field designator 'b' selects the number of the bit affected b y th e op eration, w hile the file reg ister designator 'f' represents the number of the file in which the bit is located.

The **literal** instructions may use some of the following operands:

- A literal value to be loaded into a file register (specified by 'k')
- The desired FSR register to load the literal value into (specified by 'f')
- No operand required (specified by '—')

The **control** instructions may use some of the following operands:

- A program memory address (specified by 'n')
- The mode of the `CALL` or `RETURN` instructions (specified by 's')
- The mode of the table read and table write instructions (specified by 'm')
- No operand required (specified by '—')

All i nstructions are a s ingle word, e xcept fo r fo ur double-word i nstructions. The se in structions w ere made double-word to contain the required information in 32 bits. In the second word, the 4 MSbs are '`1`'s. If this second w ord is ex ecuted as a n ins truction (b y itself), it will execute as a `NOP`.

All sin gle-word instructions are ex ecuted in a si ngle instruction cycle, unless a conditional test is true or the program counter is changed as a result of the instruction. In these cases, the execution takes two instruction cycles, with the additional instruction cycle(s) executed as a `NOP`.

The double-word instructions execute in two instruction cycles.

One instruction cycle consists of four oscillator periods. Thus, for an oscillator frequency of 4 MHz, the normal instruction execution time is 1 μs. If a conditional test is true, or the program counter is changed as a result of an i nstruction, t he i nstruction execution time i s 2 μs. Two-word branch instructions (if true) would take 3 μs.

Figure 24-1 shows the general formats that the instructions can have. All examples use the convention 'nnh' to represent a hexadecimal number.

The I nstruction Se t Sum mary, shown in Table 24-2, lists th e st andard in structions r ecognized by th e Microchip MPASM™ Assembler.

Section 24.1.1 "Standard Ins truction Set" provides a description of each instruction.

TABLE 24-1: OPCODE FIELD DESCRIPTIONS

Field	Description
a	RAM access bit a = 0: RAM location in Access RAM (BSR register is ignored) a = 1: RAM bank is specified by BSR register
bbb	Bit address within an 8-bit file register (0 to 7).
BSR	Bank Select Register. Used to select the current RAM bank.
C, DC, Z, OV, N	ALU Status bits: **C**arry, **D**igit **C**arry, **Z**ero, **Ov**erflow, **N**egative.
d	Destination select bit d = 0: store result in WREG d = 1: store result in file register f
dest	Destination: either the WREG register or the specified register file location.
f	8-bit Register file address (00h to FFh) or 2-bit FSR designator (0h to 3h).
f_s	12-bit Register file address (000h to FFFh). This is the source address.
f_d	12-bit Register file address (000h to FFFh). This is the destination address.
GIE	Global Interrupt Enable bit.
k	Literal field, constant data or label (may be either an 8-bit, 12-bit or a 20-bit value).
label	Label name.
mm	The mode of the TBLPTR register for the table read and table write instructions. Only used with table read and table write instructions:
*	No change to register (such as TBLPTR with table reads and writes)
*+	Post-Increment register (such as TBLPTR with table reads and writes)
*-	Post-Decrement register (such as TBLPTR with table reads and writes)
+*	Pre-Increment register (such as TBLPTR with table reads and writes)
n	The relative address (2's complement number) for relative branch instructions or the direct address for Call/Branch and Return instructions.
PC	Program Counter.
PCL	Program Counter Low Byte.
PCH	Program Counter High Byte.
PCLATH	Program Counter High Byte Latch.
PCLATU	Program Counter Upper Byte Latch.
$\overline{\text{PD}}$	Power-down bit.
PRODH	Product of Multiply High Byte.
PRODL	Product of Multiply Low Byte.
s	Fast Call/Return mode select bit s = 0: do not update into/from shadow registers s = 1: certain registers loaded into/from shadow registers (Fast mode)
TBLPTR	21-bit Table Pointer (points to a Program Memory location).
TABLAT	8-bit Table Latch.
$\overline{\text{TO}}$	Time-out bit.
TOS	Top-of-Stack.
u	Unused or unchanged.
WDT	Watchdog Timer.
WREG	Working register (accumulator).
x	Don't care ('0' or '1'). The assembler will generate code with x = 0. It is the recommended form of use for compatibility with all Microchip software tools.
z_s	7-bit offset value for indirect addressing of register files (source).
z_d	7-bit offset value for indirect addressing of register files (destination).
{ }	Optional argument.
[text]	Indicates an indexed address.
(text)	The contents of text.
[expr]<n>	Specifies bit n of the register indicated by the pointer expr.
→	Assigned to.
< >	Register bit field.
∈	In the set of.
italics	User defined term (font is Courier).

FIGURE 24-1: GENERAL FORMAT FOR INSTRUCTIONS

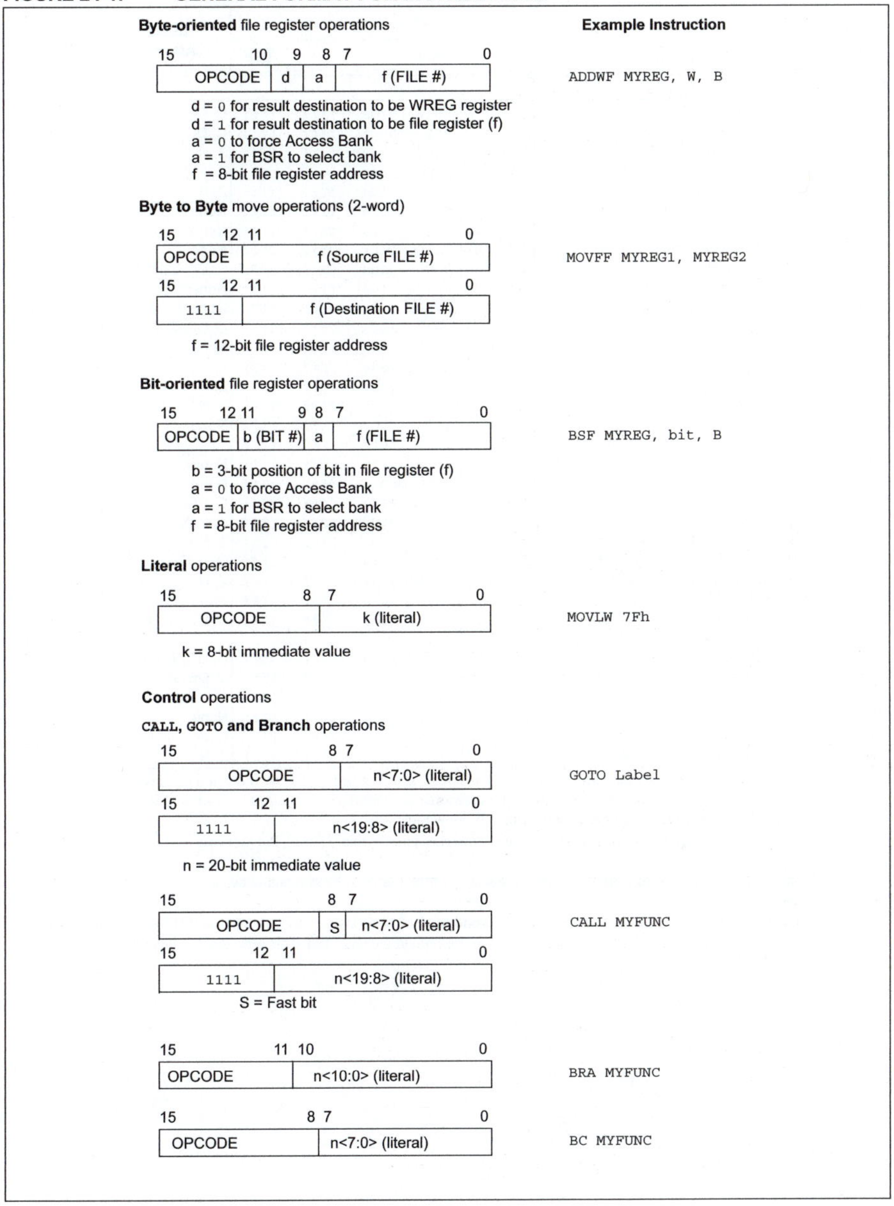

Example Instruction

Byte-oriented file register operations

15 ... 10	9	8	7 ... 0
OPCODE	d	a	f (FILE #)

`ADDWF MYREG, W, B`

d = `0` for result destination to be WREG register
d = `1` for result destination to be file register (f)
a = `0` to force Access Bank
a = `1` for BSR to select bank
f = 8-bit file register address

Byte to Byte move operations (2-word)

15 ... 12	11 ... 0
OPCODE	f (Source FILE #)

`MOVFF MYREG1, MYREG2`

15 ... 12	11 ... 0
`1111`	f (Destination FILE #)

f = 12-bit file register address

Bit-oriented file register operations

15 ... 12	11 ... 9	8	7 ... 0
OPCODE	b (BIT #)	a	f (FILE #)

`BSF MYREG, bit, B`

b = 3-bit position of bit in file register (f)
a = `0` to force Access Bank
a = `1` for BSR to select bank
f = 8-bit file register address

Literal operations

15 ... 8	7 ... 0
OPCODE	k (literal)

`MOVLW 7Fh`

k = 8-bit immediate value

Control operations

`CALL`, `GOTO` and Branch operations

15 ... 8	7 ... 0
OPCODE	n<7:0> (literal)

`GOTO Label`

15 ... 12	11 ... 0
`1111`	n<19:8> (literal)

n = 20-bit immediate value

15 ... 8		7 ... 0
OPCODE	S	n<7:0> (literal)

`CALL MYFUNC`

15 ... 12	11 ... 0
`1111`	n<19:8> (literal)

S = Fast bit

15 ... 11	10 ... 0
OPCODE	n<10:0> (literal)

`BRA MYFUNC`

15 ... 8	7 ... 0
OPCODE	n<7:0> (literal)

`BC MYFUNC`

TABLE 24-2: PIC18FXXXX INSTRUCTION SET

Mnemonic, Operands		Description	Cycles	16-Bit Instruction Word				Status Affected	Notes
				MSb			LSb		
BYTE-ORIENTED OPERATIONS									
ADDWF	f, d, a	Add WREG and f	1	0010	01da	ffff	ffff	C, DC, Z, OV, N	1, 2
ADDWFC	f, d, a	Add WREG and CARRY bit to f	1	0010	00da	ffff	ffff	C, DC, Z, OV, N	1, 2
ANDWF	f, d, a	AND WREG with f	1	0001	01da	ffff	ffff	Z, N	1,2
CLRF	f, a	Clear f	1	0110	101a	ffff	ffff	Z	2
COMF	f, d, a	Complement f	1	0001	11da	ffff	ffff	Z, N	1, 2
CPFSEQ	f, a	Compare f with WREG, skip =	1 (2 or 3)	0110	001a	ffff	ffff	None	4
CPFSGT	f, a	Compare f with WREG, skip >	1 (2 or 3)	0110	010a	ffff	ffff	None	4
CPFSLT	f, a	Compare f with WREG, skip <	1 (2 or 3)	0110	000a	ffff	ffff	None	1, 2
DECF	f, d, a	Decrement f	1	0000	01da	ffff	ffff	C, DC, Z, OV, N	1, 2, 3, 4
DECFSZ	f, d, a	Decrement f, Skip if 0	1 (2 or 3)	0010	11da	ffff	ffff	None	1, 2, 3, 4
DCFSNZ	f, d, a	Decrement f, Skip if Not 0	1 (2 or 3)	0100	11da	ffff	ffff	None	1, 2
INCF	f, d, a	Increment f	1	0010	10da	ffff	ffff	C, DC, Z, OV, N	1, 2, 3, 4
INCFSZ	f, d, a	Increment f, Skip if 0	1 (2 or 3)	0011	11da	ffff	ffff	None	4
INFSNZ	f, d, a	Increment f, Skip if Not 0	1 (2 or 3)	0100	10da	ffff	ffff	None	1, 2
IORWF	f, d, a	Inclusive OR WREG with f	1	0001	00da	ffff	ffff	Z, N	1, 2
MOVF	f, d, a	Move f	1	0101	00da	ffff	ffff	Z, N	1
MOVFF	f_s, f_d	Move f_s (source) to 1st word f_d (destination) 2nd word	2	1100 1111	ffff ffff	ffff ffff	ffff ffff	None	
MOVWF	f, a	Move WREG to f	1	0110	111a	ffff	ffff	None	
MULWF	f, a	Multiply WREG with f	1	0000	001a	ffff	ffff	None	1, 2
NEGF	f, a	Negate f	1	0110	110a	ffff	ffff	C, DC, Z, OV, N	
RLCF	f, d, a	Rotate Left f through Carry	1	0011	01da	ffff	ffff	C, Z, N	1, 2
RLNCF	f, d, a	Rotate Left f (No Carry)	1	0100	01da	ffff	ffff	Z, N	
RRCF	f, d, a	Rotate Right f through Carry	1	0011	00da	ffff	ffff	C, Z, N	
RRNCF	f, d, a	Rotate Right f (No Carry)	1	0100	00da	ffff	ffff	Z, N	
SETF	f, a	Set f	1	0110	100a	ffff	ffff	None	1, 2
SUBFWB	f, d, a	Subtract f from WREG with borrow	1	0101	01da	ffff	ffff	C, DC, Z, OV, N	
SUBWF	f, d, a	Subtract WREG from f	1	0101	11da	ffff	ffff	C, DC, Z, OV, N	1, 2
SUBWFB	f, d, a	Subtract WREG from f with borrow	1	0101	10da	ffff	ffff	C, DC, Z, OV, N	
SWAPF	f, d, a	Swap nibbles in f	1	0011	10da	ffff	ffff	None	4
TSTFSZ	f, a	Test f, skip if 0	1 (2 or 3)	0110	011a	ffff	ffff	None	1, 2
XORWF	f, d, a	Exclusive OR WREG with f	1	0001	10da	ffff	ffff	Z, N	

Note 1: When a PORT register is modified as a function of itself (e.g., `MOVF PORTB, 1, 0`), the value used will be that value present on the pins themselves. For example, if the data latch is '`1`' for a pin configured as input and is driven low by an external device, the data will be written back with a '`0`'.

2: If this instruction is executed on the TMR0 register (and where applicable, 'd' = `1`), the prescaler will be cleared if assigned.

3: If Program Counter (PC) is modified or a conditional test is true, the instruction requires two cycles. The second cycle is executed as a `NOP`.

4: Some instructions are two-word instructions. The second word of these instructions will be executed as a`NOP` unless the first word of the instruction retrieves the information embedded in these 16 bits. This ensures that all program memory locations have a valid instruction.

TABLE 24-2: PIC18FXXXX INSTRUCTION SET (CONTINUED)

Mnemonic, Operands		Description	Cycles	16-Bit Instruction Word MSb			LSb	Status Affected	Notes
BIT-ORIENTED OPERATIONS									
BCF	f, b, a	Bit Clear f	1	1001	bbba	ffff	ffff	None	1, 2
BSF	f, b, a	Bit Set f	1	1000	bbba	ffff	ffff	None	1, 2
BTFSC	f, b, a	Bit Test f, Skip if Clear	1 (2 or 3)	1011	bbba	ffff	ffff	None	3, 4
BTFSS	f, b, a	Bit Test f, Skip if Set	1 (2 or 3)	1010	bbba	ffff	ffff	None	3, 4
BTG	f, b, a	Bit Toggle f	1	0111	bbba	ffff	ffff	None	1, 2
CONTROL OPERATIONS									
BC	n	Branch if Carry	1 (2)	1110	0010	nnnn	nnnn	None	
BN	n	Branch if Negative	1 (2)	1110	0110	nnnn	nnnn	None	
BNC	n	Branch if Not Carry	1 (2)	1110	0011	nnnn	nnnn	None	
BNN	n	Branch if Not Negative	1 (2)	1110	0111	nnnn	nnnn	None	
BNOV	n	Branch if Not Overflow	1 (2)	1110	0101	nnnn	nnnn	None	
BNZ	n	Branch if Not Zero	1 (2)	1110	0001	nnnn	nnnn	None	
BOV	n	Branch if Overflow	1 (2)	1110	0100	nnnn	nnnn	None	
BRA	n	Branch Unconditionally	2	1101	0nnn	nnnn	nnnn	None	
BZ	n	Branch if Zero	1 (2)	1110	0000	nnnn	nnnn	None	
CALL	n, s	Call subroutine 1st word	2	1110	110s	kkkk	kkkk	None	
		2nd word		1111	kkkk	kkkk	kkkk		
CLRWDT	—	Clear Watchdog Timer	1	0000	0000	0000	0100	$\overline{TO}$, $\overline{PD}$	
DAW	—	Decimal Adjust WREG	1	0000	0000	0000	0111	C	
GOTO	n	Go to address 1st word	2	1110	1111	kkkk	kkkk	None	
		2nd word		1111	kkkk	kkkk	kkkk		
NOP	—	No Operation	1	0000	0000	0000	0000	None	
NOP	—	No Operation	1	1111	xxxx	xxxx	xxxx	None	4
POP	—	Pop top of return stack (TOS)	1	0000	0000	0000	0110	None	
PUSH	—	Push top of return stack (TOS)	1	0000	0000	0000	0101	None	
RCALL	n	Relative Call	2	1101	1nnn	nnnn	nnnn	None	
RESET		Software device Reset	1	0000	0000	1111	1111	All	
RETFIE	s	Return from interrupt enable	2	0000	0000	0001	000s	GIE/GIEH, PEIE/GIEL	
RETLW	k	Return with literal in WREG	2	0000	1100	kkkk	kkkk	None	
RETURN	s	Return from Subroutine	2	0000	0000	0001	001s	None	
SLEEP	—	Go into Standby mode	1	0000	0000	0000	0011	$\overline{TO}$, $\overline{PD}$	

Note 1: When a PORT register is modified as a function of itself (e.g., `MOVF PORTB, 1, 0`), the value used will be that value present on the pins themselves. For example, if the data latch is '`1`' for a pin configured as input and is driven low by an external device, the data will be written back with a '`0`'.

2: If this instruction is executed on the TMR0 register (and where applicable, 'd' = `1`), the prescaler will be cleared if assigned.

3: If Program Counter (PC) is modified or a conditional test is true, the instruction requires two cycles. The second cycle is executed as a `NOP`.

4: Some instructions are two-word instructions. The second word of these instructions will be executed as a`NOP` unless the first word of the instruction retrieves the information embedded in these 16 bits. This ensures that all program memory locations have a valid instruction.

TABLE 24-2: PIC18FXXXX INSTRUCTION SET (CONTINUED)

Mnemonic, Operands		Description	Cycles	16-Bit Instruction Word				Status Affected	Notes
				MSb			LSb		
LITERAL OPERATIONS									
ADDLW	k	Add literal and WREG	1	0000	1111	kkkk	kkkk	C, DC, Z, OV, N	
ANDLW	k	AND literal with WREG	1	0000	1011	kkkk	kkkk	Z, N	
IORLW	k	Inclusive OR literal with WREG	1	0000	1001	kkkk	kkkk	Z, N	
LFSR	f, k	Move literal (12-bit) 2nd word	2	1110	1110	00ff	kkkk	None	
		to FSR(f) 1st word		1111	0000	kkkk	kkkk		
MOVLB	k	Move literal to BSR<3:0>	1	0000	0001	0000	kkkk	None	
MOVLW	k	Move literal to WREG	1	0000	1110	kkkk	kkkk	None	
MULLW	k	Multiply literal with WREG	1	0000	1101	kkkk	kkkk	None	
RETLW	k	Return with literal in WREG	2	0000	1100	kkkk	kkkk	None	
SUBLW	k	Subtract WREG from literal	1	0000	1000	kkkk	kkkk	C, DC, Z, OV, N	
XORLW	k	Exclusive OR literal with WREG	1	0000	1010	kkkk	kkkk	Z, N	
DATA MEMORY ↔ PROGRAM MEMORY OPERATIONS									
TBLRD*		Table Read	2	0000	0000	0000	1000	None	
TBLRD*+		Table Read with post-increment		0000	0000	0000	1001	None	
TBLRD*-		Table Read with post-decrement		0000	0000	0000	1010	None	
TBLRD+*		Table Read with pre-increment		0000	0000	0000	1011	None	
TBLWT*		Table Write	2	0000	0000	0000	1100	None	
TBLWT*+		Table Write with post-increment		0000	0000	0000	1101	None	
TBLWT*-		Table Write with post-decrement		0000	0000	0000	1110	None	
TBLWT+*		Table Write with pre-increment		0000	0000	0000	1111	None	

Note 1: When a PORT register is modified as a function of itself (e.g., `MOVF PORTB, 1, 0`), the value used will be that value present on the pins themselves. For example, if the data latch is '`1`' for a pin configured as input and is driven low by an external device, the data will be written back with a '`0`'.

2: If this instruction is executed on the TMR0 register (and where applicable, 'd' = `1`), the prescaler will be cleared if assigned.

3: If Program Counter (PC) is modified or a conditional test is true, the instruction requires two cycles. The second cycle is executed as a `NOP`.

4: Some instructions are two-word instructions. The second word of these instructions will be executed as a`NOP` unless the first word of the instruction retrieves the information embedded in these 16 bits. This ensures that all program memory locations have a valid instruction.

24.1.1 STANDARD INSTRUCTION SET

ADDLW **ADD Literal to W**

Syntax: ADDLW k

Operands: $0 \le k \le 255$

Operation: (W) + k → W

Status Affected: N, OV, C, DC, Z

Encoding:

0000	1111	kkkk	kkkk

Description: The contents of W are added to the 8-bit literal 'k' and the result is placed in W.

Words: 1

Cycles: 1

Q Cycle Activity:

Q1	Q2	Q3	Q4
Decode	Read literal 'k'	Process Data	Write to W

Example: `ADDLW 15h`

Before Instruction
W = 10h
After Instruction
W = 25h

ADDWF **ADD W to f**

Syntax: ADDWF f {,d {,a}}

Operands: $0 \le f \le 255$
$d \in [0,1]$
$a \in [0,1]$

Operation: (W) + (f) → dest

Status Affected: N, OV, C, DC, Z

Encoding:

0010	01da	ffff	ffff

Description: Add W to register 'f'. If 'd' is '0', the result is stored in W. If 'd' is '1', the result is stored back in register 'f' (default).
If 'a' is '0', the Access Bank is selected. If 'a' is '1', the BSR is used to select the GPR bank (default).
If 'a' is '0' and the extended instruction set is enabled, this instruction operates in Indexed Literal Offset Addressing mode whenever $f \le 95$ (5Fh). See **Section 24.2.3 "Byte-Oriented and Bit-Oriented Instructions in Indexed Literal Offset Mode"** for details.

Words: 1

Cycles: 1

Q Cycle Activity:

Q1	Q2	Q3	Q4
Decode	Read register 'f'	Process Data	Write to destination

Example: `ADDWF REG, 0, 0`

Before Instruction
W = 17h
REG = 0C2h
After Instruction
W = 0D9h
REG = 0C2h

Note: All PIC18 instructions may take an optional label argument preceding the instruction mnemonic for use in symbolic addressing. If a label is used, the instruction format then becomes: {label} instruction argument(s).

ADDWFC **ADD W and CARRY bit to f**

Syntax: ADDWFC f {,d {,a}}

Operands: $0 \le f \le 255$
d ∈ [0,1]
a ∈ [0,1]

Operation: (W) + (f) + (C) → dest

Status Affected: N,OV, C, DC, Z

Encoding:

0010	00da	ffff	ffff

Description: Add W, the Carry flag and data memory location 'f'. If 'd' is '0', the result is placed in W. If 'd' is '1', the result is placed in data memory location 'f'.
If 'a' is '0', the Access Bank is selected. If 'a' is '1', the BSR is used to select the GPR bank (default).
If 'a' is '0' and the extended instruction set is enabled, this instruction operates in Indexed Literal Offset Addressing mode whenever $f \le 95$ (5Fh). See **Section 24.2.3 "Byte-Oriented and Bit-Oriented Instructions in Indexed Literal Offset Mode"** for details.

Words: 1

Cycles: 1

Q Cycle Activity:

Q1	Q2	Q3	Q4
Decode	Read register 'f'	Process Data	Write to destination

Example: `ADDWFC REG, 0, 1`

Before Instruction
Carry bit = 1
REG = 02h
W = 4Dh

After Instruction
Carry bit = 0
REG = 02h
W = 50h

ANDLW **AND Literal with W**

Syntax: ANDLW k

Operands: $0 \le k \le 255$

Operation: (W) .AND. k → W

Status Affected: N, Z

Encoding:

0000	1011	kkkk	kkkk

Description: The contents of W are ANDed with the 8-bit literal 'k'. The result is placed in W.

Words: 1

Cycles: 1

Q Cycle Activity:

Q1	Q2	Q3	Q4
Decode	Read literal 'k'	Process Data	Write to W

Example: `ANDLW 05Fh`

Before Instruction
W = A3h

After Instruction
W = 03h

ANDWF — AND W with f

Syntax: ANDWF f {,d {,a}}

Operands: $0 \le f \le 255$
$d \in [0,1]$
$a \in [0,1]$

Operation: (W) .AND. (f) → dest

Status Affected: N, Z

Encoding:

0001	01da	ffff	ffff

Description: The contents of W are ANDed with register 'f'. If 'd' is '0', the result is stored in W. If 'd' is '1', the result is stored back in register 'f' (default).
If 'a' is '0', the Access Bank is selected. If 'a' is '1', the BSR is used to select the GPR bank (default).
If 'a' is '0' and the extended instruction set is enabled, this instruction operates in Indexed Literal Offset Addressing mode whenever $f \le 95$ (5Fh). See **Section 24.2.3 "Byte-Oriented and Bit-Oriented Instructions in Indexed Literal Offset Mode"** for details.

Words: 1

Cycles: 1

Q Cycle Activity:

Q1	Q2	Q3	Q4
Decode	Read register 'f'	Process Data	Write to destination

Example: `ANDWF REG, 0, 0`

Before Instruction
W = 17h
REG = C2h
After Instruction
W = 02h
REG = C2h

BC — Branch if Carry

Syntax: BC n

Operands: $-128 \le n \le 127$

Operation: if Carry bit is '1'
(PC) + 2 + 2n → PC

Status Affected: None

Encoding:

1110	0010	nnnn	nnnn

Description: If the Carry bit is '1', then the program will branch.
The 2's complement number '2n' is added to the PC. Since the PC will have incremented to fetch the next instruction, the new address will be PC + 2 + 2n. This instruction is then a two-cycle instruction.

Words: 1

Cycles: 1(2)

Q Cycle Activity:
If Jump:

Q1	Q2	Q3	Q4
Decode	Read literal 'n'	Process Data	Write to PC
No operation	No operation	No operation	No operation

If No Jump:

Q1	Q2	Q3	Q4
Decode	Read literal 'n'	Process Data	No operation

Example: `HERE BC 5`

Before Instruction
PC = address (HERE)
After Instruction
If Carry = 1;
PC = addres s (HERE + 12)
If Carry = 0;
PC = addres s (HERE + 2)

BCF — Bit Clear f

Syntax:	BCF f, b {,a}
Operands:	$0 \le f \le 255$ $0 \le b \le 7$ $a \in [0,1]$
Operation:	$0 \rightarrow$ f<b>
Status Affected:	None
Encoding:	`1001` `bbba` `ffff` `ffff`
Description:	Bit 'b' in register 'f' is cleared. If 'a' is '0', the Access Bank is selected. If 'a' is '1', the BSR is used to select the GPR bank (default). If 'a' is '0' and the extended instruction set is enabled, this instruction operates in Indexed Literal Offset Addressing mode whenever $f \le 95$ (5Fh). See **Section 24.2.3 "Byte-Oriented and Bit-Oriented Instructions in Indexed Literal Offset Mode"** for details.
Words:	1
Cycles:	1

Q Cycle Activity:

Q1	Q2	Q3	Q4
Decode	Read register 'f'	Process Data	Write register 'f'

Example: `BCF FLAG_REG, 7, 0`

Before Instruction
FLAG_REG = C7h
After Instruction
FLAG_REG = 47h

BN — Branch if Negative

Syntax:	BN n
Operands:	$-128 \le n \le 127$
Operation:	if Negative bit is '1' (PC) + 2 + 2n $\rightarrow$ PC
Status Affected:	None
Encoding:	`1110` `0110` `nnnn` `nnnn`
Description:	If the Negative bit is '1', then the program will branch. The 2's complement number '2n' is added to the PC. Since the PC will have incremented to fetch the next instruction, the new address will be PC + 2 + 2n. This instruction is then a two-cycle instruction.
Words:	1
Cycles:	1(2)

Q Cycle Activity:
If Jump:

Q1	Q2	Q3	Q4
Decode	Read literal 'n'	Process Data	Write to PC
No operation	No operation	No operation	No operation

If No Jump:

Q1	Q2	Q3	Q4
Decode	Read literal 'n'	Process Data	No operation

Example: `HERE BN Jump`

Before Instruction
PC = address `(HERE)`
After Instruction
If Negative = 1;
PC = addres s `(Jump)`
If Negative = 0;
PC = addres s `(HERE + 2)`

BNC **Branch if Not Carry**

Syntax:	BNC n
Operands:	-128 ≤ n ≤ 127
Operation:	if Carry bit is '0' (PC) + 2 + 2n → PC
Status Affected:	None
Encoding:	1110 | 0011 | nnnn | nnnn
Description:	If the Carry bit is '0', then the program will branch. The 2's complement number '2n' is added to the PC. Since the PC will have incremented to fetch the next instruction, the new address will be PC + 2 + 2n. This instruction is then a two-cycle instruction.
Words:	1
Cycles:	1(2)

Q Cycle Activity:
If Jump:

Q1	Q2	Q3	Q4
Decode	Read literal 'n'	Process Data	Write to PC
No operation	No operation	No operation	No operation

If No Jump:

Q1	Q2	Q3	Q4
Decode	Read literal 'n'	Process Data	No operation

Example: `HERE BNC Jump`

Before Instruction
PC = address (HERE)
After Instruction
If Carry = 0;
PC = addres s (Jump)
If Carry = 1;
PC = addres s (HERE + 2)

BNN **Branch if Not Negative**

Syntax:	BNN n
Operands:	-128 ≤ n ≤ 127
Operation:	if Negative bit is '0' (PC) + 2 + 2n → PC
Status Affected:	None
Encoding:	1110 | 0111 | nnnn | nnnn
Description:	If the Negative bit is '0', then the program will branch. The 2's complement number '2n' is added to the PC. Since the PC will have incremented to fetch the next instruction, the new address will be PC + 2 + 2n. This instruction is then a two-cycle instruction.
Words:	1
Cycles:	1(2)

Q Cycle Activity:
If Jump:

Q1	Q2	Q3	Q4
Decode	Read literal 'n'	Process Data	Write to PC
No operation	No operation	No operation	No operation

If No Jump:

Q1	Q2	Q3	Q4
Decode	Read literal 'n'	Process Data	No operation

Example: `HERE BNN Jump`

Before Instruction
PC = address (HERE)
After Instruction
If Negative = 0;
PC = addres s (Jump)
If Negative = 1;
PC = addres s (HERE + 2)

BNOV **Branch if Not Overflow**

Syntax: BNOV n

Operands: $-128 \le n \le 127$

Operation: if Overflow bit is '0'
(PC) + 2 + 2n → PC

Status Affected: None

Encoding:

1110	0101	nnnn	nnnn

Description: If the Overflow bit is '0', then the program will branch.
The 2's complement number '2n' is added to the PC. Since the PC will have incremented to fetch the next instruction, the new address will be PC + 2 + 2n. This instruction is then a two-cycle instruction.

Words: 1

Cycles: 1(2)

Q Cycle Activity:
If Jump:

Q1	Q2	Q3	Q4
Decode	Read literal 'n'	Process Data	Write to PC
No operation	No operation	No operation	No operation

If No Jump:

Q1	Q2	Q3	Q4
Decode	Read literal 'n'	Process Data	No operation

Example: `HERE BNOV Jump`

Before Instruction
PC = address (HERE)
After Instruction
If Overflow = 0;
PC = addres s (Jump)
If Overflow = 1;
PC = addres s (HERE + 2)

BNZ **Branch if Not Zero**

Syntax: BNZ n

Operands: $-128 \le n \le 127$

Operation: if Zero bit is '0'
(PC) + 2 + 2n → PC

Status Affected: None

Encoding:

1110	0001	nnnn	nnnn

Description: If the Zero bit is '0', then the program will branch.
The 2's complement number '2n' is added to the PC. Since the PC will have incremented to fetch the next instruction, the new address will be PC + 2 + 2n. This instruction is then a two-cycle instruction.

Words: 1

Cycles: 1(2)

Q Cycle Activity:
If Jump:

Q1	Q2	Q3	Q4
Decode	Read literal 'n'	Process Data	Write to PC
No operation	No operation	No operation	No operation

If No Jump:

Q1	Q2	Q3	Q4
Decode	Read literal 'n'	Process Data	No operation

Example: `HERE BNZ Jump`

Before Instruction
PC = address (HERE)
After Instruction
If Zero = 0;
PC = addres s (Jump)
If Zero = 1;
PC = addres s (HERE + 2)

BRA **Unconditional Branch**

Syntax:	BRA n
Operands:	$-1024 \leq n \leq 1023$
Operation:	$(PC) + 2 + 2n \rightarrow PC$
Status Affected:	None
Encoding:	`1101` `0nnn` `nnnn` `nnnn`
Description:	Add the 2's complement number '2n' to the PC. Since the PC will have incremented to fetch the next instruction, the new address will be PC + 2 + 2n. This instruction is a two-cycle instruction.
Words:	1
Cycles:	2

Q Cycle Activity:

Q1	Q2	Q3	Q4
Decode	Read literal 'n'	Process Data	Write to PC
No operation	No operation	No operation	No operation

Example: `HERE BRA Jump`

Before Instruction
PC = address `(HERE)`
After Instruction
PC = addres s `(Jump)`

BSF **Bit Set f**

Syntax:	BSF f, b {,a}
Operands:	$0 \leq f \leq 255$ $0 \leq b \leq 7$ $a \in [0,1]$
Operation:	$1 \rightarrow$ f<b>
Status Affected:	None
Encoding:	`1000` `bbba` `ffff` `ffff`
Description:	Bit 'b' in register 'f' is set. If 'a' is '0', the Access Bank is selected. If 'a' is '1', the BSR is used to select the GPR bank (default). If 'a' is '0' and the extended instruction set is enabled, this instruction operates in Indexed Literal Offset Addressing mode whenever $f \leq 95$ (5Fh). See **Section 24.2.3 "Byte-Oriented and Bit-Oriented Instructions in Indexed Literal Offset Mode"** for details.
Words:	1
Cycles:	1

Q Cycle Activity:

Q1	Q2	Q3	Q4
Decode	Read register 'f'	Process Data	Write register 'f'

Example: `BSF FLAG_REG, 7, 1`

Before Instruction
FLAG_REG = 0Ah
After Instruction
FLAG_REG = 8Ah

BTFSC **Bit Test File, Skip if Clear**

Syntax: BTFSC f, b {,a}

Operands: $0 \le f \le 255$
$0 \le b \le 7$
$a \in [0,1]$

Operation: skip if (f<b>) = 0

Status Affected: None

Encoding:

1011	bbba	ffff	ffff

Description: If bit 'b' in register 'f' is '0', then the next instruction is skipped. If bit 'b' is '0', then the next instruction fetched during the current instruction execution is discarded and a NOP is executed instead, making this a two-cycle instruction.
If 'a' is '0', the Access Bank is selected. If 'a' is '1', the BSR is used to select the GPR bank (default).
If 'a' is '0' and the extended instruction set is enabled, this instruction operates in Indexed Literal Offset Addressing mode whenever $f \le 95$ (5Fh).
See **Section 24.2.3 "Byte-Oriented and Bit-Oriented Instructions in Indexed Literal Offset Mode"** for details.

Words: 1

Cycles: 1(2)
Note: 3 cycles if skip and followed by a 2-word instruction.

Q Cycle Activity:

Q1	Q2	Q3	Q4
Decode	Read register 'f'	Process Data	No operation

If skip:

Q1	Q2	Q3	Q4
No operation	No operation	No operation	No operation

If skip and followed by 2-word instruction:

Q1	Q2	Q3	Q4
No operation	No operation	No operation	No operation
No operation	No operation	No operation	No operation

Example:

```
HERE    BTFSC   FLAG, 1, 0
FALSE   :
TRUE    :
```

Before Instruction
PC = address (HERE)
After Instruction
If FLAG<1> = 0;
PC = addres s (TRUE)
If FLAG<1> = 1;
PC = addres s (FALSE)

BTFSS **Bit Test File, Skip if Set**

Syntax: BTFSS f, b {,a}

Operands: $0 \le f \le 255$
$0 \le b < 7$
$a \in [0,1]$

Operation: skip if (f<b>) = 1

Status Affected: None

Encoding:

1010	bbba	ffff	ffff

Description: If bit 'b' in register 'f' is '1', then the next instruction is skipped. If bit 'b' is '1', then the next instruction fetched during the current instruction execution is discarded and a NOP is executed instead, making this a two-cycle instruction.
If 'a' is '0', the Access Bank is selected. If 'a' is '1', the BSR is used to select the GPR bank (default).
If 'a' is '0' and the extended instruction set is enabled, this instruction operates in Indexed Literal Offset Addressing mode whenever $f \le 95$ (5Fh).
See **Section 24.2.3 "Byte-Oriented and Bit-Oriented Instructions in Indexed Literal Offset Mode"** for details.

Words: 1

Cycles: 1(2)
Note: 3 cycles if skip and followed by a 2-word instruction.

Q Cycle Activity:

Q1	Q2	Q3	Q4
Decode	Read register 'f'	Process Data	No operation

If skip:

Q1	Q2	Q3	Q4
No operation	No operation	No operation	No operation

If skip and followed by 2-word instruction:

Q1	Q2	Q3	Q4
No operation	No operation	No operation	No operation
No operation	No operation	No operation	No operation

Example:

```
HERE    BTFSS   FLAG, 1, 0
FALSE   :
TRUE    :
```

Before Instruction
PC = addres s (HERE)
After Instruction
If FLAG<1> = 0;
PC = addres s (FALSE)
If FLAG<1> = 1;
PC = addres s (TRUE)

BTG **Bit Toggle f**

Syntax:	BTG f, b {,a}
Operands:	$0 \le f \le 255$ $0 \le b < 7$ $a \in [0,1]$
Operation:	$\overline{(f<b>)} \rightarrow f<b>$
Status Affected:	None
Encoding:	`0111` `bbba` `ffff` `ffff`
Description:	Bit 'b' in data memory location 'f' is inverted. If 'a' is '0', the Access Bank is selected. If 'a' is '1', the BSR is used to select the GPR bank (default). If 'a' is '0' and the extended instruction set is enabled, this instruction operates in Indexed Literal Offset Addressing mode whenever f ≤ 95 (5Fh). See **Section 24.2.3 "Byte-Oriented and Bit-Oriented Instructions in Indexed Literal Offset Mode"** for details.
Words:	1
Cycles:	1

Q Cycle Activity:

Q1	Q2	Q3	Q4
Decode	Read register 'f'	Process Data	Write register 'f'

<u>Example:</u> `BTG PORTC, 4, 0`

Before Instruction:
PORTC = `0111 0101` [75h]
After Instruction:
PORTC = `0110 0101` [65h]

BOV **Branch if Overflow**

Syntax:	BOV n
Operands:	$-128 \le n \le 127$
Operation:	if Overflow bit is '1' (PC) + 2 + 2n → PC
Status Affected:	None
Encoding:	`1110` `0100` `nnnn` `nnnn`
Description:	If the Overflow bit is '1', then the program will branch. The 2's complement number '2n' is added to the PC. Since the PC will have incremented to fetch the next instruction, the new address will be PC + 2 + 2n. This instruction is then a two-cycle instruction.
Words:	1
Cycles:	1(2)

Q Cycle Activity:
If Jump:

Q1	Q2	Q3	Q4
Decode	Read literal 'n'	Process Data	Write to PC
No operation	No operation	No operation	No operation

If No Jump:

Q1	Q2	Q3	Q4
Decode	Read literal 'n'	Process Data	No operation

<u>Example:</u> `HERE BOV Jump`

Before Instruction
PC = address `(HERE)`
After Instruction
If Overflow = 1;
PC = addres s `(Jump)`
If Overflow = 0;
PC = addres s `(HERE + 2)`

BZ — Branch if Zero

Syntax: BZ n

Operands: $-128 \le n \le 127$

Operation: if Zero bit is '1'
(PC) + 2 + 2n → PC

Status Affected: None

Encoding:

1110	0000	nnnn	nnnn

Description: If the Zero bit is '1', then the program will branch.
The 2's complement number '2n' is added to the PC. Since the PC will have incremented to fetch the next instruction, the new address will be PC + 2 + 2n. This instruction is then a two-cycle instruction.

Words: 1

Cycles: 1(2)

Q Cycle Activity:
If Jump:

Q1	Q2	Q3	Q4
Decode	Read literal 'n'	Process Data	Write to PC
No operation	No operation	No operation	No operation

If No Jump:

Q1	Q2	Q3	Q4
Decode	Read literal 'n'	Process Data	No operation

Example: `HERE BZ Jump`

Before Instruction
PC = address (HERE)
After Instruction
If Zero = 1;
PC = addres s (Jump)
If Zero = 0;
PC = addres s (HERE + 2)

CALL — Subroutine Call

Syntax: CALL k {,s}

Operands: $0 \le k \le 1048575$
s ∈ [0,1]

Operation: (PC) + 4 → TOS,
k → PC<20:1>,
if s = 1
(W) → WS,
(STATUS) → STATUSS,
(BSR) → BSRS

Status Affected: None

Encoding:

1st word (k<7:0>)	1110	110s	$k_7$kkk	kkkk_0
2nd word(k<19:8>)	1111	k_{19}kkk	kkkk	kkkk_8

Description: Subroutine call of entire 2-Mbyte memory range. First, return address (PC + 4) is pushed onto the return stack. If 's' = 1, the W, STATUS and BSR registers are also pushed into their respective shadow registers, WS, STATUSS and BSRS. If 's' = 0, no update occurs (default). Then, the 20-bit value 'k' is loaded into PC<20:1>. CALL is a two-cycle instruction.

Words: 2

Cycles: 2

Q Cycle Activity:

Q1	Q2	Q3	Q4
Decode	Read literal 'k'<7:0>,	PUSH PC to stack	Read literal 'k'<19:8>, Write to PC
No operation	No operation	No operation	No operation

Example: `HERE CALL THERE, 1`

Before Instruction
PC = address (HERE)
After Instruction
PC = address (THERE)
TOS = address (HERE + 4)
WS = W
BSRS = BSR
STATUSS = S TATUS

CLRF **Clear f**

Syntax:	CLRF f {,a}
Operands:	$0 \le f \le 255$ $a \in [0,1]$
Operation:	000h → f 1 → Z
Status Affected:	Z
Encoding:	`0110` `101a` `ffff` `ffff`
Description:	Clears the contents of the specified register. If 'a' is '0', the Access Bank is selected. If 'a' is '1', the BSR is used to select the GPR bank (default). If 'a' is '0' and the extended instruction set is enabled, this instruction operates in Indexed Literal Offset Addressing mode whenever $f \le 95$ (5Fh). See **Section 24.2.3 "Byte-Oriented and Bit-Oriented Instructions in Indexed Literal Offset Mode"** for details.
Words:	1
Cycles:	1

Q Cycle Activity:

Q1	Q2	Q3	Q4
Decode	Read register 'f'	Process Data	Write register 'f'

Example: `CLRF FLAG_REG, 1`

Before Instruction
FLAG_REG = 5Ah
After Instruction
FLAG_REG = 00h

CLRWDT **Clear Watchdog Timer**

Syntax:	CLRWDT
Operands:	None
Operation:	000h → WDT, 000h → WDT postscaler, 1 → $\overline{TO}$, 1 → $\overline{PD}$
Status Affected:	$\overline{TO}$, $\overline{PD}$
Encoding:	`0000` `0000` `0000` `0100`
Description:	`CLRWDT` instruction resets the Watchdog Timer. It also resets the postscaler of the WDT. Status bits, $\overline{TO}$ and $\overline{PD}$, are set.
Words:	1
Cycles:	1

Q Cycle Activity:

Q1	Q2	Q3	Q4
Decode	No operation	Process Data	No operation

Example: `CLRWDT`

Before Instruction
WDT Counter = ?
After Instruction
WDT Counter = 00h
WDT Postscaler = 0
$\overline{TO}$ = 1
$\overline{PD}$ = 1

COMF — Complement f

Syntax: COMF f {,d {,a}}

Operands: $0 \le f \le 255$
d ∈ [0,1]
a ∈ [0,1]

Operation: $(\bar{f}) \rightarrow$ dest

Status Affected: N, Z

Encoding:

0001	11da	ffff	ffff

Description: The contents of register 'f' are complemented. If 'd' is '0', the result is stored in W. If 'd' is '1', the result is stored back in register 'f' (default).
If 'a' is '0', the Access Bank is selected. If 'a' is '1', the BSR is used to select the GPR bank (default).
If 'a' is '0' and the extended instruction set is enabled, this instruction operates in Indexed Literal Offset Addressing mode whenever $f \le 95$ (5Fh). See **Section 24.2.3 "Byte-Oriented and Bit-Oriented Instructions in Indexed Literal Offset Mode"** for details.

Words: 1

Cycles: 1

Q Cycle Activity:

Q1	Q2	Q3	Q4
Decode	Read register 'f'	Process Data	Write to destination

Example: `COMF REG, 0, 0`

Before Instruction
REG = 13h
After Instruction
REG = 13h
W = ECh

CPFSEQ — Compare f with W, Skip if f = W

Syntax: CPFSEQ f {,a}

Operands: $0 \le f \le 255$
a ∈ [0,1]

Operation: (f) – (W),
skip if (f) = (W)
(unsigned comparison)

Status Affected: None

Encoding:

0110	001a	ffff	ffff

Description: Compares the contents of data memory location 'f' to the contents of W by performing an unsigned subtraction.
If 'f' = W, then the fetched instruction is discarded and a NOP is executed instead, making this a two-cycle instruction.
If 'a' is '0', the Access Bank is selected. If 'a' is '1', the BSR is used to select the GPR bank (default).
If 'a' is '0' and the extended instruction set is enabled, this instruction operates in Indexed Literal Offset Addressing mode whenever $f \le 95$ (5Fh). See **Section 24.2.3 "Byte-Oriented and Bit-Oriented Instructions in Indexed Literal Offset Mode"** for details.

Words: 1

Cycles: 1(2)
Note: 3 cycles if skip and followed by a 2-word instruction.

Q Cycle Activity:

Q1	Q2	Q3	Q4
Decode	Read register 'f'	Process Data	No operation

If skip:

Q1	Q2	Q3	Q4
No operation	No operation	No operation	No operation

If skip and followed by 2-word instruction:

Q1	Q2	Q3	Q4
No operation	No operation	No operation	No operation
No operation	No operation	No operation	No operation

Example:

```
HERE      CPFSEQ REG, 0
NEQUAL    :
EQUAL     :
```

Before Instruction
PC Address = HERE
W = ?
REG = ?
After Instruction
If REG = W;
PC = Address (EQUAL)
If REG ≠ W;
PC = Address (NEQUAL)

CPFSGT **Compare f with W, Skip if f > W**

Syntax: CPFSGT f {,a}

Operands: $0 \le f \le 255$
$a \in [0,1]$

Operation: (f) – (W),
skip if (f) > (W)
(unsigned comparison)

Status Affected: None

Encoding:

0110	010a	ffff	ffff

Description: Compares the contents of data memory location 'f' to the contents of the W by performing an unsigned subtraction.
If the contents of 'f' are greater than the contents of WREG, then the fetched instruction is discarded and a `NOP` is executed instead, making this a two-cycle instruction.
If 'a' is '0', the Access Bank is selected. If 'a' is '1', the BSR is used to select the GPR bank (default).
If 'a' is '0' and the extended instruction set is enabled, this instruction operates in Indexed Literal Offset Addressing mode whenever $f \le 95$ (5Fh). See **Section 24.2.3 "Byte-Oriented and Bit-Oriented Instructions in Indexed Literal Offset Mode"** for details.

Words: 1

Cycles: 1(2)
Note: 3 cycles if skip and followed by a 2-word instruction.

Q Cycle Activity:

Q1	Q2	Q3	Q4
Decode	Read register 'f'	Process Data	No operation

If skip:

Q1	Q2	Q3	Q4
No operation	No operation	No operation	No operation

If skip and followed by 2-word instruction:

Q1	Q2	Q3	Q4
No operation	No operation	No operation	No operation
No operation	No operation	No operation	No operation

Example:

```
HERE        CPFSGT REG, 0
NGREATER    :
GREATER     :
```

Before Instruction
PC = Address (HERE)
W = ?
After Instruction
If REG > W;
PC = Address (GREATER)
If REG ≤ W;
PC = Addr ess (NGREATER)

CPFSLT **Compare f with W, Skip if f < W**

Syntax: CPFSLT f {,a}

Operands: $0 \le f \le 255$
$a \in [0,1]$

Operation: (f) – (W),
skip if (f) < (W)
(unsigned comparison)

Status Affected: None

Encoding:

0110	000a	ffff	ffff

Description: Compares the contents of data memory location 'f' to the contents of W by performing an unsigned subtraction.
If the contents of 'f' are less than the contents of W, then the fetched instruction is discarded and a `NOP` is executed instead, making this a two-cycle instruction.
If 'a' is '0', the Access Bank is selected. If 'a' is '1', the BSR is used to select the GPR bank (default).

Words: 1

Cycles: 1(2)
Note: 3 cycles if skip and followed by a 2-word instruction.

Q Cycle Activity:

Q1	Q2	Q3	Q4
Decode	Read register 'f'	Process Data	No operation

If skip:

Q1	Q2	Q3	Q4
No operation	No operation	No operation	No operation

If skip and followed by 2-word instruction:

Q1	Q2	Q3	Q4
No operation	No operation	No operation	No operation
No operation	No operation	No operation	No operation

Example:

```
HERE     CPFSLT REG, 1
NLESS    :
LESS     :
```

Before Instruction
PC = Address (HERE)
W = ?
After Instruction
If REG < W;
PC = Addr ess (LESS)
If REG ≥ W;
PC = Address (NLESS)

DAW **Decimal Adjust W Register**

Syntax:	DAW
Operands:	None

Operation:	If [W<3:0> > 9] or [DC = 1] then
	(W<3:0>) + 6 → W<3:0>;
	else
	(W<3:0>) → W<3:0>
	
	If [W<7:4> + DC > 9] or [C = 1] then
	(W<7:4>) + 6 + DC → W<7:4>;
	else
	(W<7:4>) + DC → W<7:4>

Status Affected:	C
Encoding:	0000 0000 0000 0111
Description:	DAW adjusts the eight-bit value in W, resulting from the earlier addition of two variables (each in packed BCD format) and produces a correct packed BCD result.
Words:	1
Cycles:	1

Q Cycle Activity:

Q1	Q2	Q3	Q4
Decode	Read register W	Process Data	Write W

Example 1:

```
DAW
```

Before Instruction
W = A5h
C = 0
DC = 0
After Instruction
W = 05h
C = 1
DC = 0

Example 2:

Before Instruction
W = CEh
C = 0
DC = 0
After Instruction
W = 34h
C = 1
DC = 0

DECF **Decrement f**

Syntax:	DECF f {,d {,a}}
Operands:	$0 \le f \le 255$ $d \in [0,1]$ $a \in [0,1]$
Operation:	(f) – 1 → dest
Status Affected:	C, DC, N, OV, Z
Encoding:	0000 01da ffff ffff
Description:	Decrement register 'f'. If 'd' is '0', the result is stored in W. If 'd' is '1', the result is stored back in register 'f' (default). If 'a' is '0', the Access Bank is selected. If 'a' is '1', the BSR is used to select the GPR bank (default). If 'a' is '0' and the extended instruction set is enabled, this instruction operates in Indexed Literal Offset Addressing mode whenever f ≤ 95 (5Fh). See **Section 24.2.3 "Byte-Oriented and Bit-Oriented Instructions in Indexed Literal Offset Mode"** for details.
Words:	1
Cycles:	1

Q Cycle Activity:

Q1	Q2	Q3	Q4
Decode	Read register 'f'	Process Data	Write to destination

Example: `DECF CNT, 1, 0`

Before Instruction
CNT = 01h
Z = 0
After Instruction
CNT = 00h
Z = 1

DECFSZ — Decrement f, Skip if 0

Syntax: DECFSZ f{,d {,a}}

Operands: $0 \le f \le 255$
$d \in [0,1]$
$a \in [0,1]$

Operation: (f) – 1 → dest,
skip if result = 0

Status Affected: None

Encoding:

0010	11da	ffff	ffff

Description: The contents of register 'f' are decremented. If 'd' is '0', the result is placed in W. If 'd' is '1', the result is placed back in register 'f' (default).
If the result is '0', the next instruction, which is already fetched, is discarded and a NOP is executed instead, making it a two-cycle instruction.
If 'a' is '0', the Access Bank is selected. If 'a' is '1', the BSR is used to select the GPR bank (default).
If 'a' is '0' and the extended instruction set is enabled, this instruction operates in Indexed Literal Offset Addressing mode whenever $f \le 95$ (5Fh). See **Section 24.2.3 "Byte-Oriented and Bit-Oriented Instructions in Indexed Literal Offset Mode"** for details.

Words: 1

Cycles: 1(2)
Note: 3 cycles if skip and followed by a 2-word instruction.

Q Cycle Activity:

Q1	Q2	Q3	Q4
Decode	Read register 'f'	Process Data	Write to destination

If skip:

Q1	Q2	Q3	Q4
No operation	No operation	No operation	No operation

If skip and followed by 2-word instruction:

Q1	Q2	Q3	Q4
No operation	No operation	No operation	No operation
No operation	No operation	No operation	No operation

Example:

```
HERE      DECFSZ  CNT, 1, 1
          GOTO    LOOP
CONTINUE
```

Before Instruction
PC = Address (HERE)
After Instruction
CNT = CNT – 1
If CNT = 0;
PC = Address (CONTINUE)
If CNT ≠ 0;
PC = Address (HERE + 2)

DCFSNZ — Decrement f, Skip if Not 0

Syntax: DCFSNZ f{,d {,a}}

Operands: $0 \le f \le 255$
$d \in [0,1]$
$a \in [0,1]$

Operation: (f) – 1 → dest,
skip if result ≠ 0

Status Affected: None

Encoding:

0100	11da	ffff	ffff

Description: The contents of register 'f' are decremented. If 'd' is '0', the result is placed in W. If 'd' is '1', the result is placed back in register 'f' (default).
If the result is not '0', the next instruction, which is already fetched, is discarded and a NOP is executed instead, making it a two-cycle instruction.
If 'a' is '0', the Access Bank is selected. If 'a' is '1', the BSR is used to select the GPR bank (default).
If 'a' is '0' and the extended instruction set is enabled, this instruction operates in Indexed Literal Offset Addressing mode whenever $f \le 95$ (5Fh). See **Section 24.2.3 "Byte-Oriented and Bit-Oriented Instructions in Indexed Literal Offset Mode"** for details.

Words: 1

Cycles: 1(2)
Note: 3 cycles if skip and followed by a 2-word instruction.

Q Cycle Activity:

Q1	Q2	Q3	Q4
Decode	Read register 'f'	Process Data	Write to destination

If skip:

Q1	Q2	Q3	Q4
No operation	No operation	No operation	No operation

If skip and followed by 2-word instruction:

Q1	Q2	Q3	Q4
No operation	No operation	No operation	No operation
No operation	No operation	No operation	No operation

Example:

```
HERE    DCFSNZ  TEMP, 1, 0
ZERO    :
NZERO   :
```

Before Instruction
TEMP = ?
After Instruction
TEMP = TEMP – 1,
If TEMP = 0;
PC = Address (ZERO)
If TEMP ≠ 0;
PC = Address (NZERO)

GOTO **Unconditional Branch**

Syntax:	GOTO k
Operands:	$0 \le k \le 1048575$
Operation:	$k \rightarrow$ PC<20:1>
Status Affected:	None

Encoding:

1st word (k<7:0>)	1110	1111	$k_7$kkk	kkkk$_0$
2nd word(k<19:8>)	1111	k_{19}kkk	kkkk	kkkk$_8$

Description: GOTO allows an unconditional branch anywhere within entire 2-Mbyte memory range. The 20-bit value 'k' is loaded into PC<20:1>. GOTO is always a two-cycle instruction.

Words: 2

Cycles: 2

Q Cycle Activity:

Q1	Q2	Q3	Q4
Decode	Read literal 'k'<7:0>,	No operation	Read literal 'k'<19:8>, Write to PC
No operation	No operation	No operation	No operation

Example: `GOTO THERE`

After Instruction
PC = Address (THERE)

INCF **Increment f**

Syntax:	INCF f {,d {,a}}
Operands:	$0 \le f \le 255$ $d \in [0,1]$ $a \in [0,1]$
Operation:	(f) + 1 $\rightarrow$ dest
Status Affected:	C, DC, N, OV, Z

Encoding:

0010	10da	ffff	ffff

Description: The contents of register 'f' are incremented. If 'd' is '0', the result is placed in W. If 'd' is '1', the result is placed back in register 'f' (default). If 'a' is '0', the Access Bank is selected. If 'a' is '1', the BSR is used to select the GPR bank (default). If 'a' is '0' and the extended instruction set is enabled, this instruction operates in Indexed Literal Offset Addressing mode whenever $f \le 95$ (5Fh). See **Section 24.2.3 "Byte-Oriented and Bit-Oriented Instructions in Indexed Literal Offset Mode"** for details.

Words: 1

Cycles: 1

Q Cycle Activity:

Q1	Q2	Q3	Q4
Decode	Read register 'f'	Process Data	Write to destination

Example: `INCF CNT, 1, 0`

Before Instruction
CNT = FFh
Z = 0
C = ?
DC = ?
After Instruction
CNT = 00h
Z = 1
C = 1
DC = 1

INCFSZ — Increment f, Skip if 0

Syntax: INCFSZ f {,d {,a}}

Operands: $0 \le f \le 255$
$d \in [0,1]$
$a \in [0,1]$

Operation: (f) + 1 → dest,
skip if result = 0

Status Affected: None

Encoding:

0011	11da	ffff	ffff

Description: The contents of register 'f' are incremented. If 'd' is '0', the result is placed in W. If 'd' is '1', the result is placed back in register 'f' (default).
If the result is '0', the next instruction, which is already fetched, is discarded and a `NOP` is executed instead, making it a two-cycle instruction.
If 'a' is '0', the Access Bank is selected. If 'a' is '1', the BSR is used to select the GPR bank (default).
If 'a' is '0' and the extended instruction set is enabled, this instruction operates in Indexed Literal Offset Addressing mode whenever $f \le 95$ (5Fh). See **Section 24.2.3 "Byte-Oriented and Bit-Oriented Instructions in Indexed Literal Offset Mode"** for details.

Words: 1

Cycles: 1(2)
Note: 3 cycles if skip and followed by a 2-word instruction.

Q Cycle Activity:

Q1	Q2	Q3	Q4
Decode	Read register 'f'	Process Data	Write to destination

If skip:

Q1	Q2	Q3	Q4
No operation	No operation	No operation	No operation

If skip and followed by 2-word instruction:

Q1	Q2	Q3	Q4
No operation	No operation	No operation	No operation
No operation	No operation	No operation	No operation

Example:

```
HERE     INCFSZ   CNT, 1, 0
NZERO    :
ZERO     :
```

Before Instruction
PC = Address (HERE)

After Instruction
CNT = CNT + 1
If CNT = 0;
PC = Address (ZERO)
If CNT ≠ 0;
PC = Address (NZERO)

INFSNZ — Increment f, Skip if Not 0

Syntax: INFSNZ f {,d {,a}}

Operands: $0 \le f \le 255$
$d \in [0,1]$
$a \in [0,1]$

Operation: (f) + 1 → dest,
skip if result ≠ 0

Status Affected: None

Encoding:

0100	10da	ffff	ffff

Description: The contents of register 'f' are incremented. If 'd' is '0', the result is placed in W. If 'd' is '1', the result is placed back in register 'f' (default).
If the result is not '0', the next instruction, which is already fetched, is discarded and a `NOP` is executed instead, making it a two-cycle instruction.
If 'a' is '0', the Access Bank is selected. If 'a' is '1', the BSR is used to select the GPR bank (default).
If 'a' is '0' and the extended instruction set is enabled, this instruction operates in Indexed Literal Offset Addressing mode whenever $f \le 95$ (5Fh). See **Section 24.2.3 "Byte-Oriented and Bit-Oriented Instructions in Indexed Literal Offset Mode"** for details.

Words: 1

Cycles: 1(2)
Note: 3 cycles if skip and followed by a 2-word instruction.

Q Cycle Activity:

Q1	Q2	Q3	Q4
Decode	Read register 'f'	Process Data	Write to destination

If skip:

Q1	Q2	Q3	Q4
No operation	No operation	No operation	No operation

If skip and followed by 2-word instruction:

Q1	Q2	Q3	Q4
No operation	No operation	No operation	No operation
No operation	No operation	No operation	No operation

Example:

```
HERE     INFSNZ   REG, 1, 0
ZERO
NZERO
```

Before Instruction
PC = Address (HERE)

After Instruction
REG = REG + 1
If REG ≠ 0;
PC = Address (NZERO)
If REG = 0;
PC = Address (ZERO)

IORLW **Inclusive OR Literal with W**

Syntax:	IORLW k
Operands:	$0 \leq k \leq 255$
Operation:	(W) .OR. k → W
Status Affected:	N, Z
Encoding:	0000 | 1001 | kkkk | kkkk
Description:	The contents of W are ORed with the eight-bit literal 'k'. The result is placed in W.
Words:	1
Cycles:	1

Q Cycle Activity:

Q1	Q2	Q3	Q4
Decode	Read literal 'k'	Process Data	Write to W

Example: `IORLW 35h`

Before Instruction
W = 9Ah
After Instruction
W = BFh

IORWF **Inclusive OR W with f**

Syntax:	IORWF f {,d {,a}}
Operands:	$0 \leq f \leq 255$ $d \in [0,1]$ $a \in [0,1]$
Operation:	(W) .OR. (f) → dest
Status Affected:	N, Z
Encoding:	0001 | 00da | ffff | ffff
Description:	Inclusive OR W with register 'f'. If 'd' is '0', the result is placed in W. If 'd' is '1', the result is placed back in register 'f' (default). If 'a' is '0', the Access Bank is selected. If 'a' is '1', the BSR is used to select the GPR bank (default). If 'a' is '0' and the extended instruction set is enabled, this instruction operates in Indexed Literal Offset Addressing mode whenever $f \leq 95$ (5Fh). See **Section 24.2.3 "Byte-Oriented and Bit-Oriented Instructions in Indexed Literal Offset Mode"** for details.
Words:	1
Cycles:	1

Q Cycle Activity:

Q1	Q2	Q3	Q4
Decode	Read register 'f'	Process Data	Write to destination

Example: `IORWF RESULT, 0, 1`

Before Instruction
RESULT = 13h
W = 91h
After Instruction
RESULT = 13h
W = 93h

LFSR — Load FSR

Syntax: LFSR f, k

Operands: $0 \le f \le 2$
$0 \le k \le 4095$

Operation: k → FSRf

Status Affected: None

Encoding:

1110	1110	00ff	k_{11}kkk
1111	0000	$k_7$kkk	kkkk

Description: The 12-bit literal 'k' is loaded into the File Select Register pointed to by 'f'.

Words: 2

Cycles: 2

Q Cycle Activity:

Q1	Q2	Q3	Q4
Decode	Read literal 'k' MSB	Process Data	Write literal 'k' MSB to FSRfH
Decode	Read literal 'k' LSB	Process Data	Write literal 'k' to FSRfL

<u>Example:</u> `LFSR 2, 3ABh`

After Instruction
FSR2H = 03h
FSR2L = ABh

MOVF — Move f

Syntax: MOVF f{,d{,a}}

Operands: $0 \le f \le 255$
$d \in [0,1]$
$a \in [0,1]$

Operation: f → dest

Status Affected: N, Z

Encoding:

0101	00da	ffff	ffff

Description: The contents of register 'f' are moved to a destination dependent upon the status of 'd'. If 'd' is '0', the result is placed in W. If 'd' is '1', the result is placed back in register 'f' (default). Location 'f' can be anywhere in the 256-byte bank.
If 'a' is '0', the Access Bank is selected. If 'a' is '1', the BSR is used to select the GPR bank (default).
If 'a' is '0' and the extended instruction set is enabled, this instruction operates in Indexed Literal Offset Addressing mode whenever $f \le 95$ (5Fh). See **Section 24.2.3 "Byte-Oriented and Bit-Oriented Instructions in Indexed Literal Offset Mode"** for details.

Words: 1

Cycles: 1

Q Cycle Activity:

Q1	Q2	Q3	Q4
Decode	Read register 'f'	Process Data	Write W

<u>Example:</u> `MOVF REG, 0, 0`

Before Instruction
REG = 22h
W = FFh
After Instruction
REG = 22h
W = 22h

MOVFF **Move f to f**

Syntax:	MOVFF f_s,f_d
Operands:	$0 \le f_s \le 4095$ $0 \le f_d \le 4095$
Operation:	$(f_s) \rightarrow f_d$
Status Affected:	None

Encoding:

1st word (source)	1100	ffff	ffff	$ffff_s$
2nd word (destin.)	1111	ffff	ffff	$ffff_d$

Description: The contents of source register 'f_s' are moved to destination register 'f_d'. Location of source 'f_s' can be anywhere in the 4096-byte data space (000h to FFFh) and location of destination 'f_d' can also be anywhere from 000h to FFFh.
Either source or destination can be W (a useful special situation).
`MOVFF` is particularly useful for transferring a data memory location to a peripheral register (such as the transmit buffer or an I/O port).
The `MOVFF` instruction cannot use the PCL, TOSU, TOSH or TOSL as the destination register.

Words: 2

Cycles: 2 (3)

Q Cycle Activity:

Q1	Q2	Q3	Q4
Decode	Read register 'f' (src)	Process Data	No operation
Decode	No operation No dummy read	No operation	Write register 'f' (dest)

Example: `MOVFF REG1, REG2`

Before Instruction
REG1 = 33h
REG2 = 11h
After Instruction
REG1 = 33h
REG2 = 33h

MOVLB **Move Literal to Low Nibble in BSR**

Syntax:	MOVLW k
Operands:	$0 \le k \le 255$
Operation:	$k \rightarrow$ BSR
Status Affected:	None

Encoding:

0000	0001	kkkk	kkkk

Description: The eight-bit literal 'k' is loaded into the Bank Select Register (BSR). The value of BSR<7:4> always remains '0', regardless of the value of k_7:k_4.

Words: 1

Cycles: 1

Q Cycle Activity:

Q1	Q2	Q3	Q4
Decode	Read literal 'k'	Process Data	Write literal 'k' to BSR

Example: `MOVLB 5`

Before Instruction
BSR Register = 02h
After Instruction
BSR Register = 05h

MOVLW **Move Literal to W**

Syntax: MOVLW k

Operands: $0 \le k \le 255$

Operation: $k \rightarrow W$

Status Affected: None

Encoding:

0000	1110	kkkk	kkkk

Description: The eight-bit literal 'k' is loaded into W.

Words: 1

Cycles: 1

Q Cycle Activity:

Q1	Q2	Q3	Q4
Decode	Read literal 'k'	Process Data	Write to W

Example: `MOVLW 5Ah`

After Instruction

W = 5Ah

MOVWF **Move W to f**

Syntax: MOVWF f {,a}

Operands: $0 \le f \le 255$
$a \in [0,1]$

Operation: $(W) \rightarrow f$

Status Affected: None

Encoding:

0110	111a	ffff	ffff

Description: Move data from W to register 'f'. Location 'f' can be anywhere in the 256-byte bank.
If 'a' is '0', the Access Bank is selected. If 'a' is '1', the BSR is used to select the GPR bank (default).
If 'a' is '0' and the extended instruction set is enabled, this instruction operates in Indexed Literal Offset Addressing mode whenever $f \le 95$ (5Fh). See **Section 24.2.3 "Byte-Oriented and Bit-Oriented Instructions in Indexed Literal Offset Mode"** for details.

Words: 1

Cycles: 1

Q Cycle Activity:

Q1	Q2	Q3	Q4
Decode	Read register 'f'	Process Data	Write register 'f'

Example: `MOVWF REG, 0`

Before Instruction

W = 4Fh
REG = FFh

After Instruction

W = 4Fh
REG = 4Fh

MULLW **Multiply Literal with W**

Syntax:	MULLW k
Operands:	$0 \le k \le 255$
Operation:	(W) x k → PRODH:PRODL
Status Affected:	None
Encoding:	0000 | 1101 | kkkk | kkkk
Description:	An unsigned multiplication is carried out between the contents of W and the 8-bit literal 'k'. The 16-bit result is placed in the PRODH:PRODL register pair. PRODH contains the high byte. W is unchanged. None of the Status flags are affected. Note that neither overflow nor carry is possible in this operation. A zero result is possible but not detected.
Words:	1
Cycles:	1

Q Cycle Activity:

Q1	Q2	Q3	Q4
Decode	Read literal 'k'	Process Data	Write registers PRODH: PRODL

Example: `MULLW 0C4h`

Before Instruction

W	=	E2h
PRODH	=	?
PRODL	=	?

After Instruction

W	=	E2h
PRODH	=	ADh
PRODL	=	08h

MULWF **Multiply W with f**

Syntax:	MULWF f {,a}
Operands:	$0 \le f \le 255$ $a \in [0,1]$
Operation:	(W) x (f) → PRODH:PRODL
Status Affected:	None
Encoding:	0000 | 001a | ffff | ffff
Description:	An unsigned multiplication is carried out between the contents of W and the register file location 'f'. The 16-bit result is stored in the PRODH:PRODL register pair. PRODH contains the high byte. Both W and 'f' are unchanged. None of the Status flags are affected. Note that neither overflow nor carry is possible in this operation. A zero result is possible but not detected. If 'a' is '0', the Access Bank is selected. If 'a' is '1', the BSR is used to select the GPR bank (default). If 'a' is '0' and the extended instruction set is enabled, this instruction operates in Indexed Literal Offset Addressing mode whenever $f \le 95$ (5Fh). See **Section 24.2.3 "Byte-Oriented and Bit-Oriented Instructions in Indexed Literal Offset Mode"** for details.
Words:	1
Cycles:	1

Q Cycle Activity:

Q1	Q2	Q3	Q4
Decode	Read register 'f'	Process Data	Write registers PRODH: PRODL

Example: `MULWF REG, 1`

Before Instruction

W	=	C4h
REG	=	B5h
PRODH	=	?
PRODL	=	?

After Instruction

W	=	C4h
REG	=	B5h
PRODH	=	8Ah
PRODL	=	94h

NEGF **Negate f**

Syntax:	NEGF f {,a}
Operands:	$0 \le f \le 255$ $a \in [0,1]$
Operation:	$(\bar{f}) + 1 \rightarrow f$
Status Affected:	N, OV, C, DC, Z
Encoding:	`0110` `110a` `ffff` `ffff`
Description:	Location 'f' is negated using two's complement. The result is placed in the data memory location 'f'. If 'a' is '0', the Access Bank is selected. If 'a' is '1', the BSR is used to select the GPR bank (default). If 'a' is '0' and the extended instruction set is enabled, this instruction operates in Indexed Literal Offset Addressing mode whenever $f \le 95$ (5Fh). See **Section 24.2.3 "Byte-Oriented and Bit-Oriented Instructions in Indexed Literal Offset Mode"** for details.
Words:	1
Cycles:	1

Q Cycle Activity:

Q1	Q2	Q3	Q4
Decode	Read register 'f'	Process Data	Write register 'f'

Example: `NEGF REG, 1`

Before Instruction
REG = `0011 1010` [3Ah]
After Instruction
REG = `1100 0110` [C6h]

NOP **No Operation**

Syntax:	NOP
Operands:	None
Operation:	No operation
Status Affected:	None
Encoding:	`0000` `0000` `0000` `0000` `1111` `xxxx` `xxxx` `xxxx`
Description:	No operation.
Words:	1
Cycles:	1

Q Cycle Activity:

Q1	Q2	Q3	Q4
Decode	No operation	No operation	No operation

Example:

None.

POP — **Pop Top of Return Stack**

Syntax:	POP
Operands:	None
Operation:	(TOS) → bit bucket
Status Affected:	None
Encoding:	`0000` `0000` `0000` `0110`
Description:	The TOS value is pulled off the return stack and is discarded. The TOS value then becomes the previous value that was pushed onto the return stack. This instruction is provided to enable the user to properly manage the return stack to incorporate a software stack.
Words:	1
Cycles:	1

Q Cycle Activity:

Q1	Q2	Q3	Q4
Decode	No operation	POP TOS value	No operation

Example:

```
POP
GOTO    NEW
```

Before Instruction

TOS	=	0031A2h
Stack (1 level down)	=	014332h

After Instruction

TOS	=	014332h
PC	=	NEW

PUSH — **Push Top of Return Stack**

Syntax:	PUSH
Operands:	None
Operation:	(PC + 2) → TOS
Status Affected:	None
Encoding:	`0000` `0000` `0000` `0101`
Description:	The PC + 2 is pushed onto the top of the return stack. The previous TOS value is pushed down on the stack. This instruction allows implementing a software stack by modifying TOS and then pushing it onto the return stack.
Words:	1
Cycles:	1

Q Cycle Activity:

Q1	Q2	Q3	Q4
Decode	PUSH PC + 2 onto return stack	No operation	No operation

Example:

```
PUSH
```

Before Instruction

TOS	=	345Ah
PC	=	0124h

After Instruction

PC	=	0126h
TOS	=	0126h
Stack (1 level down)	=	345Ah

RCALL	**Relative Call**			
Syntax:	RCALL n			
Operands:	$-1024 \le n \le 1023$			
Operation:	(PC) + 2 → TOS, (PC) + 2 + 2n → PC			
Status Affected:	None			
Encoding:	1101	1nnn	nnnn	nnnn
Description:	Subroutine call with a jump up to 1K from the current location. First, return address (PC + 2) is pushed onto the stack. Then, add the 2's complement number '2n' to the PC. Since the PC will have incremented to fetch the next instruction, the new address will be PC + 2 + 2n. This instruction is a two-cycle instruction.			
Words:	1			
Cycles:	2			

Q Cycle Activity:

Q1	Q2	Q3	Q4
Decode	Read literal 'n' PUSH PC to stack	Process Data	Write to PC
No operation	No operation	No operation	No operation

<u>Example:</u> `HERE RCALL Jump`

Before Instruction
PC = Address (HERE)
After Instruction
PC = Address (Jump)
TOS = Address (HERE + 2)

RESET	**Reset**			
Syntax:	RESET			
Operands:	None			
Operation:	Reset all registers and flags that are affected by a $\overline{\text{MCLR}}$ Reset.			
Status Affected:	All			
Encoding:	0000	0000	1111	1111
Description:	This instruction provides a way to execute a $\overline{\text{MCLR}}$ Reset in software.			
Words:	1			
Cycles:	1			

Q Cycle Activity:

Q1	Q2	Q3	Q4
Decode	Start Reset	No operation	No operation

<u>Example:</u> `RESET`

After Instruction
Registers = Reset Value
Flags* = Reset Value

RETFIE **Return from Interrupt**

Syntax:	RETFIE {s}
Operands:	s ∈ [0,1]
Operation:	(TOS) → PC, 1 → GIE/GIEH or PEIE/GIEL, if s = 1 (WS) → W, (STATUSS) → STATUS, (BSRS) → BSR, PCLATU, PCLATH are unchanged
Status Affected:	GIE/GIEH, PEIE/GIEL.
Encoding:	0000 | 0000 | 0001 | 000s
Description:	Return from interrupt. Stack is popped and Top-of-Stack (TOS) is loaded into the PC. Interrupts are enabled by setting either the high or low priority global interrupt enable bit. If 's' = 1, the contents of the shadow registers, WS, STATUSS and BSRS, are loaded into their corresponding registers, W, STATUS and BSR. If 's' = 0, no update of these registers occurs (default).
Words:	1
Cycles:	2

Q Cycle Activity:

Q1	Q2	Q3	Q4
Decode	No operation	No operation	POP PC from stack Set GIEH or GIEL
No operation	No operation	No operation	No operation

Example: `RETFIE 1`

After Interrupt

PC	=	TOS
W	=	WS
BSR	=	BSRS
STATUS	=	STATUSS
GIE/GIEH, PEIE/GIEL	=	1

RETLW **Return Literal to W**

Syntax:	RETLW k
Operands:	0 ≤ k ≤ 255
Operation:	k → W, (TOS) → PC, PCLATU, PCLATH are unchanged
Status Affected:	None
Encoding:	0000 | 1100 | kkkk | kkkk
Description:	W is loaded with the eight-bit literal 'k'. The program counter is loaded from the top of the stack (the return address). The high address latch (PCLATH) remains unchanged.
Words:	1
Cycles:	2

Q Cycle Activity:

Q1	Q2	Q3	Q4
Decode	Read literal 'k'	Process Data	POP PC from stack, Write to W
No operation	No operation	No operation	No operation

Example:

```
    CALL TABLE ; W contains table
               ; offset value
               ; W now has
               ; table value
    :
TABLE
    ADDWF PCL  ; W = offset
    RETLW k0   ; Begin table
    RETLW k1   ;
    :
    :
    RETLW kn   ; End of table
```

Before Instruction
W = 07h
After Instruction
W = value of kn

RETURN **Return from Subroutine**

Syntax: RETURN {s}

Operands: s ∈ [0,1]

Operation: (TOS) → PC,
if s = 1
(WS) → W,
(STATUSS) → STATUS,
(BSRS) → BSR,
PCLATU, PCLATH are unchanged

Status Affected: None

Encoding:

0000	0000	0001	001s

Description: Return from subroutine. The stack is popped and the top of the stack (TOS) is loaded into the program counter. If 's'= 1, the contents of the shadow registers, WS, STATUSS, and BSRS, are loaded into their corresponding registers, W, STATUS, and BSR. If 's' = 0, no update of these registers occurs (default).

Words: 1

Cycles: 2

Q Cycle Activity:

Q1	Q2	Q3	Q4
Decode	No operation	Process Data	POP PC from stack
No operation	No operation	No operation	No operation

Example: `RETURN`

After Instruction:
PC = TOS

RLCF **Rotate Left f through Carry**

Syntax: RLCF f {,d {,a}}

Operands: 0 ≤ f ≤ 255
d ∈ [0,1]
a ∈ [0,1]

Operation: (f<n>) → dest<n + 1>,
(f<7>) → C,
(C) → dest<0>

Status Affected: C, N, Z

Encoding:

0011	01da	ffff	ffff

Description: The contents of register 'f' are rotated one bit to the left through the Carry flag. If 'd' is '0', the result is placed in W. If 'd' is '1', the result is stored back in register 'f' (default).
If 'a' is '0', the Access Bank is selected. If 'a' is '1', the BSR is used to select the GPR bank (default).
If 'a' is '0' and the extended instruction set is enabled, this instruction operates in Indexed Literal Offset Addressing mode whenever f ≤ 95 (5Fh). See **Section 24.2.3 "Byte-Oriented and Bit-Oriented Instructions in Indexed Literal Offset Mode"** for details.

C ← register f

Words: 1

Cycles: 1

Q Cycle Activity:

Q1	Q2	Q3	Q4
Decode	Read register 'f'	Process Data	Write to destination

Example: `RLCF REG, 0, 0`

Before Instruction
REG = 1110 0110
C = 0
After Instruction
REG = 1110 0110
W = 1100 1100
C = 1

RLNCF **Rotate Left f (No Carry)**

Syntax:	RLNCF f {,d {,a}}
Operands:	$0 \le f \le 255$ $d \in [0,1]$ $a \in [0,1]$
Operation:	(f<n>) → dest<n + 1>, (f<7>) → dest<0>
Status Affected:	N, Z
Encoding:	`0100` `01da` `ffff` `ffff`
Description:	The contents of register 'f' are rotated one bit to the left. If 'd' is '0', the result is placed in W. If 'd' is '1', the result is stored back in register 'f' (default). If 'a' is '0', the Access Bank is selected. If 'a' is '1', the BSR is used to select the GPR bank (default). If 'a' is '0' and the extended instruction set is enabled, this instruction operates in Indexed Literal Offset Addressing mode whenever $f \le 95$ (5Fh). See **Section 24.2.3 "Byte-Oriented and Bit-Oriented Instructions in Indexed Literal Offset Mode"** for details.
Words:	1
Cycles:	1

register f

Q Cycle Activity:

Q1	Q2	Q3	Q4
Decode	Read register 'f'	Process Data	Write to destination

Example: `RLNCF REG, 1, 0`

Before Instruction
REG = `1010 1011`
After Instruction
REG = `0101 0111`

RRCF **Rotate Right f through Carry**

Syntax:	RRCF f {,d {,a}}
Operands:	$0 \le f \le 255$ $d \in [0,1]$ $a \in [0,1]$
Operation:	(f<n>) → dest<n – 1>, (f<0>) → C, (C) → dest<7>
Status Affected:	C, N, Z
Encoding:	`0011` `00da` `ffff` `ffff`
Description:	The contents of register 'f' are rotated one bit to the right through the Carry flag. If 'd' is '0', the result is placed in W. If 'd' is '1', the result is placed back in register 'f' (default). If 'a' is '0', the Access Bank is selected. If 'a' is '1', the BSR is used to select the GPR bank (default). If 'a' is '0' and the extended instruction set is enabled, this instruction operates in Indexed Literal Offset Addressing mode whenever $f \le 95$ (5Fh). See **Section 24.2.3 "Byte-Oriented and Bit-Oriented Instructions in Indexed Literal Offset Mode"** for details.
Words:	1
Cycles:	1

C → register f

Q Cycle Activity:

Q1	Q2	Q3	Q4
Decode	Read register 'f'	Process Data	Write to destination

Example: `RRCF REG, 0, 0`

Before Instruction
REG = `1110 0110`
C = 0
After Instruction
REG = `1110 0110`
W = `0111 0011`
C = 0

RRNCF **Rotate Right f (No Carry)**

Syntax:	RRNCF f {,d {,a}}
Operands:	$0 \le f \le 255$ $d \in [0,1]$ $a \in [0,1]$
Operation:	(f<n>) → dest<n – 1>, (f<0>) → dest<7>
Status Affected:	N, Z
Encoding:	`0100` `00da` `ffff` `ffff`
Description:	The contents of register 'f' are rotated one bit to the right. If 'd' is '0', the result is placed in W. If 'd' is '1', the result is placed back in register 'f' (default). If 'a' is '0', the Access Bank will be selected, overriding the BSR value. If 'a' is '1', then the bank will be selected as per the BSR value (default). If 'a' is '0' and the extended instruction set is enabled, this instruction operates in Indexed Literal Offset Addressing mode whenever $f \le 95$ (5Fh). See **Section 24.2.3 "Byte-Oriented and Bit-Oriented Instructions in Indexed Literal Offset Mode"** for details.
	register f
Words:	1
Cycles:	1

Q Cycle Activity:

Q1	Q2	Q3	Q4
Decode	Read register 'f'	Process Data	Write to destination

Example 1: `RRNCF REG, 1, 0`

Before Instruction
REG = 1101 0111
After Instruction
REG = 1110 1011

Example 2: `RRNCF REG, 0, 0`

Before Instruction
W = ?
REG = 1101 0111
After Instruction
W = 1110 1011
REG = 1101 0111

SETF **Set f**

Syntax:	SETF f {,a}
Operands:	$0 \le f \le 255$ $a \in [0,1]$
Operation:	FFh → f
Status Affected:	None
Encoding:	`0110` `100a` `ffff` `ffff`
Description:	The contents of the specified register are set to FFh. If 'a' is '0', the Access Bank is selected. If 'a' is '1', the BSR is used to select the GPR bank (default). If 'a' is '0' and the extended instruction set is enabled, this instruction operates in Indexed Literal Offset Addressing mode whenever $f \le 95$ (5Fh). See **Section 24.2.3 "Byte-Oriented and Bit-Oriented Instructions in Indexed Literal Offset Mode"** for details.
Words:	1
Cycles:	1

Q Cycle Activity:

Q1	Q2	Q3	Q4
Decode	Read register 'f'	Process Data	Write register 'f'

Example: `SETF REG, 1`

Before Instruction
REG = 5Ah
After Instruction
REG = FFh

SLEEP **Enter Sleep mode**

Syntax: SLEEP

Operands: None

Operation: 00h → WDT,
0 → WDT postscaler,
1 → $\overline{TO}$,
0 → $\overline{PD}$

Status Affected: $\overline{TO}$, $\overline{PD}$

Encoding:

0000	0000	0000	0011

Description: The Power-Down status bit ($\overline{PD}$) is cleared. The Time-out status bit ($\overline{TO}$) is set. Watchdog Timer and its postscaler are cleared.
The processor is put into Sleep mode with the oscillator stopped.

Words: 1

Cycles: 1

Q Cycle Activity:

Q1	Q2	Q3	Q4
Decode	No operation	Process Data	Go to Sleep

Example: `SLEEP`

Before Instruction
$\overline{TO}$ = ?
$\overline{PD}$ = ?

After Instruction
$\overline{TO}$ = 1†
$\overline{PD}$ = 0

† If WDT causes wake-up, this bit is cleared.

SUBFWB **Subtract f from W with Borrow**

Syntax: SUBFWB f {,d {,a}}

Operands: $0 \le f \le 255$
$d \in [0,1]$
$a \in [0,1]$

Operation: (W) – (f) – ($\overline{C}$) → dest

Status Affected: N, OV, C, DC, Z

Encoding:

0101	01da	ffff	ffff

Description: Subtract register 'f' and Carry flag (borrow) from W (2's complement method). If 'd' is '0', the result is stored in W. If 'd' is '1', the result is stored in register 'f' (default).
If 'a' is '0', the Access Bank is selected. If 'a' is '1', the BSR is used to select the GPR bank (default).
If 'a' is '0' and the extended instruction set is enabled, this instruction operates in Indexed Literal Offset Addressing mode whenever $f \le 95$ (5Fh). See **Section 24.2.3 "Byte-Oriented and Bit-Oriented Instructions in Indexed Literal Offset Mode"** for details.

Words: 1

Cycles: 1

Q Cycle Activity:

Q1	Q2	Q3	Q4
Decode	Read register 'f'	Process Data	Write to destination

Example 1: `SUBFWB REG, 1, 0`

Before Instruction
REG = 3
W = 2
C = 1

After Instruction
REG = FF
W = 2
C = 0
Z = 0
N = 1 ; result is negative

Example 2: `SUBFWB REG, 0, 0`

Before Instruction
REG = 2
W = 5
C = 1

After Instruction
REG = 2
W = 3
C = 1
Z = 0
N = 0 ; result is positive

Example 3: `SUBFWB REG, 1, 0`

Before Instruction
REG = 1
W = 2
C = 0

After Instruction
REG = 0
W = 2
C = 1
Z = 1 ; result is zero
N = 0

SUBLW **Subtract W from Literal**

Syntax: SUBLW k

Operands: $0 \le k \le 255$

Operation: $k - (W) \rightarrow W$

Status Affected: N, OV, C, DC, Z

Encoding:

0000	1000	kkkk	kkkk

Description: W is subtracted from the eight-bit literal 'k'. The result is placed in W.

Words: 1

Cycles: 1

Q Cycle Activity:

Q1	Q2	Q3	Q4
Decode	Read literal 'k'	Process Data	Write to W

Example 1: `SUBLW 02h`

```
Before Instruction
    W   =  01h
    C   =  ?
After Instruction
    W   =  01h
    C   =  1    ; result is positive
    Z   =  0
    N   =  0
```

Example 2: `SUBLW 02h`

```
Before Instruction
    W   =  02h
    C   =  ?
After Instruction
    W   =  00h
    C   =  1    ; result is zero
    Z   =  1
    N   =  0
```

Example 3: `SUBLW 02h`

```
Before Instruction
    W   =  03h
    C   =  ?
After Instruction
    W   =  FFh  ; (2's complement)
    C   =  0    ; result is negative
    Z   =  0
    N   =  1
```

SUBWF **Subtract W from f**

Syntax: SUBWF f {,d {,a}}

Operands: $0 \le f \le 255$
$d \in [0,1]$
$a \in [0,1]$

Operation: $(f) - (W) \rightarrow dest$

Status Affected: N, OV, C, DC, Z

Encoding:

0101	11da	ffff	ffff

Description: Subtract W from register 'f' (2's complement method). If 'd' is '0', the result is stored in W. If 'd' is '1', the result is stored back in register 'f' (default).
If 'a' is '0', the Access Bank is selected. If 'a' is '1', the BSR is used to select the GPR bank (default).
If 'a' is '0' and the extended instruction set is enabled, this instruction operates in Indexed Literal Offset Addressing mode whenever $f \le 95$ (5Fh). See **Section 24.2.3 "Byte-Oriented and Bit-Oriented Instructions in Indexed Literal Offset Mode"** for details.

Words: 1

Cycles: 1

Q Cycle Activity:

Q1	Q2	Q3	Q4
Decode	Read register 'f'	Process Data	Write to destination

Example 1: `SUBWF REG, 1, 0`

```
Before Instruction
    REG  =  3
    W    =  2
    C    =  ?
After Instruction
    REG  =  1
    W    =  2
    C    =  1    ; result is positive
    Z    =  0
    N    =  0
```

Example 2: `SUBWF REG, 0, 0`

```
Before Instruction
    REG  =  2
    W    =  2
    C    =  ?
After Instruction
    REG  =  2
    W    =  0
    C    =  1    ; esult is rero     z
    Z    =  1
    N    =  0
```

Example 3: `SUBWF REG, 1, 0`

```
Before Instruction
    REG  =  1
    W    =  2
    C    =  ?
After Instruction
    REG  =  FFh ;(2's complement)
    W    =  2
    C    =  0    ; result is negative
    Z    =  0
    N    =  1
```

SUBWFB — Subtract W from f with Borrow

Syntax:	SUBWFB f {,d {,a}}
Operands:	0 ≤ f ≤ 255 d ∈ [0,1] a ∈ [0,1]
Operation:	(f) – (W) – ($\overline{C}$) → dest
Status Affected:	N, OV, C, DC, Z
Encoding:	`0101` `10da` `ffff` `ffff`
Description:	Subtract W and the Carry flag (borrow) from register 'f' (2's complement method). If 'd' is '0', the result is stored in W. If 'd' is '1', the result is stored back in register 'f' (default). If 'a' is '0', the Access Bank is selected. If 'a' is '1', the BSR is used to select the GPR bank (default). If 'a' is '0' and the extended instruction set is enabled, this instruction operates in Indexed Literal Offset Addressing mode whenever f ≤ 95 (5Fh). See **Section 24.2.3 "Byte-Oriented and Bit-Oriented Instructions in Indexed Literal Offset Mode"** for details.
Words:	1
Cycles:	1

Q Cycle Activity:

Q1	Q2	Q3	Q4
Decode	Read register 'f'	Process Data	Write to destination

Example 1: `SUBWFB REG, 1, 0`

Before Instruction
REG = 19h (0001 1001)
W = 0Dh (0000 1101)
C = 1
After Instruction
REG = 0Ch (0000 1011)
W = 0Dh (0000 1101)
C = 1
Z = 0
N = 0 ; result is positive

Example 2: `SUBWFB REG, 0, 0`

Before Instruction
REG = 1Bh (0001 1011)
W = 1Ah (0001 1010)
C = 0
After Instruction
REG = 1Bh (0001 1011)
W = 00h
C = 1
Z = 1 ; result is zero
N = 0

Example 3: `SUBWFB REG, 1, 0`

Before Instruction
REG = 03h (0000 0011)
W = 0Eh (0000 1101)
C = 1
After Instruction
REG = F5h (1111 0100) ; [2's comp]
W = 0Eh (0000 1101)
C = 0
Z = 0
N = 1 ; result is negative

SWAPF — Swap f

Syntax:	SWAPF f {,d {,a}}
Operands:	0 ≤ f ≤ 255 d ∈ [0,1] a ∈ [0,1]
Operation:	(f<3:0>) → dest<7:4>, (f<7:4>) → dest<3:0>
Status Affected:	None
Encoding:	`0011` `10da` `ffff` `ffff`
Description:	The upper and lower nibbles of register 'f' are exchanged. If 'd' is '0', the result is placed in W. If 'd' is '1', the result is placed in register 'f' (default). If 'a' is '0', the Access Bank is selected. If 'a' is '1', the BSR is used to select the GPR bank (default). If 'a' is '0' and the extended instruction set is enabled, this instruction operates in Indexed Literal Offset Addressing mode whenever f ≤ 95 (5Fh). See **Section 24.2.3 "Byte-Oriented and Bit-Oriented Instructions in Indexed Literal Offset Mode"** for details.
Words:	1
Cycles:	1

Q Cycle Activity:

Q1	Q2	Q3	Q4
Decode	Read register 'f'	Process Data	Write to destination

Example: `SWAPF REG, 1, 0`

Before Instruction
REG = 53h
After Instruction
REG = 35h

TBLRD **Table Read**

Syntax: TBLRD (*; *+; *-; +*)

Operands: None

Operation: if TBLRD *,
(Prog Mem (TBLPTR)) → TABLAT;
TBLPTR – No Change;
if TBLRD *+,
(Prog Mem (TBLPTR)) → TABLAT;
(TBLPTR) + 1 → TBLPTR;
if TBLRD *-,
(Prog Mem (TBLPTR)) → TABLAT;
(TBLPTR) – 1 → TBLPTR;
if TBLRD +*,
(TBLPTR) + 1 → TBLPTR;
(Prog Mem (TBLPTR)) → TABLAT;

Status Affected: None

Encoding:

0000	0000	0000	10nn nn=0 * =1 *+ =2 *- =3 +*

Description: This instruction is used to read the contents of Program Memory (P.M.). To address the program memory, a pointer called Table Pointer (TBLPTR) is used.
The TBLPTR (a 21-bit pointer) points to each byte in the program memory. TBLPTR has a 2-Mbyte address range.

TBLPTR[0] = 0: Least Significant Byte of Program Memory Word

TBLPTR[0] = 1: Most Significant Byte of Program Memory Word

The TBLRD instruction can modify the value of TBLPTR as follows:

- no change
- post-increment
- post-decrement
- pre-increment

Words: 1

Cycles: 2

Q Cycle Activity:

Q1	Q2	Q3	Q4
Decode	No operation	No operation	No operation
No operation	No operation (Read Program Memory)	No operation	No operation (Write TABLAT)

TBLRD **Table Read (Continued)**

Example 1: `TBLRD *+ ;`

Before Instruction

TABLAT	=	55h
TBLPTR	=	00A356h
MEMORY (00A356h)	=	34h

After Instruction

TABLAT	=	34h
TBLPTR	=	00A357h

Example 2: `TBLRD +* ;`

Before Instruction

TABLAT	=	AAh
TBLPTR	=	01A357h
MEMORY (01A357h)	=	12h
MEMORY (01A358h)	=	34h

After Instruction

TABLAT	=	34h
TBLPTR	=	01A358h

TBLWT T able Write

Syntax:	TBLWT (*; *+; *-; +*)
Operands:	None
Operation:	if TBLWT*, (TABLAT) → Holding Register; TBLPTR – No Change; if TBLWT*+, (TABLAT) → Holding Register; (TBLPTR) + 1 → TBLPTR; if TBLWT*-, (TABLAT) → Holding Register; (TBLPTR) – 1 → TBLPTR; if TBLWT+*, (TBLPTR) + 1 → TBLPTR; (TABLAT) → Holding Register;
Status Affected:	None

Encoding:

0000	0000	0000	11nn nn=0 * =1 *+ =2 *- =3 +*

Description: This instruction uses the 3 LSBs of TBLPTR to determine which of the 8 holding registers the TABLAT is written to. The holding registers are used to program the contents of Program Memory (P.M.). (Refer to **Section 6.0 "Flash Program Memory"** for additional details on programming Flash memory.) The TBLPTR (a 21-bit pointer) points to each byte in the program memory. TBLPTR has a 2-Mbyte address range. The LSb of the TBLPTR selects which byte of the program memory location to access.

TBLPTR[0] = 0: Least Significant Byte of Program Memory Word
TBLPTR[0] = 1: Most Significant Byte of Program Memory Word

The `TBLWT` instruction can modify the value of TBLPTR as follows:

- no change
- post-increment
- post-decrement
- pre-increment

Words: 1

Cycles: 2

Q Cycle Activity:

Q1	Q2	Q3	Q4
Decode	No operation	No operation	No operation
No operation	No operation (Read TABLAT)	No operation	No operation (Write to Holding Register)

TBLWT Table Write (Continued)

Example 1: `TBLWT *+;`

Before Instruction

TABLAT =		55h
TBLPTR	=	00A356h
HOLDING REGISTER (00A356h)	=	FFh

After Instructions (table write completion)

TABLAT	=	55h
TBLPTR	=	00A357h
HOLDING REGISTER (00A356h)	=	55h

Example 2: `TBLWT +*;`

Before Instruction

TABLAT	=	34h
TBLPTR	=	01389Ah
HOLDING REGISTER (01389Ah)	=	FFh
HOLDING REGISTER (01389Bh)	=	FFh

After Instruction (table write completion)

TABLAT	=	34h
TBLPTR	=	01389Bh
HOLDING REGISTER (01389Ah)	=	FFh
HOLDING REGISTER (01389Bh)	=	34h

TSTFSZ **Test f, Skip if 0**

Syntax: TSTFSZ f {,a}

Operands: $0 \le f \le 255$
$a \in [0,1]$

Operation: skip if f = 0

Status Affected: None

Encoding:

0110	011a	ffff	ffff

Description: If 'f' = 0, the next instruction fetched during the current instruction execution is discarded and a NOP is executed, making this a two-cycle instruction.
If 'a' is '0', the Access Bank is selected. If 'a' is '1', the BSR is used to select the GPR bank (default).
If 'a' is '0' and the extended instruction set is enabled, this instruction operates in Indexed Literal Offset Addressing mode whenever $f \le 95$ (5Fh). See **Section 24.2.3 "Byte-Oriented and Bit-Oriented Instructions in Indexed Literal Offset Mode"** for details.

Words: 1

Cycles: 1(2)
Note: 3 cycles if skip and followed by a 2-word instruction.

Q Cycle Activity:

Q1	Q2	Q3	Q4
Decode	Read register 'f'	Process Data	No operation

If skip:

Q1	Q2	Q3	Q4
No operation	No operation	No operation	No operation

If skip and followed by 2-word instruction:

Q1	Q2	Q3	Q4
No operation	No operation	No operation	No operation
No operation	No operation	No operation	No operation

Example:

```
HERE    TSTFSZ  CNT, 1
NZERO   :
ZERO    :
```

Before Instruction
PC = Address (HERE)
After Instruction
If CNT = 00h,
PC = Address (ZERO)
If CNT ≠ 00h,
PC = Address (NZERO)

XORLW **Exclusive OR Literal with W**

Syntax: XORLW k

Operands: $0 \le k \le 255$

Operation: (W) .XOR. k → W

Status Affected: N, Z

Encoding:

0000	1010	kkkk	kkkk

Description: The contents of W are XORed with the 8-bit literal 'k'. The result is placed in W.

Words: 1

Cycles: 1

Q Cycle Activity:

Q1	Q2	Q3	Q4
Decode	Read literal 'k'	Process Data	Write to W

Example: `XORLW 0AFh`

Before Instruction
W = B5h
After Instruction
W = 1Ah

XORWF	**Exclusive OR W with f**
Syntax:	XORWF f {,d {,a}}
Operands:	$0 \leq f \leq 255$ $d \in [0,1]$ $a \in [0,1]$
Operation:	(W) .XOR. (f) → dest
Status Affected:	N, Z
Encoding:	`0001` `10da` `ffff` `ffff`
Description:	Exclusive OR the contents of W with register 'f'. If 'd' is '0', the result is stored in W. If 'd' is '1', the result is stored back in the register 'f' (default). If 'a' is '0', the Access Bank is selected. If 'a' is '1', the BSR is used to select the GPR bank (default). If 'a' is '0' and the extended instruction set is enabled, this instruction operates in Indexed Literal Offset Addressing mode whenever $f \leq 95$ (5Fh). See **Section 24.2.3 "Byte-Oriented and Bit-Oriented Instructions in Indexed Literal Offset Mode"** for details.
Words:	1
Cycles:	1

Q Cycle Activity:

Q1	Q2	Q3	Q4
Decode	Read register 'f'	Process Data	Write to destination

Example: `XORWF REG, 1, 0`

Before Instruction
REG = AFh
W = B5h
After Instruction
REG = 1Ah
W = B5h

APPENDIX E: PIC18F4321 SPECIAL FUNCTION REGISTERS

SPECIAL FUNCTION REGISTERS

The Special Function Registers (SFRs) are registers used by the CPU and peripheral modules for controlling the desired operation of the device. These registers are implemented as static RAM. SFRs start at the top of data memory (FFFh) and extend downward to occupy the top half of Bank 15 (F80h to FFFh). A list of these registers is given in Table E-1.

TABLE E-1: SPECIAL FUNCTION REGISTER MAP FOR PIC18F4321 FAMILY DEVICES

Address	Name	Address	Name	Address	Name	Address	Name
FFFh	TOSU	FDFh	INDF2[(1)]	FBFh	CCPR1H	F9Fh	IPR1
FFEh	TOSH	FDEh	POSTINC2[(1)]	FBEh	CCPR1L	F9Eh	PIR1
FFDh	TOSL	FDDh	POSTDEC2[(1)]	FBDh	CCP1CON	F9Dh	PIE1
FFCh	STKPTR	FDCh	PREINC2[(1)]	FBCh	CCPR2H	F9Ch	—[(2)]
FFBh	PCLATU	FDBh	PLUSW2[(1)]	FBBh	CCPR2L	F9Bh	OSCTUNE
FFAh	PCLATH	FDAh	FSR2H	FBAh	CCP2CON	F9Ah	—[(2)]
FF9h	PCL	FD9h	FSR2L	FB9h	—[(2)]	F99h	—[(2)]
FF8h	TBLPTRU	FD8h	STATUS	FB8h	BAUDCON	F98h	—[(2)]
FF7h	TBLPTRH	FD7h	TMR0H	FB7h	ECCP1DEL[(3)]	F97h	—[(2)]
FF6h	TBLPTRL	FD6h	TMR0L	FB6h	ECCP1AS[(3)]	F96h	TRISE[(3)]
FF5h	TABLAT	FD5h	T0CON	FB5h	CVRCON	F95h	TRISD[(3)]
FF4h	PRODH	FD4h	—[(2)]	FB4h	CMCON	F94h	TRISC
FF3h	PRODL	FD3h	OSCCON	FB3h	TMR3H	F93h	TRISB
FF2h	INTCON	FD2h	HLVDCON	FB2h	TMR3L	F92h	TRISA
FF1h	INTCON2	FD1h	WDTCON	FB1h	T3CON	F91h	—[(2)]
FF0h	INTCON3	FD0h	RCON	FB0h	SPBRGH	F90h	—[(2)]
FEFh	INDF0[(1)]	FCFh	TMR1H	FAFh	SPBRG	F8Fh	—[(2)]
FEEh	POSTINC0[(1)]	FCEh	TMR1L	FAEh	RCREG	F8Eh	—[(2)]
FEDh	POSTDEC0[(1)]	FCDh	T1CON	FADh	TXREG	F8Dh	LATE[(3)]
FECh	PREINC0[(1)]	FCCh	TMR2	FACh	TXSTA	F8Ch	LATD[(3)]
FEBh	PLUSW0[(1)]	FCBh	PR2	FABh	RCSTA	F8Bh	LATC
FEAh	FSR0H	FCAh	T2CON	FAAh	—[(2)]	F8Ah	LATB
FE9h	FSR0L	FC9h	SSPBUF	FA9h	EEADR	F89h	LATA
FE8h	WREG	FC8h	SSPADD	FA8h	EEDATA	F88h	—[(2)]
FE7h	INDF1[(1)]	FC7h	SSPSTAT	FA7h	EECON2[(1)]	F87h	—[(2)]
FE6h	POSTINC1[(1)]	FC6h	SSPCON1	FA6h	EECON1	F86h	—[(2)]
FE5h	POSTDEC1[(1)]	FC5h	SSPCON2	FA5h	—[(2)]	F85h	—[(2)]
FE4h	PREINC1[(1)]	FC4h	ADRESH	FA4h	—[(2)]	F84h	PORTE[(3)]
FE3h	PLUSW1[(1)]	FC3h	ADRESL	FA3h	—[(2)]	F83h	PORTD[(3)]
FE2h	FSR1H	FC2h	ADCON0	FA2h	IPR2	F82h	PORTC
FE1h	FSR1L	FC1h	ADCON1	FA1h	PIR2	F81h	PORTB
FE0h	BSR	FC0h	ADCON2	FA0h	PIE2	F80h	PORTA

Note 1: This is not a physical register.
2: Unimplemented registers are read as '0'.
3: This register is not available on 28-pin devices.

APPENDIX F: TUTORIAL FOR ASSEMBLING AND DEBUGGING A PIC18F ASSEMBLY LANGUAGE PROGRAM USING THE MPLAB

Assembling PIC18F assembly language program using MPLAB

First, download the latest versions of the MPLAB assembler and C18 compiler from the Microchip website www.microchip.com. After installing and downloading the program, you will see the following icon on your desktop:

Double click (right) on the MPLAB icon and wait until you see the following screen:

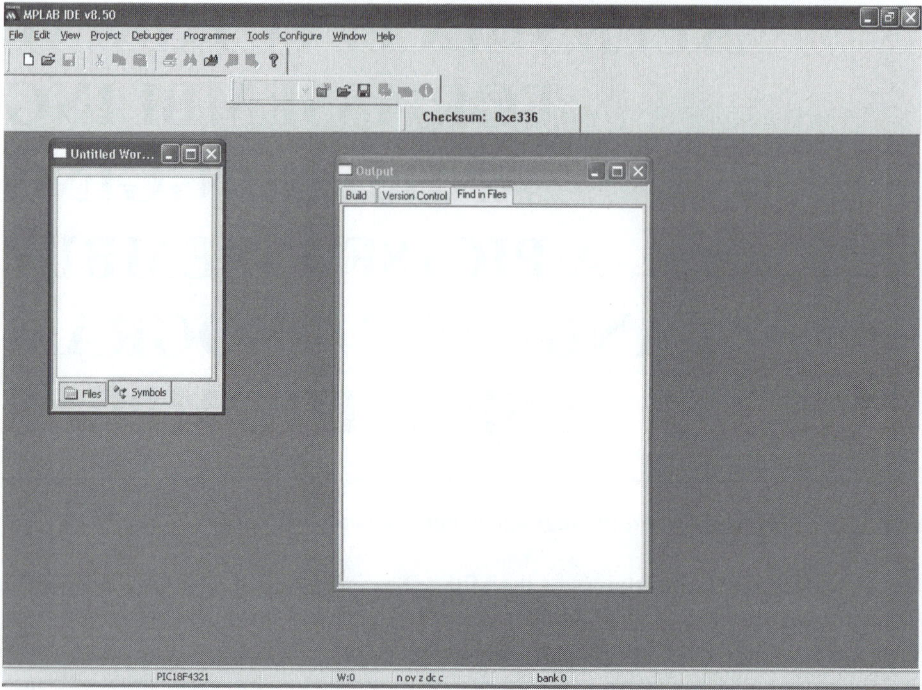

Next, click on 'Project' and then 'Project Wizard'; the following screen will appear:

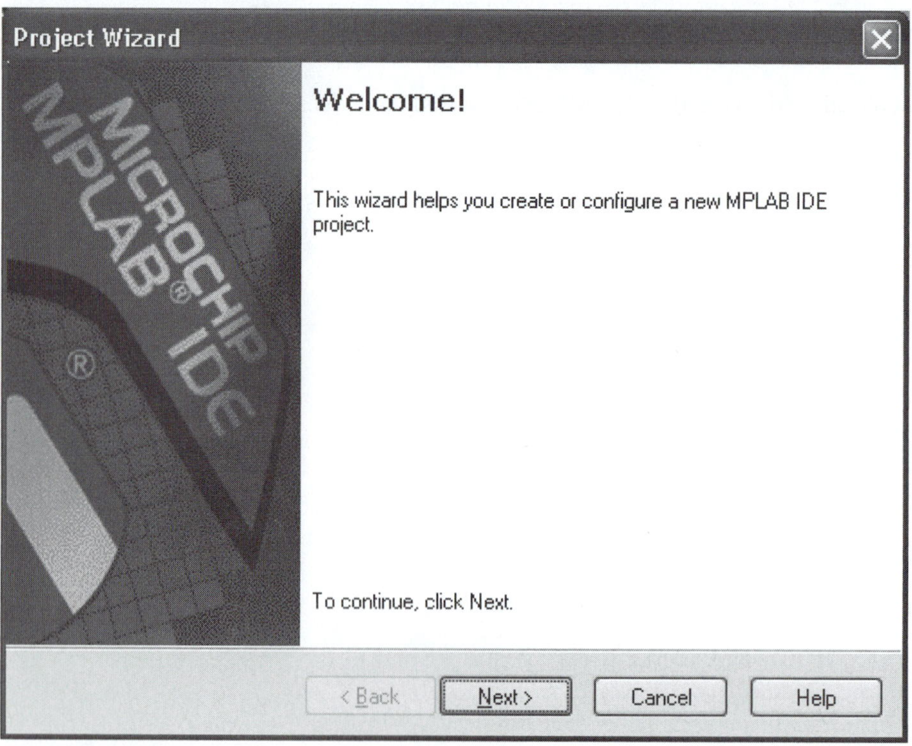

Click Next; the following screen shot will be displayed:

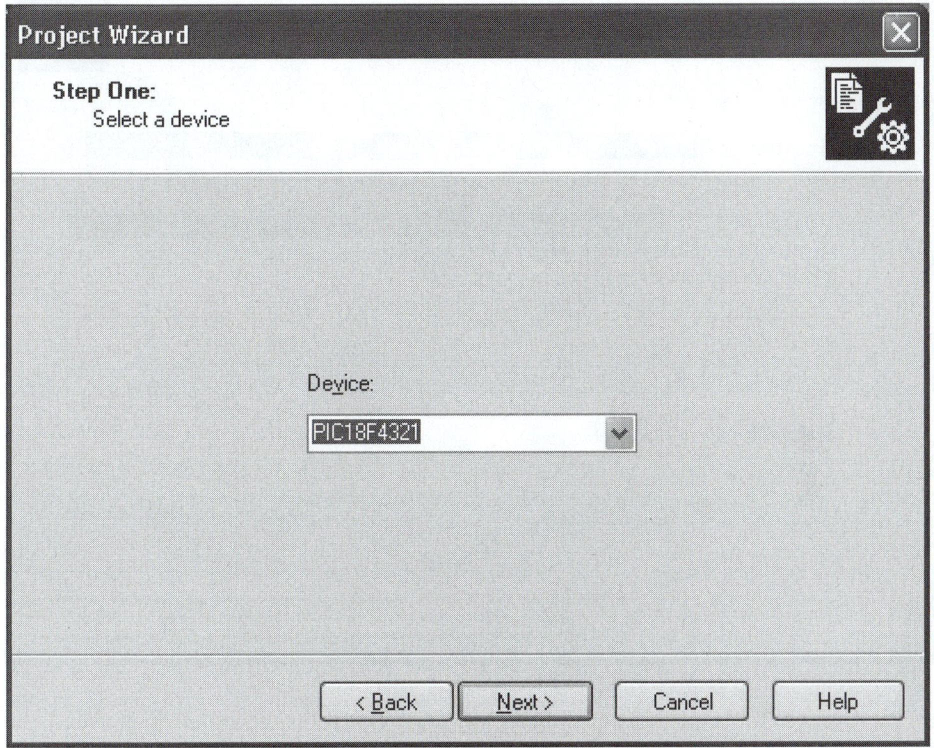

Select the device PIC18F4321, hit Next, and wait; the following will be displayed:

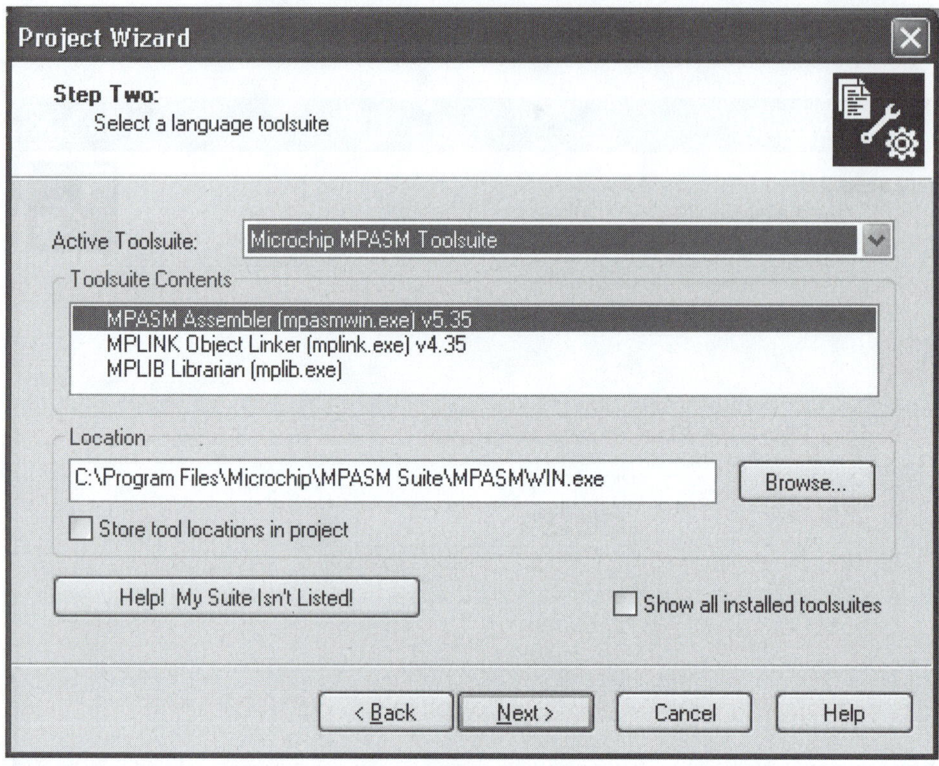

In the 'Active Toolsuite', select 'Microchip MPASM Toolsuite', and click Next; the following will be displayed:

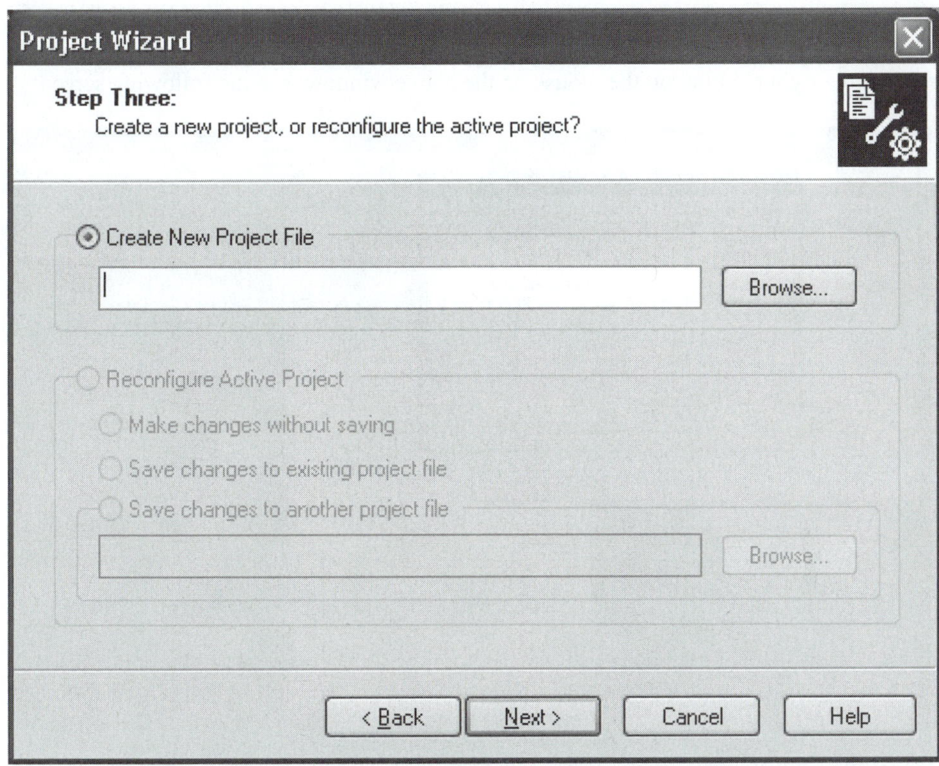

Select a location where all project contents will be placed. For this example, the folder will be placed on the desktop (arbitrarily chosen). Go to the desktop directory, make a new folder, and name the folder. In order to do this, click on 'Browse', and select desktop:

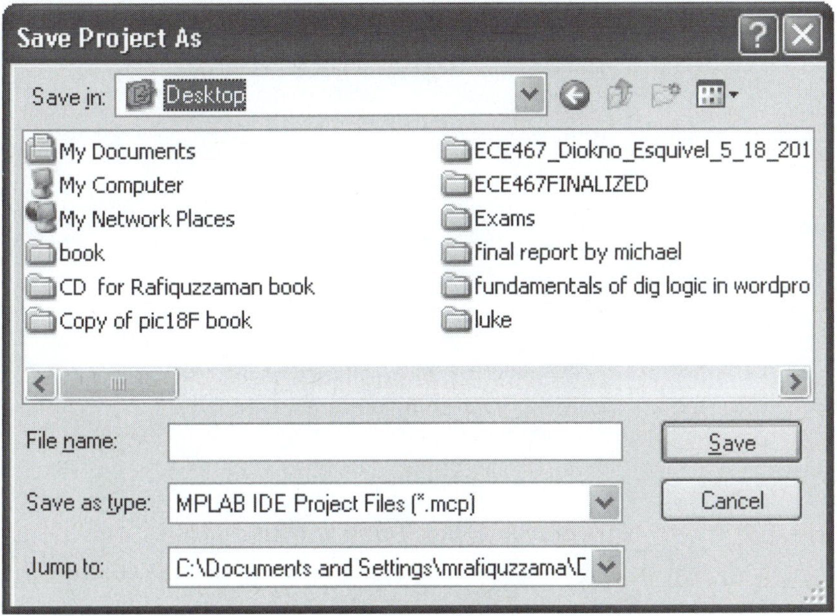

Next, create a new folder by clicking on the icon (second yellow icon from right on top row) or by right clicking on the mouse on the above window; see the following screen:

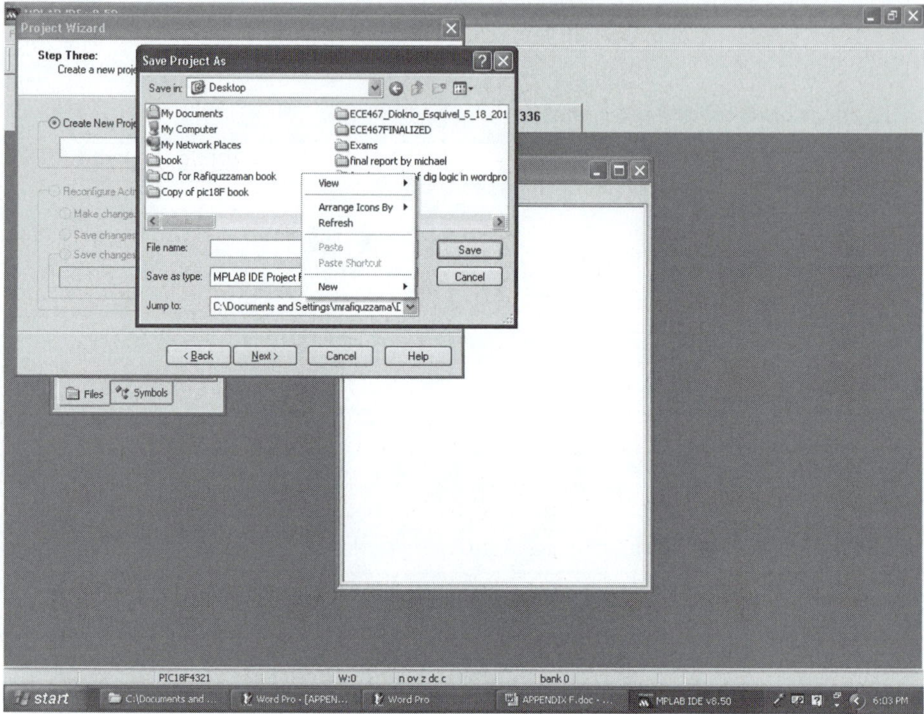

Next, click on 'New' to see the following:

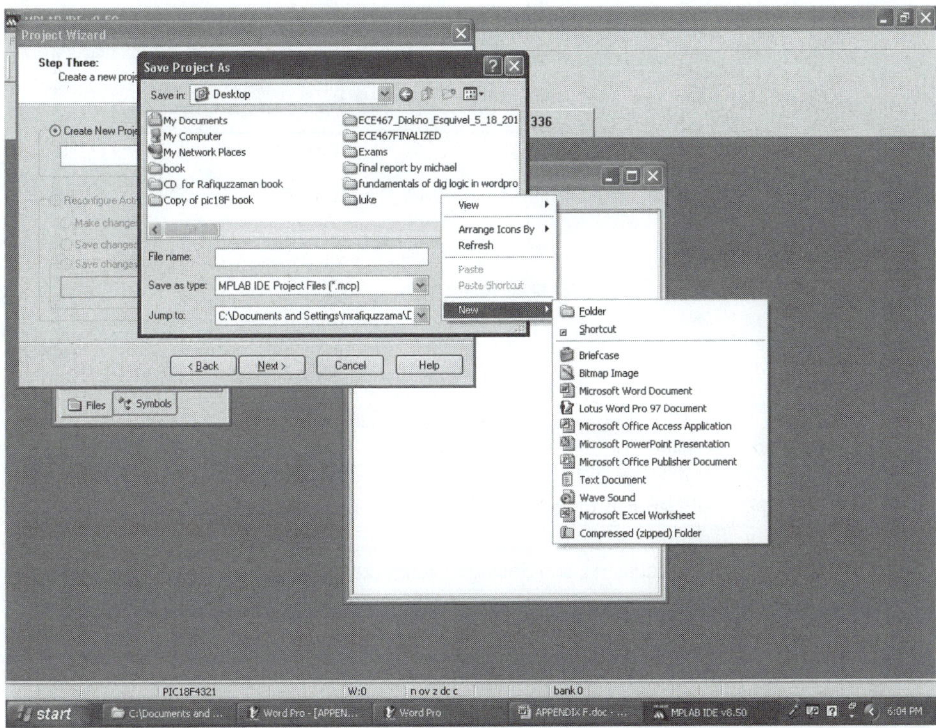

Click on folder, name it 'sum' (arbitrarily chosen name), and see the following :

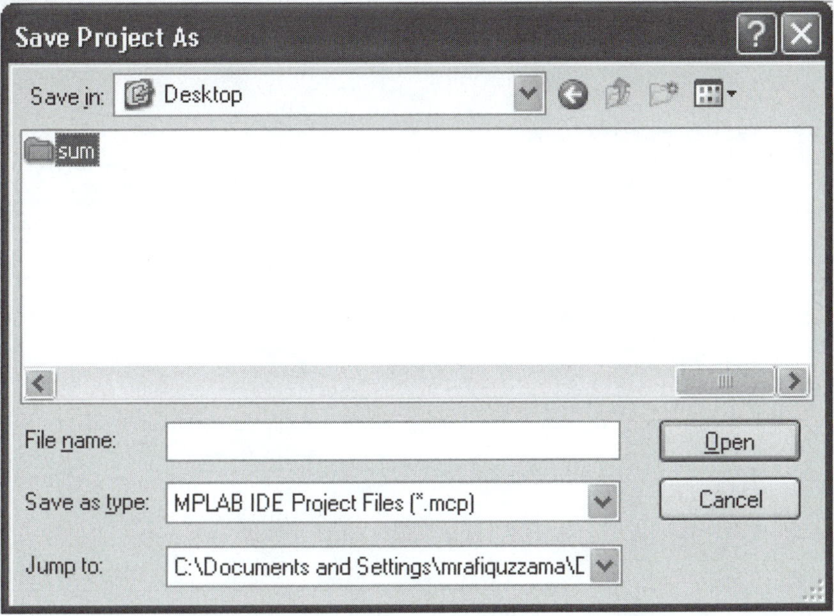

Next, type in a file name such as addition (arbitrarily chosen name), and see the following on the screen:

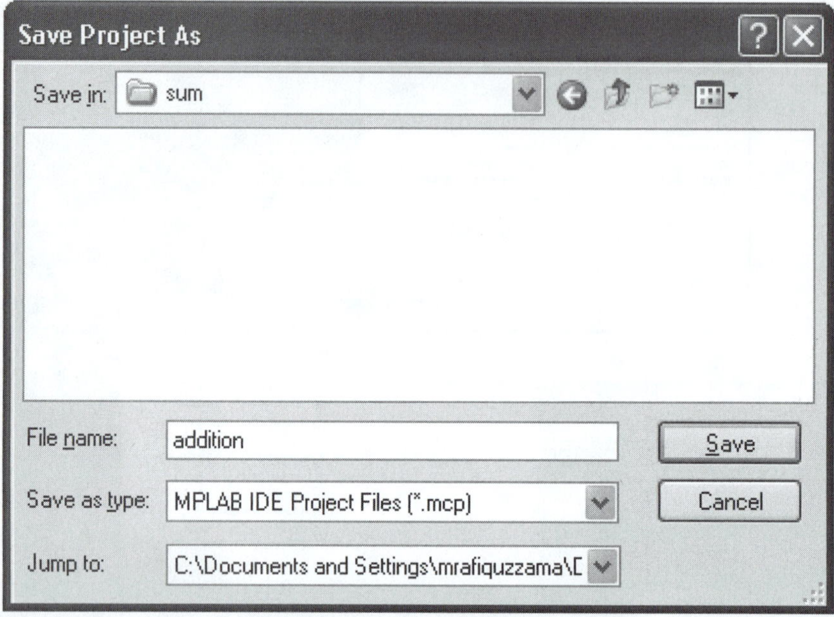

Next, click on Save; the following screen will appear:

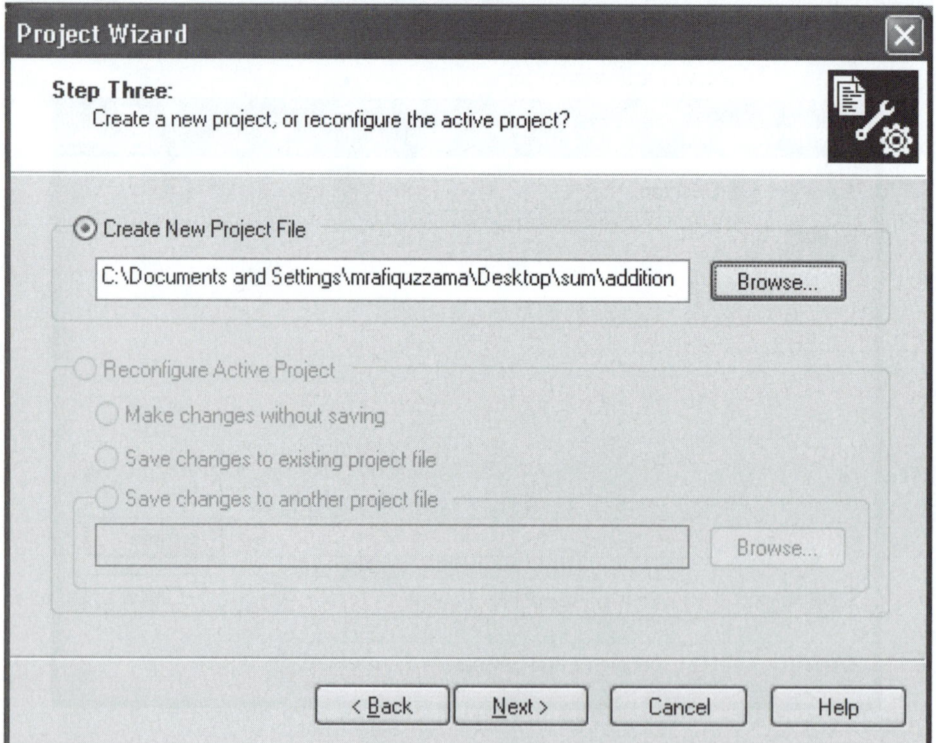

Click on Next, and see the following:

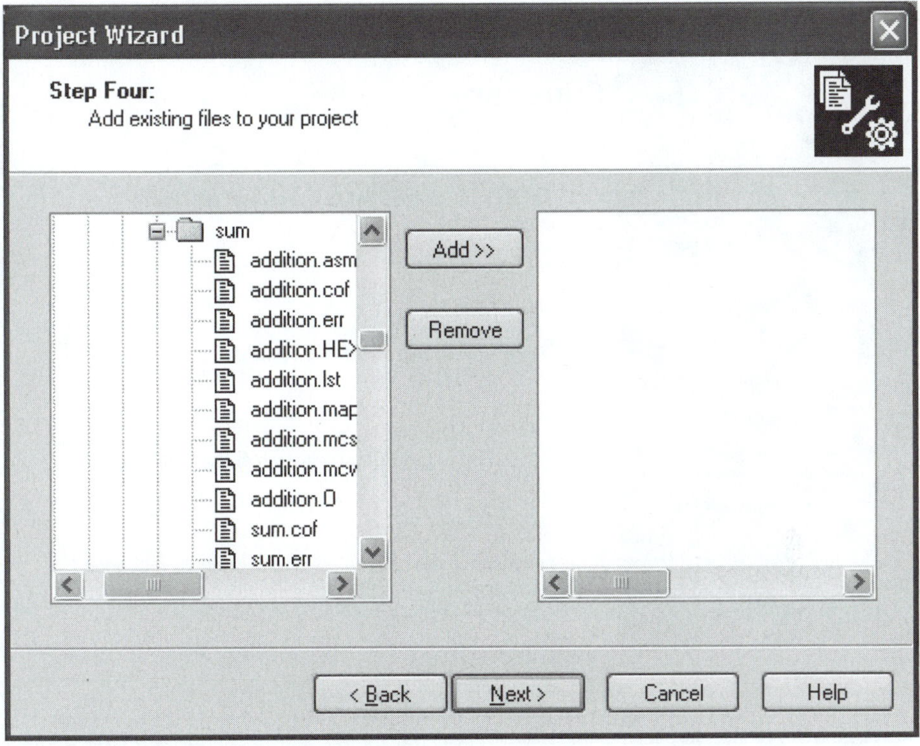

Click on Next, and see the following:

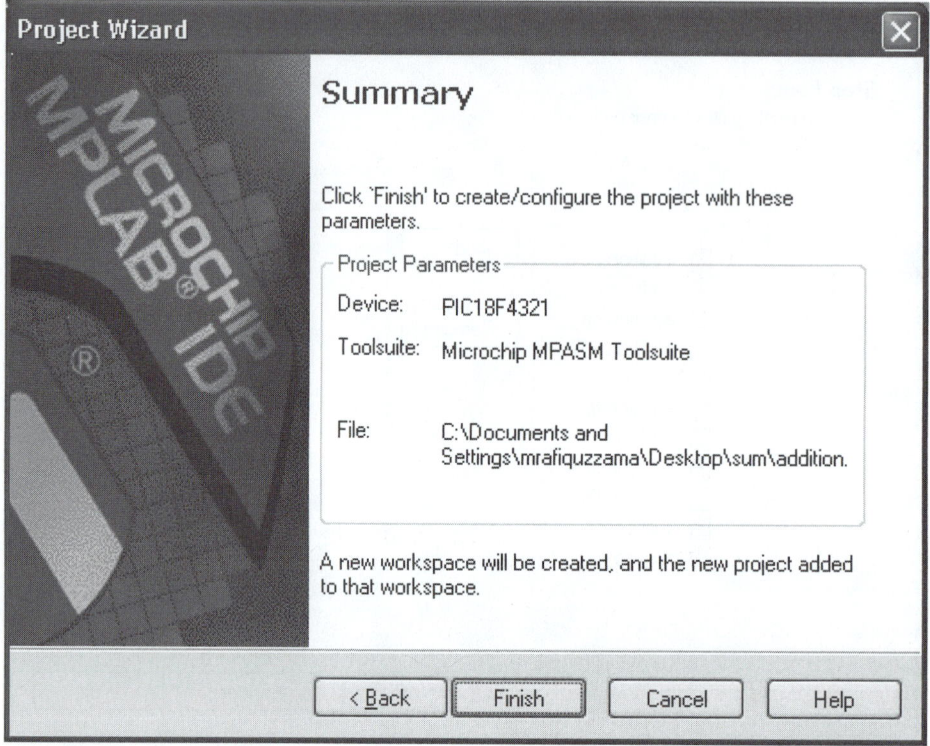

Click on Finish, and see the following:

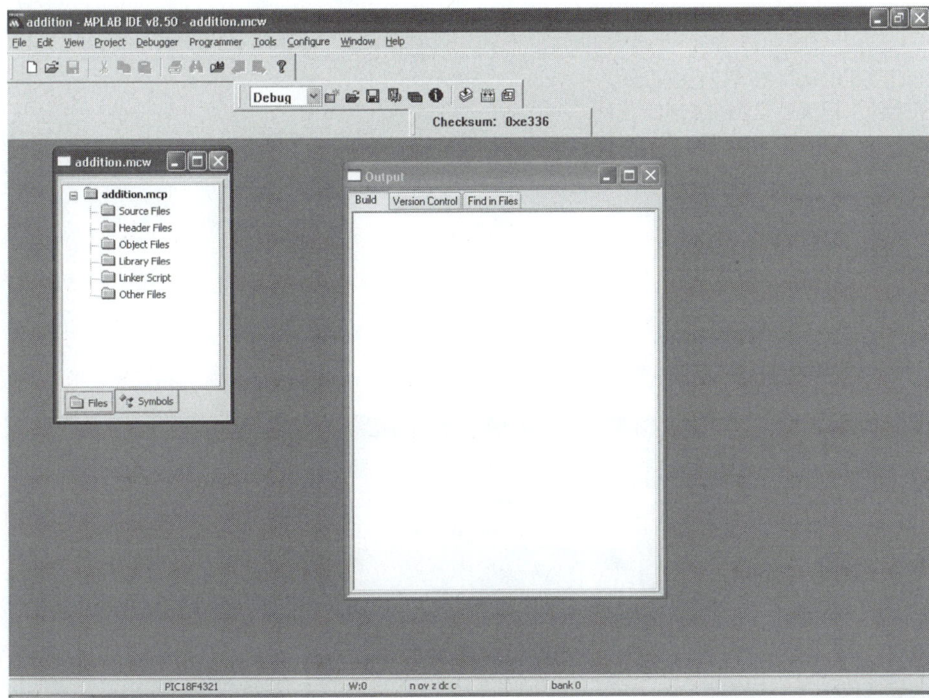

Click on File, and then New to see the following:

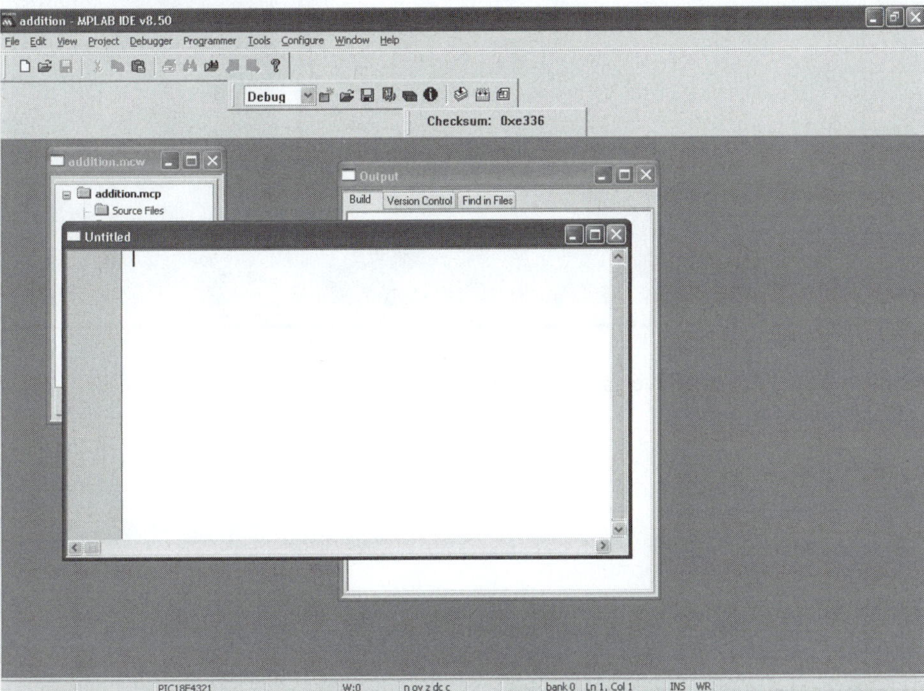

Type in the program you want to assemble. The following addition program is entered:

```
INCLUDE <P18F4321.INC>
SUM     EQU     0x50
        ORG     0x100
        MOVLW 0x02
        ADDLW 0x05
        MOVWF SUM
HERE    BRA     HERE
END
```

After entering the program, see the following:

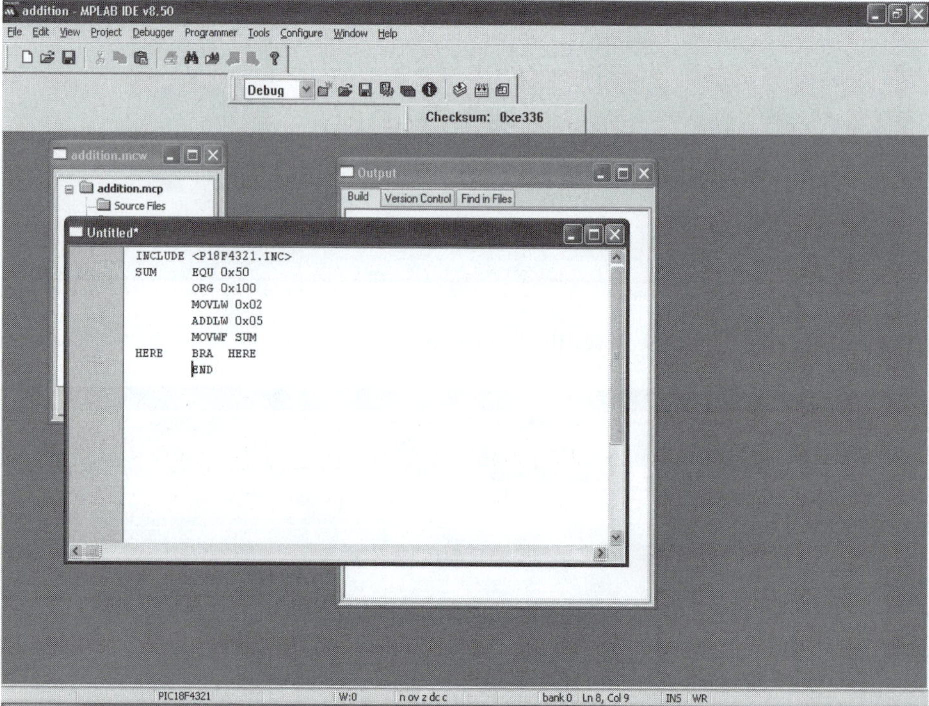

Next, click on File, and Save as, and see the following:

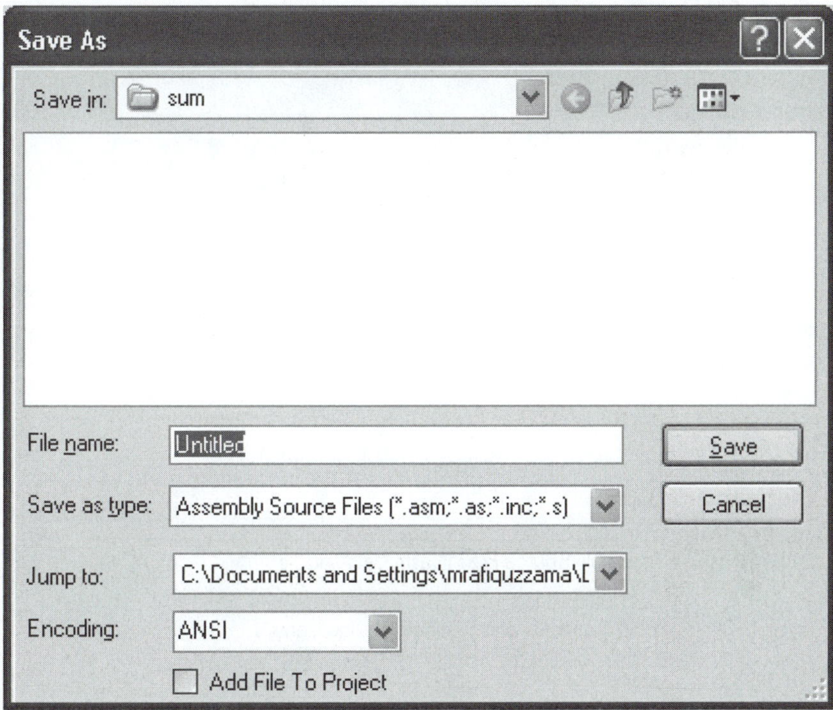

Make sure you scroll up to desktop, and then click on sum (the folder that was created before), and see the following:

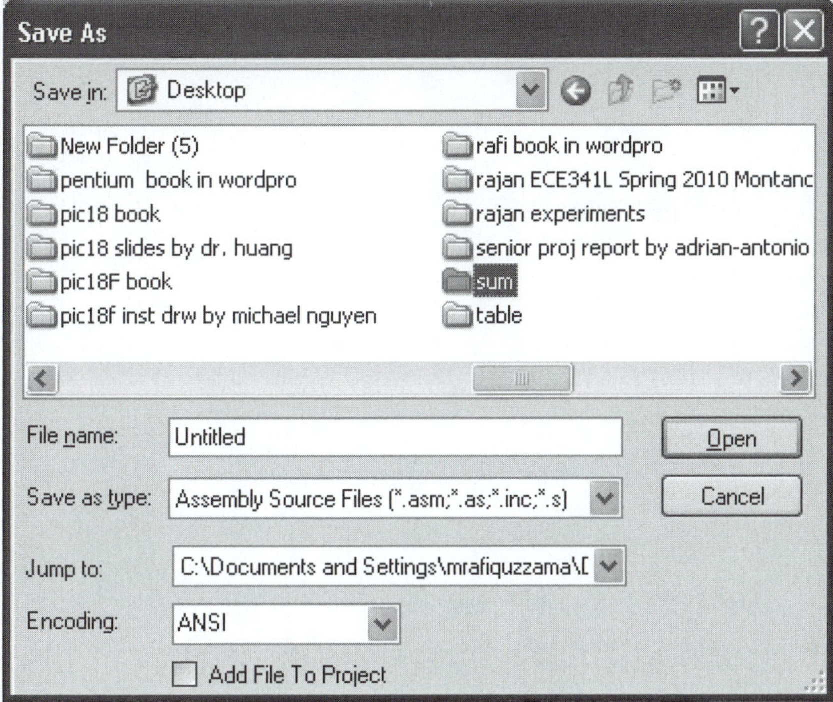

Next, double click (left) on sum to see the following:

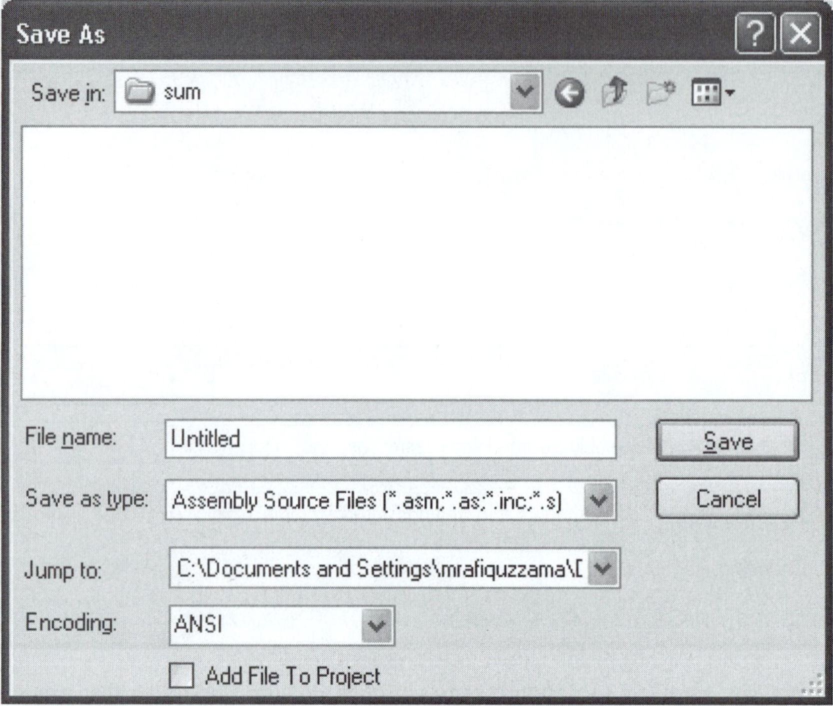

Delete Untitled, and enter the same file name 'addition' with .asm extension as File name. Click on save, and see the following screen shot (notice the display changes color):

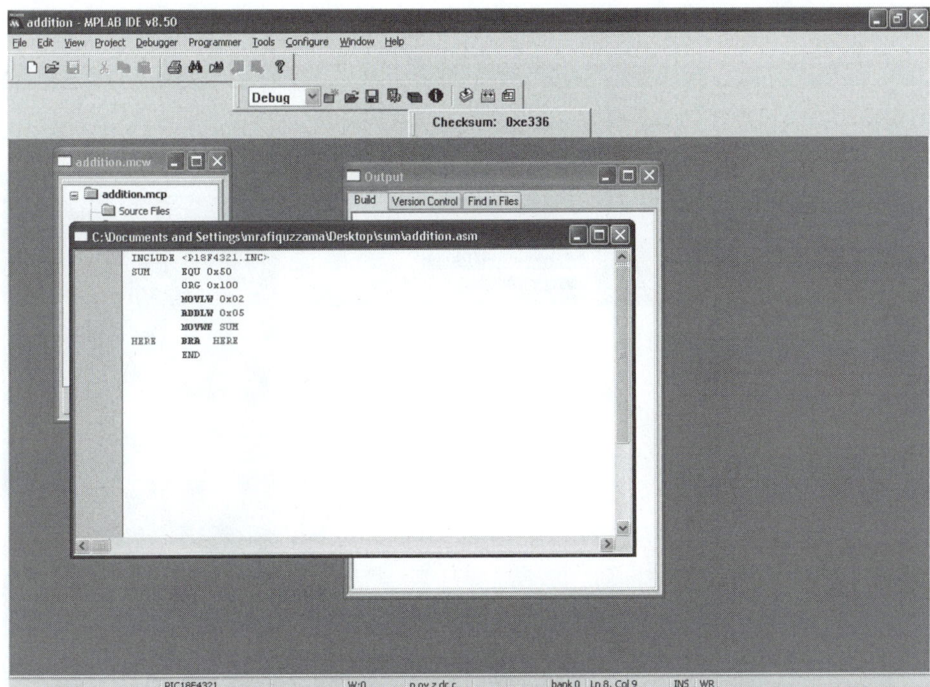

Next, highlight by clicking on the top (blue) section of addition.asm.mcp, and see the following:

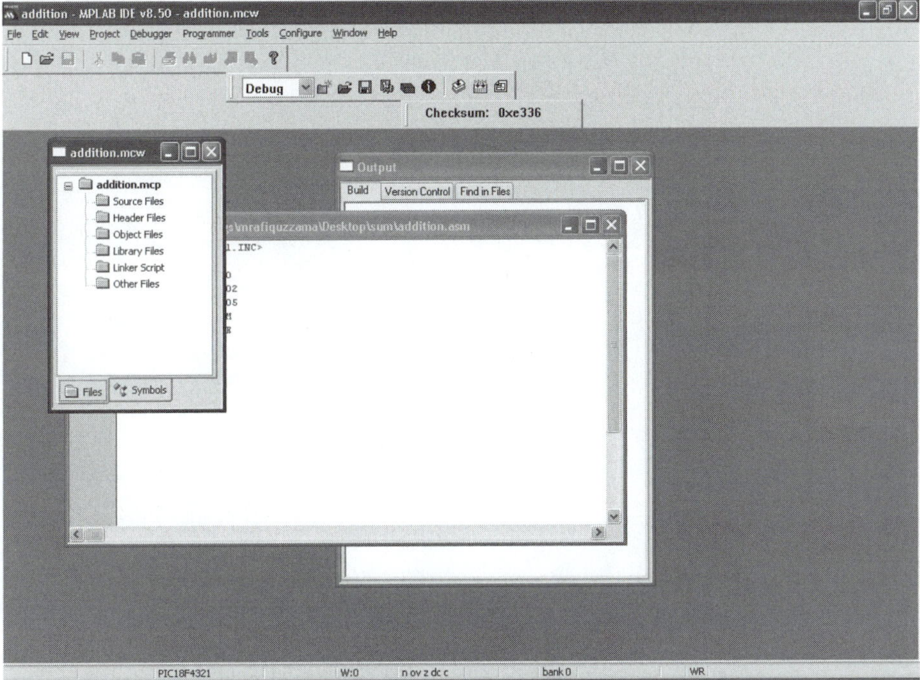

Right click on Source Files to see the following:

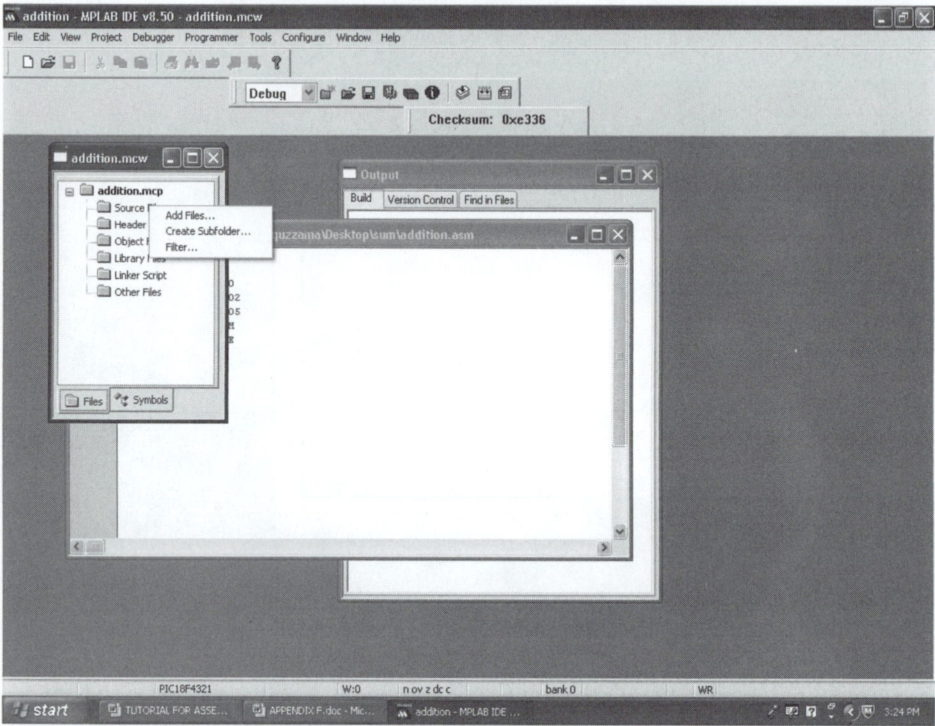

Click on Add files to see the following:

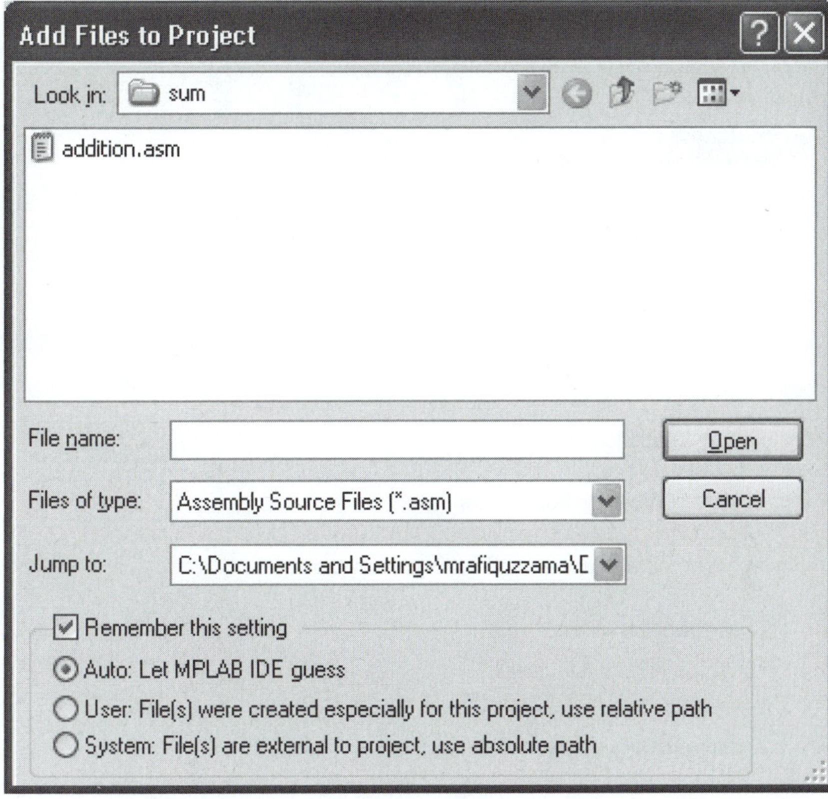

Click on addition.asm to see the following screen shot:

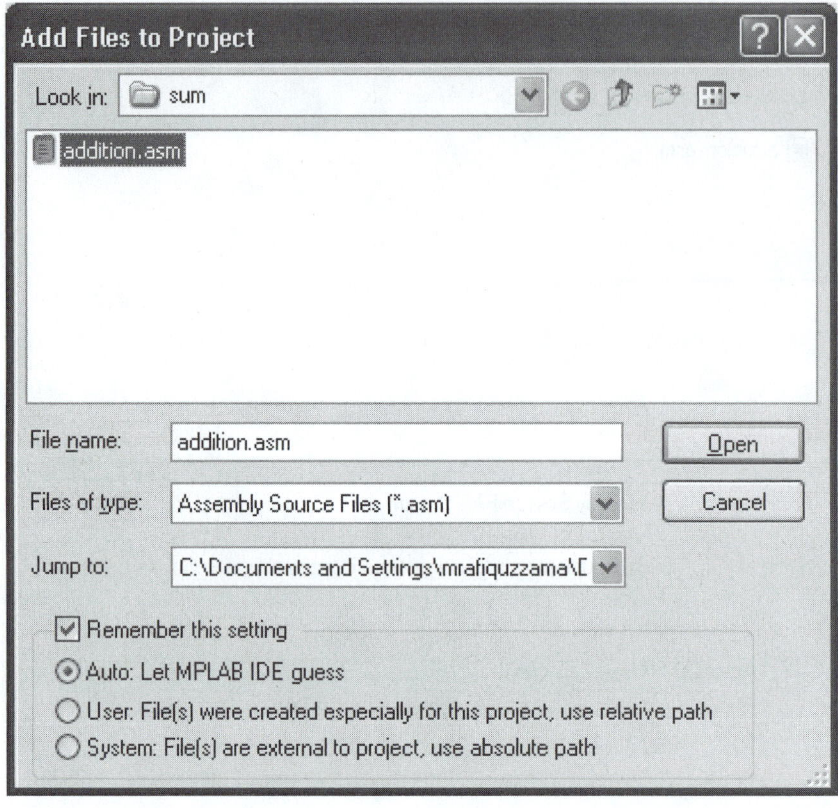

Click Open to see the following:

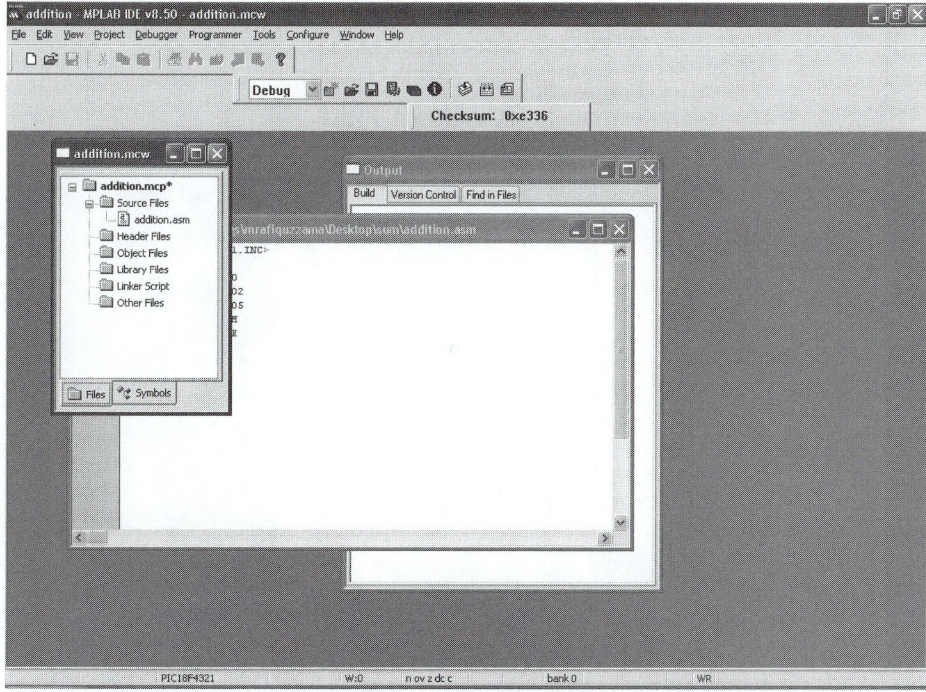

Note that addition.asm is listed under Source Files. Next, click on Project and then build all (or only the 'Build All' icon, second icon on top right of the Debug toolbar), and see the following:

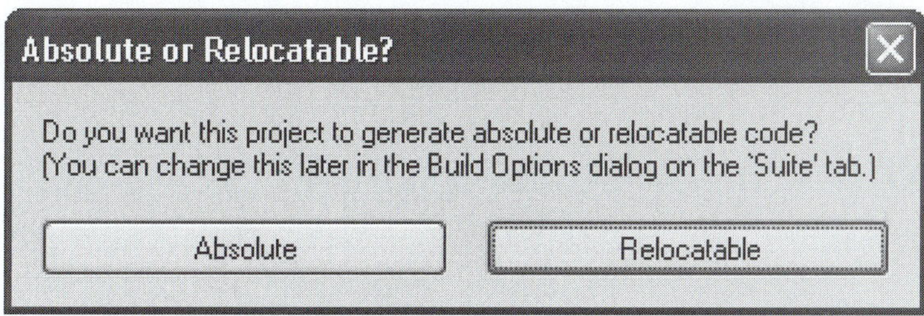

Next, click on Absolute to see the following:

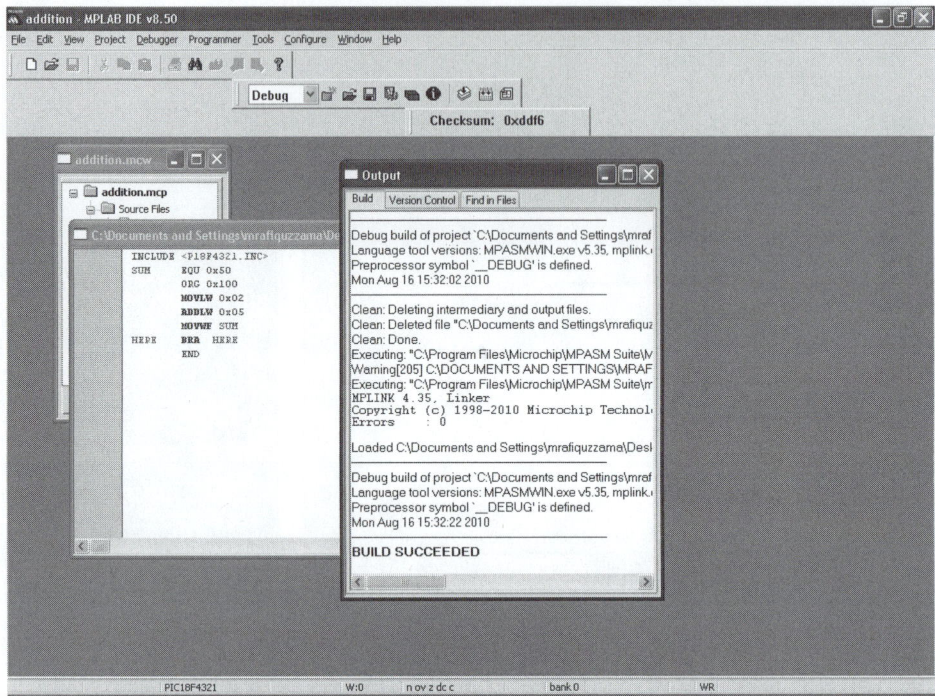

This means that the assembling the program is successful. Next the result will be verified using the debugger.

Click on Debugger, Select Tool, and then MPLAB SIM to see the following display:

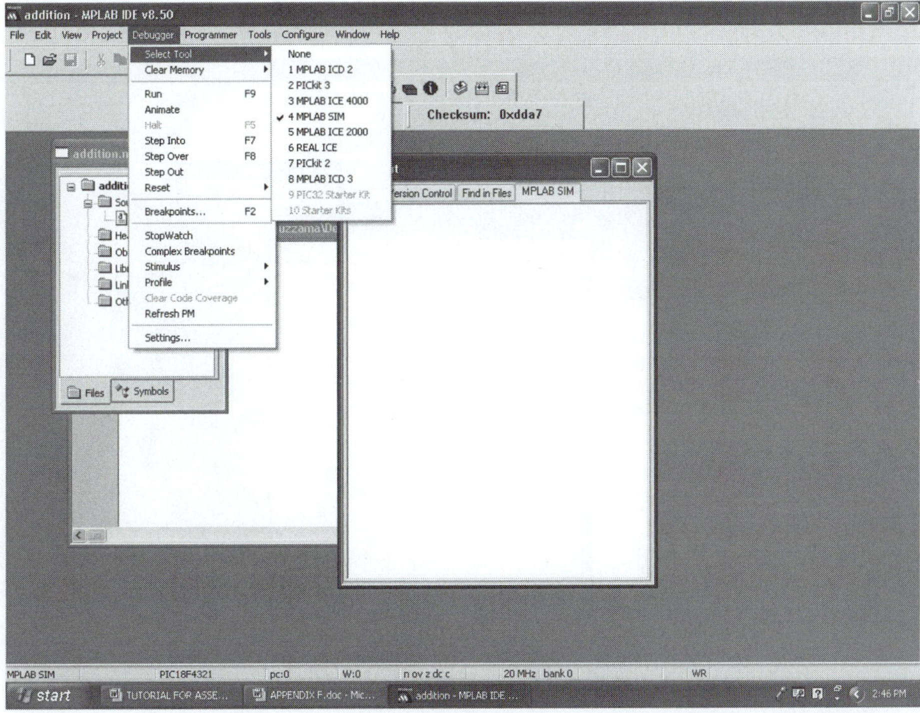

Click on MPLAB SIM to see the following:

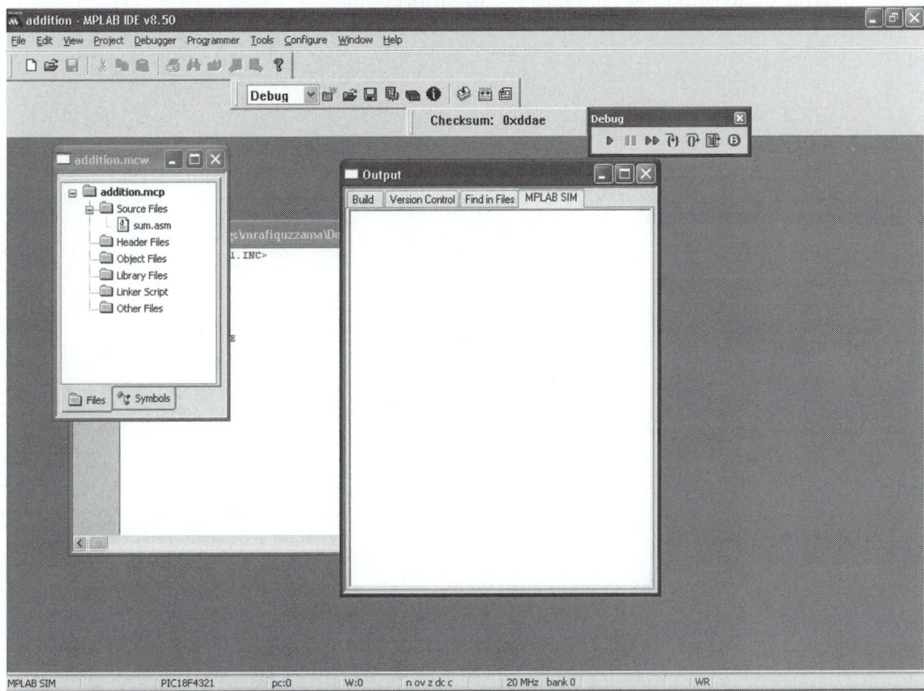

Click on View, toolbars, and Debug to see the following display with Debug toolbar:

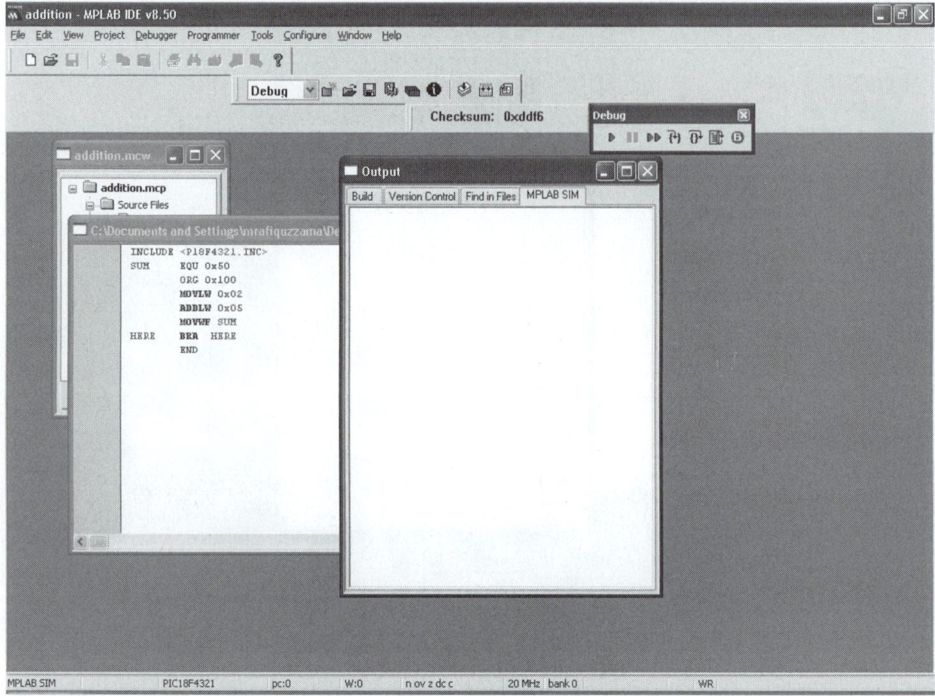

Next, click on View, and then watch to see the following:

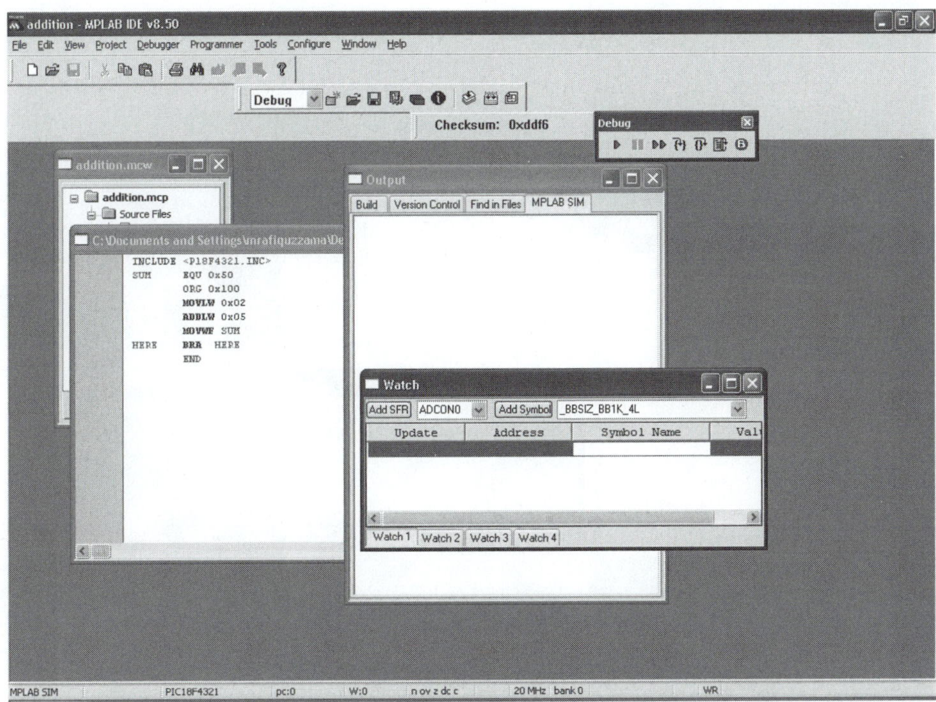

On the Watch list, you can now include WREG and SUM to monitor their contents. For example, to add WREG, scroll down to WREG by using the arrow beside ADCON0, and then click on Add SFR to see the following display:

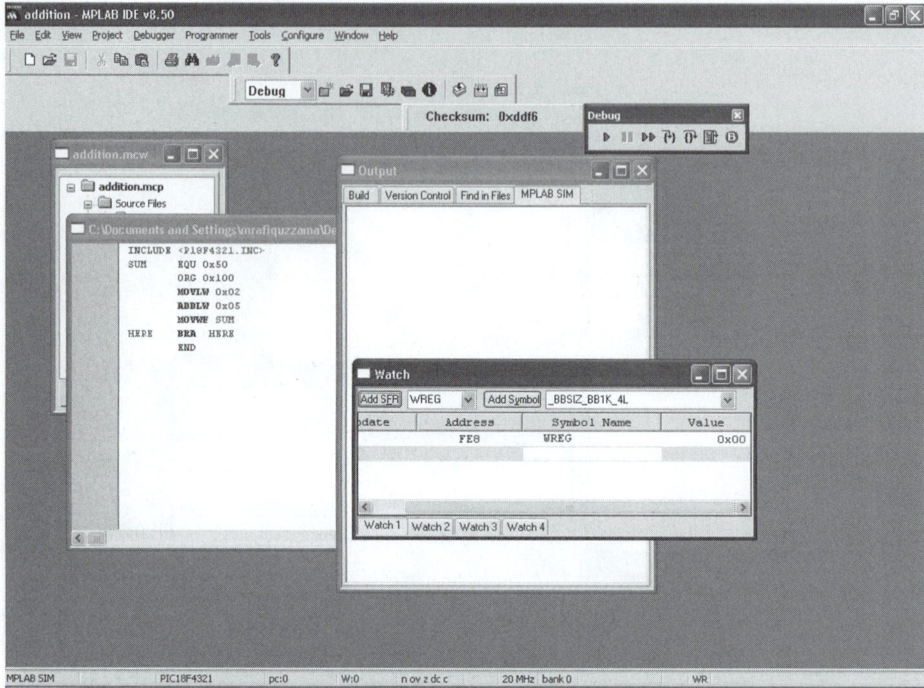

Next, scroll down using arrow beside Add Symbol, select SUM, and then click on Add Symbol to see the following display:

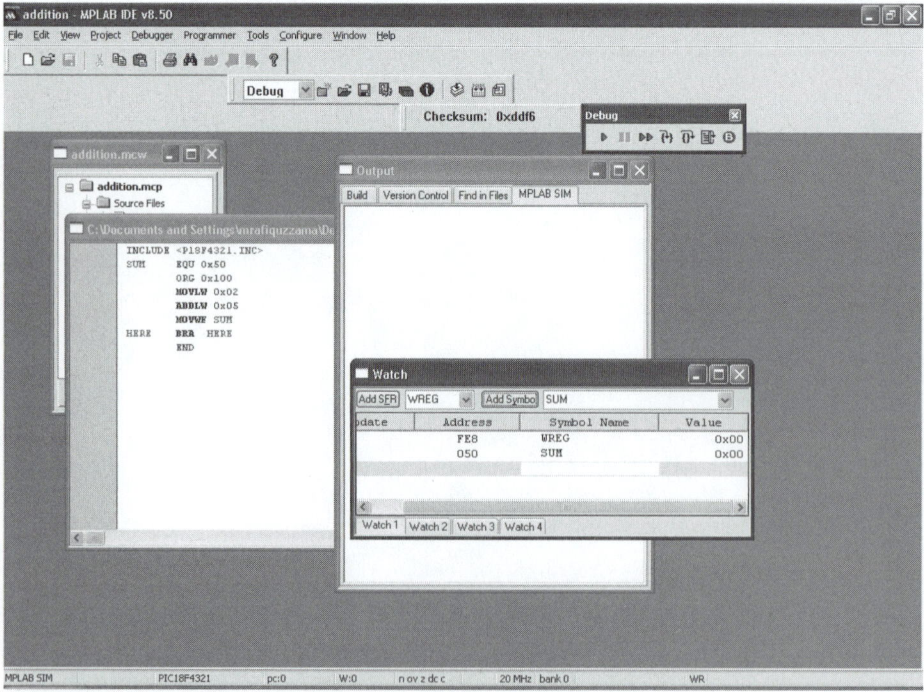

Note that SUM with address 0x50 along with contents is displayed.

In order to enter breakpoint for MOVLW 0x02, right click beside MOVLW to see:

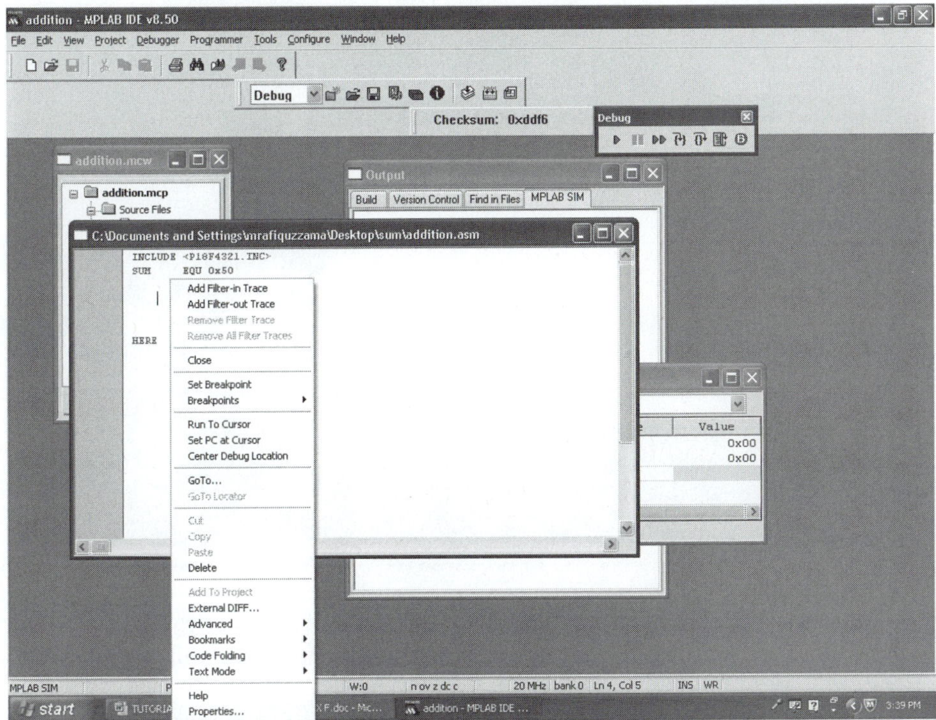

Click on set breakpoint to see the following:

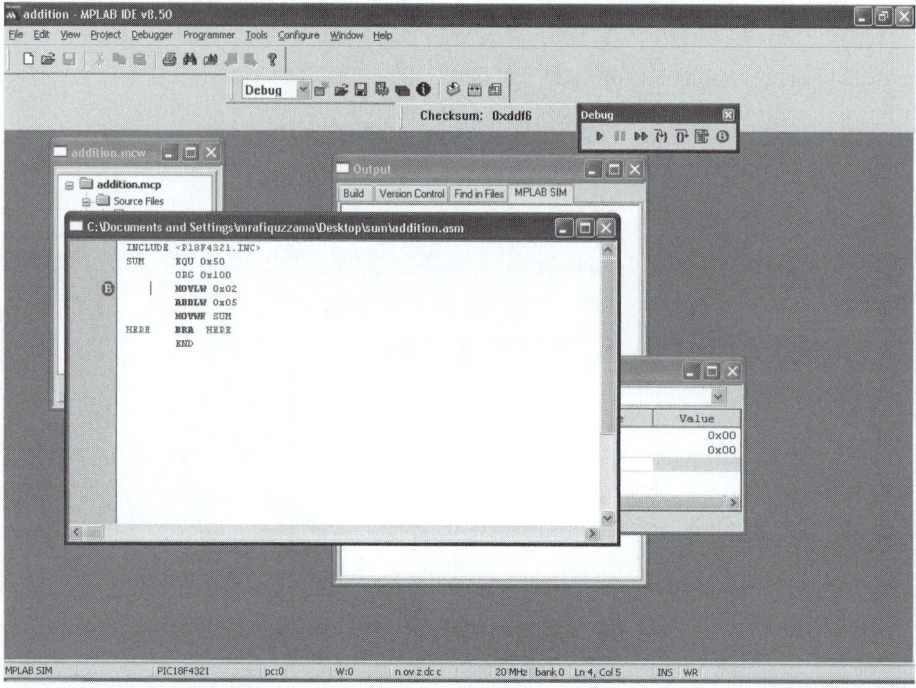

B in red means breakpoint is inserted for MOVLW0x02. Similarly, breakpoints for ADDLW 0x05 and MOVWF SUM can be inserted, and the following will be displayed:

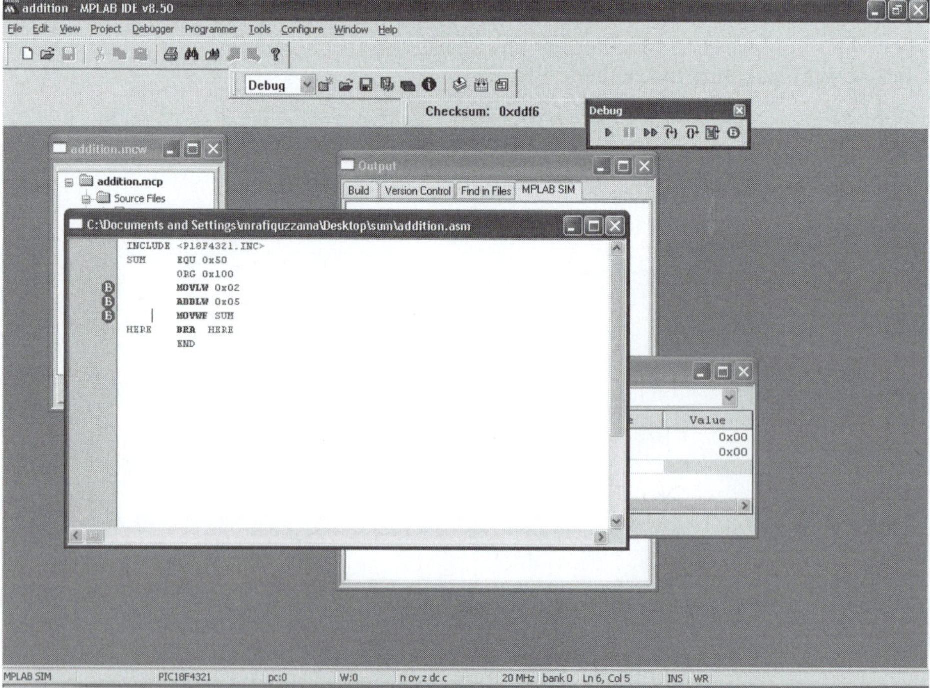

Next, locate the Debug Toolbar. If, for some reason, Debug toolbar is missing, go to view, select Toolbars, click on Debug, and see the following:

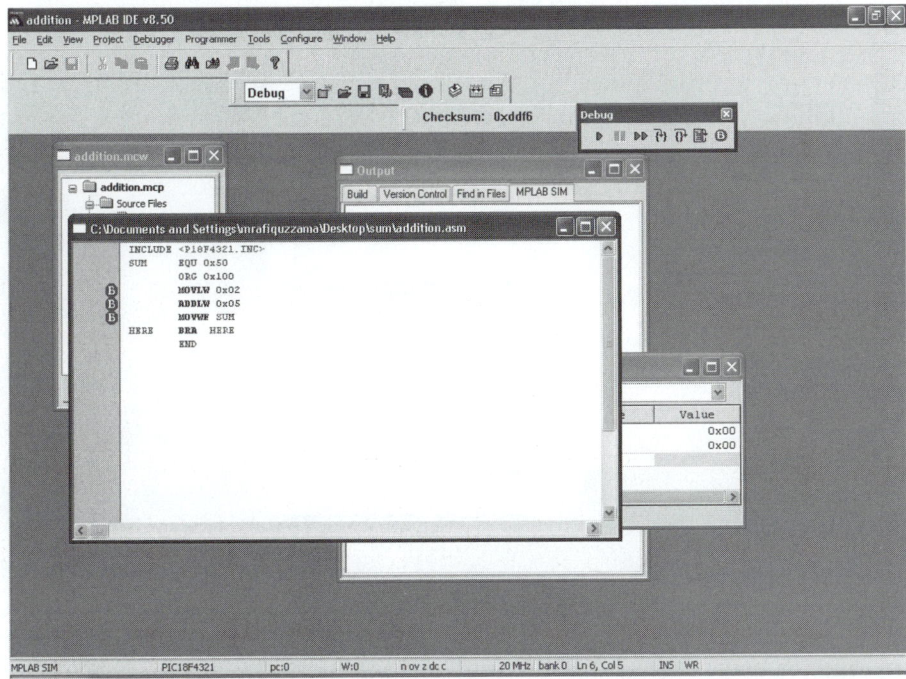

Click on the reset (first icon from right on the Debug toolbar), move the cursor to left of MOVLW 0x02, click on run (green arrow on left on the Debug toolbar) on Debug toolbar, and see that WREG is loaded with 0x02:

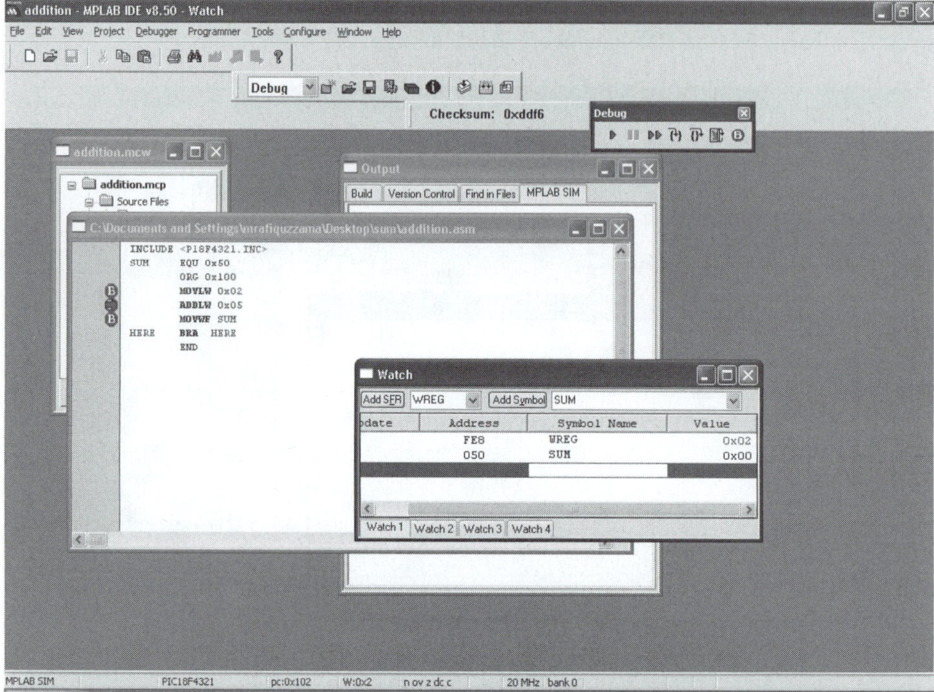

Next, click on run to execute ADDLW 0x05, and see that the result of addition0x07 is loaded into WREG as follows:

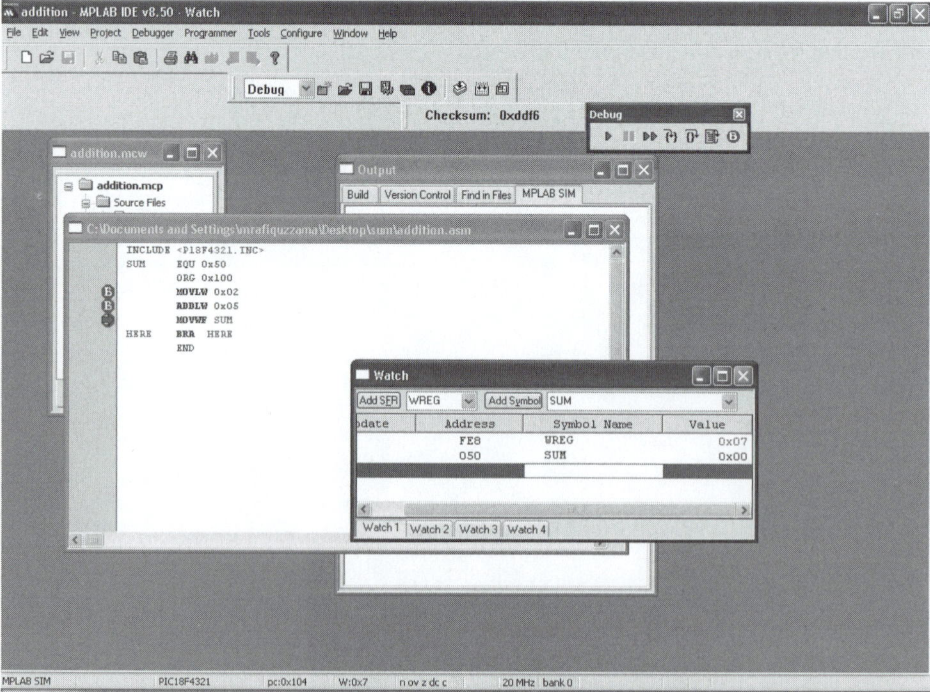

Finally, click on Animate (double green arrow on the Debug tool) to execute MOVWF SUM to see that the result 0x07 is stored in SUM (address 0x50) as follows:

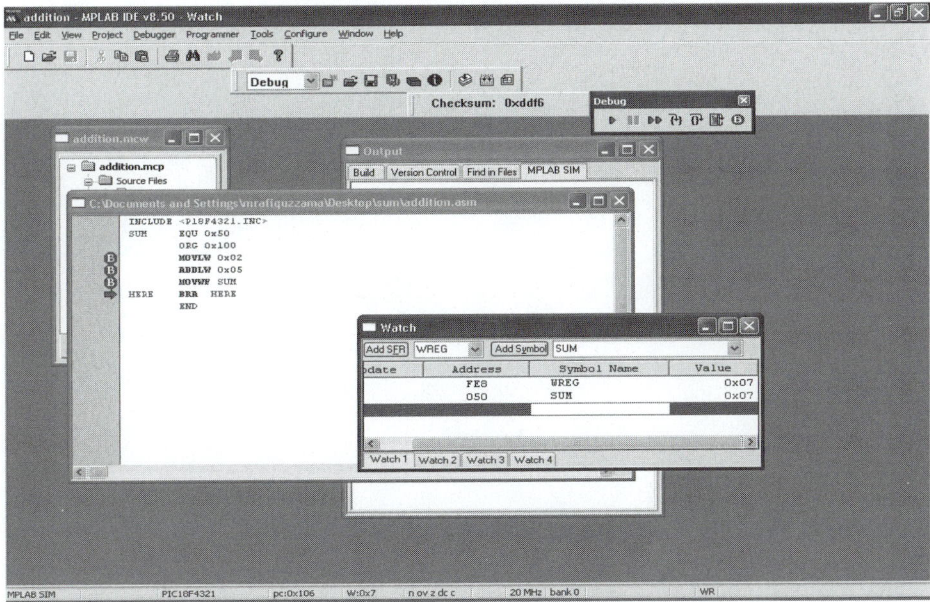

The debugging is now complete.

APPENDIX G: TUTORIAL FOR COMPILING AND DEBUGGING A C-PROGRAM USING THE MPLAB

Compiling a C-language program using MPLAB

First download the latest versions of the MPLAB assembler and C18 compiler from the Microchip website www.microchip.com. After installing and downloading the program, you will see the following icon on your desktop:

Double click (right) on the MPLAB icon and wait until you see the following screen:

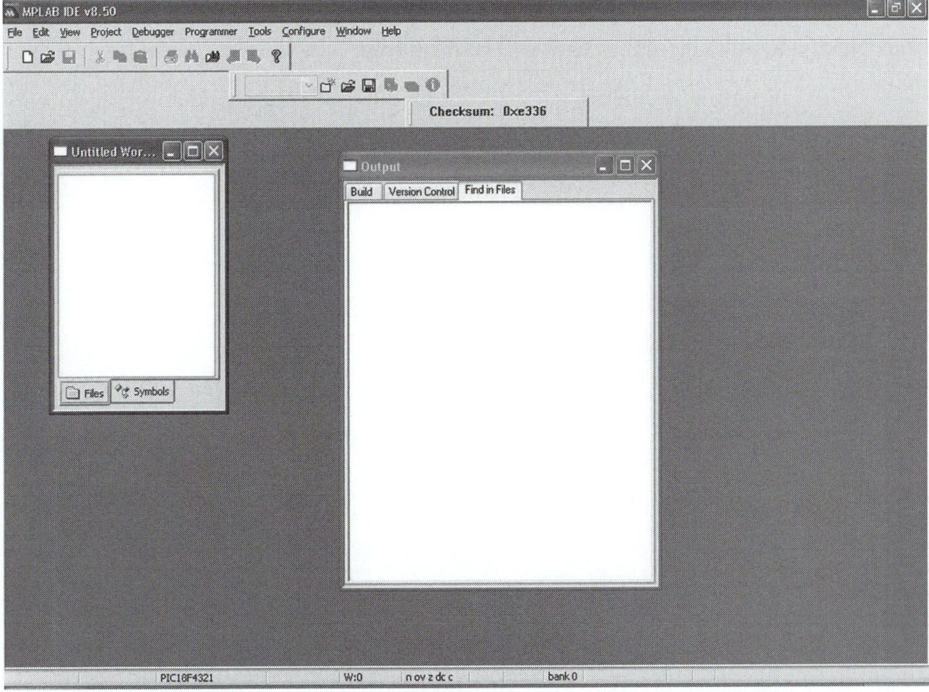

Next, click on 'Project' and then 'Project Wizard'; the following screen will appear:

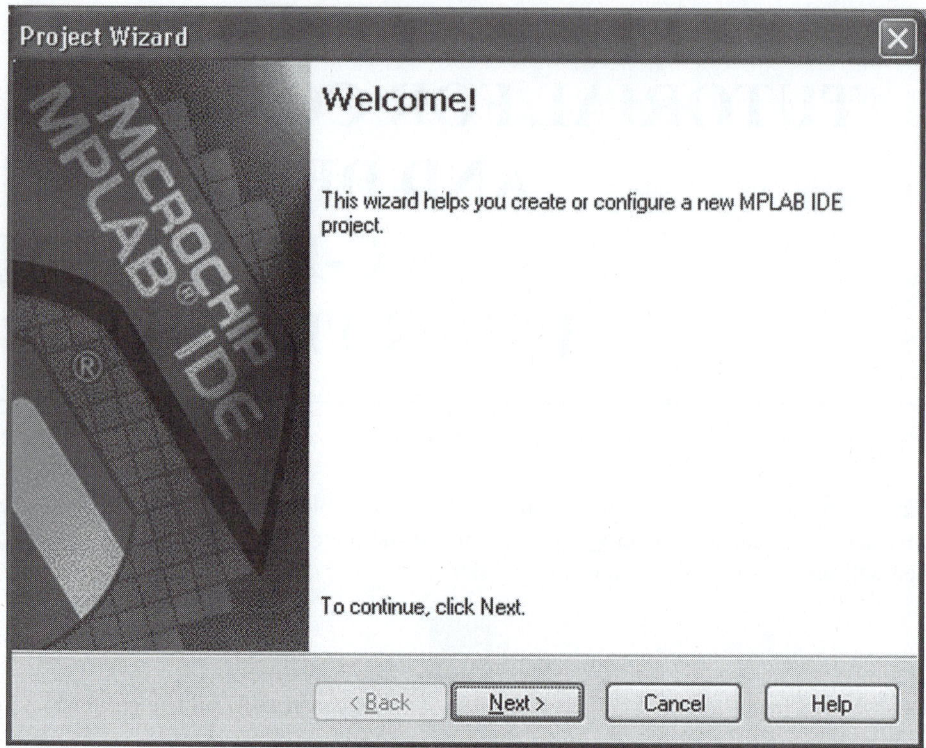

Click Next; the following screen shot will be displayed:

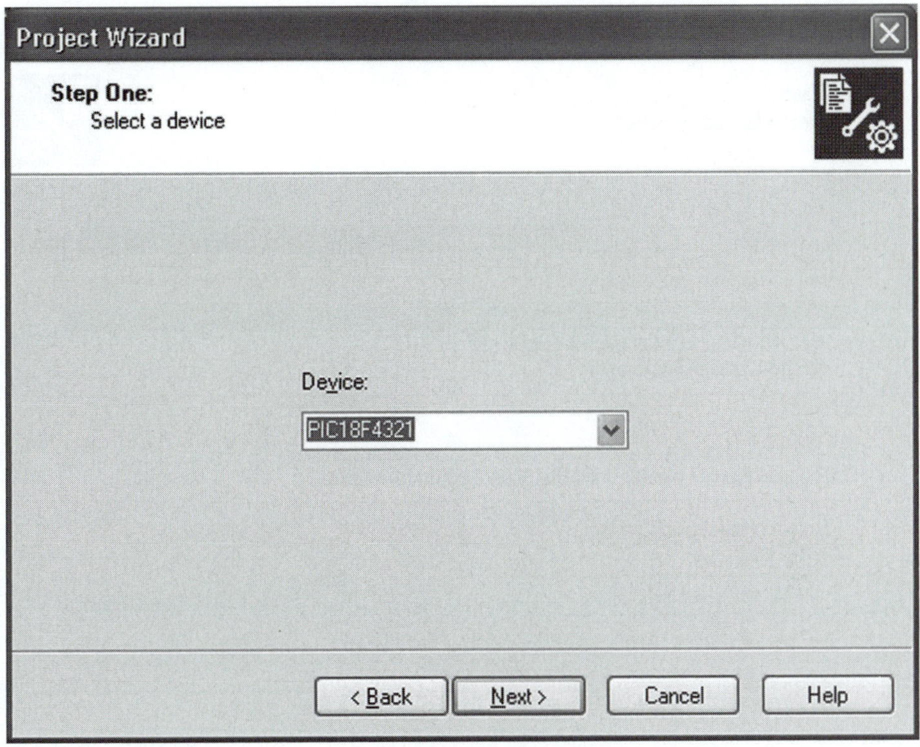

Select the device PIC18F4321, hit Next, and wait; the following will be displayed:

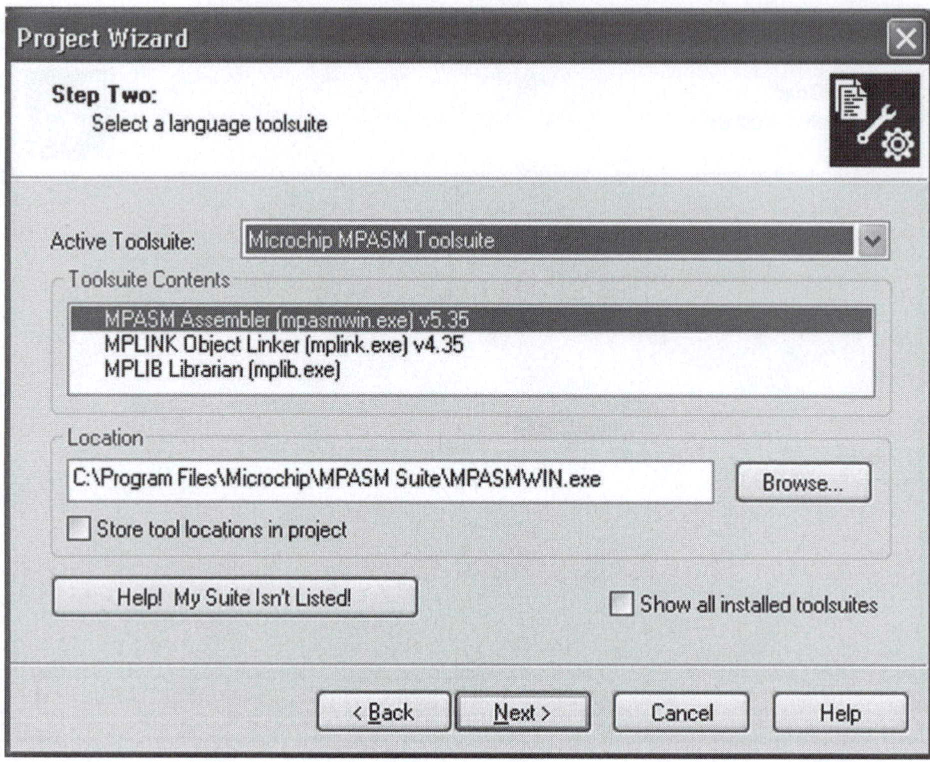

In the 'Active Toolsuite', select 'Microchip C18 Toolsuite', and click Next; the following will be displayed:

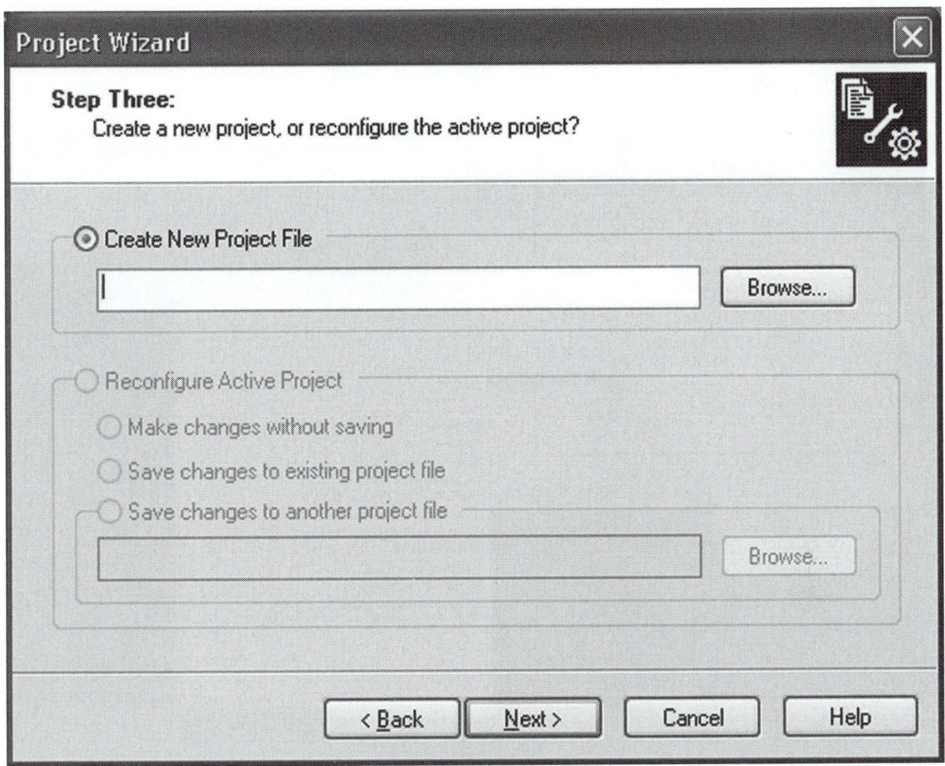

Select a location where all project contents will be placed. For this example, the folder will be placed on the desktop (arbitrarily chosen). Go to the desktop directory, make a new folder, and name the folder. In order to do this, click on 'Browse', and select desktop:

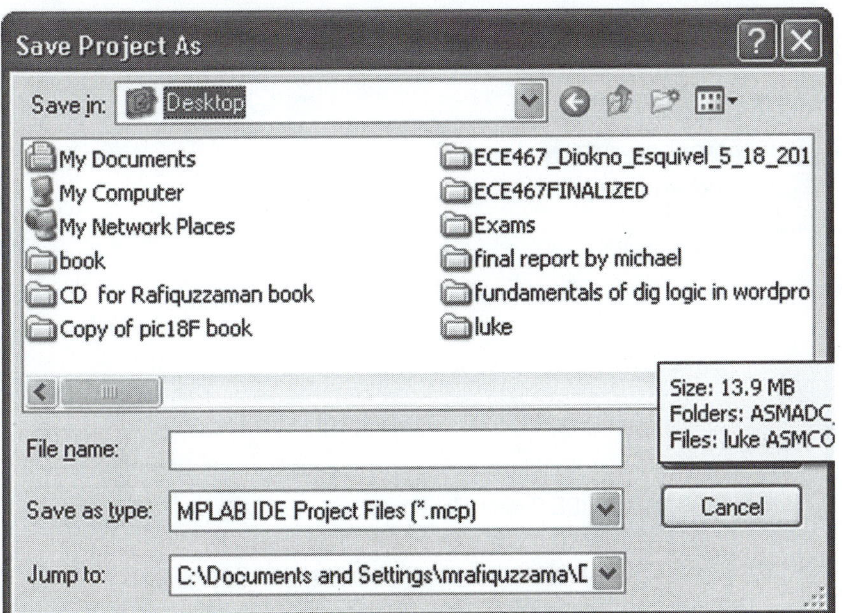

Next, create a new folder by clicking on the icon (second yellow icon from right on top row) or by right clicking on the mouse on the above window, and then go to New to see the following screen:

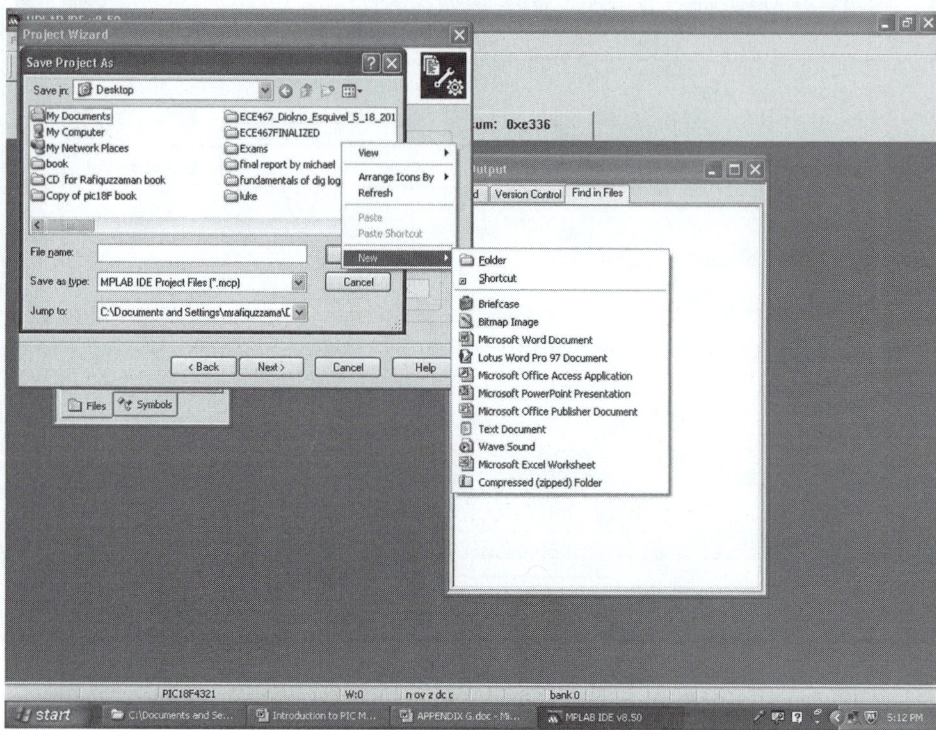

Click on Folder to see the following:

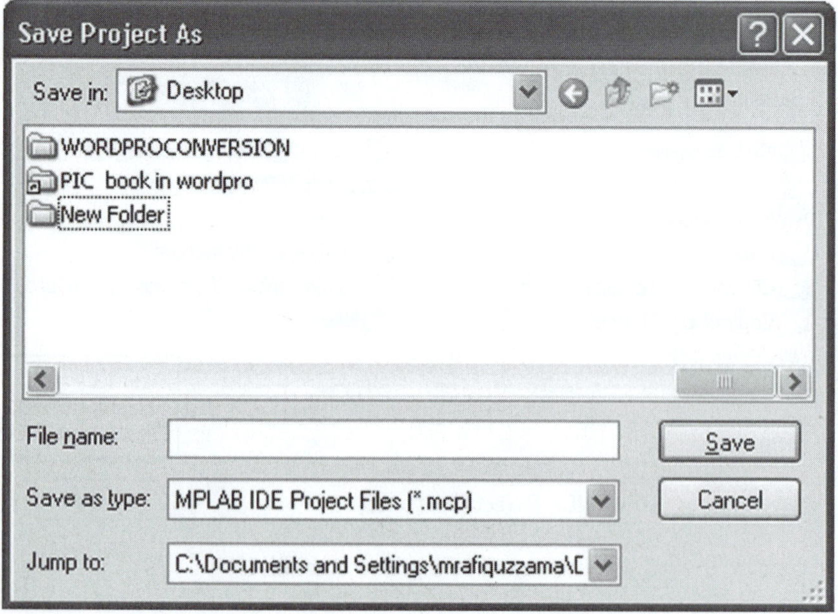

Click on folder, name it 'plus' (arbitrarily chosen name), and see the following :

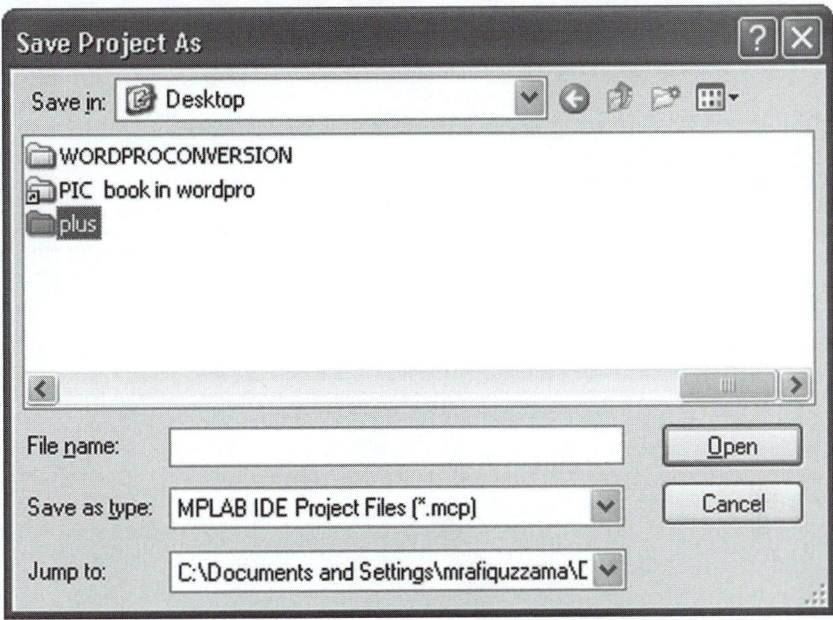

File name 'addition' is arbitrarily chosen. Type in the File name to see the following:

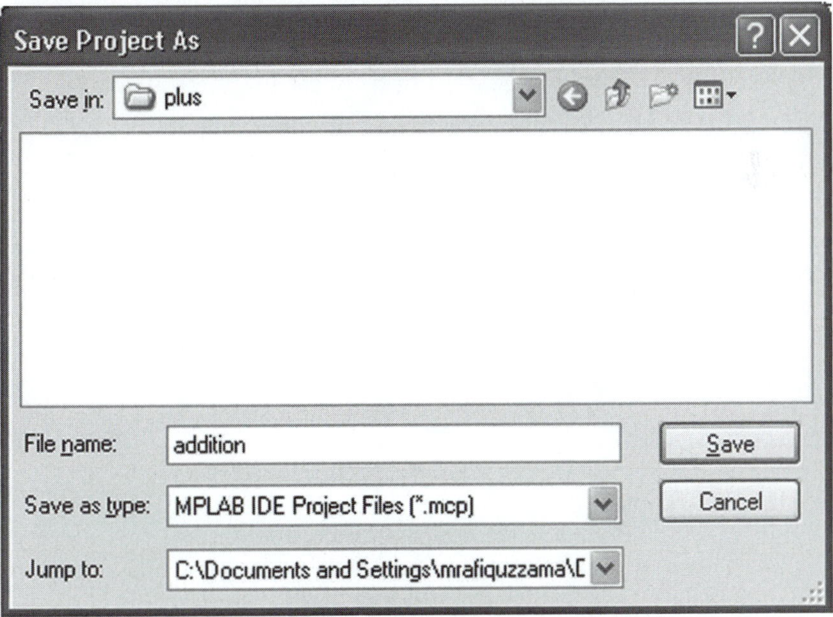

Next, click on Save; the following screen will appear:

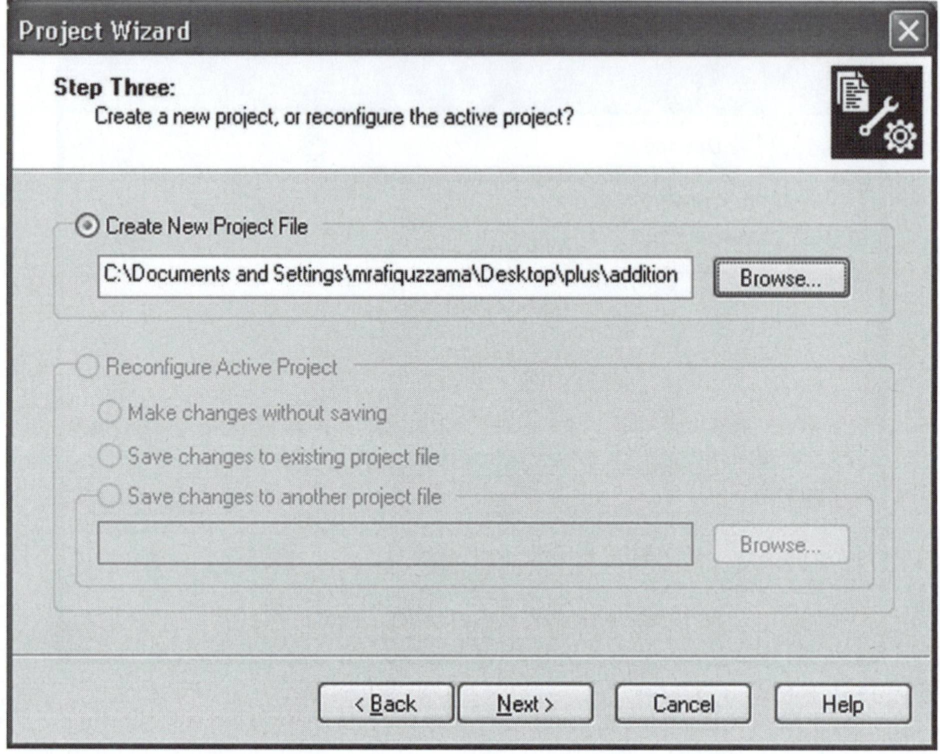

Click on Next, and see the following:

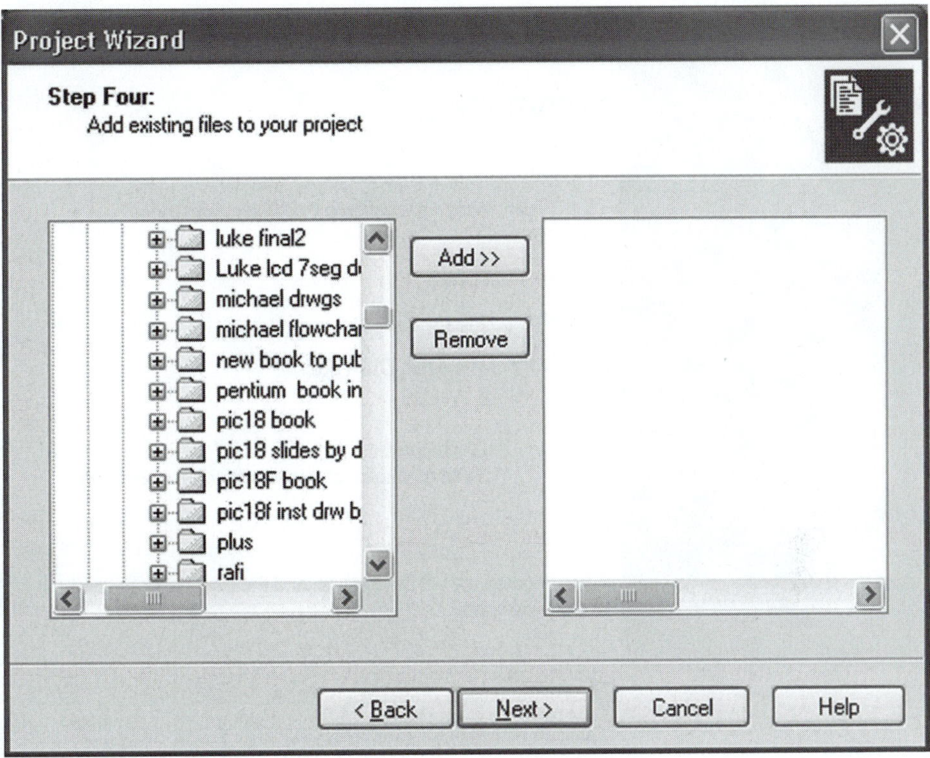

Click Next to see the following:

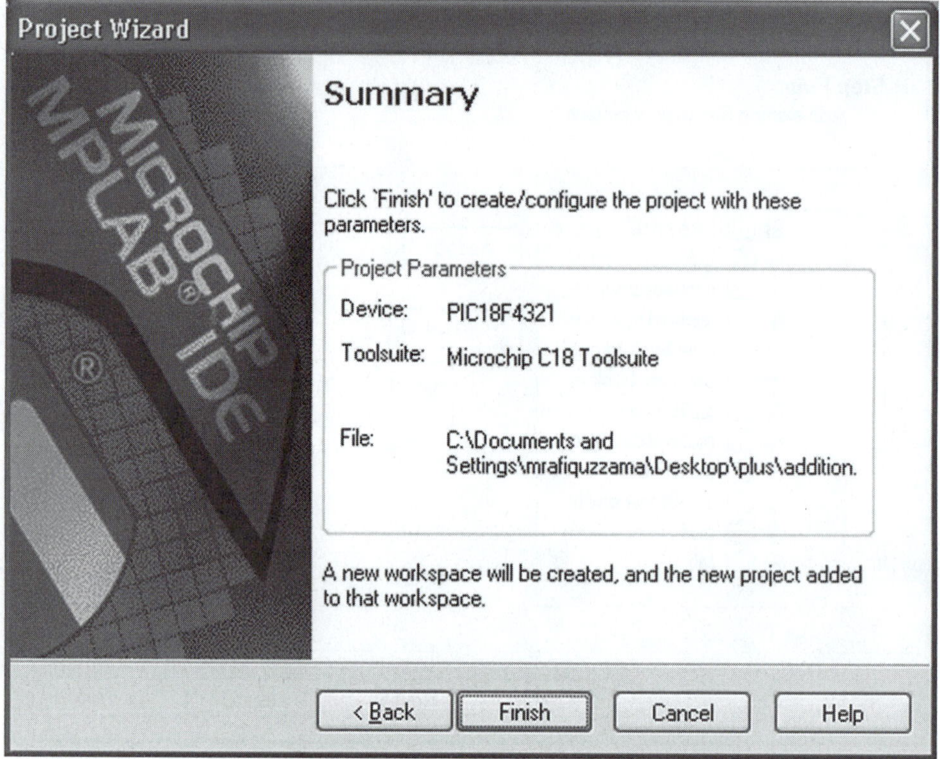

Click on Finish, and see the following:

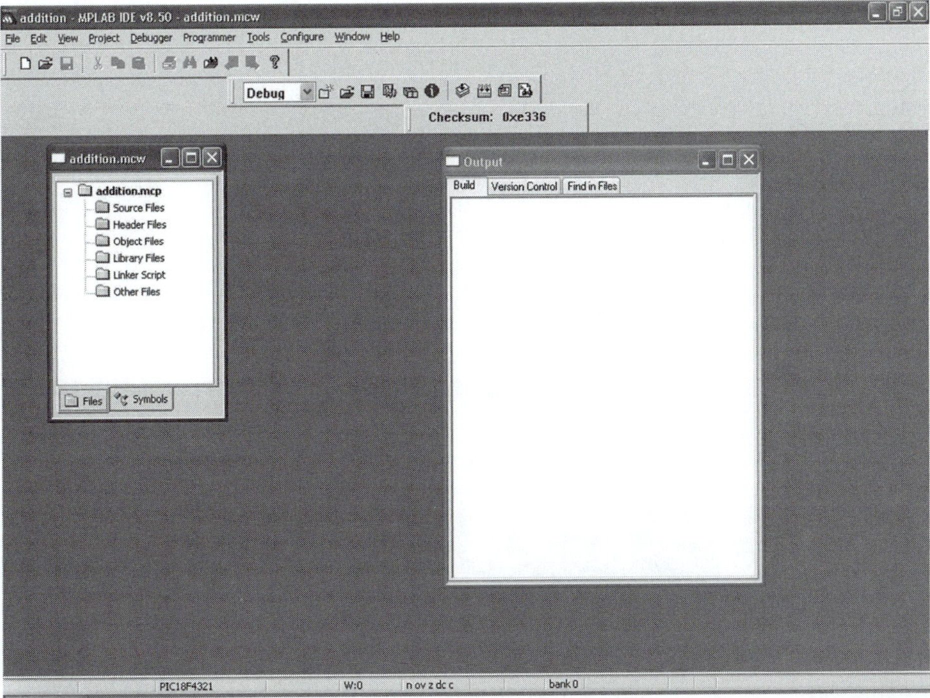

Click on File, and then New to see the following:

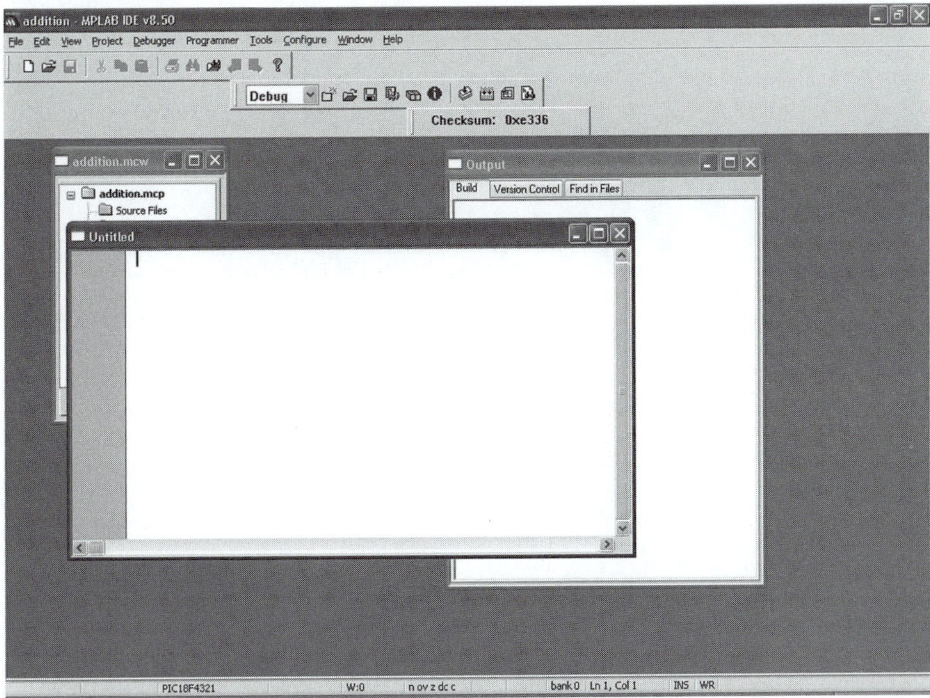

Type in the program you want to compile. The following addition program is entered:

```
#include <stdio.h>
void main (void)
{int a=5;
int b=1;
int c;
c=a+b;
while(1);
}
```

After entering the program, see the following:

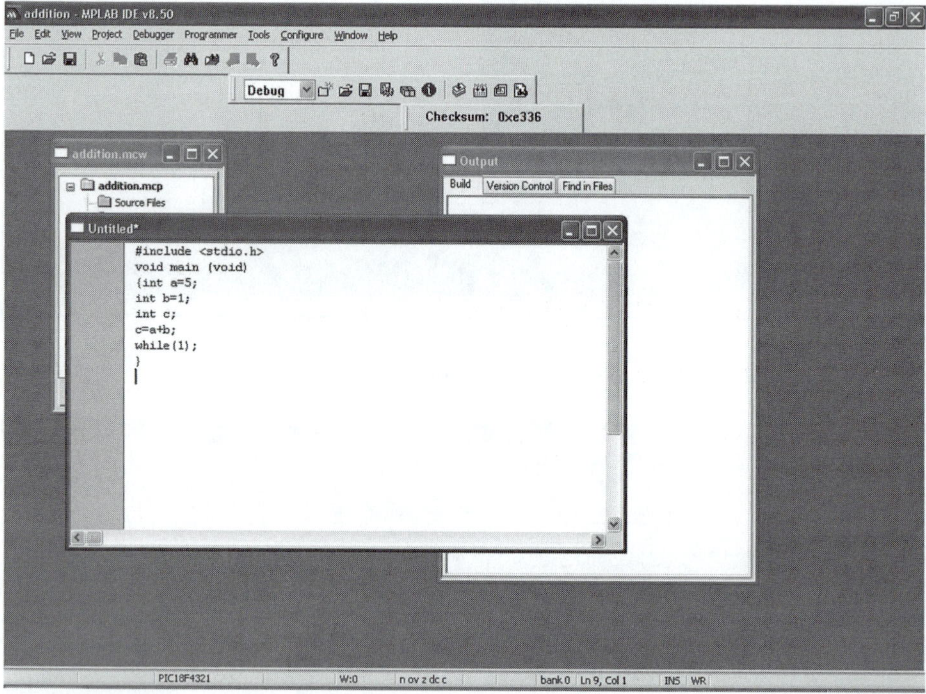

Next, click on File, and then Save as to see the following screen shot:

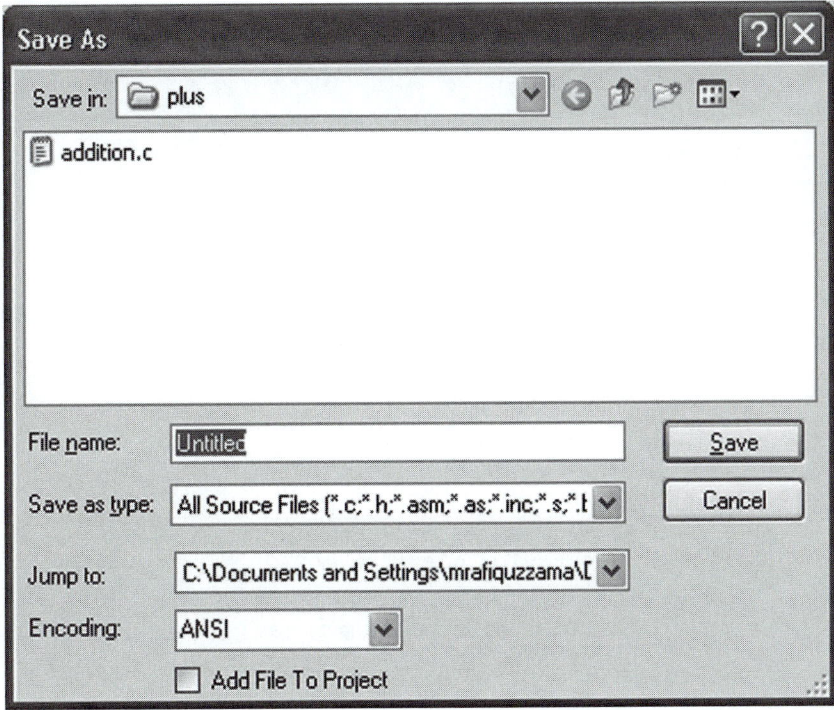

Make sure you scroll up to desktop, then click on plus (the folder which was created before), and see the following:

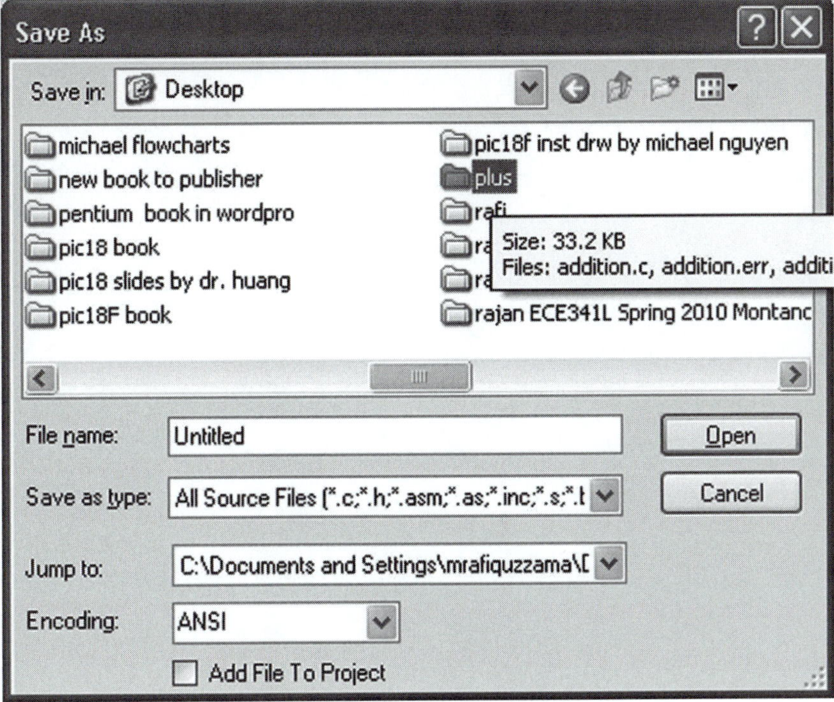

Next, double click (left) on plus to see the following:

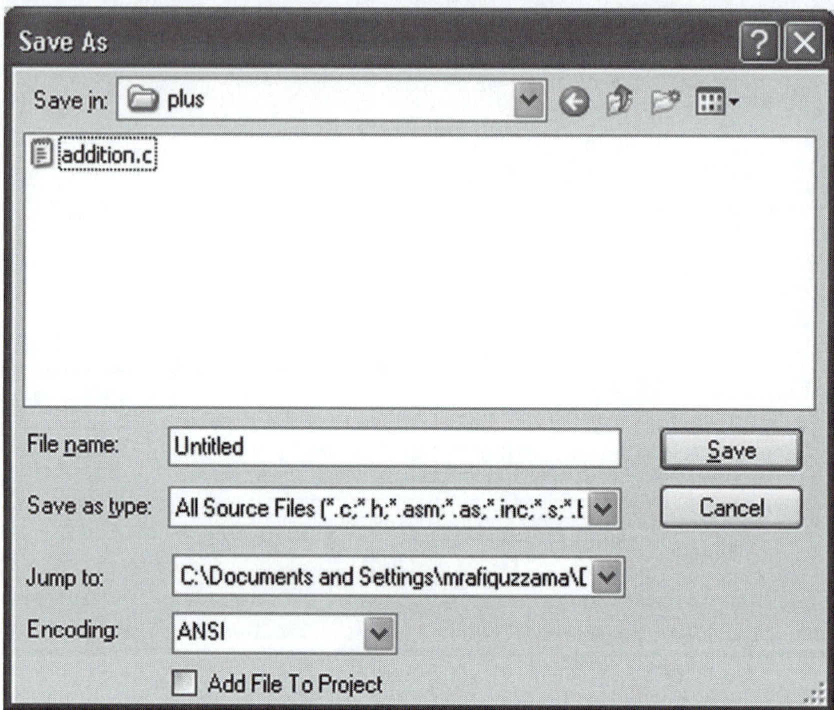

Delete Untitled, enter the same file name 'addition' with .c extension as File name. Click on save, and see the following screen shot (notice the display changes color) :

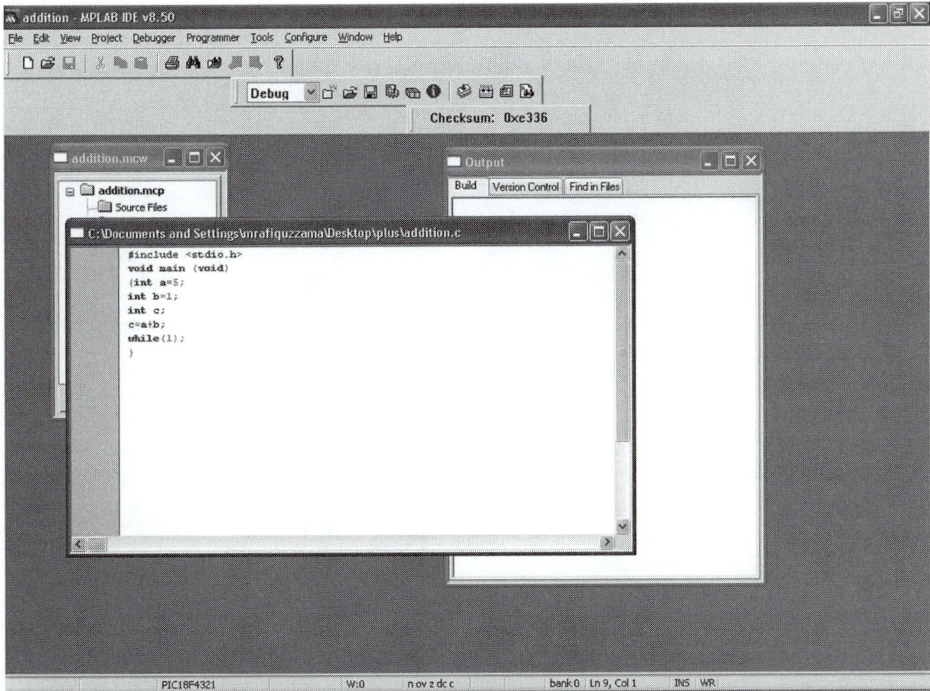

Next, highlight by clicking on the top (blue) section of addition.mcw, and see the following:

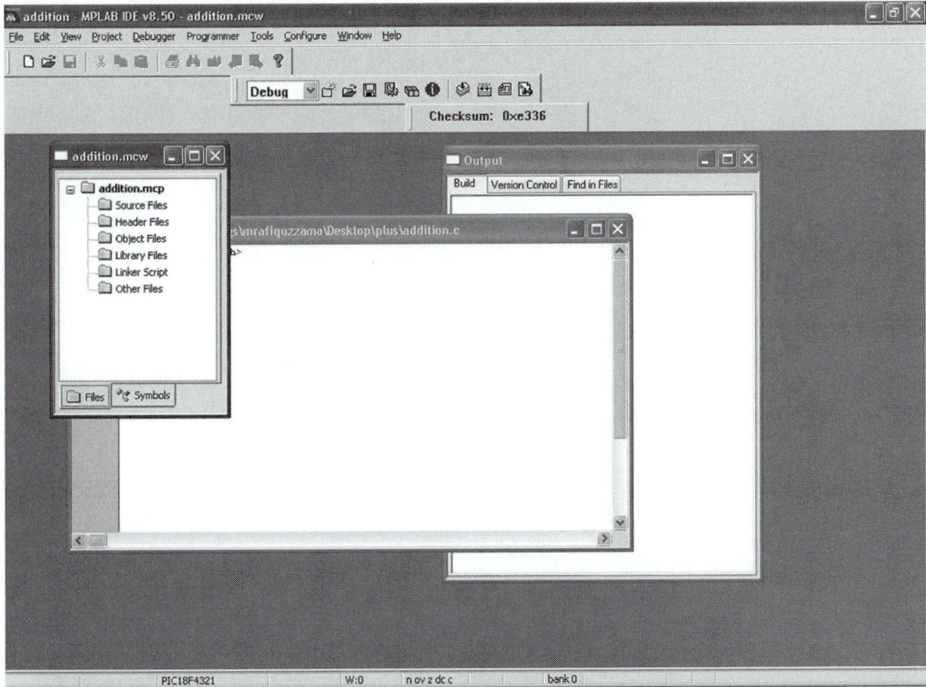

Right click on Source Files to see the following:

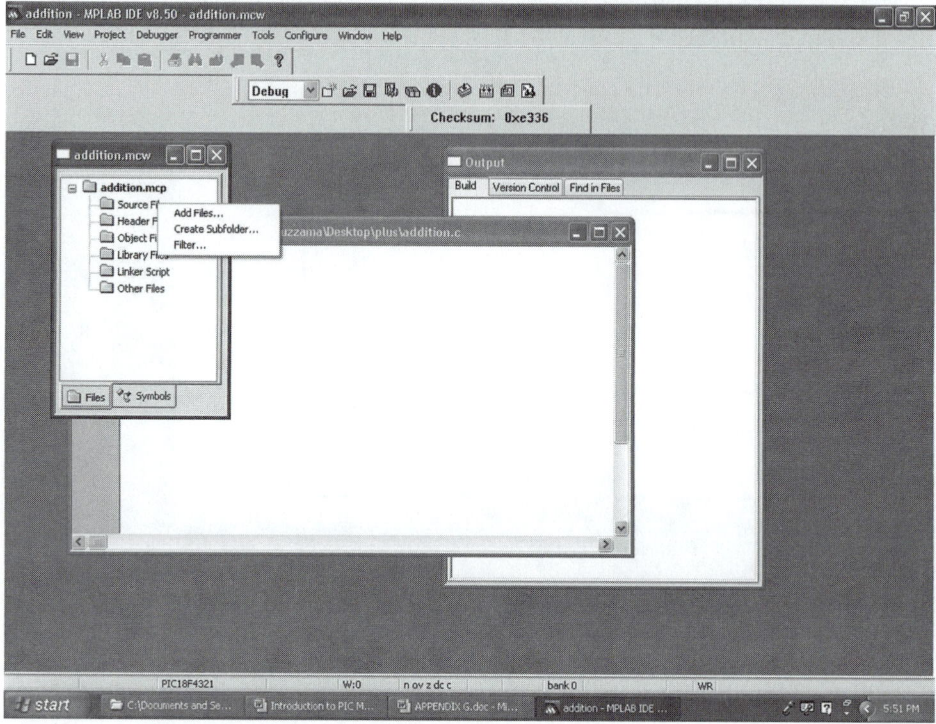

Click (left) on Add files to see the following:

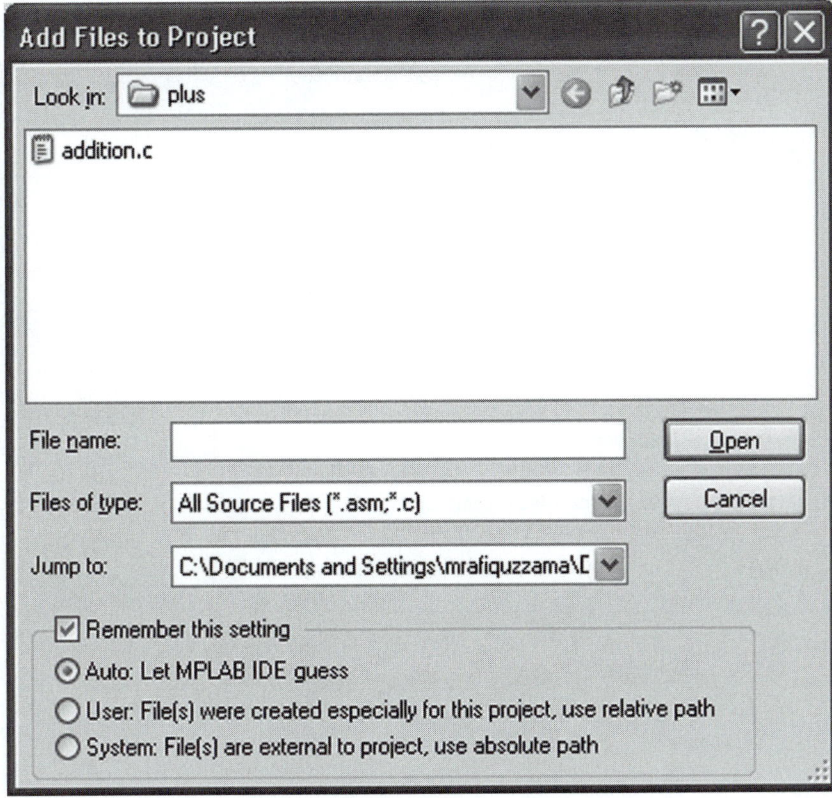

Click once (left) on addition.c on the window to see the following:

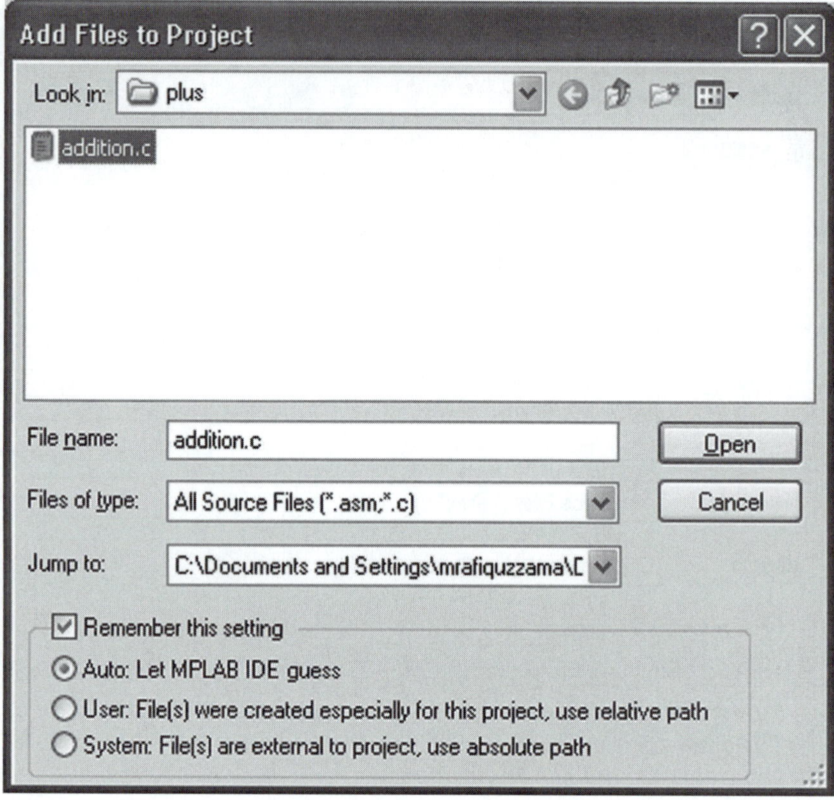

Click Open to see the following:

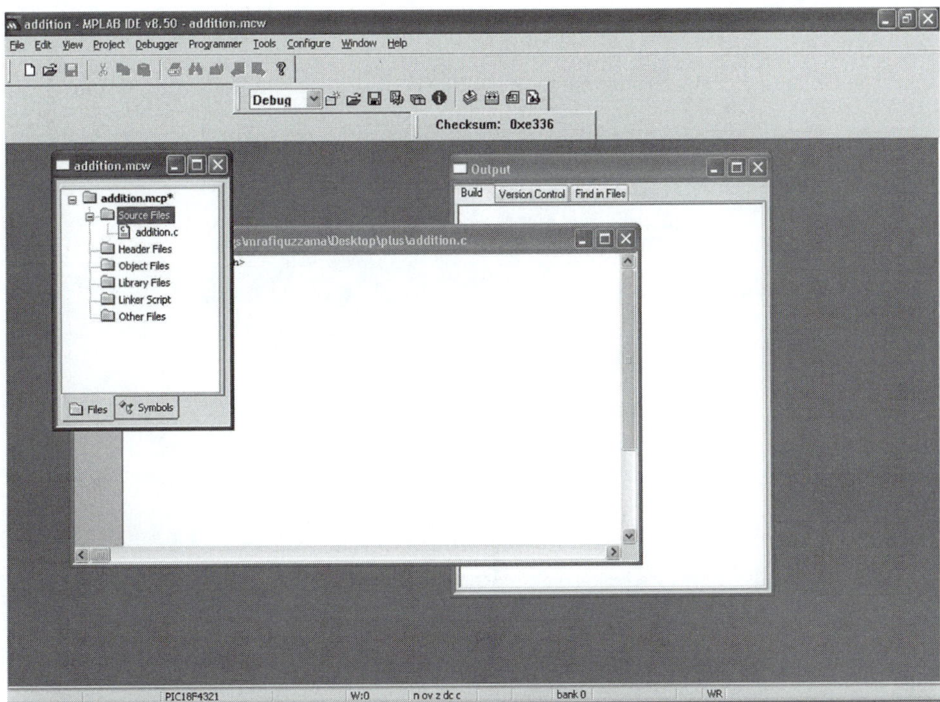

Note that addition.c is listed under Source Files. Next, click on Project and then build all (or only the 'Build All' icon, third icon on top right of the Debug toolbar), and see the following:

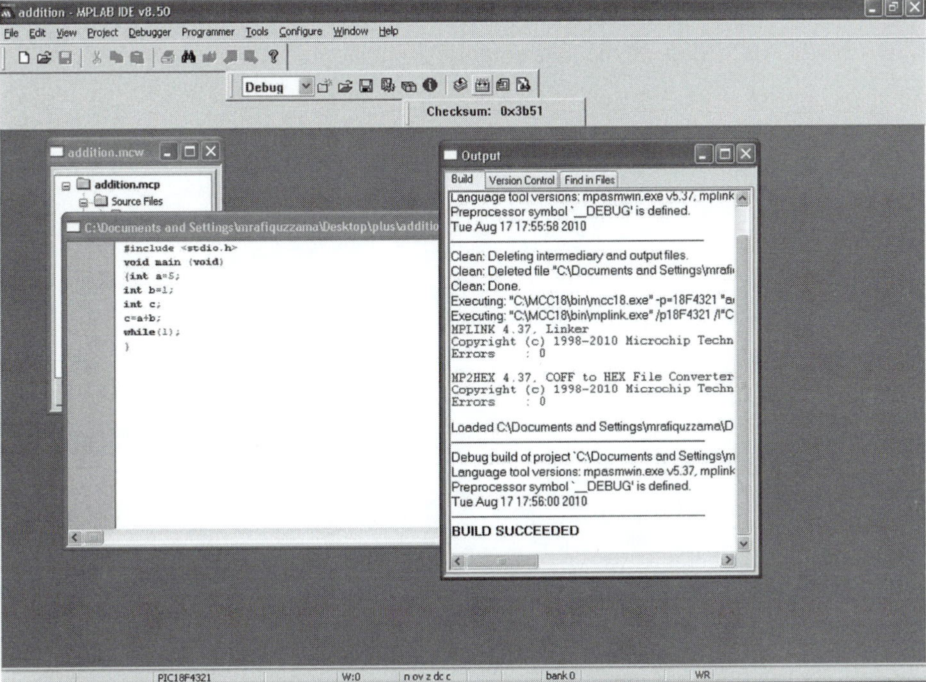

This means that compiling the C program is successful. Next, the result will be verified using the debugger.

Click on Debugger, Select Tool, and then MPLAB SIM to see the following display:

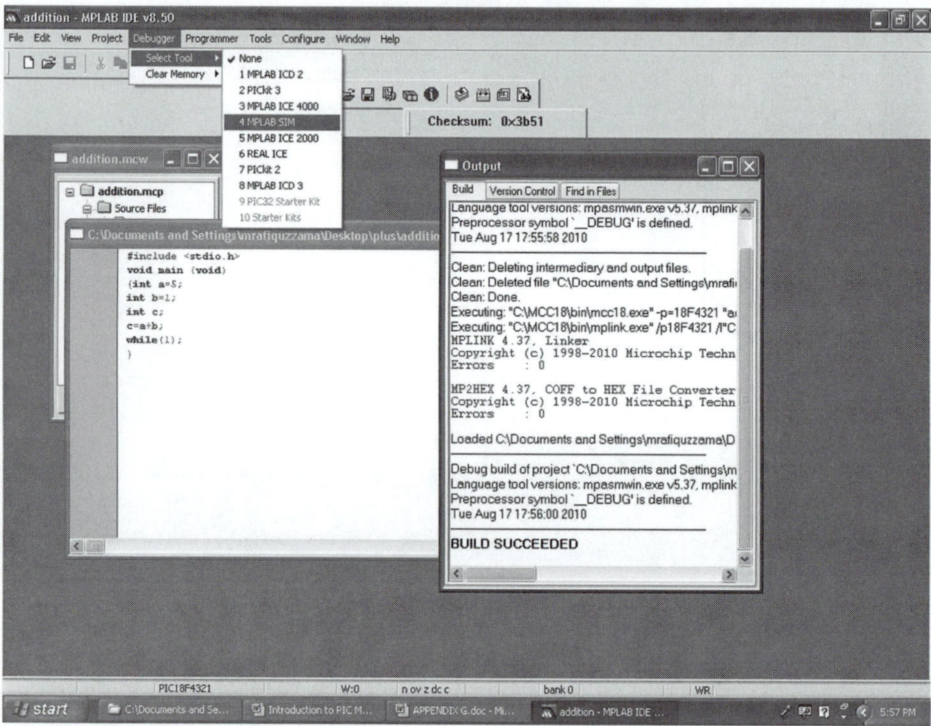

Click on MPLAB SIM to see the following:

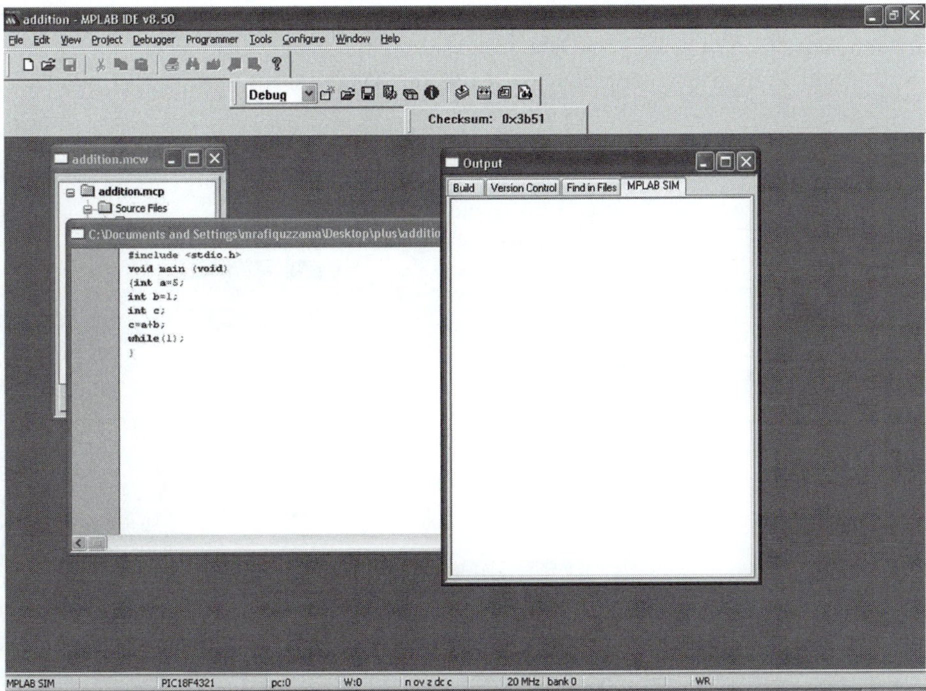

Click on View, toolbars, and Debug to see the following display with Debug toolbar:

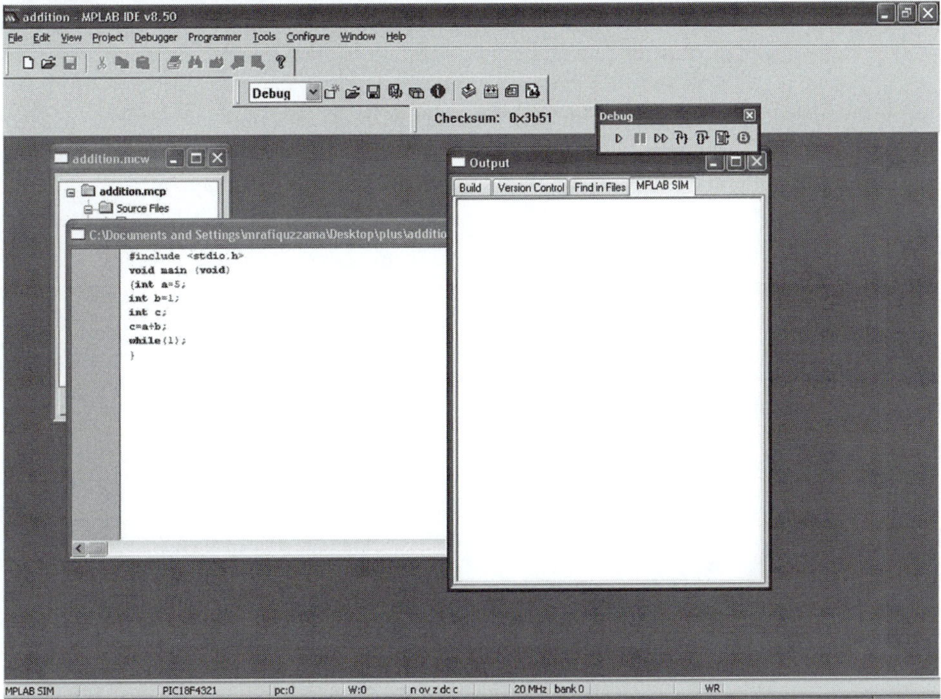

In the above, locate the Debug Toolbar. If, for some reason, Debug toolbar is missing, go to view, select Toolbars, click on Debug, and see the following:
Next, click on View, and then watch to see the following:

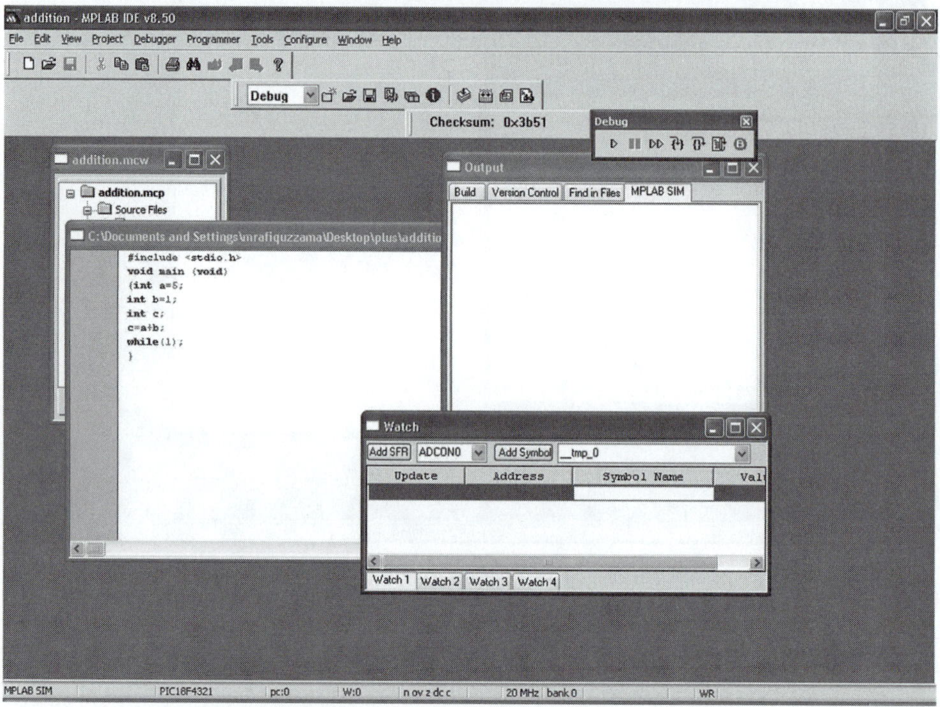

On the Watch list, you can now include locations a, b, c to monitor their contents. For example, to add 'a', simply select 'a' by scrolling down and using the arrow on the Add Symbol window, then click on Add Symbol to see the following display:

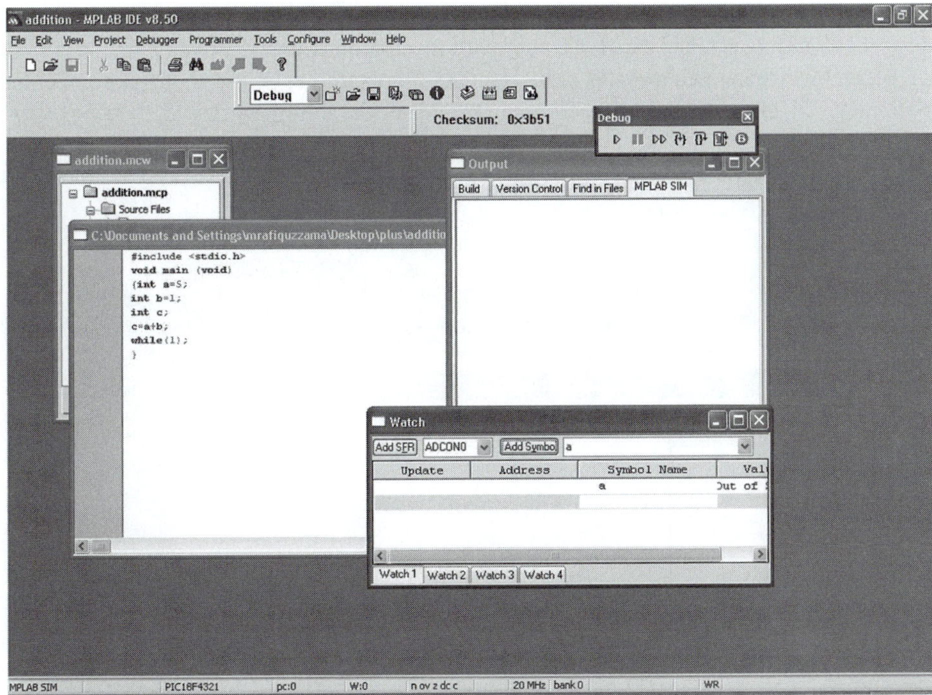

See that 'a' is displayed on the watch window. Similarly, display 'b' and 'c', and see the following screen shot:

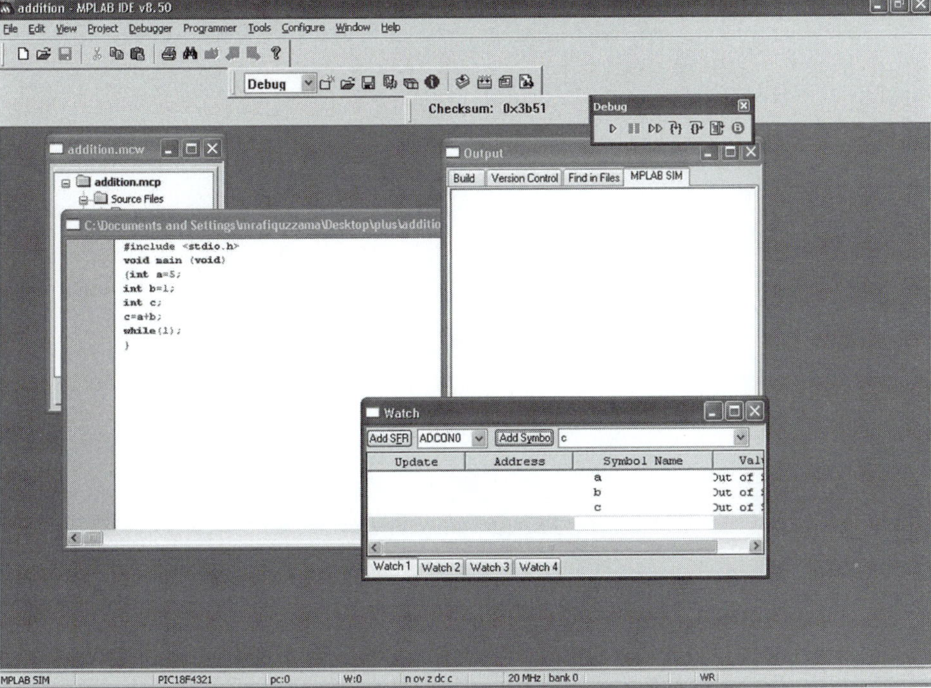

Next, insert breakpoints. Three breakpoints will be inserted for this program. One for int a = 5, one for int b = 2, and one for c = a+b. To insert a breakpoint, move the cursor to the left of the line where breakpoint is to be inserted. For example, to insert a breakpoint at int a =5, move cursor to the left of the line, click, and see the following display:

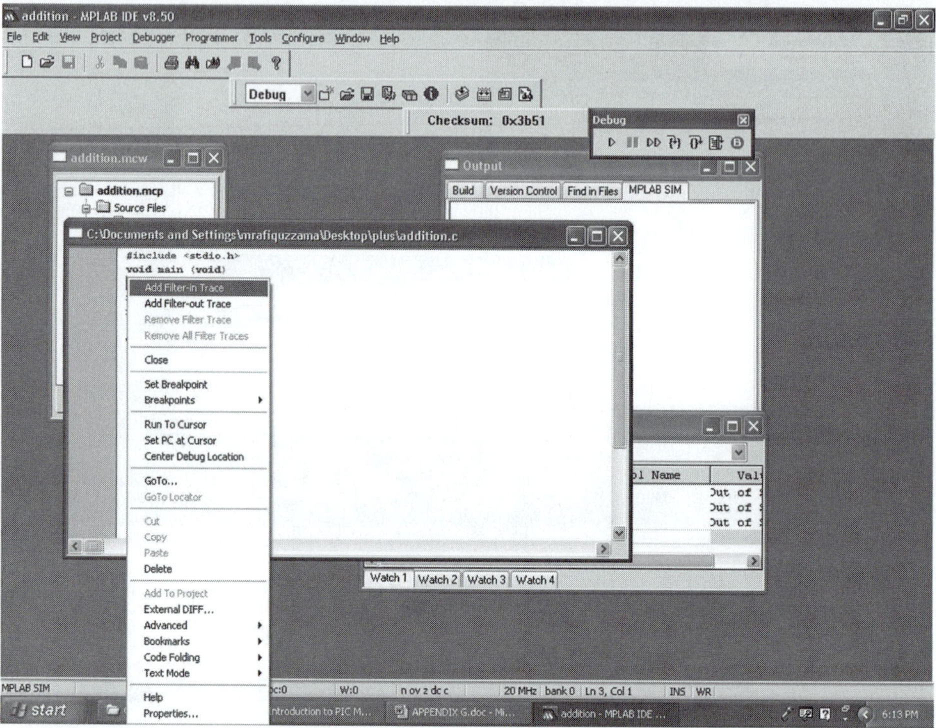

Next, click on Set Breakpoint to see the following:

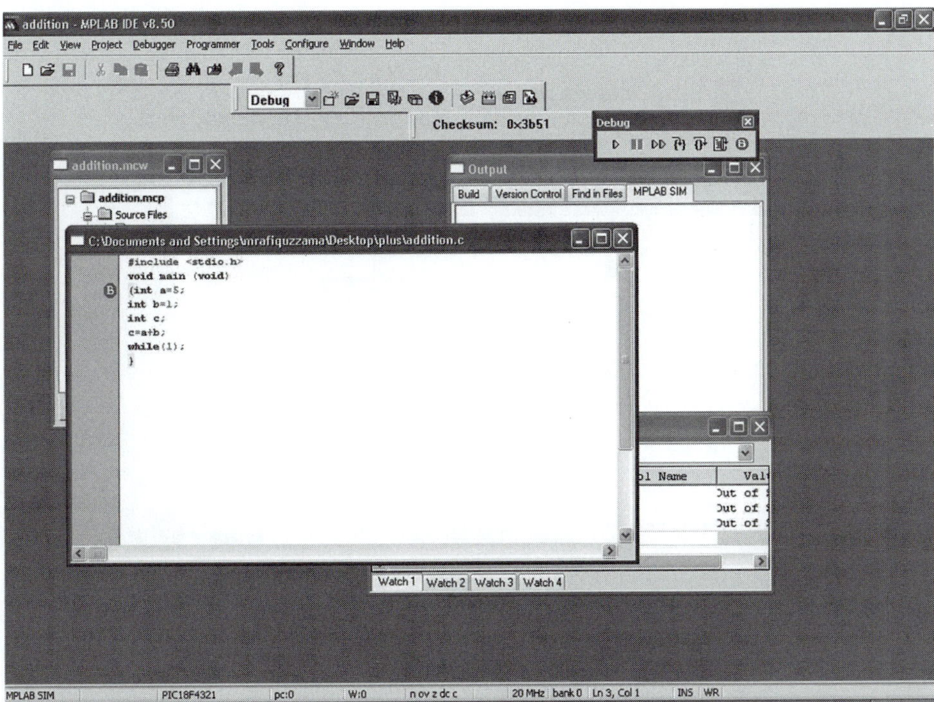

B in red on the left side of the line would indicate that the breakpoint is inserted. Similarly, insert the breakpoints for' 'b and 'c', and obtain the following display:

Next go to the Debug toolbar and Watch menu to see the contents of a, b, and c as each line is executed.

First, go to Debug toolbar, left click on reset (first symbol on right), and then click on the single arrow called the 'Run' arrow (left most arrow on the Debug menu); the code int a = 5; will be executed next. Click on single arrow again, the code is executed, and the following will be displayed:

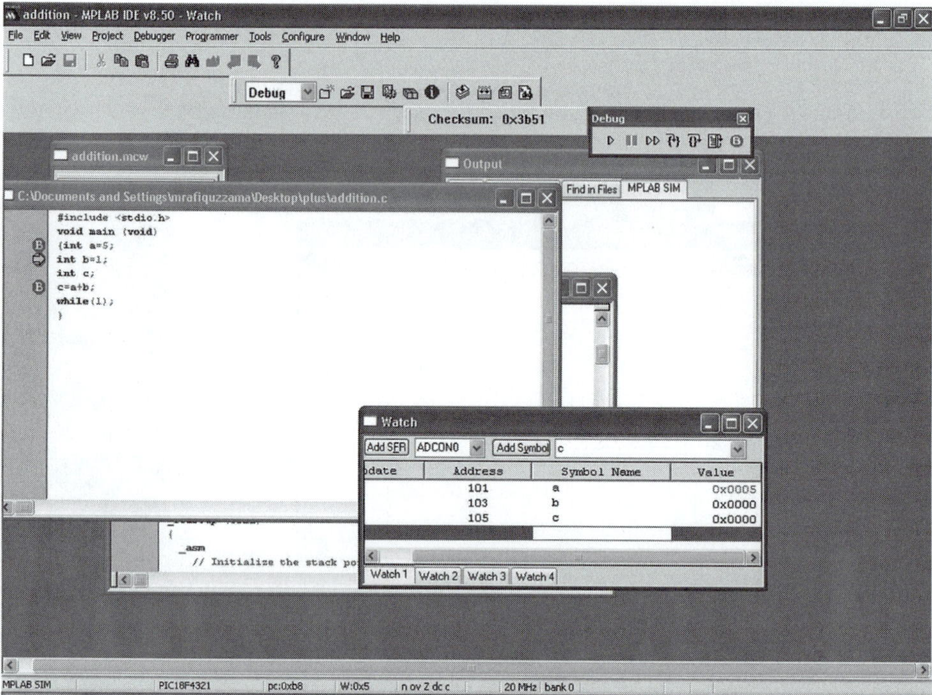

Note that 'a' contains 5. Next, left click on the single arrow; the following will be displayed:

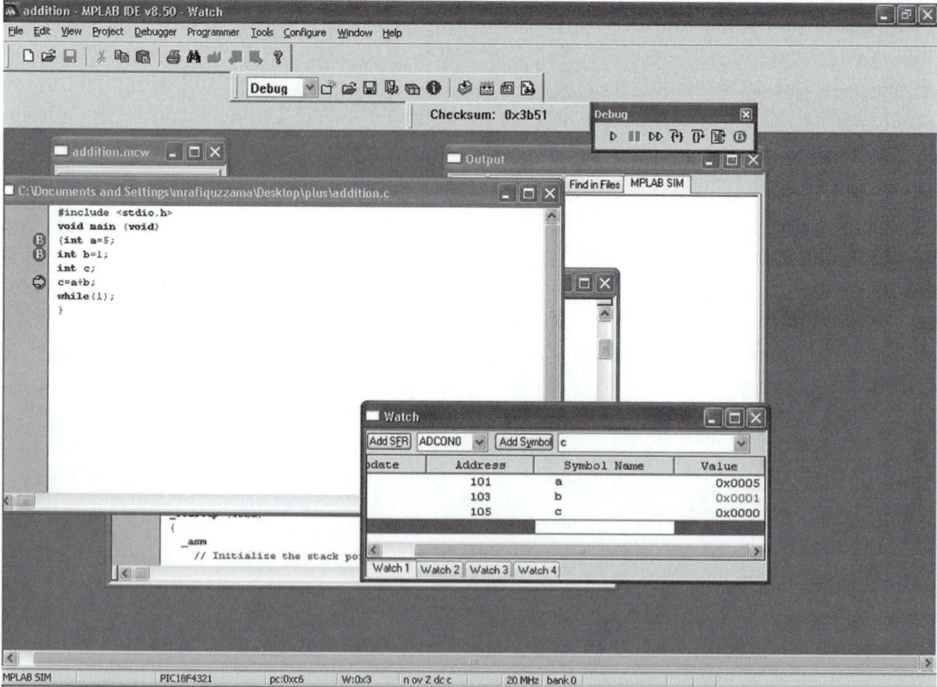

Note that ‘b’ contains 1 after execution of
Int b = 1;

Next, left click on the single arrow, and then left click on Halt (icon with two vertical lines, second from left on the Debug toolbar) to see the final result after execution of the line
C = a +b ;
as follows:

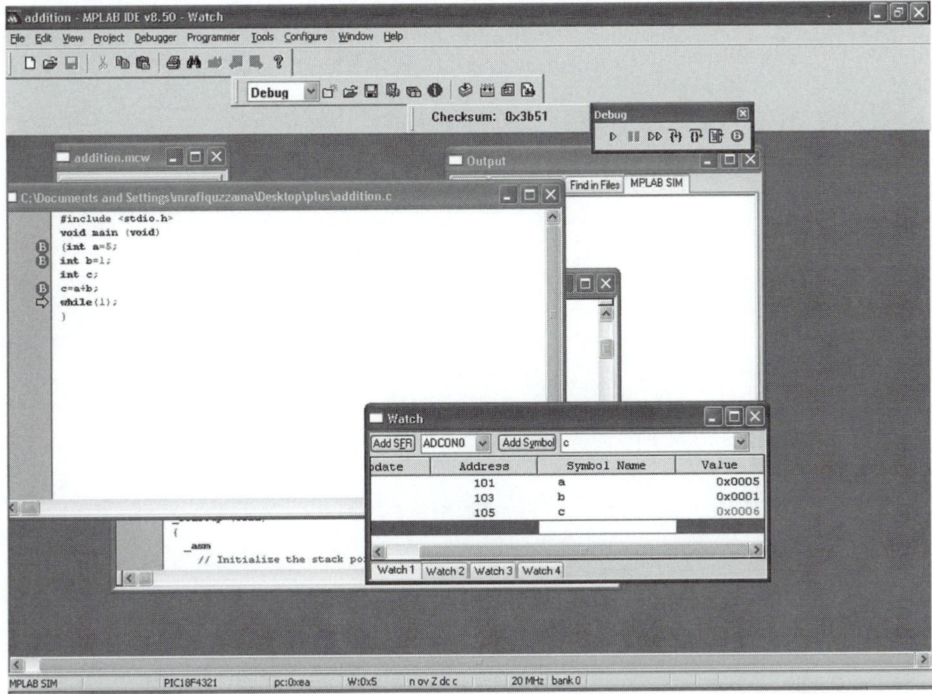

In the above, see that ‘c’ contains 6 (final answer).

The debugging is now complete.

APPENDIX H: INTERFACING THE PIC18F4321 TO PERSONAL COMPUTER USING PICkit™ 3

Appendix H contains the following:

H.1 INITIAL HARDWARE SETUP FOR THE PIC18F4321

H.2 CONNECTING THE PERSONAL COMPUTER (PC) TO THE PIC18F4321 VIA PICkit™ 3

H.3 PROGRAMMING THE PIC18F4321 FROM PERSONAL COMPUTER USING PICkit™3

H.1 INITIAL HARDWARE SETUP FOR THE PIC18F4321

Figure H.1 shows the initial set up for the PIC18F4321 microcontroller. Pin #1 of the PIC18F4321 is the RESET input for the microcontroller. The $\overline{\text{MCLR}}$ (pin #1) must be connected to the reset circuit shown in the figure. There are two pairs of pins on the

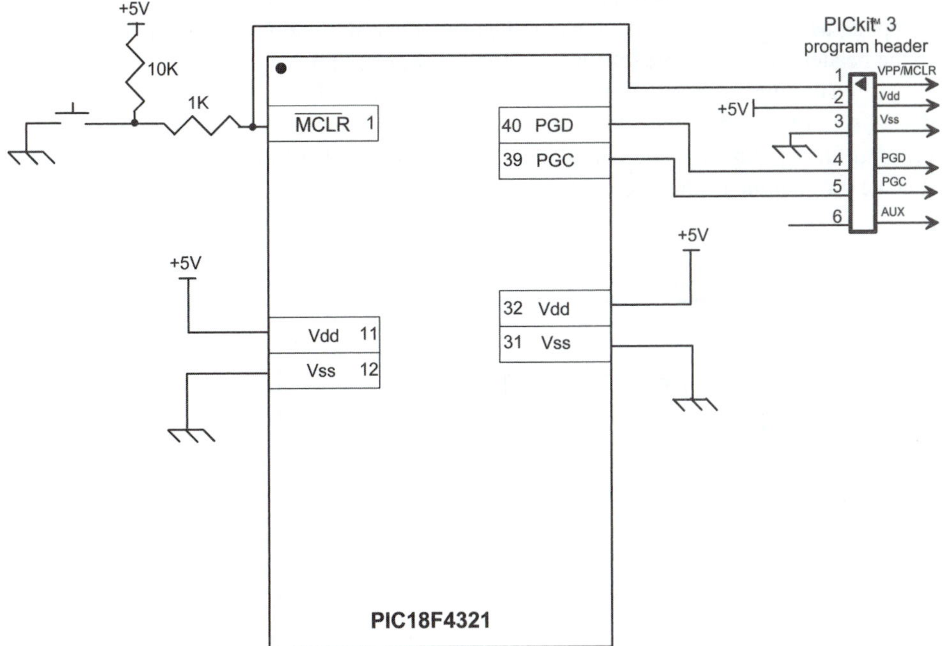

FIGURE H.1 Initial set up for the PIC18F4321

PIC18F4321 that must be connected to power and ground. For example, pins 11 and 32 must be connected directly to +5 V while pins 12 and 31 are connected directly to ground. Be sure not to connect any capacitors to these pins, connect them directly to either ground or +5 V. With any of the PIC18 family microcontrollers containing an "F" in the name, such as the PIC18F4321, the operating Vdd range is between 4.2 and 5.5 V. Figure H.1 also shows the proper connections for the header that will connect to the programmer. Note that the programmer has six pins but the sixth pin (Aux pin) makes no connections. After the PIC18F4321 is properly connected, the appropriate software must be installed.

There are two programs that must be installed in order to interface with the PIC18F. The first program is called MPLAB and the latest version can be downloaded at www.microchip.com/MPLAB. The second is called MPLAB C18 and is the C compiler for the PIC18F which can be found at www.microchip.com/c18. Note that at this site there is a link for academic use of the C18 compiler; be sure to click on the link and download the student C18 compiler. After the software has been installed, the PIC18F is now ready to be implemented.

H.2 CONNECTING THE PERSONAL COMPUTER (PC) TO THE PIC18F4321 VIA PICkit3

First, the PIC18F4321 initial setup circuit on the breadboard should be implemented. Next, the personal computer or Laptop should be connected to the PIC18F4321 using the PICkit™ 3. Figure H.2 shows a simplified block diagram of the implementation.

Figure H.3 shows a pictorial view of the implementation. The following picture shows how the initial setup along with the reset circuit for the above block diagram would look like after building it on a breadboard:

Once the circuit is built, the PICkit™ 3 can be connected to the USB port of the computer as shown in Figure H.4. Next, the header part of the PICkit™ 3 can be connected to the header pins on the breadboard as shown in Figure H.5.

The other necessary I/O devices such as switches, LEDs, LCDs, and seven-segment displays can now be connected to the PIC18F4321 on the breadboard to perform some meaningful experiments. After implementing the desired hardware, the PIC18F4321 can then be programmed using the MPLAB software.

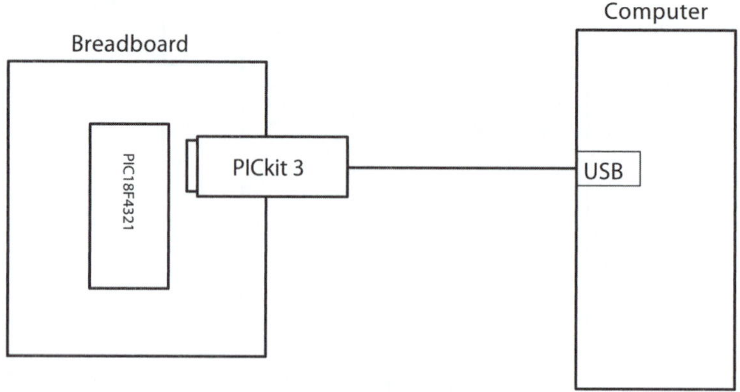

FIGURE H.2 PIC18F4321 personal computer interface using the PICkit™ 3

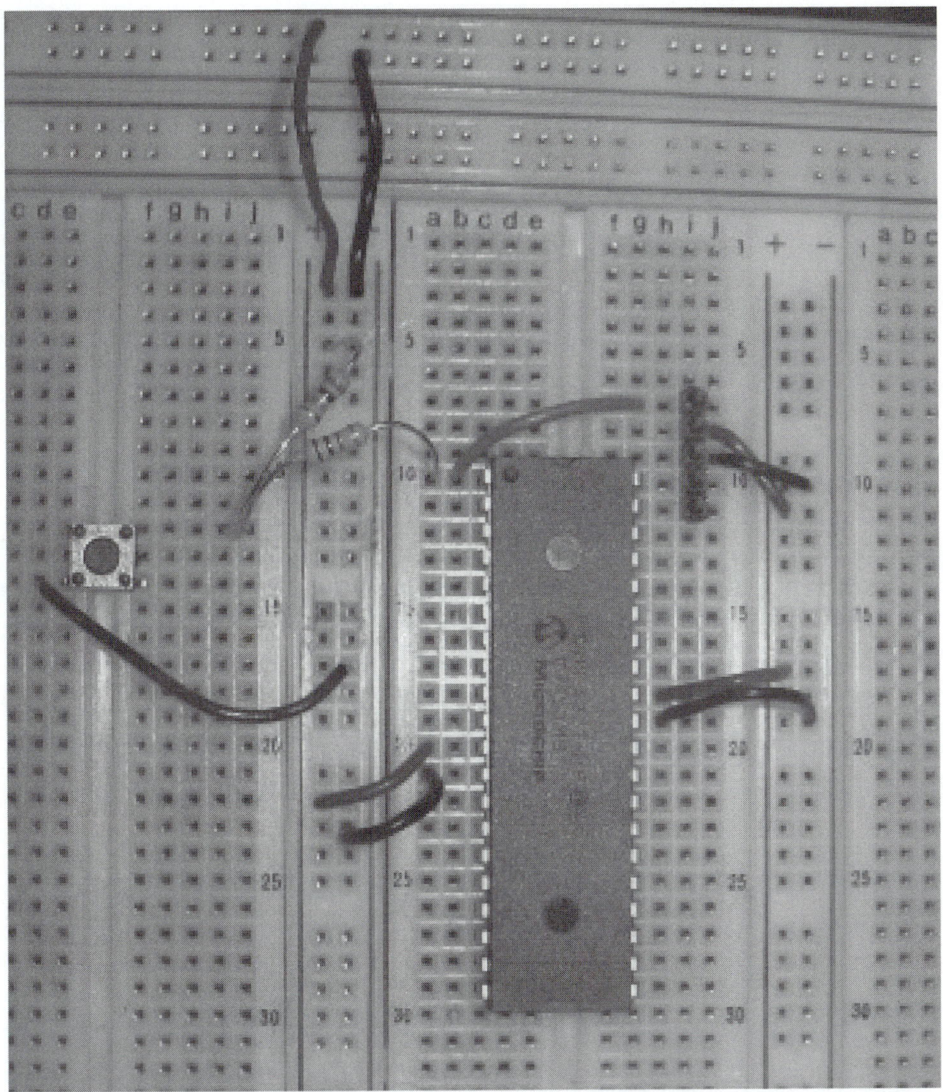

FIGURE H. 3 Pictorial view of the breadboard implementation

FIGURE H. 4 Pictorial view of connecting the PICkit™ 3 to the USB port

FIGURE H. 5 Connecting the PICkit™ 3 to the breadboard

H.3 PROGRAMMING THE PIC18F4321 FROM PERSONAL COMPUTER USING THE PICkit3

In order to configure PICkit3 from the personal computer or laptop, The user needs to click on the 'Programmer', and then select PICkit3 as follows:

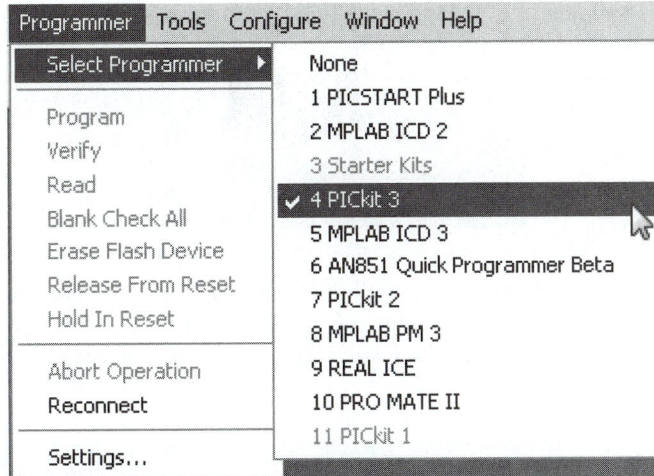

The following screen shot with the warning sign will appear. Just make sure that the PIC18F4321 microcontroller is connected to the proper voltage, and then click "OK" as follows:

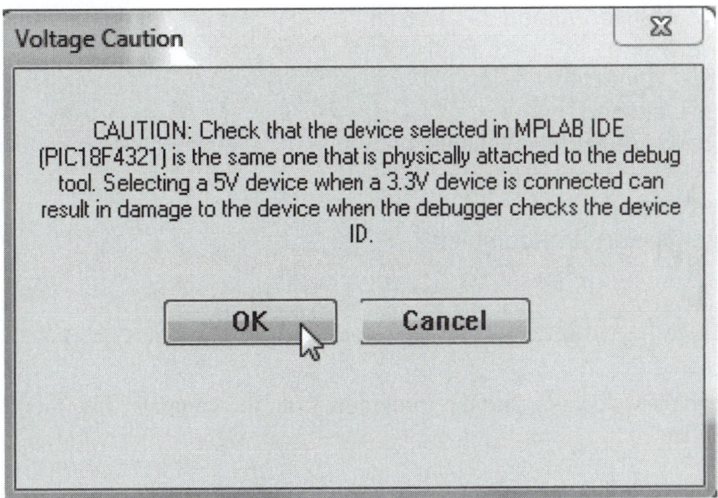

The PICkit3 is now connected.

Several options will appear at the top menu to program the PIC18F4321. The screen shot is provided below:

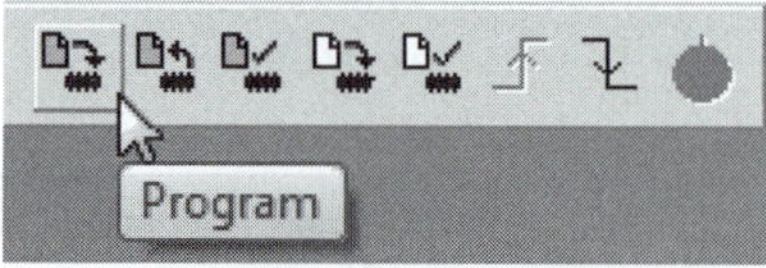

After successfully assembling or compiling a program, click the "program" option and MPLAB will download the program into the microcontroller.
The following message will apear indicating that the code was successfully programmed and verified onto the PIC18F:

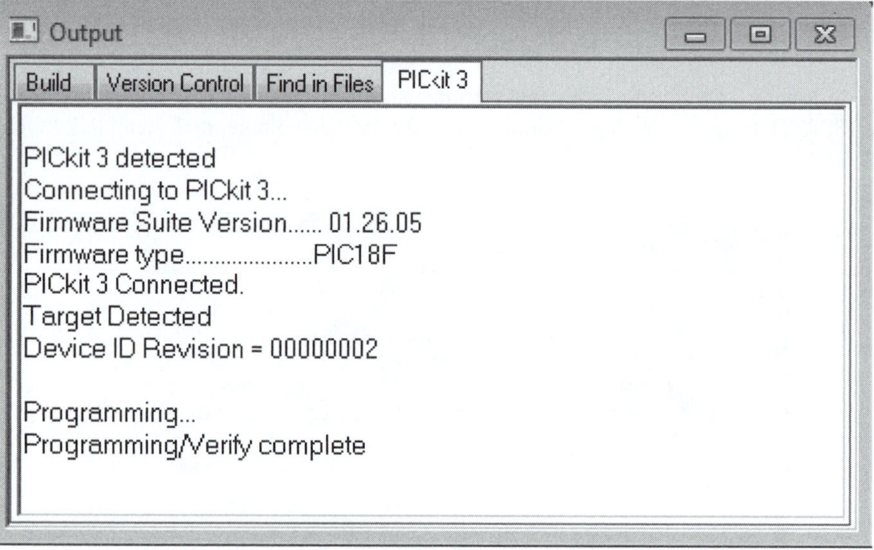

This will complete downloading the programs from the computer into the PIC18F4321 microcontroller.

BIBLIOGRAPHY

Gaonkar, Ramesh S., *Fundamentals of Microcontrollers and Applications in Embedded Systems (with the PIC18F Microcontroller Family),* Thomson Delmar Learning, 2007.

Huang, Han-Way, *PIC Microcontroller: An Introduction to Software and Hardware Interfacing,* Thomson Delmar Learning, 2005.

Johnsonbaugh, R. and Kalin., M., *C for Scientists and Engineers,* Prentice Hall, 1997.

Majidi, M. A., Mckinlay, R. D., and Causey, D., *PIC Microcontroller and Embedded Systems using assembly and C for PIC18,* Prentice Hall, 2008.

Microchip Technology, Inc., *PIC18F4321 Family Data Sheet,* 2009.

Rafiquzzaman, M., *Microprocessor Theory and Applications with 68000/68020 and Pentium,* 2008.

Rafiquzzaman, M., *Fundamentals of Digital Logic and Microcomputer Design,* 5th Edition, Wiley, 2005.

Rafiquzzaman, M., *Microprocessors and Microcomputer Development Systems - Designing Microprocessor-Based Systems*, Harper and Row, 1984.

Rafiquzzaman, M., *Microcomputer Theory and Applications with the INTEL SDK-85*, 2nd ed., John Wiley & Sons, 1987.

Rafiquzzaman, M., *Microprocessors - Theory and Applications - Intel and Motorola*, Prentice-Hall, 1992.

Rafiquzzaman, M., and Chandra, R., *Modern Computer Architecture*, West / PWS, 1988.

Rafiquzzaman, M., *Microprocessors and Microcomputer-Based System Design*, 1st ed., CRC Press, 1990.

Rafiquzzaman, M., *Microprocessors and Microcomputer-Based System Design*, 2nd ed. CRC Press, 1995.

INDEX

B

C

D

E

F

G

H

I

K

L

M

Q

R

S

T

U

V

W

X

Z